AF737428

The biology of hypogean fishes

Developments in environmental biology of fishes 21

Series Editor
EUGENE K. BALON

The biology of hypogean fishes

Guest Editor:

Aldemaro Romero

Reprinted from *Environmental biology of fishes*, Volume 62 (1–3), 2001
with addition of species and subject index

To Phil Regal
with my deepest
appreciation
A Romero

KLUWER ACADEMIC PUBLISHERS
DORDRECHT / BOSTON / LONDON

A C.I.P. Catalogue record for this book is available from the library of Congress

ISBN 1-4020-0076-6

Published by Kluwer Academic Publishers,
P.O. Box 17, 3300 AA Dordrecht, The Netherlands

Sold and distributed in the North, Central and South America
by Kluwer Academic Publishers,
101 Philip Drive, Norwell, MA 02061, U.S.A.

In all other countries, sold and distributed
by Kluwer Academic Publishers,
P.O. Box 322, 3300 AH Dordrecht, The Netherlands

Cover design by Eugene K. Balon
using posterior portion of *Amblyopsis rosae* and *Typhlichthys subterraneus*
by Loubna Bennis in the respective Threatened fishes of the world one
pagers of Aldemaro Romero (Env. Biol. Fish. 52: 434 and 53: 74).

The logo
designed by Christine Flegler-Balon, based on an idea of Emma Spertus

Printed on acid-free paper

All Rights Reserved
©2001 Kluwer Academic Publishers
No part of the material protected by this copyright notice may be reproduced or
utilized in any form or by any means, electronic or mechanical,
including photocopying, recording or by any information storage and
retrieval system, without written permission from the copyright owner.

Printed in the Netherlands

CONTENTS

How newly evolving species of cave fish learn to navigate.

Environmental Biology of Fishes **62:** 7–12, 2001.
© 2001 *Kluwer Academic Publishers. Printed in the Netherlands.*

An introduction to the special volume on the biology of hypogean fishes

Aldemaro Romero
Environmental Studies Program and Department of Biology, Macalester College, 1600 Grand Ave., St. Paul, MN 55105–1899, U.S.A. (e-mail: romero@macalester.edu)

Patterns, diversification, and something else

Hypogean fishes have always been a source of amazement and amusement throughout the history of ichthyology (Romero 2001). Controversies have ranged from evolutionary issues to behavioral adaptations, from taxonomy to ecology. The fundamental reason behind those controversies is that a number of adaptations usually referred to as 'troglomorphisms', especially blindness and depigmentation, have already been found in nearly 90 species that are largely taxonomically unrelated (Romero & Paulson 2001). Although reduction and/or loss of eyes and pigmentation seem to be common, the degree to which they occur varies and, to make matters more complicated, not all hypogean fishes have those adaptations. In some cases these species/populations freely come in and out of caves; in others, our access to them has been limited to occasional collections from wells. In some cases these species are known from a single specimen; in others they are represented by several populations. Some hypogean habitats in which these fishes are found seem to be very poor in nutrients, while others are supplied with constant and abundant amounts of organic material. In most cases, the fish in question is the only vertebrate found in its habitat, in others, they share their habitat with other hypogean fishes. Many species of non-hypogean fishes show some of the features of cave fishes. That is the case of fishes living in murky waters or the deep-sea; however, no hypogean fish has been found displaying some of the remarkable adaptations of abyssal fishes such as bioluminescence and darkening coloration.

Thus, besides being subject to (in most cases) permanent darkness, there seems to be no common rule to be applied here, or is there?

One of the pursuits of science is finding common patterns in nature; however, that goal has remained elusive for hypogean fish researchers. Some of those researchers have devoted their entire life to the study of a particular species or a particular problem among hypogean fishes; others have only visited this subject in order to test their ideas in a particular field. Studies have ranged from taxonomy to behavior, from morphology to ecology; methodologies have encompassed field studies as well as the application of the latest behavioral and molecular laboratory techniques. No wonder there is a lot of information out there but not too much of a synthesis.

So far the only attempt to provide such a synthesis was Thinès's (1969) book, which was largely a compilation of known facts. Since then, many changes have occurred; not only have a great deal of new species been discovered, but new approaches have been taken that have profoundly influenced how the scientific community looks at this subject. Hence, the idea of this special volume on the biology hypogean fishes.

Here we present 27 articles written by 38 authors from nine countries. Some of them tackle very specific issues; others provide a syntheses on specific issues. What follows is a general overview of these articles.

As a matter of introduction

Romero & Paulson (2001) provide a general list of hypogean fishes that present well-established troglomorphism. There they summarize information not previously synthesized elsewhere, with special emphasis on those features that could be used for comparative purposes to better understand the biology of these fishes. Thus, in addition to the scientific names and the (not always known) common names for these fishes, the article includes information about the etymology of their scientific name, major synonyms, history, distribution, habitat and ecology, and remarks of troglomorphisms found among these species as well as any other unusual feature described for them. Included in this article are references to two poorly-studied features that may be more common than previously thought: structural simplification or loss of both their scales and swim(gas)bladder. Remarks are also made regarding the possible phylogenetic relationships of each species.

Romero's (2001) article on the history of hypogean fish research represents an attempt to give a historical context to past and present efforts to understand the biology of these animals. It contains both qualitative and quantitative information on how research has progressed through times, defining the rather clear-cut historical periods which were largely influenced by philosophical concepts and geography. This article tries to explain why if hypogean fishes represent such an interesting model for evolutionary studies, they have yet to become a popular model for research among non-speleological investigators.

Poly (2001) provides not only a study on non-troglobitic fishes for a particular cave in West Virginia, but also synthesizes information about other non-troglobitic species from the rest of the world. This represents a new and healthy trend: students of cave fishes had until recently concentrated almost exclusively on hypogean fishes with troglomorphic features considering non-troglobitic ones as mere 'accidentals' worthy of little more than a footnote. Yet, with the gradual abandonment of typological views and the understanding that these fishes can actually provide clues to understanding hypogean fish evolution (they are sort of 'missing links'), now there is an increasing interest in them (see some other articles below).

In general, these chapters demonstrate that although a great deal of progress has been made in the last few decades, a number of issues need to be resolved, from conceptual to methodological ones, and that most hypogean species, whether they are troglomorphic or not, remain poorly known, which has hampered some of the attempts to synthesize information.

Anatomy, physiology and behavior

The relationships between morphology, physiology and behavior has always been an important component of hypogean fish research, and five articles in this volume represent current research in that tradition. Montgomery et al. (2001), for example, provide an excellent synthesis of theirs and others' research on the mechanosensory lateral line system of the hypogean form of the characid *Astyanax fasciatus*. They show how their enhanced lateral line capabilities with respect to their sighted relatives allow them to build up a 'picture' of their environment, a process that has been called active hydrodynamic imaging.

Adams & Johnson (2001) studied the metabolic rate and natural history of Ozark cavefish, *Amblyopsis rosae*, in Logan Cave, Arkansas, and found that seasonal relationships between mass and oxygen consumption may reflect alterations in environmental conditions (i.e. food availability, ambient dissolved oxygen) and changes in susceptibility to handling stress. They also report that cannibalism may not always be the rule for this amblyopsid species, as previously suggested.

Berti & Zorn (2001) report their most recent research on the locomotory responses of the cave cyprinid *Phreatichthys andruzzii* to chemical signals from conspecifics and related species. Contrary to what was previously believed, they found that there was not a positive chemotaxis to the odor of known or unknown conspecifics, but rather a strong negative chemotaxis towards the odor of unfamiliar conspecifics and of individuals belonging to the closely related species *Barbus conchonius*.

Camassa (2001) studied the responses to light in epigean and hypogean populations of the poeciliid *Gambusia affinis* from the brackish water basins of the 'Spunnulate' dolinas system of Torre Castiglione, in southeastern Italy. She found that the hypogean individuals showed, on average, scotophilia (photophobia) with little variability in their responses. On the other hand, epigean and spring-dwelling individuals of *G. affinis* from the same region showed the opposite behavior and with a great variability in their responses when compared with cave-dwelling conspecifics.

Pati (2001) presents the results of his latest experiments on the temporal organization in locomotor activity of the hypogean loach, *Nemacheilus evezardi*, and its epigean ancestor. His studies suggest that hypogean fish still possess a functional oscillator underlying their overt circadian rhythm in locomotor activity, although the ecophysiological significance of this proposition is yet to be fully understood.

Ecology and conservation

In the last decade, there has been an increased interest on the ecology and conservation status of these fishes. Thanks to the use of capture-recapture methods, which allows the gathering of data instead of the mostly casual observations of the past, new testable hypotheses have been presented and more precise information has become available. Trajano (2001a), for example, presents an overview on the ecology of hypogean fishes by synthesizing the available ecological data and comparatively analyzing that in an evolutionary context.

Brown & Johnson (2001) give us the results of their population studies and growth of the Ozark cavefish, *Amblyopsis rosae*, in Logan Cave National Wildlife Refuge, Arkansas. According to their results, this threatened species, endemic to the Springfield Plateau of the Ozark Highlands in Arkansas, Missouri, and Oklahoma, has a maximum life span of 4–5 years. They found that this fish has an average rate of growth of 0.6 mm per month, though smaller fish grew faster than larger fish

Noltie & Wicks (2001), on the other hand, took a look at how hydrogeology has shaped the ecology of Missouri's Ozark cavefish, *Amblyopsis rosae*, and the southern cavefish, *Typhlichthys subterraneus*. They explored how the differences in geography, geology, and hydrology in the habitats of both fish interact in shaping the habitat in which these two species dwell, the resulting ecological constraints that the fishes face, how their ecologies and populations reflect these constraints, and how conservation and management efforts may need to be tailored to best compliment the unique challenges that each species presents.

Trajano (2001b) studied the habitat and population biology of the armored cave catfish, *Ancistrus cryptophthalmus*, from Central Brazil and found this species to be a typical bottom-dweller with a strong preference for rocky substrates in fast-flowing stream sectors, while having relatively high population densities when compared to other cavefishes, conservatively estimated at 20 000 individuals.

Proudlove (2001) synthesized the conservation status of hypogean fishes and found that these fishes are susceptible to five main threats: habitat degradation, hydrological manipulations, environmental pollution, overexploitation, and introduced exotic species. Because they are endemic to small areas and have small populations, any threat could have serious consequences. Some species are receiving conservation attention; however, most receive none.

Kuhajda & Mayden (2001) discuss the status of the federally endangered Alabama cavefish, *Speoplatyrhinus poulsoni,* in caves of Alabama. They found that its population is relatively stable, but recommend continuous monitoring. They also observed a single specimen of *Typhlichthys subterraneus* in Key Cave, which falsifies the previous assumption that these two species were allopatric.

Three papers of the series 'Threatened fish of the world' round up this section with narratives on *Caecobarbus geertsii* (Proudlove & Romero 2001), *Ophisternon infernale* (Medina-González et al. 2001), and *Ogilbia pearsei* (Proudlove et al. 2001).

Genetics and evolution

This section contains five articles that examine genetic and evolutionary issues from different perspectives. Molecular techniques are helping tremendously to understand the true evolutionary nature of many hypogean populations while underlying the fact that drastic phenotypic differences between epigean and hypogean populations may result from very little genetic modifications. Borowsky & Mertz (2001) for example, report on the genetic differentiation among populations of the balitorid cave fish, *Schistura oedipus*, from Thailand by using DNA analysis. They found that the five populations they studied were little differentiated and that *S. oedipus* appears to be more closely related to *S. poculi* than to *S. reidi*, a proposed sister taxon.

Espinasa & Borowsky (2001) used RAPD markers to explore the origins and relationships of cave populations of the blind Mexican tetra, *Astyanax fasciatus*, in the Sierra de El Abra. Their results indicate that all cave populations are more closely related to one another than they are to the surface populations. This suggests that present day cave populations derived from a common ancestral stock, most likely due to a single colonization event, or alternatively, that strong gene flow among cave populations has occurred, raising precaution against considering each cave population as independent.

Poulson (2001a) studied the morphological and physiological correlates of evolutionary reduction of metabolic rate among amblyopsid cave fishes. He reports multiple correlates of reduction in metabolic rate during evolutionary adaptation to caves in the fish family Amblyopsidae. He found that no trait that could contribute to a lowered metabolic rate is in the same rank order as metabolic rate, though collectively the reductions were in the same order as metabolic rates. He explains that this was as expected with convergent evolution of complex traits.

Wilkens (2001) describes the convergent adaptations to cave life in the pimelodid *Rhamdia laticauda* catfish group from Central America. This species gave rise to a number of cave species. Comparison between two of them, *R. zongolicensis* and *R. reddelli*, reveals an intermediate state of reduction of the biologically functionless eyes, melanin pigmentation, and scotophilia. Given that the surface species is perfectly adapted to life in darkness, only a few constructive adaptations are developed in the cave species. For survival under conditions of low food supply in the caves, the barbels are elongated to improve the senses of taste and touch, and more fat can be deposited in the cave fish's tissue. Due to convergent evolution, the two cave species are morphologically much alike and show only minor diagnostic meristic differences. From geologic data he concluded that the two cave species started their troglobitic evolution at the end of the Pleistocene.

Finally, Parzefall (2001) reviews morphological and behavioral changes in the cave molly, *Poecilia mexicana*, from Tabasco, Mexico, which successfully colonized a sulfur cave. The eye size, melanin pigmentation and scale tuberances, as well as aggressive and schooling behavior, are reduced in the cave fish to a certain degree. Besides these regressive characters, he found some constructive ones such as an enlarged genital region in the female and an increased number of taste buds on the upper side of the head in both sexes. Nearly all the characters he studied form a genetically based gradient from the beginning to the end of the cave, which seems to be the result of gene flow from the outside to the inner part of the cave. He also reports on a special behavioral adaptation found among females which prefer bigger males during courtship. In darkness, only the cave females show this preference.

Special cases

Some of the contributing authors concentrated their efforts in taking a holistic view of some unusual populations. For example Burr et al. (2001) studied the cavernicolous sculpins of the *Cottus carolinae* species group in Perry County, Missouri. They found that this fish occupies pools and riffles of cave streams, occurring on a variety of substrates, from sediment to breakdown. A number of phenotypic characteristics separate the epigean from the hypogean form and these results are interpreted as being representative of losses associated with long-term cave habitation.

Humphreys (2001) summarized the current information on the eleotrid cavefish *Milyeringa veritas* from the arid tropics of northwestern Australia which include some of his own biogeographic studies. He found allozyme frequencies and distributions indicating significant population sub-structuring on the Cape Range peninsula. Further, he suggests that the vicariant events may have been associated with a series of eustatic low sea levels.

Hendrickson et al. (2001) present an overview of recent explorations in search of ictalurid Mexican blindcats, of the genus *Prietella*. Based on the data gathered from their visits to more than 50 sites they recommend threatened status for *Prietella phreatophila*. They located a new site for *P. lundbergi* and think that this species should probably be considered endangered.

Espinasa et al. (2001) describe a new blind cave fish population of genus *Astyanax* in Granadas Cave, in the Balsas drainage, southern México. They suggest that this new blind morph represents an independent colonization and convergent adaptation to the cave environment by this fish species. Individuals of this population display variability of their troglomorphic features, including asymmetrical degeneration of the eyes. They found that loss of pigmentation and eye reduction, although sometimes correlated, were not always linked with reduced eyes found on pigmented fish and unpigmented fish with normal eyes. Some individuals had reduced lens size or an absence of lens altogether and the retina was highly modified with photoreceptors sometimes absent.

Finally, this volume closes with an article by Poulson (2001b) in which he tries to show the importance of a broad comparative approach on morphology, physiology, behavior, life histories, and ecology that, when coupled with similar studies on deep-sea fishes, provides some interesting insights on the evolutionary history of both types of fishes, and suggests future lines of research.

As it usually happens in science, this volume provides some answers, but many more questions have yet to be elucidated. That is what makes our job so much fun.

Acknowledgements

Kelly M. Paulson, Jon Beland, and Kathryn Benz helped me proofreading many of the articles and getting in touch with some of the authors for further clarification. Ann Esson helped to prepare some of the final figures and provided extraordinary logistical support. Finally, I am deeply thankful to Eugene K. Balon who provided editorial assistance and advice in the always complicated task of putting together a volume like this.

References cited

Adams, G.L. & J.E. Johnson. 2001. Metabolic rate and natural history of Ozark cavefish, *Amblyopsis rosae*, in Logan Cave, Arkansas. Env. Biol. Fish. 62: 97–105.

Balon, E.K. 1981. Saltatory processes and altricial to precocial forms in the ontogeny of fishes. Amer. Zool. 21: 573–596.

Balon, E.K. 1989. The epigenetic mechanisms of bifurcation and alternative life-history styles. pp. 467–501. *In*: M.N. Bruton (ed.) Alternative Life-History Styles of Animals, Perspectives in Vertebrate Science 6, Kluwer Academic Publishers, Dordrecht.

Berti, R. & L. Zorn. 2001. Locomotory responses of the cave cyprinid *Phreatichthys andruzzii* to chemical signals from conspecifics and related species: new findings. Env. Biol. Fish. 62: 107–114.

Borowsky, R.B. & L. Mertz. 2001. Genetic differentiation among populations of the cave fish, *Schistura oedipus* (Cypriniformes: Balitoridae). Env. Biol. Fish. 62: 225–231.

Brown, J.Z. & J.J. Johnson. 2001. Population biology and growth of Ozark cavefish in Logan Cave National Wildlife Refuge, Arkansas. Env. Biol. Fish. 62: 161–169.

Burr, B.M., G.L. Adams, J.K., Krejca, R.J. Paul & M.L. Warren, Jr. 2001. Troglomorphic sculpins of the *Cottus carolinae* species group in Perry County, Missouri: distribution, external morphology, and conservation status. Env. Biol. Fish. 62: 279–296.

Camassa, M.M. 2001. Responses to light in epigean and hypogean populations of *Gambusia affinis* (Cyprinodontiformes: Poeciliidae). Env. Biol. Fish. 62: 115–118.

Espinasa, L. & R.B. Borowsky. 2001. Origins and relationship of cave populations of the blind Mexican tetra, *Astyanax fasciatus*, in the Sierra de El Abra. Env. Biol. Fish. 62: 233–237.

Espinasa, L., P. Rivas-Manzano & H. Espinosa Pérez. 2001. A new blind cave fish population of genus *Astyanax*: geography, morphology and behavior. Env. Biol. Fish. 62: 339–344.

Hendrickson, D.A., J.K. Krejca & J.M. Rodríguez Martinez. 2001. Mexican blindcats genus *Prietella* (Siluriformes: Ictaluridae): an overview of recent explorations. Env. Biol. Fish. 62: 315–337.

Humphreys, W.F. 2000. Background and glossary. pp. 3–14. *In*: H. Wilkens, D.C. Culver & W.F. Humphreys (ed.) Subterranean ecosystems. Elsevier, Amsterdam.

Humphreys, W.F. 2001. *Milyeringa veritas* (Eleotridae), a remarkably versatile cave fish from the arid tropics of northwestern Australia. Env. Biol. Fish. 62: 297–313.

Kuhajda, B.R. & R.L. Mayden. 2001. Status of the federally endangered Alabama cavefish, *Speoplatyrhinus poulsoni* (Amblyopsidae), in Key Cave and surrounding caves, Alabama. Env. Biol. Fish. 62: 215–222.

Medina-González, R., G. Proudlove, L. Chumba-Segura & T. Iliffe. 2001. Threatened fishes of the world: *Ophisternon infernale* (Hubbs, 1938) (Synbranchidae). Env. Biol. Fish. 62: 170.

Montgomery, J.C., S. Coombs & C.F. Baker. 2001. The mechanosensory lateral line system of the hypogean form of *Astyanax fasciatus*. Env. Biol. Fish. 62: 87–96.

Noltie, D.B & C.M. Wicks. 2001. How hydrogeology has shaped the ecology of Missouri's Ozark cavefish, *Amblyopsis rosae*, and southern cavefish, *Typhlichthys subterraneus*: insights on the sightless from understanding the underground. Env. Biol. Fish. 62: 171–194.

Parzefall, J. 2001. A review of morphological and behavioural changes in the cave molly, *Poecilia mexicana*, from Tabasco, Mexico. Env. Biol. Fish. 62: 263–275.

Pati, A.K. 2001. Temporal organization in locomotor activity of the hypogean loach, *Nemacheilus evezardi*, and its epigean ancestor. Env. Biol. Fish. 62: 119–129.

Poly, W.J. 2001. Nontroglobitic fishes in Bruffey-Hills Creek Cave, West Virginia, and other caves worldwide. Env. Biol. Fish. 62: 73–83.

Poulson, T.L. 2001a. Morphological and physiological correlates of evolutionary reduction of metabolic rate among amblyopsid cave fishes. Env. Biol. Fish. 62: 239–249.

Poulson, T.L. 2001b. Adaptations of cave fishes with some comparisons to deep-sea fishes. Env. Biol. Fish. 62: 345–364.

Proudlove, G.S. 2001. The conservation status of hypogean fishes. Env. Biol. Fish. 62: 201–213.

Proudlove, G., R. Medina-González, L. Chumba-Segura & T. Iliffe. 2001. Threatened fishes of the world: *Ogilbia pearsei* (Hubbs, 1938) (Bythitidae). Env. Biol. Fish. 62: 214.

Proudlove, G.S. & A. Romero. 2001. Threatened fishes of the world: *Caecobarbus geertsi* Boulenger, 1921 (Cyprinidae). Env. Biol. Fish. 62: 238.

Romero, A. 2001. Scientists prefer them blind: the history of hypogean fish research. Env. Biol. Fish. 62: 43–71.

Romero, A. & K.M. Paulson. 2001. It's a wonderful hypogean life: a guide to the troglomorphic fishes of the world. Env. Biol. Fish. 62: 13–41.

Thinès, G. 1969. L'evolution regressive des poissons cavernicoles et abyssaux. Masson et Cie, Paris. 394 pp.

Trajano, E. 2001b. Habitat and population data of troglobitic armoured cave catfishes, *Ancistrus cryptophthalmus* Reis, 1987, from central Brazil (Siluriformes: Loricariidae). Env. Biol. Fish. 62: 195–200.

Trajano, E. 2001a. Ecology of subterranean fishes: an overview. Env. Biol. Fish. 62: 133–160.

Wilkens, H. 2001. Convergent adaptations to cave life in the *Rhamdia laticauda* catfish group (Pimelodidae, Teleostei). Env. Biol. Fish. 62: 251–261.

Appendix. Glossary of terms frequently used in this special volume and that are unique to biospeleological studies. For a more expansive list of terms, see Humphreys (2000)

Anchi(h)aline: Aquatic habitats with a restricted exposure to open air, with one or more connections to the sea, that is influenced by both the marine and terrestrial ecosystems with which they interface. They are common in volcanic or limestone bedrock. Some examples of anchialine fishes are species of the genus *Lucifuga* in Cuba and the Bahamas as well as *Ogilbia galapagosensis* from the Galápagos Islands, *Milyeringa veritas* and *Ophisternon candidum* from northwestern Australia.

Cave: A hollow place in the earth. For cavers any natural hole, vertical, horizontal, or a combination of both which can be penetrated by a human being is a cave.

Cavernicole, cavernicolous: Any organism living in caves.

Cenote: A flooded, natural depression carved out of friable limestone with collapsed ceiling. The word *cenote* is Spanish but derived from the Mayan word *tzonot* which is somewhat loosely used in the Yucatán Peninsula of Mexico, to refer to various types of bodies of waters contained in cavities in the limestone which makes up the flat plain that constitutes that peninsula.

Crenobionts: Organisms normally found in springs and spring brooks.

Epigean: The surface environment as opposed to the hypogean one. It also applies to organisms living there.

Hypogean: The subsurface or subterranean environment as opposed to the epigean one. It also applies to organisms living there.

Karst: Soluble-rock landscape product of a combination of high rock solubility and well-developed secondary solubility. They have sinking streams, caves, enclosed depressions, flute rock outcrops, and large springs. The word derived from the name Karst (today Kars) used until early this century to describe a region in the Italian-Slovenian border with that name.

Phreatic: Underground source of water.

Scotophilia: Tendency to stay away from light exhibited by many hypogean organisms including blind ones (they use other sensory systems to detect light).

Spel(a)ean: Pertaining to cave.

Speleogenesis: The origin and development of caves.

Stygobiont: Organisms that inhabit groundwaters.

Stygobite: Hypogean organisms that show some sort of specialization to the underground environment. Adaptations may include, but are not restricted to, eye reduction, depigmentation, reduction of scale number, hyperdevelopment of some sensory organs, reduction in metabolic rates, reduction or disappearance of circadian rhythms, reduction in fecundity, increased egg size, increased life span, increased starvation resistance, reduction or disappearance of certain behaviors such as aggressiveness, schooling, and response to alarm substances. They are obligatory organisms to that environment. This term is similar in meaning to troglobite and to the precocial life-history style derived from the altricial style of the epigean ancestor (e.g., Balon 1981, 1989).

Stygofauna: Fauna inhabiting the underground water environment.

Stygophile: An organism that frequently visits and exploits hypogean resources. A fish example is the Chinese cyprinid *Varicorhinus* (*Scaphestes*) *macrolepis* which hibernates for about six months in the comparatively warm and stable cave waters (Zhang 1986). Another is *Chologaster agassizii* which occurs in springs and caves (but almost always near the surface). Lives underground but emerges above ground in springs (this term is similar in meaning to troglophile).

Stygomorphic: Organism that displays convergent phenotypic (morphological, physiological, and behavioral) characteristics of stygobites. This term is similar in meaning to troglomorphic.

Stygoxenes: Organisms that can be found accidentally in the hypogean environment. This term is similar in meaning to trogloxene.

Troglobite: (see stygobite).

Troglomorphic: (see stygomorphic).

Troglophile: (see stygophile).

Trogloxene: (see stygoxene).

Environmental Biology of Fishes **62:** 13–41, 2001.
© 2001 *Kluwer Academic Publishers. Printed in the Netherlands.*

It's a wonderful hypogean life: a guide to the troglomorphic fishes of the world

Aldemaro Romero & Kelly M. Paulson
Environmental Studies Program and Department of Biology, Macalester College, 1600 Grand Ave., St. Paul, MN 55105-1899, U.S.A. (e-mail: romero@macalester.edu)

Received 15 February 2001 Accepted 1 March 2001

Key words: hypogean fishes, biodiversity, systematics, taxonomy, geographical distribution, convergent evolution

Synopsis

There are at least 86 species of troglomorphic fishes belonging to 18 families. Some of those families are characterized by features that can be labeled as preadaptations to the hypogean life; others are not. The level of structural reduction in eye development and pigmentation is highly variable, even within some populations. Reduction in number and complexity of scales does occur but has yet to be fully documented. Reduction in the size and structure of the swim(gas)bladder may be another troglomorphic feature. There is considerable doubt on the taxonomic position of many species of troglomorphic fishes given that a number of them have been described solely on the basis of morphology while genetically they may be very closely correlated to genera different from those they have been assigned to. Geographically speaking there are no evident patterns since many of those species are not found in karstic areas but in phreatic waters. These fishes represent an excellent example of convergent evolution.

Introduction

The rate of discovery of new species/populations of troglomorphic fishes continues to increase rapidly. Although a list of these fishes has been published recently (Weber 2000), since Thinès (1969) there has been no comprehensive attempt to summarize the most important information about them. This paper aims to fill that gap and to serve as a general reference for this volume. We have not included non-troglomorphic fishes found in caves for this account because they have been summarized elsewhere in the volume (Poly 2001). At least one species, *Varicorhinus* (*Scaphestes*) *macrolepis*, hibernates for six months in caves (Zhang 1986). Also, we are aware that a number of troglomorphic species have been reported but not formally described (e.g., Trajano 1997c) and others seem to show only a few signs of troglomorphism (e.g., Burr et al. 2001). However, for the sake of simplicity, we decided to stick with those whose taxonomic status and troglomorphism has been clearly established.

Methods

We collected all available information on troglomorphic fishes published up to January 2001 following the methodology described in Romero (2001). The species were grouped by family following Nelson's (1994) systematic order. For each family a general summary of their biology is given to indicate whether or not they present generalized characters that could be understood as preadaptations to hypogean life, e.g., nocturnal habits or sensory systems that enhance information-gathering in dark waters. Some remarks are also made about the proportion of species known in comparison to how many of them are troglomorphic.

Given that for most genera and species their phylogenetic relationships with the rest of the family is uncertain, genera within families, as well as species within genera, were listed alphabetically. The species were numbered in order to facilitate the reading of the maps (Figures 1, 2). Unless otherwise indicated, the information for each species comes from the

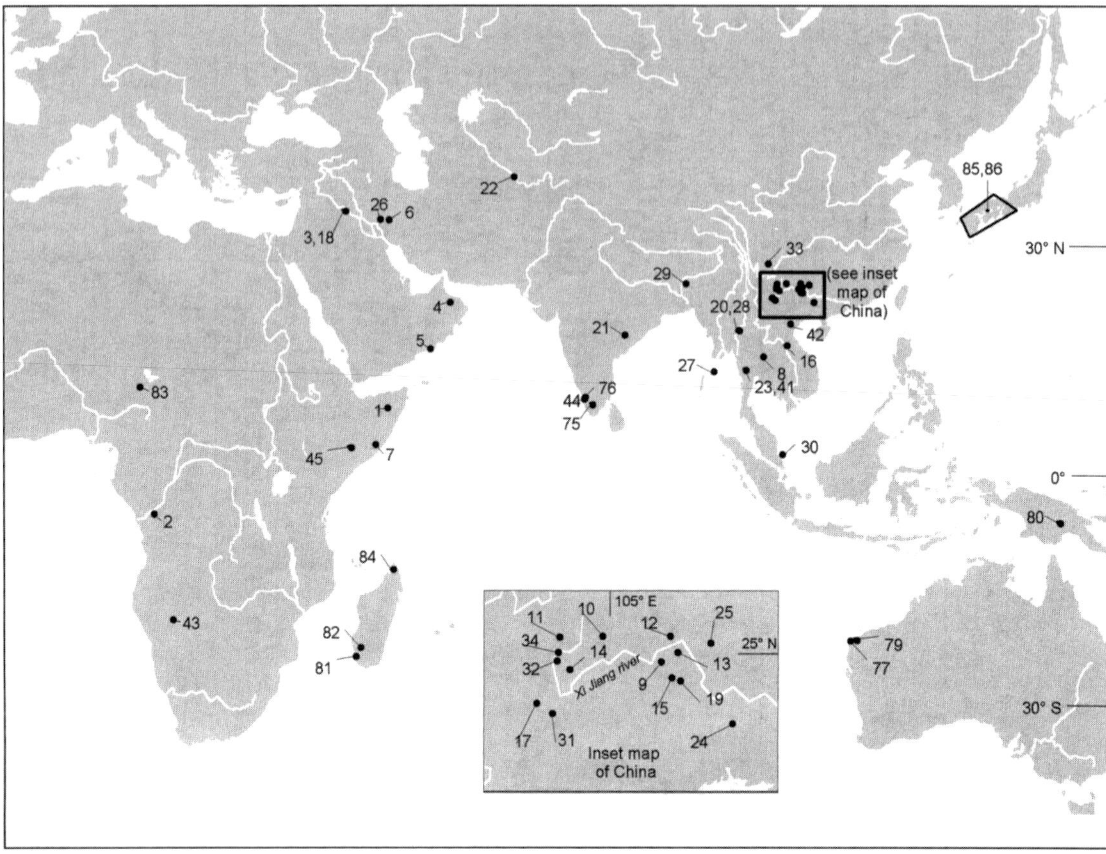

Figure 1. Localities and distributional areas for troglomorphic fishes in the Old World. Numbers represent species as cited in the text.

original description. Common name(s) are given if known. Etymology is given only for those names that were created to identify the hypogean/troglomorphic species/populations. Style for referring to etymologies follows Brown (1956) and Gotch (1995). Major synonyms are given only if they represent or have represented a source of confusion. Historical information is provided only if available. The remarks section aims to summarize information about troglomorphic features (reduction in eyes, pigmentation, and scales), behavioral features and phylogenetic relationships. Information on their conservation status is not included since it is summarized elsewhere in this volume (Proudlove 2001).

The maps accompanying this article (Figures 1, 2) were created using Arc View software in the Macalester College Cartography Laboratory in January 2001. The maps are projected with the Mercator projection, which does create some distortion in the area of landmasses towards the poles, but since these maps exclude much of the polar landmasses and that troglomorphic fishes are not found at high latitudes, this is not much of an issue. Latitudes and longitudes of troglomorphic fish distributions were determined using one of several methods: sometimes latitude and longitude information was found in the article describing the species; sometimes a description of the location was given (with varying degrees of accuracy and reproducibility), and latitude and longitude were determined using atlases and other sources; other times a range was given (as in the southeastern US and in Japan), and the best attempt was made to draw a polygon encompassing these regions; finally, some articles gave no more than a sketchy description of the location (as with some of the fishes from China) and difficulties in translation and finding the locations on maps make the latitudes and longitudes for these few species rather vague. The latitude and longitude information included in

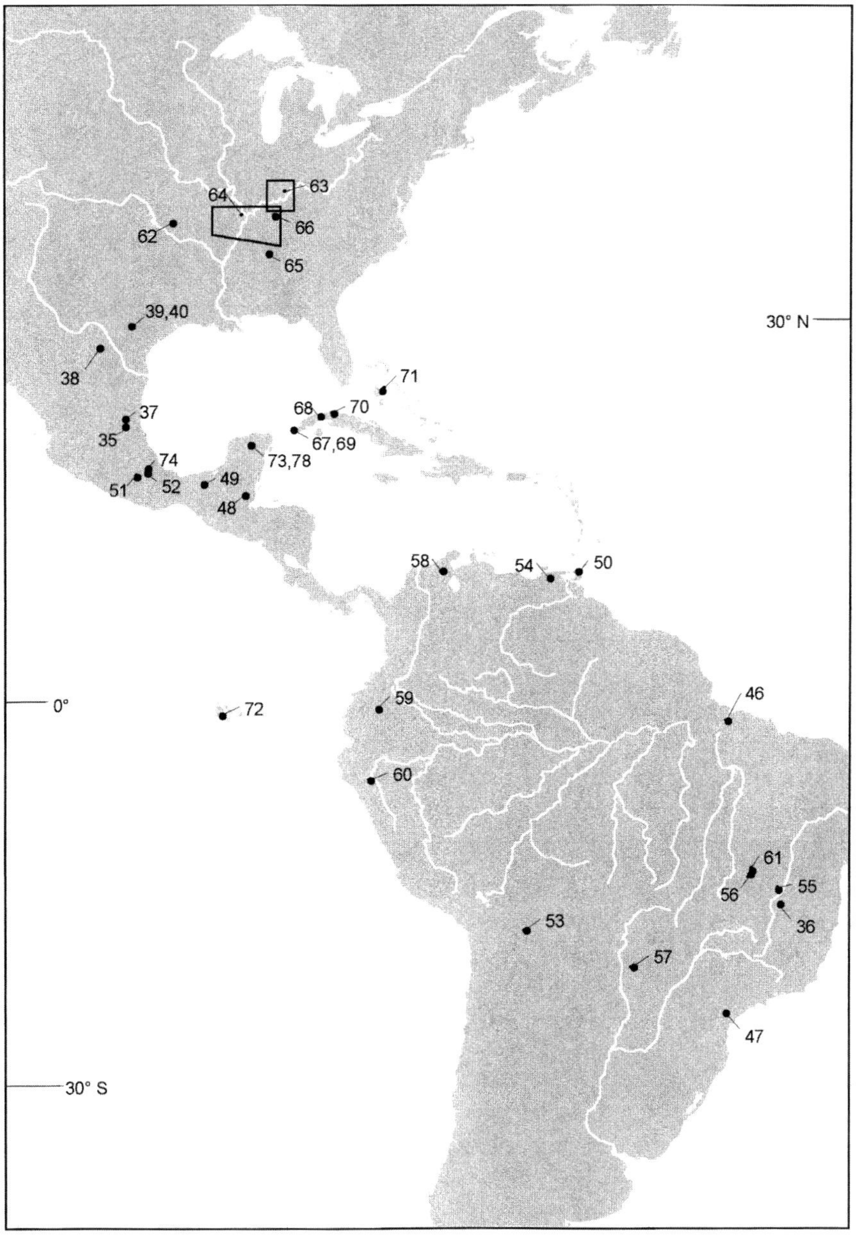

Figure 2. Localities and distributional areas for troglomorphic fishes in the American continent. Numbers represent species as cited in the text.

the article is the same as that which was used in the creation of the maps; where 'ca.' appears before the latitude and longitude, it means that it was determined using an atlas and should reflect a slightly lesser degree of accuracy. The reader should, therefore, take several things into consideration when using these maps. The maps are on a global scale and are intended to do little more than provide a general idea of where these species are located and provide a reference to the species numbers as described in the article. The nature of the methods used and the scale of the maps mean that the dots are not always (though in some cases are accurate to the tenth of a second) in the *exact* location.

Results

Family: Cyprinidae

Cyprinidae is the largest family of freshwater fishes with more than 2000 species distributed throughout Eurasia, Africa, and North America (Nelson 1994). Eighteen species/populations of troglomorphic hypogean cyprinids have been described so far. Three are from Africa and the rest are from Asia, ranging from the Middle East to Southeast Asia. Most cyprinids have scales to some degree. Nothing in their morphology, physiology or behavior seems to support the idea of a family being preadapted to the hypogean environment. Thus, the large number of hypogean representatives seems to be just the result of the fact that this is a highly speciosed family.

Most of the troglobitic species have been discovered in the last ten years in China and southeast Asia. Many of their descriptions have been published in Chinese and specimens of those species have yet to be examined by researchers others than those who described them. For those that have been more thoroughly studied, there are indications that their taxonomy may be wrong. For example, it is quite possible that *Barbopsis devecchii*, *Phreatichthys andruzzii*, *Caecobarbus geertsi* and *Iranocypris typhlops* should all be assigned to the genus *Barbus* (Ercolini et al. 1982). A very close genetic relationship between the first two is supported by allozyme studies (Sbordoni et al. 1996).

1. ***Barbopsis devecchii*** Di Caporiacco, 1926. **Common name:** Somalian blind barb (E). **Etymology:** *barb* (L) for its apparent relationship with other fishes of the genus *Barbus*; *ops* (Gr) eyes; *devecchii* for C.M. De Vecchi, governor of Italian Somalia at that time. **Major synonyms:** *Eilichthys microphthalmus* (Pellegrin, 1929), *Barbopsis stefaninii* (Gianferrari, 1930), and *Zaccarinia stefaninii* (Gianferrari, 1934). Also, the name was originally spelled once as *devecchi* and once as *devecchii*; named after De Vecchi, so *devecchii* is correct. **History:** Original description was based on 19 specimens collected in a well by two local doctors, G. Stefanini and N. Puccioni in the spring of 1925 (Di Caporiacco 1927). Because of variability in eye reduction, specimens of this species have been mistakenly assigned to several genera and species. **Distribution:** Several localities in north-central Somalia (see Ercolini et al. 1982 for map and list of specific locations) (ca. 9°10′00″ N, 48°22′10″ E). **Habitat and ecology:** In springs and wells throughout a wide and continuous phreatic layer in the Nogal Valley in an area with a wide range of salinity. **Remarks:** This species exhibits a lower degree of troglomorphism than other hypogean species, including variability in eye size; little, but still present, pigmentation; thin but existing scales; and a somewhat reduced nervous system related to visual functions. It can be best described as a fish with a variable degree of microphthalmia. It also shows marked scotophilia (photophobia) (Ercolini & Berti 1978) and does not form schools (Jankowska & Thinès 1982). It has a low metabolic rate (Ercolini et al. 1987). Allozyme studies suggest that this species is troglomorphic to a lesser degree because of a more recent isolation from its epigean ancestor (probably a *Barbus* sp.) Gene flow among the different populations seems to be rather constant when compared to that of *P. andruzzii* (Sbordoni et al. 1996; see also remarks for *P. andruzzii*).

2. ***Caecobarbus geertsii*** Boulenger, 1921. **Common name:** Congo blind barb (E); barbu aveugle (F). **Etymology:** *caeco* (L) blind; *barbus* (L) for its barbs but also because of its relationship with fishes of the genus *Barbus*; *geertsi* after M.G. Geerts who sent the first specimens to Europe on which the original description was made. **Major synonyms:** None. **History:** M. Delporte, a Belgian explorer, may have been the first European to see this fish in 1915. In 1917, during the dry season, a party of amateur cave explorers penetrated 500 m into a limestone cave situated about 700 m above sea level. This cave is located in lower Congo near Thysville (today Kanka near Mbanza-Ngungu, 5°18′ S, 14°50′ E). One of those amateur explorers, M. Geerts, waited until after the end of World War I, when it was safe for him to go back to Belgium, where he carried with him a few specimens of this fish. He sent them to George Albert Boulenger, a well known specialist on African fishes (and a Belgian). E. Randour may have found this fish in other caves in the early 1920s. Being easy to transport alive, many European scientists did some experimental work with this species early on. It became so famous that in May 1951, it was exhibited at the New York Aquarium (Romero & Benz 2000). **Distribution:** Congo, caves in the lower Congo river system (ca. 5° S, 15° E) (see Heuts & Leleup 1954 for a distributional map). **Habitat and ecology:** In caves characterized by cyclic food scarcity (Heuts 1951). **Remarks:** It has vestigial eyes that are not externally visible (Gerard 1936). It was

the first troglomorphic species described outside the American continent. After *Astyanax fasciatus*, it is one of the most-studied hypogean fish. Despite numerous attempts, breeding in captivity has failed. It tends to aggregate based on conspecific odors (Berti & Thinès 1980), is scotophilic (Thinès 1953), and responds to the alarm substance (Thinès & Legrain 1973). This species seems to be related to primitive members of the genus *Barbus* (Heuts op. cit.).

3. ***Caecocypris basimi*** Banister & Bunni, 1980. **Common name:** None given. **Etymology:** *caeco* (L) blind; *cypris* (L) because it is a cyprinid; *basimi* (L) after Basim M. Al-Azzawi, the collector of the specimens. **Major synonyms:** None. **History:** Collected for the first time by Basim M. Al-Azzawi, a collector with the Natural History Research Center of the University of Baghdad, on 21 October 1977 while visiting the locality of *T. widdowsoni* in order to collect specimens of the latter species. **Distribution:** Iraq, Al Anbar province, Al Hadithah, natural well in the Sheik Hadid shrine (ca. 34°04′ N, 42°24′ E). Same as for *T. widdowsoni*. **Habitat and ecology:** Same as for *T. widdowsoni* with which it is found syntopically. **Remarks:** Eyeless with the orbit filled with fatty tissue; vestigial eyes are present only in juveniles. Depigmented. Scales present only in large specimens.

4. ***Garra barreimiae*** Fowler & Steinitz, 1956. **Common name:** Omani blind cave fish (E). **Etymology:** *barreimiae* (L) after the type locality, Barreimei, Oman. **Major synonyms:** None. **History:** The epigean form was first known from several specimens sent to the Academy of Natural Sciences of Philadelphia between 1950 and 1951. The first hypogean, troglomorphic population was discovered in May 1980 by Andy Dunsire and M. Gallagher who collected 100 dead and 56 live individuals (Banister et al. 1992). **Distribution:** Arabian peninsula, Oman, Barreimei. From a deep cave on the south flank of the Jabal Akhdar mountains, 9 km E.S.E. of Al-Hamra (23°05′ N, 57°21′ E). **Habitat and ecology:** pH range 6.5–7.5; dH range 10.0–20.0. **Remarks:** No externally visible eyes. Depigmented. Optic lobes not fully developed. It has weak squamation. Some increased development of melanin and the optic lobes occurred among individuals exposed to light under artificial conditions.

5. ***Garra dunsirei*** Banister, 1987. **Common name:** None given. **Etymology:** *dunsirei* (L) after Andy Dunsire, the collector. **Major synonyms:** None. **History:** The first four specimens were received at the British Museum (Natural History) in 1984 and were collected by A. Dunsire and D. Green on 16 May 1980. Another 19 live specimens were collected by C. Maclelland on 26 February 1983. **Distribution:** Oman, Dhofar, Jabal Qara (variously Jabal Samhan) mountains, in pools inside caves at the bottom of a large sinkhole at Tawi Atair (17°06′ N, 54°34′ E). It seems to be very geographically isolated, even from any other species of fish; the closest species is found more than 600 km away. **Habitat and ecology:** The fish were caught in a pool in a side passage 200 m down a sink hole. **Remarks:** Eyes apparently normal. Depigmented. Less squamation than epigean *G. barreimiae*. There are only a few casual behavioral observations of this fish in captivity.

6. ***Iranocypris typhlops*** Bruun & Kaiser, 1944. **Common names:** mahi-ye kurgar (Farsi), Iran cave barb (E). **Etymology:** *iran* (L) because it is found in Iran; *cypris* (L) because it is a cyprinid; *typhlos* (Gr) blind; *ops* (Gr) eye. **Major synonyms:** None. **History:** E.W. Kaiser collected seven specimens from a well in Kaaje-Ru, valley of Ab-i-Serum, Iran, while researching fauna in the freshwaters of that country on 6 May 1937. **Distribution:** Iran, Lorestan (Khorramabad) province, Zagros mountains, Ab-i-Serum Valley near Tang-e-haft railway station, Baq-e-Loveh Oasis, a natural well at Kaaje-Ru (33°05′ N, 48°36′ E) (for full locality description see Coad 1996). **Habitat and ecology:** The well-like water resurgence in which it has been found seems to be the result of a collapsed subterranean system. This is probably part of a larger but complex, narrow, and inaccessible network of underground waters. This locality is shared by another species of hypogean fish, *Paracobitis smithi* (see below). **Remarks:** This species is rare and may be endangered due to overcollecting in its only known locality (Coad op. cit.).

7. ***Phreatichthys andruzzii*** Vinciguerra, 1924. **Common name:** Somalian cavefish (E). **Etymology:** *phreas* (L) well; *ikhthus* (Gr) fish; *andruzzii* (L) for Alcibiade Andruzzii who secured the first specimens. **Major synonyms:** None. **History:** Six specimens were collected by lieutenant Zaccarini at Bud-Bud (4°11′19″ N, 46°28′21″ E), central Somalia. He gave them to a Alcibiade Andruzzi, a local physician who, in turn, sent them to the Museo Civico de Storia Naturale di Genoa. Since then, this species has been studied

from morphological, ecological, and behavioral perspectives by a number of researchers (e.g., Berti & Zorn 2001). **Distribution:** In central Somalia (see Ercolini et al. 1982 for map and list of specific localities) (ca. 4°11′50″ N, 46°28′00″ E). **Habitat and ecology:** The different localities at which it is found have different levels of salinity within what probably is the same underground, karstic hydrographic system. Elevations of the wells range between 135 and 150 m a.s.l. **Remarks:** This species lacks visible eyes. Depigmented. Scaleless. Parts of the nervous system associated with vision are highly reduced in complexity. It exhibits specific chemical recognition (Berti 1990), responds to light (Ercolini & Berti 1975), and has low oxygen consumption (Ercolini et al. 1987). Allozyme studies have suggested that this species and *Barbopsis devecchii*, another hypogean species of the same general area but allopatric to *P. andruzzii*, may have descended from the same hypogean ancestor (probably a primitive *Barbus*). Furthermore, the genetic distance between the two does not support the contention that they belong to different genera. Gene flow among the different populations of *P. andruzzii* is probably highly restricted or even absent (Sbordoni et al. 1996).

8. ***Poropuntius speleops*** (Roberts, 1991). **Common name:** None given. **Etymology:** *speos* (Gr) cave; *ops* (Gr) eyes. **Major synonyms:** *Barbus speleops* (original description). **History:** It was collected by Phillip Chapman in April 1989. **Distribution:** Thailand, Tham Phu Khieo, Phu Khieo Wildlife Sanctuary (16°02′05″ N, 101°32′10″ E), and in the nearby Tham Phathewada (Tham means cave). **Habitat and ecology:** The cave seems to be a major flood-sink during the rainy season. A small stream flowing directly into Tham Pathewada may provide the only permanent year-round input into the underground system. The fish seems to be very abundant, with 65 to 150 individuals per 100 m over 500 m of cave length. **Remarks:** Eyes are vestigial and pigmentation is reduced among adults. Generalized squamation.

9. ***Sinocyclocheilus anatirostris*** Lin & Luo, 1986. **Common name:** duck-billed golden-line fish (E, translation from Chinese); cave dragon (E). **Etymology:** *anatis* (L) duck-like; *rostris* (L) beak. **Major synonyms:** None. **History:** No information available. **Distribution:** China, Guangxi, Ling-Yün and Le-Ye (ca. 24°45′ N, 106°30′ E). **Habitat and ecology:** Found in two caves at about 30 m from their entrance at 1.5 m in depth. **Remarks:** No externally visible eyes. Depigmented. Scaleless. Reduced swim bladder.

10. ***Sinocyclocheilus angularis*** Zheng & Wang, 1990. **Common name:** golden-line angle fish (E, translation from Chinese). **Etymology:** *angularis* (L) because at the end of the posterior portion of the head there is a bone in a form of an angle. **Major synonyms:** None. **History:** First collected in June 1986 by Zheng and Wang. **Distribution:** China, Guizhou, Panxian county, Cave is located at 25°26′ N, 104°45′ E, at 1540 m a.s.l. **Habitat and ecology:** Karstic cave. **Remarks:** Microphthalmic. Depigmented. Scales embedded.

11. ***Sinocyclocheilus anophthalmus*** Chen, Chu, Luo & Wu, 1988. **Common name:** eyeless golden-line fish (E, translation from Chinese). **Etymology:** *an* (Gr) without; *opthalmus* (Gr) eyes. **Major synonyms:** None. **History:** Collected on 18 April 1986. **Distribution:** China, Yunnan, Yiliang county, subterranean river at 25°24′ N, 103°27′ E. **Habitat and ecology:** Limestone cave. **Remarks:** Eyes not externally visible. Little pigmentation. The scales cover only part of the body.

12. ***Sinocyclocheilus cyphotergous*** (Dai, 1988). **Common name:** None given. **Etymology:** *cypho* (Gr) curved, convexed; *terg(o)us* (L) back, because of its humpback. **Major synonyms:** *Gibbibarbus cyphotergous*. **History:** No information available. **Distribution:** China, Guizhou province, Luodian county, cave at Daxiao Dong, village of Dajing (ca. 25°26′ N, 106°47′ E). **Habitat and ecology:** 300 m from the entrance of the cave. **Remarks:** Microphthalmic. Depigmented.

13. ***Sinocychlocheilus furcodorsalis*** Chen, Yang & Lan, 1997. **Common name:** crossed-fork back golden-line fish (E, translation from Chinese). **Etymology:** *furco* (L) forked; *dorsalis* (L) on the back, because the anterior part of the humpback is bifurcated. **Major synonyms:** None. **History:** Collected in September 1993 by the authors who described it. **Distribution:** China, Guangxi, in an underground river in Tiane County (ca. 25°00′ N, 107°00′ E). **Habitat and ecology:** No information available. **Remarks:** No externally visible eyes. Depigmented. Scales on the body but not on the head.

14. ***Sinocyclocheilus hyalinus*** Chen & Yang in Chen, Yang & Zhu, 1994. **Common name:** translucent golden-line barbel (E, translation from Chinese); hyaline fish (E, translation from Chinese). **Etymology:** late Latin *hyalinus*, from *hyalinos* (Gr) of glass, referring to the transparency of the fish. **Major synonyms:** None. **History:** Two specimens were collected in June 1991. This may have been the first troglomorphic fish ever reported in print (Romero 2001). **Distribution:** China, Yunnan, Luxi county, Alu limestone caves (24°33′ N, 103°45′ E). **Habitat and ecology:** In an underground river (Yusun River) of a cave of 625 m in length at an elevation of 1712.87 m a.s.l. Feeds on microscopic animals and aquatic insects. **Remarks:** No externally visible eyes. Depigmentated. Scaleless. May be related to the epigean *S. angustiporus*.

15. ***Sinocyclocheilus microphthalmus*** Li, 1989. **Common name:** Small eye golden-line fish (E, translation from Chinese). **Etymology:** *micro* (Gr) small; *ophthalmus* (Gr) eyes. **Major synonyms:** None. **History:** Li collected it in 1986, between April and May. **Distribution:** China, Guangxi Zhuang autonomous region, Ling Yun county, karst cave at Luo-lou Village (24°20′ N, 106°49′ E). **Habitat and ecology:** Karst cave from which a river flows. **Remarks:** Microphthalmic. Depigmented. Covered by large scales.

16. ***Troglocyclocheilus khammouanensis*** Kottelat & Bréhier, 1999. **Common name:** paa kham khom (Laotian for 'fish which crosses the streams'). **Etymology:** *trogle* (Gr) hole; *cyclocheilus* (L) a generic name proposed for cyprinid fishes with which this genus is supposedly related; *khammouanensis* (L) after the limestone formation where it was found. **Major synonyms:** None. **History:** Holotype collected 24 February 1998 by F. Bréhier. Only one specimen has been collected but more have been observed. **Distribution:** Laos, Khammouan province, Khoung Nam Dôn, resurgence of Nam Dôn (stream) near Ban Phôndou (village) (17°33′50″ N, 104°52′20″ E). May have a wider distribution. **Habitat and ecology:** Cave. Swims close to the bottom and hides near the bottom. **Remarks:** No externally visible eyes. Depigmented. Body entirely scaled. Several rows of sensory papillae on lateral and dorsal sides of the head. May be related to species of the genus *Cyclocheilichthys*.

17. ***Typhlobarbus nudiventris*** Chu & Chen, 1982. **Common name:** None given. **Etymology:** *typhlo* (Gr) blind; *barbus* (L) for its relationship with the genus *Barbus*; *nudi ventris* (L) for the naked ventral portion. **Major synonyms:** None. **History:** Known from only three specimens. No other information available. **Distribution:** China, Yunnan province, Jian Jianshui county, known only from a sinkhole about 100 m below the surface at an altitude of 1450 m, in Yangjieba (ca. 23°39′ N, 102°46′ E). **Habitat and ecology:** No information available. **Remarks:** No externally visible eyes. No known phylogenetic relationships except for the fact that it has been placed in the subfamily Barbinae.

18. ***Typhlogarra widdowsoni*** Trewavas, 1955. **Common name:** Iraq blind barb (E). **Etymology:** *typhlo* (Gr) blind; *garra* (L) because of its close relationship to the genus *Garra*; *widdowsoni* (L) after A.G. Widdowson who secured the first specimens. **Major synonyms:** None. **History:** Dr. Donald and Mr. Khosraf, by request of Mr. A.G. Widdowson, all of them with the Iraq Petroleum Co., collected two specimens of these fish which ended up in the British Museum (Natural History) in July 1953. Three living individuals were later captured and also sent to that institution. **Distribution:** Iraq, in two caves, one about 8 m below ground, 10 km north of Haditha (34°04′ N, 42°24′ E), and the other, known as Pigeon Hole, is found about 8 km south of Haditha, near the Euphrates. **Habitat and ecology:** Little is known except that the type localities are large caves fed by underground streams. It shares its habitat with *Caecocypris basini*. **Remarks:** No externally visible eyes. Depigmented. Has scales only along the lateral line. The nervous system associated with vision is very much reduced. A few melanophores are present in the tissue between the optic lobes and the cerebellum. This species is slightly scotophilic (Marshall & Thinès 1958). It is closely related to *Garra rufa*.

Family: Cobitidae

This is a family composed of more than 110 species of freshwater fishes from Eurasia and Morocco. They have a wormlike to fusiform body and are mostly bottom dwellers (Nelson 1994). Only one troglomorphic species has been described so far.

19. ***Protocobitis typhlops*** Yang, Chen & Lan, 1994. **Common name:** None given. **Etymology:** *proto*

(Gr) primitive; *cobitis* (L) a generic name for members of the subfamily Cobitinae; *typhlos* (Gr) blind; *ops* (Gr) eye. **Major synonyms:** None. **History:** Specimens collected in August 1991 by Lan Jia-Hu. **Distribution:** China, Guangxi, Duan county, Xiaao town (24°15′ N, 107°05′ E). **Habitat and ecology:** In a cave at 210 m a.s.l. **Remarks:** No externally visible eyes. Depigmented. Scales rudimentary and only along the midline of the sides of the body. Lacks lateral line. It also lacks a bony swimbladder capsule, which may be a primitive character.

Family: Balitoridae

This is a family of freshwater species from Eurasia composed by more than 600 species. They have three or more pair of barbels. Some species are scaleless. Many tend to hide underneath rocks (Nelson 1994). They have 14 troglobitic representatives. Yang et al. (1994) mentioned a *Oreonectes* sp. for a limestone cave near Xiaao, Duan county, Guanxi, China, yet to be described.

20. ***Cryptotora thamicola*** (Kottelat, 1988). **Common name:** waterfall climbing cave fish of Thailand (E). **Etymology:** *Cryptotora* (Gr) is a contraction of *kryptos* (hidden, an allusion to the hidden cave life of this monospecific genus) and *balitora*, the name-sake of the family; *tham* (L) transcription of the Thai word for cave; *cola* (L) *colere*, to inhabit. **Major synonyms:** *Homaloptera thamicola* (original description) (Kottelat 1998). **History:** First collected by J. Taylor on 7 May 1985. **Distribution:** Thailand, Mae Hong Son province, Susa Cave (19°28′ N, 98°08′ E). **Habitat and ecology:** The holotype was found on a ledge, apparently climbing up a small waterfall about 0.6 km from the cave entrance. **Remarks:** No visible eyes. Depigmented except for some appendages. Scaleless.

21. ***Nemacheilus evezardi*** Day, 1872. **Common name:** hill stream loach (E). **Etymology:** not applicable. **Major synonyms:** Kottelat (1990) changed the generic status of this species to *Indoreonectes* but without explaining why. Singh & Yazdani (1993) gave it the name of *Oreonectes evezardi*. **History:** At the British Museum there are two specimens of *N. evezardi* (BMNH 1960, p. 6.2, pp. 1–2) collected from 'a cave in India', one of which is fully pigmented and the other somewhat depigmented but apparently with normal eyes. Menon (1987) mentioned the existence of 13 specimens collected in Bhaje Cave, Karla, Poona, but without saying whether or not these fish displayed any troglomorphic characters. **Distribution:** India, Madhya Pradesh state, Bastar district, Kotumsar Cave (18°52′09″ N, 81°56′05″ E) at 560 m a.s.l. (Kottelat 1990). **Habitat and ecology:** A cave subject to frequent flooding during the monsoon season. **Remarks:** Hypogean forms exhibit lower oxygen consumption than the epigean forms (Biswas 1991). The hypogean individuals also exhibit circadian and circannual rhythmicity (Biswas et al. 1990).

22. ***Nemacheilus starostini*** Parin, 1983. **Common name:** Kughitang blind loach (E). **Etymology:** *starostini* (L) after I.V. Starostin, a fish researcher and collector specialized in Turkmenia. **Major synonyms:** *Noemacheilus* (*Troglobitis*) *starostini* (original description), *Paracobitis starostini* (Weber, 2000). **History:** Seven live individuals and four preserved specimens were captured in December 1981 by V. Dolzansky, a member of a Moscow speleological expedition. The head of that expedition, Vladimir Maltsev, had first found the fish in that very sinkhole five years earlier and caught several specimens which were lost during shipment. **Distribution:** Turkmenistan, Chardzhou province, Khrebet Kughitang (mountains) (several sinkholes) (ca. 37°55′ N, 66°23′ E). **Habitat and ecology:** The type locality is a sinkhole of about 19 m in depth with a seasonally fluctuating water level. A large portion of the pool is exposed to light part of the time. The pool is connected to an underwater stream. Up to 40 fish can be observed at a time in the pool. They feed on insect larvae and small benthic crustaceans (Smirnov 1983). **Remarks:** No externally visible eyes. Depigmented. Scaleless. No swimbladder.

23. ***Nemacheilus troglocataractus*** Kottelat & Géry, 1989. **Common name:** None known. **Etymology:** *trogle* (Gr) hole; *cataracta* (L) waterfall. **Major synonyms:** None. **History:** Collected in February 1986 by R. Brouquisse and in June 1986 by the Thais-Maros expedition. **Distribution:** Thailand, Kanchanaburi province, Tham Sai Yok Noi [=Tham Nam Tok], 3 km north-northwest of Nam Tok (14°15′ N, 99°04′ E). **Habitat and ecology:** Underwater stream. Syntopic with *Pterocryptis buccata*. **Remarks:** No externally visible eyes. Depigmented. Scales embedded.

24. ***Oreonectes anophthalmus*** Zheng in Anonymous, 1981. **Common name:** None given. **Etymology:** *an* (Gr) without; *opthalmus* (Gr) eyes. **Major synonyms:** *Nemacheilus anophthalmus* (Kottelat, 1988). **History:** No information available. **Distribution:** China, Guangxi, Wuming county, Qifeng Shan (hill), Taiji Cave (ca. 23°06′ N, 108°40′ E). **Habitat and ecology:** No information available. **Remarks:** No externally visible eyes.

25. ***Paracobitis longibarbatus*** Chen, Yang, Sket & Aljancic, 1998. **Common name:** None given. **Etymology:** *longi* (L) long; *barbatus* (L) barbels. **Major synonyms:** None. **History:** The first three specimens were collected during February and March 1995 by a joint Slovenian-Chinese expedition. **Distribution:** China, Guizhou province, Libo county, in a cave 13 km northeast of the town of Libo (ca. 25°15′ N, 108°00′ E). **Habitat and ecology:** The cave contains both streams and lakes with a sandy/rocky bottom. **Remarks:** No externally visible eyes. Depigmented. Rostral and maxillary barbels are well developed. It is related to *Paracobitis variegatus longidorsalis*, which occurs in the same basin of Xijiang. They appear to be scotophilic.

26. ***Paracobitis smithi*** (Greenwood, 1976). **Common name:** blind loach (E). **Etymology:** *smithi* (L) after Anthony Smith who collected the first specimens. **Major synonyms:** *Noemacheilus smithi* (original description) (Nalbant & Bianco, 1998). **History:** First collected in April 1976. **Distribution:** Iran, Lorestan (Khorramabad) province, Zagros mountains, Ab-i-Serum Valley near Tang-e-haft railway station, Baq-e-Loveh Oasis, a natural well at Kaaje-Ru (33°05′ N, 48°36′ E) (for full locality description see Coad 1996). **Habitat and ecology:** The well-like water resurgence in which it has been found seems to be the result of a collapsed subterranean system. This is probably part of a larger but complex, narrow, and inaccessible network of underground waters. Syntopic with *Iranocypris typhlops* (see above). **Remarks:** No externally visible eyes. Depigmented.

27. ***Schistura jarutanini*** Kottelat, 1990. **Common name:** Srisawat blind cave loach (E). **Etymology:** *jarutanini* (L) after Kitipong Jarutanin, who collected the first specimens. **Major synonyms:** None. **History:** First collected by K. Jarutanin in November 1987. **Distribution:** Thailand, Kanchanaburi province, Amphoe Sri Sawat, Tham (cave) Ba Dan (ca. 14°02′ N, 94°32′ E). **Habitat and ecology:** Underground stream. **Remarks:** Eyes reduced to absent. Slightly depigmented. Scales embedded.

28. ***Schistura oedipus*** (Kottelat, 1988). **Common name:** None given. **Etymology:** *schistura* (Gr) or skhizeîn to split; *oura* (Gr) tail, an allusion to forked caudal fin; *oedipus* for Oedipus, a mythic Theban king who tore out his eyes. **Major synonyms:** *Nemacheilus oedipus* (original description). **History:** The holotype and paratype were collected by J. Dunkley on 7 May 1985. **Distribution:** Thailand, Mae Hong Son province, Tham (cave) Nam Lang (19°31′ N, 98°09′ E) (see Kottelat 1990 for distributional map). **Habitat and ecology:** An outflow cave with a stream which is probably part of Nam Lang, part of a karstic endoreic basin. **Remarks:** Microphthalmic. Depigmented. Scales embedded.

29. ***Schistura sijuensis*** Menon, 1987. **Common name:** None given. **Etymology:** *sijuensis* (L) after Siju, the name of the cave where they were discovered. **Major synonyms:** *Noemacheilus sijuensis* original combination/description; *Noemacheilus Mesonoemachilus sijuensis* on plate caption of original description; *Mesonoemacheilus sijuensis*. Pillai & Yazdani (1977) refer to this hypogean fish as *Nemacheilus multifasciatus*, yet, but Talwar & Jhingran (1991) consider them separate species, and when referring to *N. multifasciatus* make no mention of being found in a cave. **History:** The first cave individuals belonging to this species may have been reported by Hora (1924) who described three specimens as '*Nemacheilus* sp.' from the 'Siju Cave, Assam, India'. One specimen, caught within 35 m of the cave mouth, showed coloration comparable to the epigean forms. Two others, netted about 550 m from the entrance of the cave, were paler in color and apparently had reduced eyes. **Distribution:** India, Meghalaya, Garo Hills, Siju Cave (ca. 25°25′ N, 90°30′ E). **Habitat and ecology:** No information available. **Remarks:** Eye size and pigmentation variable.

30. ***Sundoreonectes tiomanensis*** Kottelat, 1990. **Common name:** None given. **Etymology:** *tiomanensis* (L) after Pulau Tioman, the type locality. **Major synonyms:** None. **History:** First collected by Mark

Day on 17 August 1988. **Distribution:** Malaysia, Pulau Tioman (island), cave in Mount Gunung Kajang at 1000 m a.s.l. (ca. 2°50′ N, 104°15′ E). **Habitat and ecology:** In shallow pools of a small granitic cave system. **Remarks:** Reduced, partly embedded eyes. Partially depigmented.

31. ***Triplophysa gejiuensis*** (Chu & Chen, 1979). **Common name:** Gejiu blind loach (E). **Etymology:** *gejiuensis* (L) for the province of Gejiu. **Major synonyms:** *Noemacheilus gejiuensis* (original description) (Zhou & Cui, 1997), *Schistura gejiuensis* (Zhu, 1989). **History:** The seven original specimens were captured on 30 March 1978. **Distribution:** China, Yunnan province, Gejiu county, Bajiaojing near Qiafang (23°16′ N, 103°09′ E). **Habitat and ecology:** In subterranean waters at a depth of 400 m underground. **Remarks:** No externally visible eyes. Depigmented. Scaleless. Has no swimbladder. It was classified as a new species based only on troglomorphic characters. Paired fins and barbels are greatly elongated.

32. ***Triplophysa shilinensis*** (Chu & Yang in Chen, Yang & Xu, 1992). **Common name:** blind loach (E). **Etymology:** *shilinensis* (L) for the words Shi Lin which in Chinese means ‘stone forest’, the name of the general area where this species is found. **Major synomyms:** None. **History:** The first two specimens were collected by the authors describing the species on 18 March 1991. **Distribution:** China, Yunnan, Lunan county, cave at Weiboyi village near Stone Forest (24°47′ N, 103°22′ E). **Habitat and ecology:** The cave has a funnel-like opening and goes down vertically for 109 m before it reaches a horizontal river where this fish can be found. Insects have been found in their stomachs. **Remarks:** Eyes completely degenerated. Completely depigmented; body described as ‘translucent’. Scaleless. Posterior chamber of air bladder reduced, and anterior chamber enclosed in a bony capsule.

33. ***Triplophysa xiangxiensis*** (Yang, Yuan & Liao, 1986). **Common name:** None given. **Etymology:** *xiangxiensis* (L) because it is found in the area of Xiangxi. **Major synonyms:** *Noemacheilus xiangxiensis* (original description), *Schistura xiangxiensis* (in Zhu, 1989). **History:** No information available. **Distribution:** China, Xiangxi, Hunan province, Longshan county, Feihu Cave (ca. 27°52′ N, 102°15′ E). **Habitat and ecology:** Subterranean waters, Yuan River system, 20 m underground. **Remarks:** No externally visible eyes.

34. ***Triplophysa yunnanensis*** Yang, 1990. **Common name:** None given. **Etymology:** *yunnanensis* (L) because it is found in the province of Yunnan. **Major synonyms:** None. **History:** No information available. **Distribution:** China, Yunnan province, Yiliang (ca. 25°00′ N, 103°23′ E). **Habitat and ecology:** No information available. **Remarks:** Microphthalmic but pigmented.

Family: Characidae

This is a highly diversified (ca. 900 species) family of freshwater fishes from America and Africa (Nelson 1994). They show no apparent morphological and/or behavioral preadaptation to the hypogean environment. It is somewhat surprising that more representatives of this family have not been found in caves given its ample distribution and high diversity. Most characids are completely covered with cycloid scales. There are at least two species of troglomorphic characids.

35. ***Astyanax fasciatus* species complex** (Cuvier, 1819). **Common name:** Mexican blind/cave tetra (E). **Etymology:** not applicable. **Major synonyms:** The epigean form of this species was originally described by Cuvier (1819, p. 352, pl. 26 (figure 2) as *Chalceus fasciatus* based on material from Brazil. The genus *Astyanax* was created by Baird & Girard (1854) and then the species was reassigned to it. The specific status of the epigean form in the areas surrounding the cave localities for the troglobitic ones is in dispute. The whole family Characidae is very complex, with many species that are difficult to differentiate from each other and are not always allopatric (Géry 1977, Nelson 1994). New species of the genus *Astyanax* have been described in Mexico using morphological characters only (e.g., Contreras-Balderas & Lozano-Vilano 1988, Lozano-Vilano & Contreras-Balderas 1990). Common synonyms are *A. mexicanus*, a rearrangement of *Tetragonopterus mexicanus* (De Filippi, 1853), and *Astyanax fasciatus mexicanus*. The use of *mexicanus* has been most commonly employed by US-based authors, especially when dealing with epigean individuals from Texas. However, clear taxonomic differences and geographic limits between *A. fasciatus* and *A. mexicanus* have yet to be established. Until the taxonomy of the entire genus is fully clarified, the most parsimonious approach is to use the name *Astyanax*

fasciatus, clarifying when necessary whether we refer to the epigean or troglobitic populations. We avoid the use of the term 'cave populations' because some hypogean individuals are actually eyed and pigmented. **History:** The first cave, troglobitic population (from La Cueva Chica in the San Luis de Potosí area) was described as *Anoptichthys jordani* (Hubbs & Innes, 1936), the second one (from La Cueva El Pachón) was named *Anoptichthys antrobius* (Alvarez, 1946), and a third one (from La Cueva de Los Sabinos) was named *Anoptichthys hubbsi* (Alvarez, 1947). Most if not all of the papers that still use either those names, or a combination of *Astyanax* and the specific names given to cave populations, are written by experimentalists, not field or taxonomic researchers. As more cave populations were discovered it became evident that this typological approach was creating a lot of confusion (Romero 2001). **Distribution:** The epigean *A. fasciatus* is a freshwater fish with a broad distribution in the New World. Although obligatory and facultative cave populations of the eyed form have been reported for Belize and Costa Rica (Romero 1984), the Yucatán Peninsula (Hubbs 1938), and Brazil (Trajano 2001a), the only region in which they have developed into blind, depigmented individuals is the San Luis de Potosí area, east central Mexico, where at least thirty-one caves supporting subterranean populations of this fish can be found (Espinasa et al. 2001) (ca. 22°05′ N, 99°00′ W). Not all the cave populations of this area display the same degree of morphological divergence from the surface forms, however. Some are completely blind and depigmented while others are only partially so. Three caves contain only individuals with full eyes and pigmentation. Eleven of these populations include blind, eyed, as well as phenotypically intermediate forms (Mitchell et al. 1977, Romero 1983, Espinasa et al. op. cit.). **Habitat and ecology:** In numerous caves. Some of those caves appear to be very poor from a trophic viewpoint while others (e.g., La Cueva Chica) have a very rich input of nutrients thanks to large bat colonies producing guano on which these fish feed. **Remarks:** Troglomorphic individuals have reduced eyes (most are not externally visible), pigmentation (including totally depigmented), and a larger number of taste buds (Schemmel 1967). The surface and cave forms interbreed in both natural and laboratory conditions, producing fertile hybrids with a phenotypically intermediate form in the F_1 generation, and with an F_2 generation whose individuals range from an almost completely blind and depigmented form to an almost eyed and pigmented one (Peters & Peters 1973, Sadoglu 1957, Wilkens 1969). Breeding, electrophoretic, and karyotypic studies support the contention that the cave and epigean forms are the same species (Avise & Selander 1972, Kirby et al. 1977). The two forms are not only different in their morphology, but also in their behavior: unlike the surface/eyed form, the blind form never schools, is active all the time, and is not aggressive (Boucquey et al. 1965, Breder 1942, Breder & Gresser 1941, Erckens & Weber 1976). Although the blind form does produce an alarm substance, it does not respond to it (Pfeiffer 1966). Schemmel (1980) also reported differences between the two forms in the angle of inclination used when feeding from the bottom. Differences in the level of phototactic responses among different populations have also been reported (Romero 1985).

36. ***Stygichthys typhlops*** Brittan & Böhlke, 1965. **Common name:** Brazilian blind characid (E). **Etymology:** *stygos* (Gr) from the underworld in the Greek mythology; *ichthyos* (Gr) fish; *typhlos* (Gr) blind; *ops* (Gr) eye. **History:** Known from only one specimen captured on 16 May 1962 by Joseph A. Tosi, Jr., an ecologist with the Instituto de Agricultura of the Organization of American States. The fish was captured in water brought up from a depth of 30 m during a well-drilling operation. **Distribution:** Brazil, Minas Gerais state, in one well location at Jaiba. The precise location is unknown. Jaiba may refer to an area (Serra de Jaíba) or to a town, near to which the well could be situated. It is the area between Januaria and Janaúba, in northern Minas Gerais State, central-eastern Brazil (Trajano personal communication) (ca. 16°15′ N, 44°00′ W). **Habitat and ecology:** Groundwater, obtained from a well drilled at 30 m. Although the tooth structure seems to indicate a mostly vegetarian diet, it is unlikely that they maintain such a diet in the underground environment. This may be a relictual morphological feature. **Remarks:** No externally visible eyes. There is no distinct lateral-line system. Totally depigmented. Unlike the Mexican cave tetra, the ancestral form of this fish has yet to be determined. The Brazilian blind characid has a number of features typical of the Tetragonopterinae. More specifically, it may be related to the *Hyphessobrycon-Hasemania* complex, because of this group's reduction in the circumorbital bones; however that just may be a coincidence given that the reduction of those bones is a convergent feature among cave fishes. They also differ from other Tetragonopterinae in the large number of inner

premaxillary and maxillary teeth, as well as the small number of anal rays which is unusual for this group of characids (Romero & McLeran 2000).

Family: Ictaluridae
About 45 species of freshwater fishes found in North America from southern Canada to Guatemala. They are scaleless (Nelson 1994). Their possession of barbels as well as their nocturnal habits might be argued to be preadaptive features. Four unrelated eyeless species are known from deep wells and caves. All four species show a remarkable convergence of features that include the lack of externally visible eyes, depigmentation, and reduced lateral line (Walsh & Gilbert 1995).

37. ***Prietella lundbergi*** Walsh & Gilbert, 1995. **Common name:** phantom blindcat (E); bagre ciego duende (S). **Etymology:** *lundbergi* (L) after John G. Lundberg, an American ichthyologist who has greatly contributed to the systematics and paleontology of ictalurid and neotropical catfishes. **Major synonyms:** None. **History:** Collected by T. Morris on 17 April 1989. **Distribution:** Mexico, Tamaulipas state, Río Tamesí drainage, Nacimiento (spring) de San Rafael de Los Castros, subsurface thermal spring (ca. 22°40′ N, 99°00′ W). **Habitat and ecology:** Cave subject to periodic flooding. **Remarks:** No visible eyes. Depigmented. Reduced lateral line.

38. ***Prietella phreatophila*** Carranza, 1954. **Common name:** Mexican blindcat (E), bagre ciego de Múzquiz (S). **Etymology:** *prietella* (L) after Carlos Prieto, sponsor of the expedition to collect these specimens; *phreas* (L) well; *philos* (Gr) loving, with affinity to. **Major synonyms:** None. **History:** Collected on August 1954 by J. Carranza after hearing of the presence of a blind catfish in a well in that area. **Distribution:** Several localities in northern Mexico. Original specimens collected at El Potrero de Doña Mariana, Municipio Múzquiz, Cohauila state, in a well at the base of Sierra de Santa Rosa (ca. 27°55′ N, 101°07′ W). **Habitat and ecology:** It has been kept under starvation in the laboratory for up to 44 months and it is believed that it may last longer under natural conditions, which suggests that its environment is very poor in nutrients (Hendrickson et al. 2001). **Remarks:** No externally visible eyes. Depigmented. Reduced lateral line. It has a swimbladder.

39. ***Satan eurystomus*** Hubbs & Bailey, 1947. **Common name:** widemouth blindcat (E). **Etymology:** *satan* (Heb) prince of darkness; *eury* (Gr) broad; *stomus* (Gr) mouth ('widemouth prince of darkness') (Hubbs liked to name cave fishes with diabolic names based on the idea of hell being underground; see etymology for *Ophisternon infernale*). **Major synonyms:** None. **History:** Carl L. Hubbs visited the Witte Memorial Museum in San Antonio, Texas in 1938 when the director of that museum, Ellen S. Quillin, made available to him two blind catfishes which had been pumped from deep artesian wells in the vicinity of San Antonio. One of them was *Trogloglanis pattersoni* but the other one was of an unknown species. Apparently these fish had been pumped out of wells since the 1920s. **Distribution:** Texas, Edwards Aquifer. Holotype collected at an artesian well (depth 381 m), near San Antonio, Bexar county (ca. 29°30′ N, 98°30′ W). **Habitat and ecology:** An opportunistic predator. **Remarks:** No externally visible eyes. Depigmented. Exhibits degeneration of the swimbladder.

40. ***Trogloglanis pattersoni*** Eigenmann, 1919. **Common name:** toothless blindcat (E). **Etymology:** *troglo* (Gr) cave; *glanis* (Gr) catfish, originally from Glanis, the name of a river; *pattersoni* (L) after J.T. Patterson who secured the first specimen. **Major synonyms:** None. **History:** The holotype was collected on an unknown date by G.W. Brackenridge of San Antonio, Texas, the owner of the well. He gave the specimen to Patterson who, in turn, sent it to Eigenmann. A second specimen was caught in June 1934 by Josef Boecke in a ditch fed by an artesian well on his farm NE to San Antonio, Texas, and a third one was collected sometime before 1955 near Von Ormy, Texas. **Distribution:** Texas, Edwards Aquifer (ca. 29°30′ N, 98°30′ W). **Habitat and ecology:** Found in artesian waters up to 1300 m deep. This species is a herbivore, feeding on fungal growth and detritus found in the wells. Temperature may be a limiting factor in the distribution of this fish (Longley & Karnei 1979). It may be a scavenger. **Remarks:** Blind. Depigmented. Exhibits rudimentation in a number of regions of the nervous system related to vision.

Family: Siluridae
About 100 species of freshwater in Europe and Asia. Most are scaleless. Silurids have one or two pairs of barbels (Nelson 1994). The presence of barbels and

nocturnal habits could be considered preadaptations. One, maybe two troglomorphic species from southeast Asia.

41. ***Pterocryptis buccata*** Ng & Kottelat, 1998. **Common name:** None given. **Etymology:** *bucca* (L) mouth, because their cheeks look inflated due to their musculature. **Major synonyms:** None. **History:** The hypogean individuals were collected by L. Deharveng and colleagues in June 1986. **Distribution:** Thailand, Kanchanaburi province, Amphoe Sai Yok, Mae Khlong basin (ca. 14°15′ N, 99°04′ E). **Habitat and ecology:** Syntopic with *Nemacheilus troglocataractus*. **Remarks:** There are both epigean and hypogean populations. The hypogean individuals are depigmented and have sunken eyes.

42. ***Pterocryptis cucphuongensis*** (Mai, 1978). **Common name:** Cá niêt Cúc phurong (Vietnamese). **Etymology:** *cucphuongensis* (L) after Cúc phurong National Park where the fish was found. **Major synonyms:** *Silurus cucphuongensis* (in original description). New name given by Ng & Kottelat (1998). **History:** Collector and date unknown. **Distribution:** Vietnam, Cuc Phurong National Park, foothills of the northern Annamite Mountains, some 100 km southwest of Hanoi. The park comprises parts of Ha Nam Ninh, Ha Son Binh and Thanh Hoa provinces (20°19′ N, 105°22′ E). **Habitat and ecology:** There is no information in the description that this fish is a hypogean one. **Remarks:** It is microphthalmic. Description based on a single individual that may have not been well preserved. Based on the original description, it is not clear why this fish has been labeled as a 'cave fish' (e.g., Weber 2000).

Family: Clariidae

This family includes about 100 species of freshwater fishes found in Africa, Syria and southern and western Asia (Philippines to Java). They have four pairs of barbels and are scaleless. Airbreathing is accomplished with a labyrinthic organ arising from the gill arches ('labyrinth catfishes'). Some species are capable of traveling over short distances on land ('walking catfishes'). Some are burrowers with small eyes (Nelson 1994). These could be considered as preadaptations to the hypogean life. Three troglomorphic species have been described so far.

43. ***Clarias cavernicola*** Trewavas, 1936. **Common name:** cave catfish (E); spelonkbaber (Afrikaans); blinder höhlenwels (G). **Etymology:** *cavernicola* (L) living in a cave. **Major synonyms:** None. **History:** It may have been encountered first between 1917 and 1919 by geologists Jaeger & Waibel (1921). First collected by Karl Jordan who visited that cave in 1921 but not described until 1936. **Distribution:** Namibia, known only from a underground lake at Aigamas Cave, north of Otavi and west of Grootfontein (19°25′ S, 17°18′ E). **Habitat and ecology:** Lives in caves over shelves in open, clear water. Usually swims near the surface of the water. Feeds on the remains of insects (mostly beetles), bat and baboon guano, and insects associated with animal carcasses (Bruton 1995). **Remarks:** Eyes not visible but sunken and with variability in development. Depigmented.

44. ***Horaglanis krishnai*** Menon, 1950. **Common name:** Indian blind catfish (E). **Etymology:** *hora* (L) after S.L. Hora, Director of the Zoological Survey of India; *glanis* (L) catfish; *krishnai* (L) after the collector N. Krishna. **Major synonyms:** None. **History:** Four specimens were collected in July 1948 by N. Krishna Pillay after draining a well. He gave the specimens to K. Gopinath, who in turn sent it to Hora. The two specimens were received in August 1949 and the other two were lost. **Distribution:** India, Kerala state, Travancore, in two dugout wells at Kottayam (between 8°04′ N, 76°13′ E and 10°21′ N, 77°38′ E) (Mercy et al. 1982). **Habitat and ecology:** No information available. Syntopic with *Monopterus indicus* (Eapen, 1963). **Remarks:** No externally visible eyes. Free, not enclosed, swimbladder. It has a bulbous stomach that helps store food and the ileo-sphincter helps retain the digested food to maximize absorption (Mercy & Pillai 1984–5).

45. ***Uegitglanis zammaranoi*** Gianferrari, 1923. **Common name:** None given. **Etymology:** *uegit* (L) after the type locality, a pool in Uegit, El Uegit, Somalia; *glanis* (L) eponymous spirit of the Sacred springs in the town of Glanum in Provence, then there was the Glanicae, a triad of local Mother Goddesses associated with healing springs (all this makes reference to the fact that the this fish was found in a pool associated with a spring); *zammaranoi* (L) after the collector, V.T. Zammaranoi. **Major synonyms:** None. **History:** The first specimens were collected by an Italian army officer, V.T. Zammaranoi at an unspecified

date. **Distribution:** Somalia, several localities in southern Somalia (see Ercolini et al. 1982 for distributional map) (ca. 3°48′30″ N, 43°14′47″ E). **Habitat and ecology:** Tolerates remarkable variations in salinity and chemicals in the water. It reproduces between January and March, in the dry season (Ercolini et al. op. cit.). **Remarks:** No externally visible eyes. Depigmented. Scaleless. It lacks optic nerves, chiasm, and has a reduced brain size, particularly in the optic lobe and cerebellum (Bianchi 1975, Ercolini & Berti 1977). The epidermis has cells that contain an alarm substance. It does not present any specialization in olfactory structures (Bianchi et al. 1978). Displays aggressive behavior similar to epigean Siluriformes (Berti & Ercolini 1979). Studies on their responses to light have been inconclusive (Ercolni & Berti op. cit.).

Family: Pimelodidae

This is a family of about 300 species of freshwater fishes found in Central and South America (north to southernmost Mexico). They have three pairs of barbels and a naked body. Barbels and nocturnal habits make members of this family preadapted to the hypogean environment (Nelson 1994).

Seven troglomorphic species have been described so far. Additionally, Hubbs (1936, 1938) reported specimens of *R. guatemalensis* with certain degree of depigmentation and eye reduction and called them *Rhamdia guatemalensis decolor* and *Rhamdia guatemalensis stygaea*, respectively. The four other species/populations of *Rhamdia* described for southern Mexico and Belize represent a puzzle not only because of their close similarity to epigean *Rhamdia*, but also because of their great variability in troglomorphic characters and the lack of molecular studies that could shed some light on their evolution. There may be at least two more species of stygobiontic representatives for this family. One reported as *Imparfinis* sp. by Trajano (1997c) for the caves in the Chapada Diamantina, Itaete County, Bahia State, Brazil, and a *Taunayia* sp. also reported by Trajano (op. cit.) for Brazil, Bahia state, Campo Formoso County, Toca do Gonçalo Cave.

46. ***Phreatobius cisternarum*** Goeldi, 1905. **Common name:** None given. **Etymology:** *phreatos* (Gr) from a spring; *bius* (Gr) alive; *cisternarum* (L) from a deposit; all this in reference to the phreatic origin of this fish. **Major synonyms:** *Heptaterus cisternarum* (Andreas Buckup, 1988). The date of publication is usually cited as 1904, year in which Goeldi presented the paper describing this species before the Sixth International Congress of Zoology at Berne. However, the proceedings of that meeting were not published until 1905 and, according to article 21.1 of the International Code of Scientific Nomenclature it is the date of publication that must be taken in consideration. **History:** No information available. **Distribution:** Brazil, Pará and Amapá states, at the Rio Amazonas delta, Belém city, Pará, and Macapá city. Original locality: Marajo I., at the mouth of the Amazon, Brazil (in a cistern) (ca. 1°27′ S, 48°29′ W). **Habitat and ecology:** Phreatic. **Remarks:** Rudimentary eyes. Reduced pigmentation, mostly in the head and anterodorsal parts of the body. Its taxonomic position is very much in question (de Pinna 1998).

47. ***Pimelodella kronei*** (Miranda-Ribeiro, 1907). **Common name:** ceguinho (P). **Etymology:** *kronei* (L) after Ricardo Krone, who first reported this fish to Miranda-Ribeiro. **Major synonyms:** *Typhlobagrus kronei* (original description); *Caecorhamdella brasiliensis* (Borodin, 1927), *Pimelodella lateristriga* var. *kronei* (Haseman, 1911). **History:** This fish may have been known for some time, since the paper describing it has two pictures of the type locality dated 1898 and taken by Krone. **Distribution:** Brazil, southern São Paulo state, Upper Rio Ribeira basin, Iporanga, Areiras, ÁguasQuentes, Bombas, and Córrego Seco caves, on the right margin of the Rio Betari, and Alambari de Cima and Guruvuta caves, on the left margin (24°25′ S, 48°35′ W). **Habitat and ecology:** Headwater streams, with rocky beds and rapids intercalated with soft-bottomed pools. **Remarks:** Variable vestigial eyes. Varying pigmentation from dark to almost totally depigmented with most individuals being intermediate. They display agonistic behavior (Trajano 1991). Telencephalon is enlarged probably due to the improvement of olfactory and integrative centers related to social communication. Its putative ancestor is *Pimelodella transitoria*.

48. ***Rhamdia laticauda typhla*** Greenfield, Greenfield & Woods, 1982. **Common name:** cave chulín (E). **Etymology:** *typhla* (Gr) blind. **Major synonyms:** None. **History:** First specimens were collected on 10 July 1972. **Distribution:** Belize, Mountain Pine Ridge, Las Cuevas (16°45′ N, 89°00′ W). **Habitat and ecology:** Found in pools together with troglomorphic shrimps. **Remarks:** Variable eye size and

pigmentation. The *R. guatemalensis* reported by Hubbs (1936, 1938) showed a certain degree of depigmentation and eye reduction and was called *Rhamdia guatemalensis decolor* and *Rhamdia gualtemalensis stygaea*, respectively; yet these subspecies show a much lesser degree of depigmentation and eye reduction than *R. laticauda typhla*.

49. ***Rhamdia macuspanensis*** Weber & Wilkens, 1998. **Common name:** None given. **Etymology:** *macuspanensis* (L) after the town of Macuspana, about 20 km from the type locality. **Major synonyms:** None. **History:** The first specimen was provided by V. Sbordoni and specimens used for the species description were collected on 11 April 1996 by Ulrike Strecker, Axel Werber and Horst Wilkens. **Distribution:** Mexico, Tabasco state, ca. 20 km southeast of Macuspana, Grutas de Agua Blanca, Balneario de Agua Blanca at 64.9 km of the road (carretera federal 186) from Villahermosa to Chetumal (17°37′15.5″ N, 92°28′22.6″ W). **Habitat and ecology:** Crustaceans and epigean *Rhamdia* have been observed in the same cave. **Remarks:** Sunken eyes. Reduced pigmentation.

50. ***Rhamdia quelen*** (Quoy & Gaimard, 1824). **Common name:** eyeless catfish of Trinidad (E). **Etymology:** not applicable. **Major synonyms:** *Caecorhamdia urichi* (description of the cave population, Norman, 1926); *R. quelen urichi*, *Caecorhamdella urichi* (Hubbs, 1938). **History:** Cave individuals were first collected by Friederick William Urich in the 'Guacharo' cave (actually Oropouche or Urumaca cave). By July 1924, the British Museum of Natural History had received from Urich a specimen of 'blind' fish. The fish was examined by John Roxborough Norman who, fearing that the specimen could represent an accident of nature rather than a normal fish species, requested two more specimens and Urich complied. In October of 1926 Norman published his paper and named the fish *Caecorhamdia urichi* [*caeco* (L) blind; *rhamdia* (L) the genus of a catfish to which this cave fish seemed most related to; *urichi* (L) honoring Urich, the collector] (Romero & Creswell 2000). **Distribution:** Trinidad and Tobago, Trinidad, northern range, Pool at about 70 m from the entrance to the Oropouche or Urumaca cave (10°42′53″ N, 61°10′29″ W). **Habitat and ecology:** The cave is linear and limestone with an emergent stream running its full length. It is at an altitude of 185–215 m a.s.l. Fish are only found in the dark section of the cave. They feed on the droppings of oil birds, *Steatornis caripensis*. **Remarks:** There is some variability in eye size and pigmentation but even the ones with the least visible eyes cannot be considered anything but microphthalmic. Fish are extremely scotophilic (Romero & Creswell op. cit.).

51. ***Rhamdia reddelli*** Miller, 1984. **Common name:** juil ciego (S). **Etymology:** *reddelli* (L) after James R. Reddell who donated all the type specimens. **Major synonyms:** Considered just a population of *Rhamdia laticauda* by Silfvergrip (1996). **History:** Holotype collected on 9 March 1973. **Distribution:** Mexico, Oaxaca state, ca. 9 km southwest of Acatlán, Cueva del Nacimiento del Río San Antonio (ca. 18°12′ N, 98°02′ W). **Habitat and ecology:** In a deep lake about 120 m from the entrance of the cave. The cave is at 100 m a.s.l. A rich invertebrate troglomorphic fauna is found in the same cave. **Remarks:** No externally visible eyes. Depigmented. Individuals will darken under light conditions.

52. ***Rhamdia zongolicensis*** Wilkens, 1993 **Common name:** None given. **Etymology:** *zongolicensis* (L) after Zongolica, the geographic region where it occurs. **Major synonyms:** According to Silfvergrip (1996) it is a local population of *Rhamdia laticauda*. **History:** Collected on 17 December 1992 by Ulrike Strecker and Horst Wilkens. **Distribution:** Mexico, Veracruz state, Sierra de Zongolica, Cueva del Ostoc or Cueva de El Túnel (ca. 18°30′ N, 97°10′ W). **Habitat and ecology:** No other aquatic fauna has been reported for the same cave. **Remarks:** Slightly depigmented. Individuals will darken under light conditions. This fish is extremely similar to *R. laticauda* but also to *R. reddelli* reason by which some authors (e.g., Silfvergrip op. cit.) do not consider it a valid species.

Family: Trichomycteridae

This is a family of about 160 species of freshwater fishes found in Costa Rica, Panama and South America. They have a naked and elongate body. They usually have two pairs of barbels on maxilla, but lack barbels on the chin (Nelson 1994). The presence of barbels and nocturnal habits of some species make them preadapted to the hypogean environment. Of the ca. 100 species of the genus *Trichomycterus*, three species/populations have been found to display troglomorphic characters while other non-troglomorphic, hypogean populations have been reported for the Colombian Andes (Sket 1988) and two more for Brazil (Trajano & Souza 1994, Trajano 1997c).

53. ***Trichomycterus chaberti*** Durand, 1968. **Common name:** None given. **Etymology:** *chaberti* (L) after J. Chabert, collector of the first specimens. **Major synonyms:** *Pygidium chaberti* (Ringuelet, 1974). **History:** 30 specimens were collected by Chabert and his companions during a visit to the type locality on September 1966. An earlier specimen may have been captured by B. Antelo in August and sent to R.A. Ringulet but not described until 1974 (Ringuelet op. cit.). **Distribution:** Bolivia, Charcas province, Potosí department, Cave of Umayalanta, 8 km northwest of the town of Torotoro, 130 km south of Cochabamba (ca. 18°20′ S, 65°30′ W). **Habitat and ecology:** The cave is at an altitude of 2850 m. The entrance is a sinkhole of 140 m in depth through which a body of water penetrates for 1620 m. **Remarks:** Eye size and pigmentation are variable, from almost normal for a trichomycterid (which in comparison with other fishes appears to be microphthalmic to somewhat reduced) to a typical troglomorphic fish with very reduced, not externally visible eyes.

54. ***Trichomycterus conradi*** (Eigenmann, 1912). **Common name:** bagre ciego de la cueva del Guácharo (S). **Etymology:** not applicable. **Major synonyms:** *Pygidium conradi* (original description); *T. guianensis* (Galán, 1995). **History:** First reference to fishes from this cave is by Schultz (1949). **Distribution:** Venezuela, Monagas state, Cueva del Guácharo (10°10′33″ N, 63°33′06″ W). **Habitat and ecology:** This is a large cave in which fish can be found throughout most of its waters. This fish has the tendency to burrow in the sediment, which may make its observation difficult (Romero 1987). **Remarks:** It has reduced eyes and variable pigmentation (Nalbant & Linares 1987). Many of the morphological differences reported by Andreani Armas (1990) may be influenced by the natural variability of this species; some of those are typical of epigean species whose hypogean population is in status nascendi. Agonistic behavior for this species has been reported (Romero op. cit.). The definite taxonomy of this species has yet to be clarified. Part of the problem is the poor knowledge on the systematics of the genus *Trichomycterus* in general. Although morphological differences have been used to suggest specific differentiation between the hypogean population and the epigean one in nearby waters (Andreani Armas op. cit., Galán 1995), such contention is not supported by genetic studies (Pérez & Moodie 1993).

55. ***Trichomycterus itacarambiensis*** Trajano & de Pinna, 1996. **Common name:** None given. **Etymology:** *itacarambiensis* (L) after the name of the county, Itaracambi, in the state of Minas de Gerais, where it was found. **Major synonyms:** None. **History:** Although specimens of this fish had been collected since 1984 and even papers published about it (Carvalho & de Pinna 1986), it was not formally described until 1996. **Distribution:** Brazil, Minas Gerais state, Município (county) de Itacarambi, creek inside Olhos d'Agua Cave (15°06′06″ S, 44°09′30″ W). **Habitat and ecology:** Found throughout most of the cave. Generally, depigmented individuals increase in number as one goes farther into the cave. The estimated population size is between 1500 and 2000 (Trajano 1997b). This fish feeds mostly on living animals such as small arthropods and oligochaetes (Trajano 1997a). **Remarks:** This population consists of both depigmented and pigmented individuals with variable eye size and variable pigmentation.

Family: Loricariidae

This is a large family (ca. 550 species) of freshwater fishes from Costa Rica, Panama and South America. Their body lacks true scales but is generally covered by bony plates. Their mouth is ventral, with or without conspicuous barbels. They are usually found in swift-flowing streams from lowlands up to 3000 m a.s.l. They tend to be detritivorous (Nelson 1994). No remarkable characteristic seems to make fish of this family preadapted to the hypogean environment. The genus *Ancistrus* has 60 species, three of which are hypogean.

56. ***Ancistrus cryptophthalmus*** Reis, 1987. **Common name:** None given. **Etymology:** *cryptos* (Gr) hidden, covered, concealed; *ophthalmos* (Gr) eye. **Major synonyms:** None. **History:** This fish was known since at least 1972 when a specimen was collected, possibly by Rene Ginet, but then lost. The first specimens in museum collections today are paratypes collected by E. Ferrari and Claude Vidal in 1972. The holotype was collected in 1980 by a group of the Clube Alpino Paulita. **Distribution:** Brazil, Goiás state, Sao Domingos, Rio Sao Vicente system, Caverna Passa

Três, ca. 300 m into the cave (13°20′ to 13°52′ S, 46°16′ to 46°30′ W). **Habitat and ecology:** It is a bottom-dweller with a strong preference for rocky substrates in fast-flowing stream sectors. It is found in relatively high population densities (Trajano 2001b). **Remarks:** It coexists with a *Trichomycterus* sp. and *Astyanax* sp. Eyes in different degrees of reduction, sometime sunken under the skin. Small specimens have externally visible eyes which are reduced in size and become completely covered by dermal ossifications and skin during growth. Body almost entirely depigmented.

57. ***Ancistrus formoso*** Sabino & Trajano, 1997. **Common name:** None given. **Etymology:** *formoso* (L) after the type locality, Rio Formoso. **Major synonyms:** None. **History:** The first specimen was collected in 1995 and the second in 1996. **Distribution:** Brazil, Mato Grosso do Sul state, Município (county) de Bonito, Serra da Bodoquena, Buraco do Ducho cave (21°08′ S, 56°28′ W). **Habitat and ecology**: It is found in resurgences of caves of difficult access. **Remarks:** Reduced eyes not externally visible. It is totally depigmented. Abdomen naked.

58. ***Ancistrus galani*** Perez & Viloria, 1994. **Common name:** None given. **Etymology:** *galani* (L) after Carlos Galán who collected the holotype. **Major synonyms:** None. **History:** The holotype and the paratype were collected on 10 March 1990 and 25 March 1991, respectively. **Distribution:** Venezuela, Zulia state, Sierra de Perijá, Río Socuy, Cueva de Los Laureles (10°45′04″ N, 72°27′42″ W). **Habitat and ecology:** This cave is inhabited by a number of common epigean fishes, including *Ancistrus brevifilis bodenhameri*, none of which show true troglomorphic characters. *A. galani* is found only in remote gallery of the cave. **Remarks:** Reduced eyes and pigmentation. Description made based on only two specimens.

Family: Astroblepidae

This is a family of about 40 species of freshwater fishes found in the Andean region (South America) and Panama. Astroblepids have two pairs of barbels. They are characterized by their naked body and the presence of a mouth disc which allows some of them to climb the faces of waterfalls of torrental mountain streams. They also have maxillary and nasal barbels (Nelson 1994). These could be considered preadapted characters. There are two hypogean species within the genus *Astroblepus*. The vague reference to the Colombian *Astroblepus cyclopus* (originally described as *Pimelodus cyclopum*) by Humboldt (1805) as a cave fish cannot be substantiated (Romero & Paulson 2001).

59. ***Astroblepus pholeter*** Collette, 1962. **Common name:** None given. **Etymology:** *pholeter* (Gr) means 'one who lurks in hole'. **Major synonyms:** None. **History:** Collected in 1961 and 1962 by Pedro I. Porras who sent the specimens to the U.S. National Museum. **Distribution:** Ecuador, Napo province, Jumandi Cave in Latas, 4 km north of Archidona (0°20′–0°40′ S, 77°50′ W). **Habitat and ecology:** Found between 300 and 2000 m from the entrance of the cave. Hides in holes of rocks. Insect remains have been found in the stomachs of captured specimens. **Remarks:** Microphthalmic and slightly depigmented. It has a reduced swimbladder (Chardon 1966). Precise relationship to other members of the genus *Astroblepus* has yet to be established.

60. ***Astroblepus riberae*** Cardona & Guerao, 1994. **Common name:** None given. **Etymology:** *riberae* (L) after C. Ribera, the collector of the first specimens. **Major synonyms:** None. **History:** Collected on 20 August 1977 by Ribera. **Distribution:** Peru, Cajamarca region, Río Reque basin, Pacific drainage, Grutas (caves) de Ninabamba (6°19′48″ S, 78°30′36″ W). **Habitat and ecology:** No information available. **Remarks:** Vestigial eyes under the skin. Depigmented. Scaleless. Precise relationship with other members of the genus *Astroblepus* has yet to be established.

Family: Sternopygidae

This is a family made up of 15 species found in the freshwaters of South America. They have an elongated, compressed body. Like other members of the order Gymnotiformes, they posses electric organs that are the result of transformed muscles. They use these organs to search for food, obstacles, other fishes (electrogenic or not) and for defense. They are mainly nocturnal fishes many of which are found only in the deep waters of major rivers of the Amazon basin (Nelson 1994). These characters could be considered preadaptations to the hypogean environment. Only one hypogean species has been described so far.

61. ***Eigenmannia vicentespelaea*** Triques, 1996. **Common name:** None given. **Etymology:** *vicente* (L)

from the name of the cave; *spelaea* (L) cave. **Major synonyms:** None. **History:** First collected by P. Slavec in July 1978. **Distribution:** Brazil, Goiás state, Rio Sao Vicente system, Rio Tocantins basin, at Sao Domingos, Gruta (cave) Sao Vicente II (13°35′00″ S, 46°22′30″ W). **Habitat and ecology:** No information available. **Remarks:** The main difference between this species and other members of the same genus is its reduced eyes which are covered by thick skin. It is somewhat depigmented. The body is covered by uniformly cycloid scales whose size decreases from lateral line to dorsal and ventral regions. Scales in the dorsal and ventral regions are two or three times smaller in diameter than those near the lateral line. It is not known whether this fish utilizes its electric organs in the hypogean environment. Even if this was the case, and despite its obvious usefulness in this habitat, this cannot be considered a phylogenetic novelty since this is a character shared by all members of the order Gymnotiformes.

Family: Amblyopsidae
This is a family of freshwater fishes found in the southern and eastern United States. It is composed of six species, four of which are clearly troglomorphic; one is a trogloxene (facultative cavernicole) and one is found in swamps. Scales are cycloid but head is naked. Eyes range from small to vestigial. They have rows of sensory papillae on the head, body, and tail (Nelson 1994). The fact that the only epigean species of the family lives in dark swamp waters, one is a facultative cavernicole, and the other four are troglomorphic species makes this family the archetypical hypogean fish group.

62. ***Amblyopsis rosae*** (Eigenmann, 1897). **Common name:** Ozark cavefish (E). **Etymology:** *rosae* (L) after C.H. Eigenmann's wife and ichthyologist Rosa Smith. **Major synonyms:** *Typhlichthys rosae*, original description; *Troglichthys rosae* (Eigenmann, 1899). **History:** First reported from specimens collected from caves near Sarcoxie, Missuori by Garman (1889) as *Typhlichthys subterraneus*. The description was first published by Eigenmann (1897) with brief characterization involving eyes, and then described in more detail in Eigenmann (1898). **Distribution:** U.S.A., it is presently known from 41 caves in the Ozark Plateau, distributed over seven counties in three states, northwestern Arkansas, northeastern Oklahoma, and in the southern corner of Missouri (ca. 36°53′ N, 95°03′ W). The verified historic range was larger (Romero 1998a). **Habitat and ecology:** Found mostly in small cave streams with chert or rubble bottom. Also occasionally found in pools over silt and sand bottom. Stomachs examined contained copepods which constituted about 70–90% by volume with the balance being primarily small salamanders, crayfish, isopods, amphipods and young of their own species (Poulson 1963). **Remarks:** Rudimentary eyes hidden under the skin. Depigmented body with few nonfunctional pigmented cells. It is a very old and highly genetically distinct clade of amblyopsid fishes. It has been proposed that there are two subspecies of this fish: *A. rosae whitae* for the White River drainage and *A. rosae arkansasus* for the Middle Arkansas River drainage (Romero op. cit.).

63. ***Amblyopsis spelaea*** DeKay, 1842. **Common name:** northern cavefish (E). **Etymology:** *amblus* (Gr) blunt, point taken off; *ops* (Gr) eyes; *spelaea* (L) cave. **Major synonyms:** Originally described as *Amblyopsis spelaeus*. **History:** This was the first troglobitic species of fish ever described in the scientific literature. **Distribution:** U.S.A., in about 2500 caves in Kentucky and about 1800 caves in southern Indiana (ca. 37°–39° N, 85°–87° W). Its distribution may be limited by competition with another amblyopsid cavefish, *Typhlichthys subterraneus*. **Habitat and ecology:** Found only in caves and subterranean passages of well-developed karst terrain whose water comes from the precipitation diverted underground through sinkholes and sinking streams. It is most often found in caves with uniform silt-sand substrates. It is a top predator (Romero & Bennis 1998). **Remarks:** Rudimentary eyes hidden under the skin. Body depigmented with non-functional pigment cells. It has small cycloid scales. It has a well-defined annual cycle: breeding occurs during high water from February to April, the females carry the eggs in their gill cavities until hatching and carry the young until their lose their yolksacs, a total period of 4–5 months, then the young appear in late summer and early fall. Low reproductive rate. It is scotophilic (Green & Romero 1997).

64. ***Chologaster agassizii*** Putnam, 1872. **Common name:** spring cavefish (E). **Etymology:** *aggassizzi* (L) after Louis Agassiz who allowed Putnam to study the specimens. **Major synonyms:** *Forbesella agassizi* (Page & Burr, 1991), *Forbesichthys agassizi* (Boschung, 1992, Etnier & Starnes, 1993).

Preoccupied by *Forbesella* in tunicates, replaced by *Forbesichthys* Jordan, 1929. **History:** First collected in a well in Lebanon, Tennessee, and presented to the Museum of Comparative Zoology at Harvard, by J.M. Safford in January 1854. **Distribution:** U.S.A., central and western Kentucky (west to Tennessee River) to southern central Tennessee; west across southern Illinois to southeastern Missouri (ca. 35°00′–37°30′ N, 86°–92° W). **Habitat and ecology:** Occurs in springs and caves (but almost always near surface). Lives underground but emerges above ground in springs (Page & Burr op. cit.). **Remarks:** Active in springs at night and usually retreats underground during the day. The few individuals which venture into the spring portions of their habitat may show strong rheotaxis for half to one-minute periods but usually show strong thigmotaxis and hide under rocks or debris (Poulson 1963). Feeds at night on amphipods, midge larvae, tiny worms and microcrustaceans. Fecundity averages about 100 eggs per female (Etnier & Starnes op. cit.).

65. ***Speoplatyrhinus poulsoni*** Cooper & Kuehne, 1974. **Common name:** Alabama cavefish (E). **Etymology:** *speos* (Gr) cave; *platys* (Gr) flat; *rhinus* (Gr) nose, referring to the greatly flattened snout; *poulsoni* (L) after American cave fish researcher T.L. Poulson. **Major synonyms:** None. **History:** Captured by J.E. Cooper and his spouse on 19 March 1967 while collecting crayfish in Kay Cave, Alabama. **Distribution:** U.S.A., Alabama, apparently restricted to Key Cave, Lauderdale county, in the north bank of the Tennessee River. **Habitat and ecology:** The habitat is mostly lacustrine (lentic, phreatic groundwater) (for more details see Romero 1998c). Population is estimateed to be no more than 100 individuals, which would made of this one the most endangered fish species in the world. **Remarks:** No externally visible eyes. Body depigmented with few scattered nonfunctional melanophores. Body covered by embedded cycloid scales.

66. ***Typhlichthys subterraneus*** Girard, 1859. **Common name:** southern cavefish (E). **Etymology:** *typhlops* (Gr) blind, *ichthys* for iktus (Gr) fish; *subterraneus* (L) underground. **Major synonyms:** *Typhlichthys wyandotte* (Eigenmann, 1905), *T. osborni* (Eigenmann, 1905), *T. eigenmanni* (Hubbs, 1938). **History:** The first specimen was taken from a well near Bowling Green, Warren Co., Kentucky, and sent to the Smithsonian Institution by J.E. Younglove. **Distribution:** U.S.A., subterranean waters in two major disjunct ranges being separated by the Mississippi River, Ozark Plateau of central and southeastern Missouri and northeastern Arkansas, and Cumberland and Interior Low plateaus of northwest Alabama, northwest Georgia, central Tennessee and Kentucky and southern Indiana (ca. 37°00′ N, 86°29′ W). **Habitat and ecology:** Mostly lentic but also in pools of streams at or near water table (for more details see Romero 1998b). **Remarks:** Rudimentary eyes hidden under the skin. Body depigmented with few non-functional pigment cells. Scales minute and embedded. It does not respond to light (Green & Romero 1997).

Family: Bythitidae

This is a family composed of about 90 species of mostly marine fishes with a few representatives living in brackish and fresh waters. Their distribution includes the Atlantic, Indian, and Pacific Oceans. They usually have scales and the swimbladder is present. Their opercles usually have a strong spine. All members of this family are viviparous (Nelson 1994). They have no characters that can be labeled as preadaptaed to the hypogean environment, yet seven species belonging to two genera of the family are hypogean. Of the six species of *Lucifuga*, five are hypogean. *Lucifuga* is divided into two subgenera; the difference between the two is that *Lucifuga* (*Lucifuga*) has palatine bones but without teeth with pectoral fin rays numbering 12–14, usually 12–13, while *Lucifuga* (*Stygicola*) has palatine teeth and pectoral fin rays numbering 15–20. The species of this genus need to be studied using genetic techniques. Given that some of these species can be found together as well as the large overlap of characters among them, it raises questions of whether they represent just populations of no more than two species. Hybridizations may also occur. Until these species are studied from a population genetics perspective, rather than purely morphological, we will not know for sure. Of the five species of *Ogilbia,* two are hypogean.

67. ***Lucifuga (Lucifuga) subterraneus*** Poey, 1858. **Common name:** Cuban cusk-eel (E). **Etymology:** *fugere lucem* (L) because it is photophobic; *subterraneus* (L) underground. **Major synonyms:** None. **History:** The first person to report this fish was Tranquilino Sandalio de Noda in 1831, which predates the first reports of *Amblyopsis spelaea*. Through him and

others, Poey secured the specimens he used in his first published description (Romero 2001). **Distribution:** Cuba, in several anchialine caves of the Pinar del Rio and Habana provinces (ca. 21°50′ N, 84°57′ W). **Habitat and ecology:** In caves and sink-holes formed by solution of underlying limestone. Frequently found syntopically with *L. dentatus*. It can tolerate brackish waters. It feeds on cirolanid isopods. **Remarks:** Eyes poorly developed and not externally visible in many preserved specimens. Yellowish-white pigmentation. The body is covered with small, imbricate scales.

68. ***Lucifuga (Lucifuga) teresinarum*** Díaz Perez, 1988. **Common name:** None given. **Etymology:** *teresinarum* (L) after Maria Teresa del Valley Portilla and Maria Teresita de la Hoz Gonzalez, both faculty at the Universidad de la Habana. **Major synonyms:** None. **History:** Both the holotype and the paratype were captured on 20 October 1986 by A. Montoto Lima and E. García Machado, both skin divers. **Distribution:** Cuba, Habana province, Cuevas de Ashton (Cueva de la Lechuza and Cueva Baño II), Barrios de las Cañas Artemisa (ca. 22°50′ N, 82°40′ W). **Habitat and ecology:** In the Cueva de la Lechuga it is found with isopods and introduced guppies. In the Cueva Baño II is found with *L. subterraneus* and *L. dentatus* as well as several species of crustaceans. This fish species is prey of crustaceans, *Epilobocera* sp. **Remarks:** Eyes not visible externally. Depigmented. Cycloid scales on parts of the body. It has a number of characteristics that place it as a composite of *L.* (*L.*) *subterraneus* (upper profile of the head, lack of palatine teeth, number of pectoral rays) and *L.* (*S.*) *dentatus* (free caudal fin and male intromitent organ).

69. ***Lucifuga (Stygicola) dentatus*** Poey, 1858. **Common name:** toothed Cuban cusk-eel (E). **Etymology:** *dentatus* (L) with teeth. **Major synonyms**: *Stygicola dentatus* (original description). **History:** Same as for *L.* (*L.*) *subterraneus*. **Distribution:** Same as for *L.* (*L.*) *subterraneus* (ca. 21°50′ N, 84°57′ W). **Habitat and ecology:** Same as for *L. subterraneus*. Found syntopically with *L.* (*L.*) *subterraneus* and *L.* (*L.*) *teresinarum*. Feeds on cirolanid isopods. **Remarks:** Eyes not externally visible. Depigmented. Viviparous.

70. ***Lucifuga (Stygicola) simile*** Nalbant, 1981. **Common name:** None given. **Etymology:** *simile* (L) because of its intermediate features between *L. subterraneus* and *L. dentatus*. **Major synonyms:** None. **History:** Collected by C. Jubertie on 7 November 1970. **Distribution:** Cuba, Matanzas province, Grieta Punta de Guana, ca. 100 m from shoreline of Atlantic Ocean (ca. 23°04′ N, 81°35′ W). **Habitat and ecology:** Found in a sinkhole near the ocean. The only stomach contents reported so far for this species have been plant remains. **Remarks:** Completely depigmented, 57–60 anal rays. This species has palatine teeth as the other representatives of the subgenus *Stygicola*, but at the same time its caudal fin is confluent with both dorsal and anal fins, which is unique to *L. subterraneus*.

71. ***Lucifuga (Stygicola) spelaeotes*** Cohen & Robins, 1970. **Common name:** New Providence cusk-eel (E). **Etymology:** *spelaeotes* (Gr) cave dweller. **Major synonyms:** None. **History:** The first specimens were collected by Carleton Ray on 24 October 1967. **Distribution:** Bahamas, New Providence Island, Mermaid's Pool, SW Nassau (ca. 25°05′ N, 77°20′ W). **Habitat and ecology:** This fish lives in anchialine caves (Iliffe 1992) on several Bahamian Islands. Some of them (e.g., Mermaids Hole, Uncle Charlie's Blue Hole) are typical inland holes of the cenote type, circular openings between 30 and 250 m across and from 2 to 110 m depth. **Remarks:** Eyes not very well developed and small. The entire body is covered by small, rounded imbricate scales. It has more coloration than *L. dentatus* and seems more primitive than *L. dentatus*.

72. ***Ogilbia galapagosensis*** (Poll & LeLeup, 1965). **Common name:** Galapagos cuskeel (E). **Etymology:** *galapagosensis* (L) because it is found in the Galápagos Islands. **Major synonyms:** *Caecogilbia galapagosensis* (original description). **History:** Collected on 4 December 1964 by A. De Roy and N. Leleup. **Distribution:** Chile, Galápagos Archipelago, Santa Cruz Island (ca. 1°00′ S, 90°55′ W). **Habitat and ecology:** In crevices of a lagoon. Rare and difficult to observe (Illife 1991). **Remarks:** Eyes and pigmentation reduced and variable. However, in the pictures of this fish (Allen & Robertson 1994) examined by us there is no noticeable difference between *O. galapagosensis* and other species of the same genus living in the same area. It has small scales. Hardly a 'cave' fish, despite the fact that it is usually listed as such.

73. ***Ogilbia pearsei*** (Hubbs, 1938). **Common name:** Dama ciega blanca (S). **Etymology:** *pearsei* (L) after A.S. Pearse who collected the first specimens. **Major synonyms:** *Typhlias pearsei* (original description),

Typhliasina pearsi (e.g., Wilkens 1973, Chumba-Segura 1983). **History:** Collected by Pearse on 22 June 1936. **Distribution:** Mexico, Yucatán peninsula (see Chumba-Segura 1983 for distributional map) (ca. 20°40′ N, 88°32′ W). **Habitat and ecology:** Stable environments, clear waters. Does not seem to be very abundant (Chumba-Segura op. cit). **Remarks:** Eyes and pigmentation variable. Small scales covering only parts of the body. Sex ratio may be skewed towards females.

Family: Poecilidae
This family is represented by about 300 species of freshwater and brackish water fish found from eastern United States to South America and in Africa, including Madagascar (Nelson 1994). No characters that could be assigned as preadapted. Only one hypogean species.

74. ***Poecilia mexicana*** Steindachner, 1863. **Common name:** cave shortfin molly (E). **Etymology:** Not applicable. **Major synonyms:** *Poecilia sphenops* (Valenciennes, 1846), *Mollienisia sphenops* (Regan, 1913), *Mollienisia mexicana* (Hubbs, 1961). **History:** The first cave population was described by Gordon & Rosen (1962). **Distribution:** Mexico, Estado de Tabasco, Cueva de las Sardinas, near Tapijulapa Village, about 26 km south of Teapa (ca. 18°51′ N, 97°08′ W). **Habitat and ecology:** The environment is a calcium sulfate stream. **Remarks:** There is a lot of variation in the troglomorphic characters: the deeper you go into the cave, the more the eyes are reduced in size and the lower the pigmentation, while there is an increase in the thickness of the lips of the mouths (Parzefal 2001).

Family: Synbranchidae
This family is made up of about 17 species of freshwater and occasionally brackish water fishes. Found in tropical and subtropical waters in Asia, Indo-Australian Archipelago, West Africa (Liberia), Mexico, Central and South America. They are eellike fishes with no pectoral or pelvic fins; rudimentary dorsal and anal fins; and a caudal fin that is small or rudimentary, or lacking altogether. Scales are absent or confined to the caudal region. They have small eyes and lack swimbladders or ribs. Mostly air breathers. Many are burrowers (Nelson 1994). All these characters can be considered preadaptive to the hypogean environment.

There are two genera with two species each that are hypogean. *Ophisternon* is a genus of six species of circumtropical distribution with a disjunct distribution of species; they all lack scales. *Monopterus* is an Old World genus represented by seven species that also lack scales; two of those species are hypogean. The contention by Galán (1982, 1995) that *M. boueti* was found in caves is erroneous. Given the geographic proximity of the two hypogean species of *Monopterus*, it would be interesting to investigate the genetic relationships between both as well with their putative ancestor *M. albus*. A depigmented individual of *Synbranchus marmoratus* was collected in a cave in northeastern Venezuela in 1979 (Galán 1982), but this fish recovered its pigmentation when exposed to light. Further studies in this locality are needed in order to confirm the troglomorphic nature of this potential hypogean population.

75. ***Monopterus (Monopterus) eapeni*** Talwar in Talwar & Jhingran, 1991. **Common name:** Malabar swampeel (E). **Etymology:** *eapeni* (L) after K.C. Eapen, who first described the species under a preoccupied name. **Major synonyms:** *Monopterus indicus* (Eapen, 1963). Secondarily preoccupied in *Monopterus* by *Amphipnous indicus* Silas & Dawson, 1961, replaced by *Monopterus eapeni* Talwar, 1991. **History:** No information available. **Distribution:** India, Kerala state, well at Kottayam (9°30′ N, 76°33′ E). **Habitat and ecology:** No information available except that another troglomorphic fish, *Horaglanis krishnai*, is found in the same well. **Remarks:** Sunken eyes. Scaleless. It may have derived from *M. albus*.

76. ***Monopterus roseni*** Bailey & Gans, 1998. **Common name:** None given. **Etymology:** *roseni* (L) after Donn E. Rosen, an American ichthyologist. **Major synonyms:** None. **History:** Collected in July 1986 by Carl Gans, S.K. Saraswat and C. Rajasunderam. **Distribution:** India, Kerala state, from a well, Periyam village (10°38′ N, 76°22′ E). This location is about 102 km slightly west of north from Kottayam, the type locality of *M. eapeni*. **Habitat and ecology:** Collected in the village water supply, 50 m a.s.l. **Remarks:** Blind. Depigmented. Scaleless. The non-troglomorphic differences between these two species are mostly in body proportions and number of vertebrae.

77. ***Ophisternon candidum*** (Mees, 1962). **Common name:** blind cave eel (E). **Etymology:** *candidum* (L)

whitish, for the pale appearance of the skin. **Major synonyms:** *Anommatophasma candidum* (original description). **History:** First observed by Alf Snell who mentioned the existence of a 'blind eel' sometime in 1959, and in that year A.N. Douglas and G.F. Mees collected the fist specimen. **Distribution:** Australia, western and northeastern coastal plain of the Cape Range peninsula. Found in 11 locations (two now destroyed) in numbers ranging from one to three individuals (for a distributional map see Humphreys 1999) (ca. 22°07′ S, 113°57′ E). **Habitat and ecology:** In wells, sinkholes, and caves, and possibly in groundwater. This species feeds on invertebrates, both hypogean and epigean, that accidentally fall into their habitat (Humphreys & Feinberg 1995). **Remarks:** No externally visible eyes. Whitish coloration in the skin (Romero & Vanselow 2000b). Scaleless. It may have derived from *Synbranchus bengalensis*, the only other synbranchid known to occur in Western Australia.

78. ***Ophisternon infernale*** (Hubbs, 1938). **Common name:** blind swamp eel (E), anguila ciega de los cenotes (S). **Etymology:** *infernale* (L) from hell (Hubbs liked to name cave fishes with diabolic names based on the idea of hell being underground; see etymology for *Satan eurystomus*). **Major synonyms:** *Pluto infernalis* (original description); *Furmastix infernalis* (Whitley, 1951b); *Synbranchus infernalis* (Mees, 1962). **History:** The holotype was collected by A.S. Pearse on 7 July 1936. **Distribution:** Mexico, in the cenotes of central and northern Yucatan peninsula (see Chumba-Segura 1984 for distributional map) (ca. 20°40′ N, 88°30′ W). **Habitat and ecology:** Found in shallow waters of sinkholes with sediments that are nutrient-rich. **Remarks:** No externally visible eyes. Depigmented. Scaleless. Most likely is derived from *Synbranchus marmoratus*.

Family: Eleotridae
This is a family of mostly tropical and subtropical fishes, both marine and freshwater, represented by about 150 species (Nelson 1994). No particular characteristics that could be seen as preadapted.

79. ***Milyeringa veritas*** Whitley, 1945. **Common name:** blind cave gudgeon (E). **Etymology:** *Milyeringa* (L) after Milyering, Yardie, Western Australia, the type locality; *veritas* (L) true, 'because, like Truth, it was found at the bottom of a well' (Whitley 1951a). **Major synonyms:** None. **History:** The first twelve specimens were collected from a freshwater well by its owner, Eric Payne, on 18 October 1944. It is believed that the existence of this fish was known by the locals since the 1920's. **Distribution:** Australia, western and northeastern coastal plain of the Cape Range peninsula and Barrow Island. Found in 19 locations (now absent from two) in numbers of up to more than 100 individuals (Humphreys 2001) (ca. 22°00′ S, 114°00′ E). **Habitat and ecology:** In wells, sinkholes, caves, and bores and probably widely in the anchialine and fresh groundwater. It is sympatric with *Ophisternon candidum* but seems much more abundant. **Remarks:** Eyes not externally visible. Whitish coloration in the skin. Scales are reduced and entirely absent from the head; there are about 28 rows in a longitudinal line. *M. veritas* closest affinity is possibly with the eleotrid genus *Butis*, a mangrove dwelling genus of brackish and freshwaters on the Indo-Australian Archipelago (Romero & Vanselow 2000a).

80. ***Oxyeleotris caeca*** Allen, 1996. **Common name:** ihaaribi (blind, in Yagi language). **Etymology:** *caeca* (L) blind. **Major synonyms:** None. **History:** Rumors of the existence of this fish date back to 1978 but it was not captured until June 1995 by inhabitants of Kafa Village. **Distribution:** Papua New Guinea, creek near Kafa, tributary of the Mubi River, upper Kikori River system (ca. 6°34′30″ S, 143°29′30″ E). **Habitat and ecology:** In a creek connected to a small cave 100–200 m upstream, at an elevation of 650 m a.s.l. **Remarks:** Eyes not externally visible. Mostly depigmented except for the dorsal region. Most of the scales embedded. Description based on a single specimen. Seems to be closely related to *O. fimbriata*, one of the most widely distributed gudgeons in New Guinea and the only freshwater-restricted member of the genus occurring on both sides of the Central Dividing Range.

81. ***Typhleotris madagascariensis*** Petit, 1933. **Common name:** toho (in Mihafale). **Etymology:** *typhlos* (Gr) blind; *eleotris* (L) because of its relation to the genus *Eleotris*; *madagascariensis* (L) after the island where it was found. **Major synonyms:** None. **History:** Collected by H. Perrier de la Bàthie at an unspecified date. **Distribution:** Madagascar, province of Toliara: south of Onilahy River, at Mitoho caves, underground waters in Mahafaly, ca. 1 km north of the natural reserve of Lake Tsimanampetsotsa (ca. 24°08′ S, 43°48′ E). **Habitat and ecology:** In underground lakes below

what is now a desert where it feeds on crustaceans and insects. **Remarks:** Sunken eyes. Variable levels of depigmentation (Angel 1949). Degenerated ctenoid scales (Poisson & Besairie 1947).

82. ***Typhleotris pauliani*** Arnoult, 1959. **Common name:** None given. **Etymology:** *pauliani* (L) after R. Pauliani who secured the first specimens. **Major synonyms:** None. **History:** Pauliani collected the first specimens in July 1956. **Distribution:** Madagascar, Andalambezo region near Morombe, north of Onilahy River (ca. 23°00′ S, 44°30′ E). **Habitat and ecology:** In underground lakes below what is now a desert where it feeds on bat guano and aquatic insects. **Remarks:** No externally visible eyes. Depigmented. Irregular (degenerated?) scales that sometimes look cycloid, sometimes ctenoid.

Family: Gobiidae
This is a very large family (possibly >2000 spp.) of mostly marine and brackish species. Mostly of small size, they tend to be cryptic bottom dwelling carnivores feeding on small benthic invertebrates. Very common in fresh waters on oceanic islands (Nelson 1994). Thus, it is not surprising that all four hypogean species described thus far are from islands: two from Japan, one from the Phillippines and one from Madagascar.

The two species of *Luciogobius* seem to be valid based on the differences in their meristic, non-troglomorphic differences and allopatry; both of them most probably descended from *Luciogobius guttatus*. The third species also seems to be valid: the monotypic genus *Caecogobius* does not appear to be related to any other extant species. More information is needed regarding the fourth, *Glossogobius ankaranensis,* given the fact that it has been found syntopically with the congeneric *Glossogobius callidus*.

83. ***Caecogobius cryptophthalmus*** Berti & Ercolini, 1991. **Common name:** None given. **Etymology:** *caeco* (L) blind; *gobius* (L) for its relationship to species of the genus *Gobius*; *crypto* (L) hidden; *ophthalmus* (L) eyes. **Major synonyms:** None. **History:** The first four specimens were collected by members of the Federazione Speleologica Veneta in January and February 1987. **Distribution:** Philippines, Samar Island, Calbiga Cave system (12°00′ N, 125°00′ E). **Habitat and ecology:** No information available. **Remarks:** No externally visible eyes. Strongly reduced pigmentation. Phylogenetic relationships are unknown.

84. ***Glossogobius ankaranensis*** Banister, 1994. **Common name:** None given. **Etymology:** *ankaranensis* (L) after the region of Ankarana where this fish was first collected. **Major synonyms:** *Gobius ankaranensis* (Wilson, 1996). **History:** First collected by Jane Wilson and her colleagues in 1986. **Distribution:** Madagascar, Ankarana reserve in a subterranean river (see Wilson op. cit. for a map of the area) (ca. 12°50′ S, 49°10′ E). **Habitat and ecology:** In a cave where they feed on blind and eyed shrimps. They may also feed on insects and bat guano. It is found syntopically with eyed *Glossogobius callidus*. **Remarks:** Eyes reduced to small spots. Body depigmented. Thin and weak ctenoid scales.

85. ***Luciogobius albus*** Regan, 1940. **Common name:** dôkutsu-mimizuhaze (J), cave earthworm goby (E). **Etymology:** *albus* (L) white, referring to the color of the skin. **Major synonyms:** None. **History:** First reported by Masatoshi Iwata in 1934 (Torii 1951). Based on two specimens of *Luciogobius guttatus guttatus* of Tomiyama (1936, p. 51, figure 10C). **Distribution:** Asia, Japan, Prefectures of Wakayama, Kochi, Shimane, Nagasaki and the Gotô islands (Masuda et al. 1984). Originally collected in the cave of Daikon-zima in Simani-ken (ca. 32°–36° N, 128°–136° E; but not throughout the entire square defined by these coordinates). **Habitat and ecology:** Inhabits underground waters near the sea. **Remarks:** Vestigial eyes. Depigmented (pinkish due to the color of the blood). Like *L. pallidus*, it lacks free rays on the pectoral fin.

86. ***Luciogobius pallidus*** Regan, 1940. **Common name:** ido-mimizuhaze (J), well earthworm goby (E). **Etymology:** *pallidus* (L) for pale referring to the color of the skin. **Major synonyms:** None. **History:** First reported by Masatoshi Iwata in 1934 (Torii 1951). Based on artesian well specimens of *Luciogobius guttatus guttatus* of Tomiyama (1936, p. 51, figure 10D). **Distribution:** Asia, Japan, prefectures of Mie, Wakayama, Kochi, Ehime, Kumamoto, and Yamaguchi (Masuda et al 1984, Uéno & Morimoto 1970). Original specimens collected from an artesian well (ca. 32°–36° N, 128°–136° E; but not throughout the entire square defined by these coordinates). **Habitat and ecology:** Inhabits underground waters near

the sea. **Remarks:** Reduced, sunken eyes. Slightly depigmented. Like *L. albus*, it lacks free rays on the pectoral fin.

Acknowledgements

Jerry Reedy helped with the etymologies. Giang Ho helped with the translation of literature in Vietnamese. Qin Zheng and Janet Ebaugh were helpful translating texts from Chinese. Carol Gersmehl and the Macalester College Cartography Laboratory assisted with the maps.

References cited

Allen, G.R. 1996. *Oxyeleotris caeca*, a new species of blind cave fish (Eleotridae) from Papua New Guinea. Rev. Fr. Aquariol. 23: 43–46.

Allen, G.R. & D.R. Robertson. 1994. Fishes of the tropical eastern Pacific. Crawford House Press, Bathurst. 332 pp.

Alvarez, J. 1946. Revisión del género *Anoptichthys* con descripción de una especie nueva (Pisc., Characidae). An. Esc. Nac. Cienc. Biol. Mex. 4: 263–282.

Alvarez, J. 1947. Descripción de *Anoptichthys hubbsi* caracínido ciego de la cueva de los Sabinos, S.L.P. Rev. Soc. Mex. Hist. Nat. 8: 215–219.

Andreani Armas, L.E. 1990. Estudio comparativo de dos poblaciones, una hipogea y otra epigea de *Trichomycterus* sp. (Siluriformes, Trichomycteridae). Bol. Soc. Venez. Speleol. (24): 7–11.

Andreas Buckup, P. 1988. The genus *Heptapterus* (Teleostei, Pimelodidae) in southern Brazil and Uruguay with the description of a new species. Copeia 1988: 641–653.

Angel, F. 1949. Contribution a l'etude du *Typhleotris madagascariensis*, poisson aveugle, cavernicole, du Sud-Ouest de Madagascar. Bull. Mus. Hist. Nat. Paris 21: 56–59.

Anonymous. 1981. Freshwater fishes of Guangxi. Fisheries Research Institute, Guangxi, Remnin Press, Nanking. 257 pp. (in Chinese).

Arnoult, J. 1959. Une nouvelle espèce de poisson aveugle de Madagascar: *Typhleotris pauliani* n. sp. Mem. Inst. Sci. Madagascar 13A: 133–138.

Avise, J.C. & R.K. Selander. 1972. Evolutionary genetics of cave-dwelling fishes of the genus *Astyanax*. Evol. 26: 1–19.

Bailey, R.M. & C. Gans. 1998. Two new synbranchid fishes, *Monopterus roseni* from Peninsular India and *M. desilvai* from Sri Lanka. Occas. Pap. Mus. Zool. Univ. Mich. (726): 1–18.

Baird, S.F. & C.F. Girard. 1854. Descriptions of new species of fishes collected in Texas, New Mexico and Sonora, by Mr. John H. Clark, on the U.S. and Mexican Boundary Survey, and in Texas by Capt. Stewart Van Vliet, U.S.A. Proc. Acad. Nat. Sci. Phil. 7: 24–29.

Banister, K.E. 1987. Two new species of *Garra* (Teleostei-Cyprinidae) from the Arabian peninsula. Bull. Brit. Mus. Nat. Hist. 52: 59–70.

Banister, K.E. 1994. *Glossogobius ankaranensis*, a new species of blind cave goby from Madagascar (Pisces: Gobioidei: Gobiidae). Aqua, J. Ichthy. Aquat. Biol. 1: 25–28.

Banister, K.E., J. Bell & M. Crumpler. 1992. Omani blind cave fish. Aquarist and Pondkeeper 1992: 38–40.

Banister, K.E. & M.K. Bunni. 1980. A new blind cyprinid fish from Iraq. Bull. Brit. Mus. Nat. Hist. (Zool.). 38: 151–158.

Berti, R. 1990. Specific chemical recognition in *Phreatichthys andruzzii* (Pisces, Cyprinidae). Mem. Biospeleol. 17: 31.

Berti, R. & A. Ercolini. 1979. Aggressive behaviour in the anophthalmic phreatic fish *Uegitglanis zammaranoi* Gianferrari (Claridae, Siluriformes). Monit. Zool. Ital 13: 197.

Berti, R. & A. Ercolini 1991. *Caecogobius, cryptophthalmus* n. gen. n. sp. (Gobiidae Gobiinae), the first stygobic fish from Philippines. Trop. Zool. 4: 129–138.

Berti, R. & G. Thinès. 1980. Influence of chemical signals on the topographic orientation of the cave fish *Caecobarbus geertsi* – Boulenger (Pisces, Cyprinidae). Experientia 36: 1384–1385.

Berti, R. & L. Zorn. 2001. Locomotory responses of the cave cyprinid *Phreatichthys andruzzii* to chemical signals from conspecifics and related species: new findings. Env. Biol. Fish. 62: 107–114 (this volume).

Bianchi. S. 1975. Preliminary observations on the epidermis of *Uegitglanis zammaranoi* Gianferrari (Claridae: Siluriformes). Monit. Zool. Ital. 5: 91–101.

Bianchi, S., G. Delfino & A. Ercolini. 1978. Morphology and structure of the olfactory organ in *Uegitglanis zammaranoi* Gianferrari (Claridae, Siluriformes), anophthalmic phreatic fish from Somalia. Monit. Zool. Ital. 10: 157–171.

Biswas, J. 1991. Annual modulation of diel motor activity rhythm of the dusk active loach *Nemacheils evezardi* (Day). A correlation between day length and circadian parameters. Proc. Indian Nat. Acad. Sci. B57: 339–346.

Biswas, J., A.K. Pati, R.K. Pradhan & R.S. Kanoje. 1990. Comparative aspects of reproductive phase dependent adjustments in behavioural circadian rhythms of epigean and hypogean fish. Comp. Physiol. Ecol. 15: 134–139.

Borodin, N.A. 1927. A new blind catfish from Brazil. Amer. Mus. Novit. (263): 1–5.

Boschung, H.T. 1992. Catalogue of freshwater and marine fishes of Alabama. Bull. Alabama Mus. Nat. Hist. (14): 1–266.

Boucquey, C., G. Thines & C. Van Der Borght. 1965. Étude comparative de la capacité photopathique et de l'activité chez le poisson cavernicole *Anoptichthys antrobius*, chez la forme epigee ancestrale *Astyanax mexicanus*, et chez les hybrides F_1 (*Astyanax* x *Anoptichthys*) et F_2. pp. 79–103. *In*: J. Mendioni (ed.) La Distribution Temporelle des Activités Animales et Humaines, Masson et Cie, Paris.

Boulenger, G. 1921. Description d'un poisson aveugle decouvert par M.G. Geerts dans la grotte de Thysville (Bas-Congo). Rev. Zool. Afr. 9: 252–253.

Breder, C.M. 1942. Descriptive ecology of La Cueva Chica, with especial reference to the blind fish, *Anoptichthys*. Zoologica 27: 7–15.

Breder, C.M. & E.B. Gresser. 1941. Correlations between structural eye defects and behavior in the Mexican blind characin. Zoologica 26: 123–131.

Brittan, M.R. & J.E. Böhlke. 1965. A new blind characid fish from southeaster Brazil. Notulae Naturae (380): 1–4.

Brown, R.W. 1956. Composition of scientific words. Smithsonian Institution Press, Washington, D.C. 882 pp.

Bruton, M.N. 1995. Threatened fishes of the world: *Clarias cavernicola* Trewavas, 1936 (Clariidae). Env. Biol. Fish. 43: 162.

Bruun, A.F. & E.W. Kaiser. 1944. *Iranocypris typhlops*, the first true cave fish from Asia. Danish Scient. Invest. Iran (4): 1–8.

Burr, B.M., G.L. Adams, J.K. Krejca, R.J. Paul & M.L. Warren, Jr. 2001. Troglomorphic sculpins of the *Cottus carolinae* species group in Perry County, Missouri: distribution, external morphology, and conservation status review. Env. Biol. Fish. 62: 279–296 (this volume).

Cardona, L. & G. Guerao. 1994. *Astroblepus riberae*, una nueva especie de siluriforme cavernícola del Peru (Osteichthyes: Astroblepidae). Mém. Biospéléol. (21): 21–24.

Carranza, J. 1954. Descripcion del primer bagre anoftalmo y depigmentado encontrado en aguas mexicanas (Pisc., Ameiur.). Ciencia (Mexico) 14: 129–136.

Carvalho, A.M. & M.C.C. de Pinna. 1986. Estudo de uma populacao hipogea de *Trichomycterus* (Ostariophysi, Siluroidei, Trichomucteridae) da Gruta de Olhos d'Agua, MG. Espeleo-Tema 15: 53–64.

Chardon, M. 1966. Specialisation anatomique de l'appareil de Weber d'*Astroblepus pholeter*, silure cavernicole microphtalme de la Republique de l'Equateur. Acad. Roy. Belg. Bull. Cl. Sci. 52: 846–863.

Chen, Y.-Y., X.-L. Chu, Z.-Y. Luo & J.-Y. Wu. 1988. A new blind cyprinid fish from Yunnan, China with a reference to the evolution of its characters. Acta Zool. Sin. 34: 64–70 (in Chinese).

Chen, Y.-R., J.-X. Yang & J.-H. Lan. 1997. One new species of blind cavefish from Guangxi with comments on its phylogenetic status (Cypriniformes: Cyprinidae: Barbinae). Acta Zootaxon. Sin. 22: 219–223 (in Chinese).

Chen, Y.-R., J.-X. Yang, B. Sket & G. Aljancic. 1998. A new blind cave loach of *Paracobitis* with comments on its characters evolution. Zool. Res. 19: 59–63 (in Chinese).

Chen, Y.-R., J.-X. Yang & G.-C. Xu. 1992. A new blind loach of *Triplophysa* from Yunnan Stone Forest with comments on its phylogenetic reiationship [sic]. Zool. Res. 13: 17–23 (in Chinese).

Chen, Y.-R., J.-X. Yang & Z.-G. Zhu. 1994. A new fish of the genus *Sinocyclocheilus* from Yunnan with comments on its characteristic adaptation (Cypriniformes: Cyprinidae). Acta Zool. Sin. 19: 246–253 (in Chinese).

Chu, S.-L. & Y.-R. Chen. 1979. A new blind cobitid fish (Pisces, Cypriniformes) from subterranean waters in Yunnan, China. Acta Zool. Sin. 25: 285–287 (in Chinese).

Chu, X.-L. & Y.-R. Chen. 1982. A new genus and species of blind cyprinid fish from China with special reference to its relationships. Acta Zool. Sin. 28: 383–388 (in Chinese).

Chumba-Segura, L. 1983. Brotulidae: *Typhliasina pearsei*. Fauna de Los Cenotes de Yucatan (4): 1–9.

Chumba-Segura, L. 1984. Synbranchidae: *Ophisternon infernale*. Fauna de Los Cenotes de Yucatán (6): 1–9.

Coad, B.W. 1996. Threatened fishes of the world: *Iranocypris typhlops* Bruun & Kaiser, 1944 (Cyprinidae). Env. Biol. Fish. 46: 374.

Collette, B.B. 1962. *Astroblepus pholeter*, a new species of cave-dwelling catfish from eastern Ecuador. Proc. Biol. Soc. Wash. 75: 311–314.

Cohen, D.M. & C.R. Robins. 1970. A new ophidioid fish (genus *Lucifuga*) from a limestone sink, New Providence Island, Bahamas. Proc. Biol. Soc. Wash. 83: 133–144.

Contreras-Balderas, S. & M. de L. Lozano-Vilano. 1988. Problemas nomenclaturales de las formas mexicanas del género *Astyanax* (Pisces:Characidae). Zoología Informa (38): 1–13.

Cooper, J.E. & R.A. Kuehne. 1974. *Speoplatyrhinus poulsoni*, a new genus and species of subterranean fish from Alabama. Copeia 1974: 486–493.

Cuvier, G. 1819. Sur les poissons du sous-genre *Hydrocyon*, sur deux nouvelles espèces de *Chalceus*, sur trois nouvelles espèces du serrasalmes, et sur l'*Argentina glossodonta* de Forskahl, qui est l'*Albula gonorhynchus* de Bloch. Mem. Mus. Natl. Hist. Nat. 5: 351–379.

Dai, D.-Y. 1988. Un nouveau poisson cavernicole. Spelunca Mem. (16): 88–89.

Day, F. 1872. Monograph of Indian Cyprinidae, Part IV. J. Asiatic Soc. 41: 171–198.

DeKay, J.E. 1842. Zoology of New York or the New-York fauna, Part IV, fishes. W. & A. White & J. Visscher, Albany. 566 pp.

De Filippi, F. 1853. Nouvelles espèces de poissons. Rev. Mag. Zool. (Ser. 2) v. 5: 164–171.

de Pinna, M.C.C. 1998. Phylogenetic relationships of neotropical Siluriformes (Teleostei: Ostariophysi): historical overview and synthesis of hypotheses. pp. 279–330. *In*: L.R. Malabarba, R.E. Reis, R.P. Vari, Z.M. Lucena & C.A.S. Lucena (ed.) Phylogeny and Classification of Neotropical Fishes, Edipucrs, Porto Alegre.

Díaz Perez, P.A. 1988. *Lucifuga teresinarum* sp. n., nueva especie cubana de peces troglobios (Ophidiiformes, Bythitidae). Rev. Biol. (Habana) 2: 37–43.

Di Caporiacco, L. 1926. Un nuovo genere di ciprinide somalo delle acque di pozzo (One new cyprinid genus from a well in Somalia). Monit. Zool. Ital. 37: 23–25.

Di Capporiacco, L. 1927. Pesci recolti in Somalia nel 1924, della spedizione Stefanini e Puccini. Monit. Zool. Ital. 38: 84–89.

Durand, J. 1968. Etude des poissons récoltés dans la grotte de Umayalanta (Bolivie), *Trichomycterus chaberti* sp. n. Ann. Spéléol. 23: 343–353.

Eapen, K.C. 1963. A new species of *Monoptereus* from South India. Bull. Dept. Mar. Biol. Ocean. Univ. Kerala 1: 129–132.

Eigenmann, C.H. 1897. The Amblyopsidae and eyes of blind fishes. Proc. Indiana Acad. Sci. 1897: 230–231.

Eigenmann, C.H. 1898. On the Amblyopsidae. Science 7: 227.

Eigenmann, C.H. 1899. A case of convergence. Proc. Indiana Acad. Sci. 1899: 247–251.

Eigenmann, C.H. 1905. Divergence and convergence in fishes. Biol. Bull. Woods Hole 8: 59–66.

Eigenmann, C.H. 1912. The freshwater fishes of British Guiana, including a study of the ecological grouping of species, and the relation of the fauna of the plateau to that of the lowlands. Mem. Carnegie Mus. 5: 1–103.

Eigenmann, C.H. 1919. *Trogloglanis pattersoni* a new blind fish from San Antonio, Texas. Proc. Am. Philos. Soc. 58: 397–400.

Erckens, W. & F. Weber. 1976. Rudiments of an ability for time measuremens in cavernicole fish *Anoptichthys jordani* Hubbs & Innes (Pisces: Characidae). Experientia 32: 1297–1299.

Ercolini, A. & R. Berti. 1975. Light sensitivity experiments and morphology studies of the blind phreatic fish *Phreatichthys andruzzii* Vinciguerra from Somalia. Monit. Zool. Ital. 6: 29–43.

Ercolini, A. & R. Berti. 1977. Morphology and response to light of *Uegitglanis zammaranoi* Gianferrari, anophthalmic phreatic fish from Somalia. Monit. Zool. Ital. 9: 183–199.

Ercolini, A. & R. Berti. 1978. Morphology and response to light of *Barbopsis devecchii* di Caporiacco (Cyprinidae), microphthalmic phreatic fish from Somalia. Monit. Zool. Ital. 10: 299–314.

Ercolini, A., R. Berti, L. Chelazzi & G. Messana. 1982. Researches on the phreatobic fishes of Somalia: achievements and prospects. Monit. Zool. Ital. 17: 219–241.

Ercolini, A., R. Berti, L. Chelazzi & G. Messana. 1987. Oxygen consumption in hypogean and epigean cyprinids (Pisces). Monit. Zool. Ital. 22(Suppl.): 23–30.

Espinasa, L., P. Rivas-Manzano & H. Espinosa Pérez. 2001. A new blind cave fish population of genus *Astyanax*: geography, morphology and behavior. Env. Biol. Fish. 62: 339–344 (this volume).

Etnier, D.A. & W.C. Starnes. 1993. The fishes of Tennessee. University of Tennessee Press, Knoxville. 681 pp.

Fowler, H.W. & H. Steinitz. 1956. Fishes from Cyprus, Iran, Iraq, Israel and Oman. Bull. Res. Counc. Israel. 5B: 260–292.

Galán, C. 1982. Fauna troglobia de Venezuela: sinopsis, biología, ambiente, distribución y evolución. Bol. Soc. Venez. Speleol. (29): 20–38.

Galán, C. 1995. Fauna troglobia de Venezuela: sinopsis, biología, ambiente, distribución y evolución. Bol. Soc. Venez. Speleol. 29: 20–38.

Gerard, P. 1936. Sur l'existence des vestiges oculaires chez *Caecobarbus geertsi*. Mus. Roy. Hist. Nat. Melages P. Pelseneer 1936: 549–552.

Géry, J. 1977. Characoids of the world. T.F.H. Publications, Neptune City. 672 pp.

Gianferrari, L. 1923. *Uegitglanis zammaranoi* un nuovo siluride cieco africano (*Uegitglanis zammaranoi*, a new African cave silurid). Soc. Ital. Sci. Nat. Milan. 62: 1–3.

Gianferrari, L. 1930. Un nuovo ciprinide Somalo (*Barbopsis Stefaninii* Gianf.). Atti. Soc. Ital. Sci. Nat. (Milano) 69: 106–111.

Gianferrari, L. 1934. I pesci cavernicoli ciechi africani. Ross. Fauna 2: 33–36.

Girard, C.F. 1859. Ichthyological notes. Proc. Acad. Nat. Sci. Phila. 1859: 63–64.

Goeldi, E.A. 1905. Nova zoologica aus der Amazonas-Region. Neue Wirbeltiere. C.R. 6. Congr. Internat. Zool., Berne 1905: 542–549.

Gordon, M.S. & G.E. Rosen. 1962. A cavernicolous form of the poeciliid fish *Poecilia sphenops* from Tabasco, Mexico. Copeia 1962: 360–368.

Gotch, A.F. 1995. Latin names explained. A guide to the scientific classification of reptiles, birds & mammals. Facts on File, New York. 714 pp.

Green, S. & A. Romero. 1997. Responses to light in two blind cave fishes (*Amblyopsis spelaea* and *Typhlichthys subterraneus*) (Pisces: Amblyopsidae). Env. Biol. Fish. 50: 167–174.

Greenfield, D.W., T.A. Greenfield & R.L. Woods. 1982. A new subspecies of cave-dwelling pimelodid catfish, *Rhamdia laticauda typhla* from Belize, Central America. Brenesia (19/20): 563–576.

Greenwood, P.H. 1976. A new and eyeless cobitid fish (Pisces, Cypriniformes) from the Zagros Mountains, Iran. J. Zool. 180: 129–137.

Haseman, J.D. 1911. Descriptions of some new species of fishes and miscellaneous notes on others obtained during the expedition of the Carnegie Museum to central South America. Ann. Carnegie Mus. 7: 315–328.

Hendrickson, D.A., J.K. Krejca & J.M. Rodríguez Martinez. 2001. Mexican blindcats genus *Prietella* (Siluriformes: Ictaluridae): an overview of recent explorations. Env. Biol. Fish. 62: 315–337 (this volume).

Heuts, M.-J. 1951. Ecology, variation and adaptation of the blind African cave fish *Caecobarbus geerstsii* Blgr. Ann. Soc. Roy. Zool. Belg. 82: 115–230.

Heuts, M.J. & N. Leleup. 1954. La geographie et l'ecologie des grottes du Bas-Congoles habitats de *Caecobarbus geertsi* Blgr. Ann. Mus. Roy. Congo Belge 35: 1–71.

Hora, S.L. 1924. Fish of the Siju Cave, Garo Hills, Assam. Rec. Indian Mus. 26: 27–31.

Hubbs, C.L. 1936. XVII. Fishes of the Yucatan Peninsula. Carnegie Inst. Wash. Publ. (457): 157–287.

Hubbs, C.L. 1938. Fishes from the caves of Yucatan. Carnegie Inst. Wash. Publ. (491): 261–295.

Hubbs, C.L. 1961. Isolating mechanisms in the speciation of fishes. pp. 5–23. *In*: W.F. Blair (ed.) Vertebrate Speciation: A symposium, University of Texas Press, Austin.

Hubbs, C.L. & R.M. Bailey. 1947. Blind catfishes from artesian waters of Texas. Occas. Pap. Mus. Zool. Univ. Mich. (499): 1–15

Hubbs, C.L. & W.T. Innes. 1936. The first known blind fish of the family Characidae: a new genus from Mexico. Occas. Pap. Mus. Zool. Univ. Mich. (342): 1–7.

Humboldt, A. von. 1805. Mémoire sur une nouvelle espèce de pimelode, jetée par les volcans du Royaume de Quito. Voyage de Humboldt et Bonpland, Deuxième partie, Observations de Zoologie et d'Anatomie comparée, Paris 1: 21–25.

Humphreys, W.F. 1999. The distribution of Australian cave fishes. Rec. Western Austr. Mus. 19: 469–472.

Humphreys, W.F. 2001. *Milyeringa veritas* (Eleotridae), a remarkably versatile cave fish from the arid tropics of northwestern Australia. Env. Biol. Fish. 62: 297–313 (this volume).

Humphreys, W.F. & M.N. Feinberg. 1995. Food of the blind cave fishes of northwestern Australia. Rec. Western. Austr. Mus. 17: 29–33.

Iliffe, T.M. 1991. Archialine fauna of the Galapagos islands. pp. 209–231. *In*: M.J. James (ed.) Galapagos Marine Invertebrates, Plenum Press, New York.

Iliffe, T.M. 1992. Anchialine cave biology. pp. 613–636. *In*: A.I. Camacho (ed.) The Natural History of Biospeleology, Museo Nacional de Ciencias Naturales, Madrid.

Jaeger, F. & L. Waibel. 1921. Beiträge zur Landeskunde von Südwestafrika. E.S. Mittler, Berlin. 80 pp.

Jankowska, M. & G. Thinès. 1982. A comparative study of group density in cave and epigean fishes. Behav. Proc. 7: 281–294.
Kirby, R.F., K.W. Thompson & C. Hubbs. 1977. Karyotypic similarities between the Mexican and blind tetras. Copeia 1977: 578–580.
Kottelat, M. 1988. Two species of cavefishes from northern Thailand in the genera *Nemacheilus* and *Homaloptera* (Osteichthyes: Homalopteridae). Rec. Aust. Mus. 40: 225–231.
Kottelat, M. 1990. New species and populations of cave nemacheilines in south and south-east Asia (Osteichthyes, Balitoridae). Mém. Biospéleol. 17: 49–55.
Kottelat, M. 1998. *Homaloptera yuwonoi*, a new species of hillstream loach from Borneo, with a new generic name for *H. thamicola* (Teleostei, Balitoridae). Ichthy. Expl. Freshw. 9: 267–272.
Kottelat M. & F. Bréhier. 1999. *Troglocyclocheilus khammouanensis*, a new genus and species of cave fish from the Khammouan karst, Laos (Teleostei, Cyprinidae). Rev. Suiss. Zool. 106: 347–359.
Kottelat, M. & J. Géry. 1989. *Nemacheilus troglocataractus*, a new blind cavefish from Thailand. (Osteichthyes, Balitoridae). Spixiana 11: 273–277.
Li, G.-L. 1989. On a new fish of the genus *Sincyclocheilus* from Guangxi, China (Cypriniformes: Cyprinidae: Barbinae). Acta Zootaxon. Sin. 14: 123–126 (in Chinese).
Lin, R.-D. & Z.-F. Luo. 1986. A new blind barbid fish (Pisces, Cyprinidae) from subterranean water in Guangxi, China. Acta Hydrobiol. Sin. 10: 380–382 (in Chinese).
Longley, G. & H. Karnei. 1979. Status of *Trogloglanis pattersoni* Eigenmann, the toothless blindcat and status of *Satan eurystomus* Hubbs and Bailey, the widemouth blindcat. U.S. Fish and Wildlife Service, Albuquerque. 54 pp.
Lozano-Vilano, M. de L. & S. Contreras-Balderas. 1990. *Astyanax armandoi*, n. sp. from Chiapas, mexico (Pisces, Ostariophysi: Characidae) with a comparison to the nominal species *A. aeneus* and *A. mexicanus*. Universidad Y Ciencia 7: 95–107.
Mai, D.Y. 1978. Identification of the fresh-water fishes of North Viet Nam. Scientific & Technology Publisher, Hanoi. 340 pp. (in Vietnamese).
Marshall, N.B. & G.L. Thines. 1958. Studies of the brain, sense organs and light sensitivity of a blind cave fish (*Typhlogarra widdowsoni*) from Iraq. Proc. Zool. Soc. London 131: 441–456.
Masuda, H., K. Amaoka, C. Araga, T. Uyeno & T. Yoshino 1984. The fishes of the Japanese Archipelago. Tokai University Press, Tokyo. 437 pp.
Mees, G.F. 1962. The subterranean freshwater fauna of Yardie Creek Station, North West Cape, Western Australia. J. R. Soc. West. Aust. 45: 24–32.
Menon, A.G.K. 1950. On a remarkable blind siluroid fish of the family Clariidae from Kerala (India). Rec. Indian Mus. 48: 59–66.
Menon, A.G.K. 1987. The fauna of India and the adjacent countries. Amra Press, Madras. 259 pp.
Mercy, T.V.A., K.G. Padmanabhan & N.K. Pillai. 1982. Morphological studies on the oocytes of the blind catfish *Horaglanis krishnai* Menon. Zool. Anz. 209: 221–223.
Mercy, T.V.A. & N.K. Pillai. 1984–5. The anatomy and histology of the alimentary tract of the blind catfish *Horaglanis krishnai*. Int. J. Speleol. 14: 1–4.
Miller, R.R. 1984. *Rhamdia reddelli*, new species, the first blind pimelodid catfish from middle America, with a key to the Mexican species. Trans. San Diego Soc. Nat. Hist. 20: 135–144.
Miranda-Ribeiro, A. de 1907. Uma novidade ichthyologica. Kosmos 4: 1–3.
Mitchell, R.W., W.H. Russel & W.R. Elliot. 1977. Mexican eyeless characin fishes, genus *Astyanax*: environment, distribution, and evolution. Special Publications No. 12, Texas University Press, Lubbock. 89 pp.
Nalbant, T.T. 1981. A study of the subterranean species of *Lucifuga* from Cuba, with the description of *Lucifuga simile* sp. n. (Pisces, Ophidiiformes, Bythitidae). Résult. Expéd. Biospéol. Cubano-Roum. Cuba 3: 185–190.
Nalbant, T.T. & P.G. Bianco. 1998. The loaches of Iran and adjacent regions with description of six new species (Cobitoidea). Ital. J. Zool. 65: 109–123.
Nalbant, T.T. & O. Linares. 1987. A study of a subterranean population of *Trichomycterus guianense* (Eigenmann, 1909) from Venezuela (Pisces, Siluriformes, Trichomycteridae). pp. 211–218. *In*: V. Decu (ed.) Fauna Hipogea y Hemiedáfica de Venezuela y de Otros Paises de América del Sur, Editura Academiei Republicii Sociliste Romania, Bucareşti.
Nelson, J.S. 1994. Fishes of the world. John Wiley & Sons, New York. 600 pp.
Ng, H.H. & M. Kottelat.1998. *Pterocryptis buccata*, a new species of catfish from western Thailand (Teleostei: Siluridae) with epigean and hypogean populations. Ichth. Research 45: 393–399.
Nielsen, J.G. 1974. Fish types in the Zoological Museum of Copenhagen. Zoological Museum, University of Copenhagen, Copenhagen. 115 pp.
Norman, J.R. 1926. A new blind catfish from Trinidad, with a list of the blind cave-fishes. Ann. Mag. Nat. Hist. 18: 324–331.
Page, L.M. & B.M. Burr. 1991. A field guide to freshwater fishes, North America, north of Mexico. The Peterson Field Guide Series, Boston. 432 pp.
Parin, N.V. 1983. *Noemacheilus* (*Troglocobitis*) *starostini* sp. n. (Osteichthyes, Cobitidae) a new blind fish from subterranean waters of Kugitangtau (Turkmenia). Zool. Zh. 62: 83–89 (in Russian).
Parzefal, J. 2001. A review of morphological and behavioural changes in the cave molly, *Poecilia mexicana*, from Tabasco, Mexico. Env. Biol. Fish. 62: 263–275 (this volume).
Pellegrin, J. 1929. L'*Eilichthys microphthalmus* Pellegrin, poisson cavernicole de la Somalie Italienne. Bull. Mus. Hist. Nat. Paris 1: 363–364.
Pérez, J.E. & G.E.E. Moodie. 1993. Genetic variation in a cave-dwelling Venezuelan catfish. Acta Cient. Venez. 44: 28–31.
Pérez, A. & A. Viloria.1994. *Ancistrus galani* n. sp. (Siluriformes: Loricariidae), with comments on biospeleological explorations in western Venezuela. Mém. Biospéléol. 21: 103–107.
Peters, N. & G. Peters. 1973. Problemes genetiques de l'evolution regressive des cavernicoles. pp. 187–201. *In*: L.H. Schroeder (ed.) Genetics and Mutagenesis in Fish, Springer-Verlag, New York.
Petit, G. 1933. Un poisson cavernicole aveugle des eaux douces de Madagascar: *Typhleotris madagascariensis* gen. et sp. nov. C.R. Hebd. Seances Acad. Sci. 197: 347–348.

Pfeiffer, W. 1966. Über die Vererbung des Schreckreaktion bei *Astyanax* (Characidae, Pisces). Z. Vererbungel. 98: 97–105.

Pillai, R.S. & G.M. Yazdani. 1977. Ichthyo-fauna of Garo Hills, Meghalaya (India). Rec. Zool. Surv. India 72: 1–22.

Poey, F. 1858. Memorias sobre la historia natural de la isla de Cuba, 2 volumes. Barcina, Habana. 439 pp.

Poisson, H. & H. Bessairie. 1947. Note sur une novelle station d'un poisson cavernicole aveugle du plateau calcaire Mahafaly. Bull. Acad. Malgache 26: 167–168.

Poll, M. & N. LeLeup. 1965. Un poisson aveugle nouveau de la famille des Brotulidae provenant des îles Galapagos. Bull. Acad. R. Belg. (Cl. Sci.) 51: 464–474.

Poly, W.J. 2001. Nontroglobitic fishes in Bruffey-Hills Creek Cave, West Virginia, and other caves worldwide. Env. Biol. Fish. 62: 73–83 (this volume).

Poulson, T.L. 1963. Cave adaptation in amblyopsid fishes. Amer. Midl. Nat. 70: 257–290.

Proudlove, G.S. 2001. The conservation status of hypogean fishes. Env. Biol. Fish. 62: 201–203 (this volume).

Putnam, F.W. 1872. The blind fishes of the Mammoth Cave and their allies. Amer. Nat. 6: 6–30.

Quoy, J.R.C. & J.P. Gaimard 1824. Description des poissons. pp. 1–328. *In*: L. De Freycinet (ed.) Voyage autour du Monde. . .exécuté sur les corvettes de L.M. 'L'Uranie' et 'La Physicienne,' pendant les années 1817, 1818, 1819 et 1820. Chez Pillet Aîné, Paris (7 volumes).

Regan, C.T. 1913. A revision of the cyprinodont fishes of the subfamily Poeciliinae. Proc. Zool. Soc. Lond. 1913: 977–1018.

Regan, C.T. 1940. The fishes of the gobiid genus *Luciogobius* Gill. Ann. Mag. Nat. Hist. 5: 462–465.

Reis, R.E. 1987. *Ancistrus cryptophthalmus* sp. n., a blind mailed catfish from the Tocantins River basin, Brazil (Pisces, Siluriformes, Loricariidae). Rev. Fr. Aquariol. 14: 81–84.

Ringuelet, R.A. 1974. Breves comentarios sobre un pez cavernicola de Bolivia (*Pygidium chaberti* Durand, 1968). Neotropica 20: 65–66.

Roberts, T.R. 1991. *Barbus speleops* new species, a blind cavefish from Tham Phu Khieo, Mekong Basin, Thailand. Nat. Hist. Bull. Siam Soc. 39: 103–109.

Romero, A. 1983. Introgressive hybridization in a population of *Astyanax fasciatus* (Pisces: Characidae) at La Cueva Chica. Nat. Speleol. Soc. Bull. 45: 81–85.

Romero, A. 1984. Behavior in an 'intermediate' population of the subterranean-dwelling characid *Astyanax fasciatus*. Env. Biol. Fish. 10: 203–207.

Romero, A. 1985. Ontogenetic change in phototactic responses of surface and cave populations of *Astyanax fasciatus* (Pisces: Characidae). Copeia 1985: 1004–1011.

Romero, A. 1987. Observaciones preliminares sobre la conducta en individuos cavernícolas de *Trichomycterus conradi* (Pisces: Trichomycteridae). Bol. Soc. Ven. Cienc. Nat. 41: 203–216.

Romero, A. 1998a. Threatened fishes of the world: *Amblyopsis rosae* (Eigenmann, 1842) (Amblyopsidae). Env. Biol. Fish. 52: 434.

Romero, A. 1998b. Threatened fishes of the world: *Typhlichthys subterraneus* (Girard, 1860) (Amblyopsidae). Env. Biol. Fish. 53: 74.

Romero, A. 1998c. Threatened fishes of the world: *Speoplatyrhinus poulsoni* Cooper and Kuehne, 1974 (Amblyopsidae). Env. Biol. Fish. 53: 293–294.

Romero, A. 2001. Scientists prefer them blind: the history of hypogean fish research. Env. Biol. Fish. 62: 43–71 (this volume).

Romero, A. & L. Bennis. 1998. Threatened fishes of the world: *Amblyopsis spelaea* De Kay, 1842 (Amblyopsidae). Env. Biol. Fish. 51: 420.

Romero, A. & K. Benz. 2000. The unsung heroes of speleology. Nat. Speleol. Soc. News 58: 106, 126.

Romero, A. & J.E. Creswell. 2000. In search of the elusive 'eyeless' cave fish Trinidad, W.I. Nat. Speleol. Soc. News 58: 282–283.

Romero, A. & A. McLeran. 2000. Threatened fishes of the world: *Stygichthys typhlops* Brittan & Böhlke, 1965 (Characidae). Env. Biol. Fish. 57: 270.

Romero, A. & K.M. Paulson. 2001. Humboldt's alleged cave fish from Ecuador. J. Spelean Hist. (in press).

Romero, A. & P.B.S. Vanselow. 2000a. Threatened fishes of the world: *Milyeringa veritas* Whitley, 1945 (Eleotridae). Env. Biol. Fish. 57: 36.

Romero, A. & P.B.S. Vanselow. 2000b. Threatened fishes of the world: *Ophisternon candidum* (Mees 1962) (Synbranchidae). Env. Biol. Fish. 58: 214.

Sabino, J. & E. Trajano. 1997. A new species of blind armoured catfish, genus *Ancistrus*, from caves of Bodoquena region, Mato Grosso do Sul, southwestern Brazil (Siluriformes, Loricariidae, Ancistrinae). Rev. Fr. Aquariol. 24: 73–78.

Sadoglu, P. 1957. Mendelian inheritance in the hybrids between the Mexican blind cave fishes and their overground ancestor. Verh. Dtsch. Zool. Ges., Graz 1957: 432–439.

Sbordoni, M.C., E. De Matthaeis, M. Mattoccia, R. Berti & V. Sbordoni. 1996. Genetic variability and differentiation of hypogean cyprinid fishes from Somalia. J. Zool. Syst. Evol. Res. 34: 75–84.

Schemmel, C. 1967. Vergleichende Untersuchungen an den Hautsinnesorganen ober – und unterirdisch Lebender *Astyanax*–Formen, ein Beitrag zur evolution der Cavernicolen. Z. Morph. Tiere 61: 255–316.

Schemmel, C. 1980. Studies on the genetics of feeding behaviour in the cave fish *Astyanax mexicanus* f. *Anoptichthys*. An example of apparent monofactorial inheritance by polygenes. Z. Tierpsychol. 53: 9–22.

Schultz, L.P. 1949. A further contribution to the ichthyology of Venezuela. Proc. U.S. Nat. Mus. 99: 1–211.

Silas, E.G. & E. Dawson. 1961. *Amphipnous indicus*, a new synbranchoid eel from India, with a redefinition of the genus and a synopsis to the species of *Amphipnous* Müller. J. Bombay Nat. Hist. Soc. 58: 366–378.

Silfvergrip, A.M.C. 1996. A systematic revision of the neotropical catfish genus *Rhamdia* (Teleostei, Pimelodidae). Swedish Museum of Natural History, Stockholm. 156 pp.

Singh, D.F. & G.M. Yazdani. 1993. Studies on the ichthyofauna of Nasik District, Maharashtra, India. Rec. Zool. Surv. India 90: 195–201.

Sket, B. 1988. Speleobiological investigations in the Colombian Andes. Vest. Leningr. Univ. Ser. 3, Biol. 36: 53–62.

Smirnov, S. 1983. A blind cave loach from southwestern Asia, *Noemacheilus starostini* Parin. Trop. Fish Hobb. (3): 54–58.

Steindachner, F. 1863. Beiträge zur Kenntniss der Sciaenoiden Brasiliens und der Cyprinodonten Mejicos. Sitzungsber. Math.-Naturwiss., Classe K. Akad. Wiss., Wien 48: 162–185.

Talwar, P.K. & A.G. Jhingran. 1991. Inland fishes of India and adjacent countries. Oxford & IBH Publishing Co., Calcutta. 2 volumes, 1158 pp.

Thinès, G. 1953. Recherches experimentales sur la photosensibilité du poisson aveugle *Caecobarbus geertsii* Blgr. Ann. Soc. Roy. Zool. Belg. 84: 231–265.

Thinès, G. 1969. L'évolution régressive des poissons cavernicoles et abyssaux. Masson et Cie., Paris. 394 pp.

Thinès, G. & J.-M. Legrain. 1973. Effects de la substance d'alarme sur le comportament des poissons cavernicoles *Anoptichthys jordani* (Characidae) et *Caecobarbus geeretsi* (Cyprinidae). Ann. Speleol. 28: 291–297.

Tomiyama, I. 1936. Gobiidae of Japan. Japan. J. Zool. 7: 37–112.

Torii, H. 1951. Die Hohleen-Vertebraten aus Japan und seinen Nachbarschaften. Annot. Zool. Japan. 24: 170–179.

Trajano, E. 1991. The agonistic behaviour of *Pimelodella kronei*, a troglobitic catfish from southeastern Brazil (Siluriformes, Pimelodidade). Behav. Proc. 23: 113–124.

Trajano, E. 1997a. Food and reproduction of *Trichomycterus itacarambiensis*, cave catfish from south-eastern Brazil. J. Fish Biol. 51: 53–63.

Trajano, E. 1997b. Population ecology of *Trichomycterus itacarambiensis*, a cave catfish from eastern Brazil (Siluriformes, Trichomycteridae). Env. Biol. Fish. 50: 357–369.

Trajano, E. 1997c. Synopsis of Brazilian troglomorphic fishes. Mem. Biospeleol. 24: 119–126.

Trajano, E. 2001a. Ecology of subterranean fishes: an overview. Env. Biol. Fish. 62: 133–160 (this volume).

Trajano, E. 2001b. Habitat and population data of troglobitic armoured cave catfishes, *Ancistrus cryptophthalmus* Reis, 1987, from central Brazil (Siluriformes: Loricariidae). Env. Biol. Fish. 62: 195–200 (this volume).

Trajano, E. & M.C.C. de Pinna. 1996. A new cave species of *Trichomycterus* from eastern Brazil (Siluriformes, Trichomycteridae). Rev. Fr. Aquariol. 23: 85–90.

Trajano, E. & A.M. Souza. 1994. The behavior of *Ancistrus cryptophthalmus*, an armoured blind catfish from caves of central Brazil, with notes on synopic *Trichomycterus* sp. (Siluriformes, Loricariidae, Trichomycteridae). Mém. Biospeléol. 21: 151–159.

Trewavas, E. 1936. Dr. Karl Jordan's expedition to South-West Africa and Angola: the fresh-water fishes. Novit. Zool. 40: 63–74.

Trewavas, E. 1955. A blind fish from Iraq, related to *Garra*. Ann. Mag. Nat. Hist. 8: 551–555.

Triques, M.L. 1996. *Eigenmannia vicentespelaea*, a new species of cave dwelling electrogenic neotropical fish (Ostariophisi: Gymnotiformes: Sternopygidae). Rev. Fr. Aquariol. 23: 1–4.

Uéno, S.-I. & Y. Morimoto. 1970. The fauna of the insular lava caves in west Japan. I. General account. Bull. Nat. Sci. Mus (Japan) 13: 443–454.

Valenciennes, A.in G. Cuvier & A. Valenciennes. 1846. Histoire naturelle des poissons, Tome dix-huitième, Suite du livre dix-huitième, Cyprinoïdes. Livre dix-neuvième, Des Ésoces ou Lucioïdes 18: 1–505.

Vinciguerra, D. 1924. Descrizione di un ciprinide cieco proveniente dalla Somalia Italiana (Description of a blind cyprinid from the Italian Somalia). Ann. Mus. Civ. Stor. Nat. Genoa 51: 239–243.

Walsh, S.J. & C.R. Gilbert. 1995. New species of troglobitic catfish of the genus *Prietella* (Siluriformes: Ictaluridae) from northeastern México. Copeia 1995: 850–861.

Weber, A. 2000. Fish and amphibia. pp. 109–132. *In*: H. Wilkens, D.C. Culver & W.F. Humphrey (ed.) Subterranean Ecosystems, Elsevier, Amsterdam.

Weber, A. & H. Wilkens. 1998. *Rhamdia macuspanensis*: a new species of troglobitic pimelodid catfish (Siluriformes; Pimelodidae) from a cave in Tabasco, Mexico. Copeia 1998: 998–1004.

Whitley, G.P. 1945. New sharks and fishes from Western Australia, Part 2. Aust. Zool. 11: 1–42.

Whitley, G.P. 1951a. A naturalist at north-west Cape. Austr. Mus. Mag. 5: 158–164.

Whitley, G.P. 1951b. New fish names and records. Proc. R. Zool. Soc. N. S. W. 1949–50: 61–68.

Wilkens, H. 1969. Beitrage zur Degeneration des Auges bei Cavernicolen, Genzahl und Manifestationsart. Zool. Anz. 180: 454–464.

Wilkens, H. 1973. Über das phylogenetische Alter von Hohlentieren. Untersuchungen über die Cavernicole Susswasserfauna Yucatans. Sond. Z. Zool. Syst. Evol. 11: 49–60.

Wilkens, H. 1993. A new species of *Rhamdia* (Pisces: Pimelodidae) from a cave in the Sierra de Zongolica (Veracruz, Mexico). Mitt. Hamb. Zool. Mus. Inst. 90: 375–378.

Wilson, J.M. 1996. Conservation and ecology of a new blind fish *Glossogobius ankaranensis* from the Ankarana caves, Masagascar. Oryx: 30: 218–221.

Yang, J.-X. 1990. Nemacheilinae (Cypriniformes: Cobitidae). pp. 12–63. *In*: X. Chu & Y. Chen (ed.) The Fishes of Yunnan, China, Part 2. Science Press, Beijing (in Chinese).

Yang, J.-X., Y.-R. Chen & J.-H. Lan. 1994. *Protocobitis typhlops*, a new genus and species of cave loach from China (Cypriniformes: Cobitidae). Ichthyol. Explor. Freshwaters 5: 91–96.

Yang, G.-Y., F.-X. Yuan & Y.-M. Liao. 1986. A new blind Cobitidae fish from the subterranean water in Xiangxi, China. J. Huazhong Agric. Univ. 5: 219–223 (in Chinese).

Zhang, C. 1986. On the ecological adaptation and geographical distribution of the barbine fish *Varicorhinus (Scaphestes) macrolepis*. Acta Zool. Sin. 32: 266–272.

Zheng, J. & J. Wang. 1990. Description of a new species of the genus *Sinocyclocheilus* from China (Cypriniformes: Cyprinidae). Acta Zootaxon. Sin. 15: 251–254 (in Chinese).

Zhou, W. & G.-H. Cui. 1997. Fishes of the genus *Triplophysa* (Cypriniformes: Balitoridae) in the Yuanjiang (upper Red River) basin of Yunnan, China, with description of a new species. Ichthyol. Explor. Freshw. 8: 177–183.

Zhu, S.-Q. 1989. The loaches of the subfamily Nemacheilinae in China (Cypriniformes: Cobitidae). Jiangsu Science and Technology Publishing House, Nanjing. 150 pp. (in Chinese).

Environmental Biology of Fishes **62**: 43–71, 2001.
© 2001 *Kluwer Academic Publishers. Printed in the Netherlands.*

Scientists prefer them blind: the history of hypogean fish research

Aldemaro Romero
Environmental Studies Program and Department of Biology, Macalester College, 1600 Grand Ave., St. Paul, MN 55105-1899, U.S.A. (e-mail: romero@macalester.edu)

Received 22 December 2000 Accepted 24 January 2001

Key words: neo-Lamarckism, orthogenesis, typological thinking, neo-Darwinism, evolution, evolutionary progress, regressive evolution

Synopsis

The history of hypogean fish research has been strongly influenced by neo-Lamarckism (including orthogenesis) and typological thinking. Only in the last few decades neo-Darwinism has made any inroads in the research approach to this subject. The majority of the most distinguished and productive hypogean fish researchers have used their research subjects to confirm their own views on evolution rather than to use those subjects as a spring of knowledge to enrich mainstream biological thought. Of these views, I found that the most perversive of all is the notion of evolutionary 'progress' that has led many researchers to envision hypogean fishes as prime examples of 'regressive' evolution. I propose that the utilization of hypogean fishes for the study of convergent evolution should catapult these subjects of research into prime objects of evolutionary ideas.

Introduction

The development of a particular area of science is deeply influenced by its history. Through historical studies, one can better perceive the impacts of fashion, error, and the effects of conceptual inertia (Horder 1998). By doing that the researcher can also project the best venues to develop future lines of research.

In this paper I analyze the scientific views on the nature of hypogean fishes. I will contend that both neo-Lamarckism (including orthogenesis) and typological thinking have prevented the use of hypogean fishes as one of the prime examples to explain some general biological phenomena in evolutionary biology as it is the case for convergent evolution. Another question is how hypogean fish research fits within the broader context of the history of three sub-disciplines: ichthyology, speleology, and evolutionary biology. This is a rather difficult task due to lack of current contextual material.

For example, a comprehensive history of ichthyology has yet to be written. Although some attempts have been made, none of these give us the entire picture. They are either too old (Cuvier 1828 [1995], Günther 1880, Jordan 1905) or limited in scope geographically (Myers 1964, Hubbs 1964) or conceptually (Pietsch & Anderson 1997). Thus, there is not an appropriate background to refer to in this regard. The history of speleology has faced similar problems. The only attempt to summarize the history of this field emphasized geomorphology (Shaw 1992). Barr's (1966) short article on the history of cave research is confined to the United States. The role played by the study of cave organisms in general on the neo-Lamarckian movement in the United States as well as the idea of progress is better known (e.g., Ruse 1996).

Methods

I compiled 1607 published references on hypogean fishes. They were obtained by using standard bibliographical sources, including the *Zoological Record* up to 1999, the *Biological Abstracts* up to June 2000, Dean's *Bibliography of Fishes* (Dean 1916–1923),

Current References in Fish Research (Cvancara[1]), as well as other bibliographical data that I have accumulated in the last 20 years. These references were analyzed by author, topic, year of publication, and by country/institution using ProCite 5.[2] I emphasized studies on hypogean fishes that show some degree of troglomorphy (blindness, depigmentation) The study of non-troglomorphic hypogean fishes, despite their tremendous potential for understanding the processes of cave colonization and evolution, has been largely overlooked by scientists and, thus, have remained mostly anecdotal (Poly 2001). Although this type of compilation can never be fully complete, I am confident that I have studied and analyzed all relevant sources.

In order to provide adequate historical and geographical context, I have added the dates and places of birth and death in parentheses after mentioning, for the first time, all major players in the history of hypogean fish research. Secondary players are only given the years of birth and death, when known. All scientific names of fishes are given following Romero (2001a). I used this instead of Eschmeyer's standard *Catalogue of Fishes*[3] because that list is sometimes ambiguous when it comes to specific names for certain hypogean populations and it is not entirely complete regarding hypogean species. Sometimes those names are followed by parentheses to mention the name with which the species was originally described. For referential purposes, I present the chronology of the discovery of valid species/populations of troglomorphic fishes in Table 1.

Results

I identified six periods of hypogean fish research: (1) pre-Linnean (1541–1752), (2) first discoveries and research (1805–1864), (3) American neo-Lamarckism (1868–1919), (4) dominance of typological thinking (1921–1940), (5) American renaissance (1936–1960), and (6) times of philosophical conflict (1960–1990). Notice that between (1), (2), (3), and (4) there are gaps; this is because during those in-between years nothing significant happened in the field. On the other hand, there is an overlap between (5) and (6); this is because the American and European schools that dominated

[1] Cvancara, V.A. 1976–1998. Current references in fish research, 23 volumes.

[2] 2000. ProCite Version 5.0. ISI ResearchSoft, Berkeley.

[3] http://www.calacademy.org/research/ichthyology/catalog/

Table 1. Chronological order of discoveries of troglomorphic species/populations of hypogean fishes (sensu Romero 2001a). Following an asterisk is the author that reported the first hypogean/troglomorphic population of an already described epigean species.

Amblyopsis spelaea DeKay, 1842
Lucifuga (Lucifuga) subterraneus Poey, 1858
Lucifuga (Stygicola) dentatus Poey, 1858
Typhlichthys subterraneus Girard, 1859
Chologaster agassizii Putnam, 1872
Amblyopsis rosae (Eigenmann, 1897)
Phreatobius cisternarum Goeldi, 1905
Pimelodella kronei (Miranda-Ribeiro, 1907)
Trogloglanis pattersoni Eigenmann, 1919
Caecobarbus geertsii Boulenger, 1921
Uegitglanis zammaranoi Gianferrari, 1923
Phreatichthys andruzzii Vinciguerra, 1924
Rhamdia quelen (Quoy & Gaimard, 1824) (*Norman 1926)
Barbopsis devecchii Di Caporiacco, 1926
Typhleotris madagascariensis Petit, 1933
Clarias cavernicola Trewavas, 1936
Astyanax fasciatus (Cuvier, 1819) (*Hubbs & Innes 1936)
Ogilbia pearsei (Hubbs, 1938)
Ophisternon infernale (Hubbs, 1938)
Luciogobius pallidus Regan, 1940
Luciogobius albus Regan, 1940
Iranocypris typhlops Bruun & Kaiser, 1944
Milyeringa veritas Whitley, 1945
Satan eurystomus Hubbs & Bailey, 1947
Trichomycterus conradi (Eigenmann, 1912) (*Schultz 1949)
Horaglanis krishnai Menon, 1950
Prietella phreatophila Carranza, 1954
Typhlogarra widdowsoni Trewavas, 1955
Garra barreimiae barreimiae Fowler & Steinitz, 1956
Typhleotris pauliani Arnoult, 1959
Ophisternon candidum (Mees, 1962)
Astroblepus pholeter Collette, 1962
Poecilia mexicana Steindachner, 1863 (*Gordon & Rosen 1962)
Ogilbia galapagosensis (Poll & LeLeup, 1965)
Stygichthys typhlops Brittan & Böhlke, 1965
Trichomycterus chaberti Durand, 1968
Nemacheilus evezardi Day, 1872 (*Thines 1969?)
Lucifuga (Stygicola) spelaeotes Cohen & Robins, 1970
Speoplatyrhinus poulsoni Cooper & Kuehne, 1974
Paracobitis smithi (Greenwood, 1976)
Pterocryptis cucphuongensis (Mai, 1978)
Triplophysa gejiuensis (Chu & Chen, 1979)
Caecocypris basimi Banister & Bunni, 1980
Oreonectes anophthalmus Zheng, 1981
Lucifuga (Stygicola) simile Nalbant, 1981
Rhamdia laticauda (Kner, 1858) (*Greenfield et al. 1982)
Typhlobarbus nudiventris Chu & Chen, 1982
Nemacheilus starostini Parin, 1983
Rhamdia redelli Miller, 1984
Triplophysa xiangxiensis (Yang, Yuan & Liao, 1986)
Sinocyclocheilus anatirostris Lin & Luo, 1986
Garra dunsirei Banister, 1987

Table 1. Continued.

Nemacheilus sijuensis Menon, 1987
Ancistrus cryptophthalmus Reis, 1987
Sinocyclocheilus anophthalmus Chen, Chu, Luo & Wu, 1988
Sinocyclocheilus cyphotergous (Dai, 1988)
Schistura oedipus (Kottelat, 1988)
Cryptotora thamicola (Kottelat, 1998)
Lucifuga (Lucifuga) teresinarum Díaz Perez, 1988
Sinocyclocheilus microphthalmus Li, 1989
Nemacheilus troglocataractus Kottelat & Géry, 1989
Sinocyclocheilus angularis Zheng & Wang, 1990
Triplophysa yunnanensis Yang in Wu, 1990
Schistura jarutanini Kottelat, 1990
Sundoreonectes tiomanensis Kottelat, 1990
Monopterus eapeni Talwar in Talwar & Jhingran, 1991
Poropuntius speleops (Roberts, 1991)
Caecogobius cryptophthalmus Berti & Ercolini, 1991
Triplophysa shilinensis Chen & Yang, 1992
Rhamdia zongolicensis Wilkens, 1993
Glossogobius ankaranensis Banister, 1994
Sinocyclocheilus hyalinus Chen & Yang in Chen, Yang & Zhu, 1994
Protocobitis typhlops Yang, Chen & Lan, 1994
Ancistrus galani Perez & Viloria, 1994
Astroblepus riberae Cardona & Guerao, 1994
Prietella lundbergi Walsh & Gilbert, 1995
Oxyeleotris caeca Allen, 1996
Eigenmannia vicentespelaea Triques, 1996
Trichomycterus itacarambiensis Trajano & de Pinna, 1996
Ancistrus formoso Sabino & Trajano, 1997
Sinocyclocheilus furcodorsalis Chen, Yang & Lan, 1997
Paracobitis longibarbatus Chen, Yang, Sket & Aljancic, 1998
Monopterus roseni Bailey & Gans, 1998
Rhamdia macuspanensis Weber & Wilkens, 1998
Pterocryptis buccata Ng & Kottelat, 1998
Troglocychlocheilus khammouanensis Kottelat & Brehier, 1999

each of those periods, respectively, were very much independent of each other, not only geographically but also conceptually (see below).

Pre-Linnean times (1541–1752)

The first known printed reference of a hypogean fish is that of the hyaline fish, *Sinocyclocheilus hyalinus*, reported by Jie[4] (1541) for the Alu Limestone caves (103°45′ E, 24°33′ N) in what is today Luxi County, Yunnan, China. However, specimens of the hyaline fish were not collected for research until 1991 and the description of the species was not published in the scientific literature until three years later (Chen et al. 1994).

There are at least three more pre-Linnean printed references to hypogean fishes, all of them from Europe. Given that they have been cited as precursors in the history of biospeleology, it is important to analyze their validity as true scientific developments.

The first was by Jacques Besson (b. Colombières, France, 1530?; d. Orléans, France, 1573). He was an engineer and mathematician with no formal training in the natural sciences (Romero & Lomax 2000). Besson (1569, p. 41) reported little eels (‘*petites anguilles*’) in a cave stream. Although Shaw (1992, p. 227) claims that such observation took place ‘in a cave stream in France’, Besson did not give a locality of where he made that observation. Besson did not describe the fish as being blind and/or depigmented (extraordinary characteristics even to the casual observer). He may have seen common eels, *Anguilla anguilla*, or a species of some of the European freshwater fishes with eel-like bodies that are sympatric with the areas he used to travel (France and Switzerland). Those fish families include Petromizontidae, Cobitidae, Siluridae, and Clariidae (Blanc et al. 1971).

The second was Athanasius Kircher (b. Geisa, Germany, 2 May 1602; d. Rome, 28 November 1680). This prolific Jesuit priest polymath wrote, in what is probably the first printed work on speleology (Figure 1), that ‘There is also in the landscape of Krain [Carniola?] close to the town Haubach a huge field from which each year during Spring time a large body of water containing fish bursts forth with the result that in a few days it transforms the field into a lake teeming with fish (. . .) in Switzerland rivers rise from the caves of the mountains, that flow from May until September, but stop the rest of the time (. . .) as they come out of the mountains, are full of fish, which is clear proof that they [the fish] emerge from subterranean waters along the rivers (. . .) it is not implausible that, as under the earth all kind of fishes occur and live’ (Kircher 1665, 2, p. 85). These references to subterranean fishes, however, are vague, unsubstantiated, and given Kircher’s reputation as an uncritical repeater of other people’s tales, highly suspect (Romero 2000). Furthermore, he makes no reference to features associated with troglomorphic fishes – blindness and depigmentation.

The third pre-Linnean reference to subterranean fishes in Europe was by Marc-René Marquis de Montalembert (b. Angoulême, France, 16 July 1714; d. Paris, France, 29 March 1800) (Figure 2). He was

[4] Jie, Yi-Jing. 1541. Report on the Alu Cave. Publisher’s name and place of publication unknown (in Chinese).

Figure 1. Athanasius Kircher.

Figure 2. The only known portrait of Marc René de Montalembert.

an aristocrat, military man, and engineer known for his design of fortifications. He reported a blind, subterranean fish in a spring at Gabard, Angoumois, near one of his estates in southwestern France. He noted: 'it is common to fish either blind or one-eyed pike; one-eyed ones always miss the right eye and among the blind ones, the right eye seems further reduced than the left eyes' (Montalembert 1748, p. 28). He left no drawings, much less preserved specimens. He said that what he saw was a pike. That, by itself, is not surprising. The pike, *Esox lucius*, is by far the most common freshwater fish of the Northern Hemisphere. The fact that this fish can be identified as a pike despite being blind is also not surprising. Many hypogean fishes are almost identical to their surface, epigean forms except for the reduction of eyes and pigmentation. But Montalembert never mentioned depigmentation in his description. Furthermore, he says that some of the fish lacked one eye and, when that was the case, it was always the right eye. Troglomorphic fish generally show the same degree of reduction in both eyes. Finally, the location mentioned by Montalembert cannot be found today (Romero 1999a) nor has any true blind cave fish ever been described from Europe.

Thus, all pre-Linnean reports of European troglomorphic hypogean fishes are unsupported by scientific evidence. Two of them (Besson's 1569, Kircher's 1665) do not even describe them with the typical troglomorphic features while the third (Montalembert 1748) is suspect. The most compelling evidence that the reports by these three authors are inaccurate is the fact that no true troglomorphic fish has been found in Europe to this date. All the European citations display the characteristic of their times: vague (Besson's), uncritical and scholastic (Kircher's), or fascinated with monstrosities

(Montalembert). Among all pre-Linnean references, only Jie's (1541) has been substantiated by facts.

First discoveries and research (1805–1864)

This period was characterized by two factors: The first is the discovery and publication in the scientific literature of the first species of troglomorphic fishes. The second is the use of morphological studies aimed to describe in detail their different features. During this period all troglomorphic fishes were found either in the United States or Cuba. In the United States, the years preceding the American Civil War (1861–1865) were marked by explorations by both individuals acting independently and state-sponsored surveys.

An early published post-Linnean record of a possible hypogean fish is that of Alexander von Humboldt (1769–1859) who described *Astroblepus cyclopus* (= *Pimelodus cyclopum*) which, according to him, could be found in the subterranean waters in the Andes of Quito in the basin of Río Esmeraldas, Ecuador (Humboldt 1805). However, he never saw that fish in any hypogean environment (Romero & Paulson 2001). Furthermore, this is actually an eyed, pigmented species with ample distribution in northwestern South America (Burgess 1989, p. 448).

The first published record of a confirmed troglomorphic fish in the Western Hemisphere is probably that of James Flint. He was a Scotchman who, while visiting the United States, lived for several months in Jeffersonville, Indiana, in 1820 and recorded that 'a Colonel C – [sic] of Indiana told me that a settler in his neighbourhood [sic] digging a well, penetrated into a stream of water, and found blind fishes in it.' He added as a footnote that 'Since the above was written, a notice of blind fishes has appeared (if I mistake not) [sic] in the memoirs of the Wernerian Society of Edinburgh' (Flint 1822, p. 256). I reviewed the entire collection of this journal, which consists of eight volumes published between 1808 and 1838, and did not find any reference to that blind fish.

The first published report of a troglomorphic fish sighted in its natural environment in the Western Hemisphere was, probably, by Robert Davidson (1808–1876). In October 1836 he visited Mammoth Cave in Kentucky accompanied by Stephen Bishop (1780–1850), a self-educated black slave who guided visitors through the cave. He reported that '*white fish*[5] were found here without eyes' whose existence was already known by some of the locals (Davidson 1840, pp. 54–56). Others have pointed out that it was on 20 September 1838 that the Echo River in Mammoth Cave was discovered, and in it, a blind fish (Soule 1982).

The first time that a troglomorphic fish was mentioned in the scientific literature was in a short note in the Proceedings of the Academy of Natural Sciences of Philadelphia (Anonymous 1842). There it was reported that a W.T. Craigie donated to the Academy at the 24 May 1842 meeting a specimen of 'a small white fish, also eyeless (presumed to belong to a subgenus of *Silurus*), taken from a small stream called the "River Styx" in the Mammoth Cave, Kentucky, about two and one-half miles from the entrance'. Today, at the collection of the Academy there are three specimens of *Amblyopsis spelaea* in alcohol, that appear linked to this donation. Two are catalogued as ANSP 7964 collected by W.T. Craige, and the other, ANSP 7964, collected by 'Mrs. C.H. Graff, Messrs. Craige & Lambert'. All three specimens were captured in Mammoth Cave, but no dates are given.

The first scientific description of a troglomorphic fish was carried out by James Ellsworth DeKay (b. Lisbon, Portugal, 12 October 1792; d. Oyster Bay, Long Island, New York, 21 November 1851) (Figure 3). Son of an American captain stationed in Lisbon, DeKay studied medicine at Yale and Edinburgh, Scotland. He was one of the founders of the Academy of Medicine and an active member of both the New York Lyceum of Natural History and the American Association for the Advancement of Science. He was the first of the first-generation of American-born naturalists making important contributions in ichthyology (Hubbs 1964).

DeKay's work can be framed within the movement that started in the 1940's when several states of the United States inaugurated natural history surveys and published catalogues of the local faunas (Coe 1918). He was selected to contribute a book on the zoology of New York state. In it, DeKay (1842, pp. 187–188) gave the first recognized scientific description of a troglomorphic fish, the northern cavefish, *Amblyopsis spelaea* (= *Amblyopsis spelaeus*). The description was not very detailed nor of a great quality. This could have been due to the fact that it was based on a poor specimen in the Cabinet of the Lyceum of Natural History of New York (Putnam 1872) or to the fact that DeKay was not a trained ichthyologist (Smallwood 1941, pp. 163–164).

Although the northern cavefish was captured in the River Styx, Mammoth Cave, Kentucky, DeKay included that and many other non-New York species

[5] Italics in the original.

Figure 3. James Ellsworth DeKay, the first who described in the scientific literature a troglomorphic fish, *Amblyopsis spelaea*.

Figure 4. Jeffries Wyman (courtesy of the Ernst Mayr Library of the Museum of Comparative Zoology, Harvard University ©President and Fellows of Harvard College).

because 'It cannot therefore fail to be perceived that the Ichthyology of New-York will embrace a very large proportion of the Fishes of the United States' (DeKay 1842, p. iv). He actually placed this new species in a list of fishes under the subheading '(EXTRA-LIMITAL)' [sic]. The specimen, which originally belonged to the Cabinet of the Lyceum of Natural History of New York, cannot be located today and is presumed lost.

DeKay would never write about *Amblyopsis* again; however, this fish caught the attention of a number of anatomists who immediately began to study it. The first of those was Jeffries Wyman (b. Chelmsford, Middlesex, Massachusetts, 11 August 1814; d. Bethlehem, New Hampshire, 4 September 1874) (Figure 4). He graduated in medicine from Harvard University in 1837 and studied under George Cuvier and Richard Owen (Gifford 1967). Wyman helped to lay the foundations of comparative anatomy in the U.S. He was responsible – together with Louis Agassiz and Asa Gray – for making Harvard the most important center for the study of natural history in the United States. He is little remembered today mostly because he was a very modest man who avoided generalizations.

Wyman, Agassiz, and Gray, embraced the philosophical, or transcendental anatomy, i.e., the search for ideal patterns of structure in nature (Appel 1988). Thus, the discovery of a blind cave fish attracted Wyman's attention and he described *A. spelaea* in great detail (Wyman 1843), to the point that some have mistakenly referred to him as the first person who described a blind cave fish (e.g., Gurnee 1992).

In his first paper on *A. spelaea*, he reported that 'On the most careful dissection no traces of eyes were found' (Wyman 1843, p. 96). Later he wrote that 'The optic lobes existed; according to the general rules of physiology these should not exist; as they bare strict relation to the sense of sight, which receives its nerve from them (. . .) Here the optic lobes were not so large as the allies fishes, but yet they were of good size, and nearly as large as the cerebral lobes' (Wyman 1851, p. 349). He later re-examined three specimens and found imperfect eyes covered by tissue and, hence, unable to see. He proposed that this imperfection of the eyes 'might be owing to a want of stimulus through a series of generations' and that the organ of vision, however imperfect, 'it is more like the eyes of other vertebrates' (Wyman 1854a, p. 19). He also pointed out numerous structures without evident functions, organs that were of morphological rather than physiological value (Wyman 1854b). He produced very detailed drawings of the internal anatomy of *A. spelaea* (Wyman 1872).

For Wyman, *A. spelaea* was an excellent subject of study in his quest for evidence of a common plan underlying the differences caused by adaptive modifications. Although he quickly converted to evolutionism, he did not accept natural selection as its mechanism. He even regarded Agassiz as backwards for his refusal to accept evolution (Appel 1988). Yet, all of his papers on *A. spelaea* were devoid of evolutionary speculations, something that characterized most of his writings.

The next naturalist to study *A. spelaea* was August Otto Theodor Tellkampf (b. Heinde, Germany, 27 April 1812; d. Hannover, Germany, 7 September 1883) (Romero 2001b). He 'Americanized' his name by adding either 'A.'or 'G.' as his middle initial. In 1838 he received a doctorate in Medicine from the University of Wurzburg, Bavaria, and immigrated into the United States in 1839, where he practiced medicine in Cleveland and New York. He had some interests in cave fauna, having visited Mammoth Cave in October 1842 (Tellkampf 1844a,b), and described several species of invertebrates. He was a member of the Lyceum of Natural History of New York.

Tellkampf contributed detailed descriptions of *A. spelaea* and concluded that its eyes and those of blind cave crayfishes had become rudimentary as a result of disuse: 'While it is true, in general, that all animals retain their essential form, and that no species passes over into another by transformation, we know that less material changes of form are produced by external influences such as changes in climate or food, lasting though many generations of the same species'. But if the lack of light could only produce change within a species, then where was the original unmodified species? Tellkampf remained cautious on this point: the relationship of the blind fauna to unmodified species could not be settled until 'such species, corresponding with them in all essential points, are found' (Tellkampf 1844b, p. 393).

European-based researchers also started to show interest in *A. spelaea*. By January 1844 a specimen had arrived in Edinburgh, Scotland, that was collected by townsman Gordon A. Thompson. William Thompson (1805–1852), then president of the Belfast Natural History and Philosophical Society, reported it as 'perhaps the first examples of their respective species brought thence to Europe' (Thompson 1844, p. 111).

Thus, during the first years of research into *A. spelaea*, this fish was considered to be largely a curiosity with varying significance. It was not until 1847 that the first published insight on the potential importance of troglomorphic fishes to biological research was published. Interestingly enough, it came from an unrepentant anti-evolutionist: Jean Louis Rodolphe Agassiz (b. Motieren-Vuly, Switzerland, 28 May 1807; d. Cambridge, Massachusetts, 14 December 1873) (Figure 5). During the 5 October 1847 meeting of the American Academy of Arts and Sciences, Agassiz

Figure 5. Louis Agassiz (Courtesy of the Ernst Mayr Library of the Museum of Comparative Zoology, Harvard University ©President and Fellows of Harvard College).

proposed a 'Plan for an investigation of the embryology, anatomy and effect of light on the blind-fish of the Mammoth Cave, *Amblyopsis spelaeus*' (Agassiz 1847, p. 180). There he suggested that by studying this fish 'there was an opportunity to settle, by actual experiment, the extent of physical influences in causing organized beings to assume their peculiar and distinctive characteristics in relation to the media in which they live'. Agassiz, the creationist, was not thinking in terms of the environment influencing evolution, but rather the effects of the environment on development. He proposed to raise individuals of *A. spelaea* under different light conditions (darkness, moderate, and intense light) and see if 'there is an eye formed in the dark to ascertain when and how (the pigmentation) disappears, as it is entirely wanting in the full-grown individuals, and again notice the differences in this respect between specimens growing under the influence of light' (Agassiz op. cit., p. 180).

He never carried out those experiments, yet he kept insisting on the importance of *A. spelaea* in biological research: 'You asked me to give my opinion, respecting the primitive state of the eyeless animals of the Mammoth Cave. This is one of the most important questions to settle in natural history, and I have several years ago, proposed a plan for its investigation which, if well conducted would lead to as important results, for it might settle, once for ever, the question, in what condition and where the animals now living on the earth, were first called into existence. But the investigation would involve such long and laborious researches, that I doubt it will ever be undertaken. (...) If physical circumstances ever modified organized beings, it should be easily ascertained here.' Despite these difficulties he remained optimist: 'Whoever would settle the question by direct experiment might be sure to earn the everlasting gratitude of men of science, and here is a great aim for the young American naturalist who would not shrink from the idea of devoting his life to the solution of one great question' (Agassiz 1851, p. 255). These words may not have fallen into a vacuum since, as we will see below, several of his students showed a great deal of interest in cave fishes.

Despite his insight, there is no question that Agassiz maintained an anti-evolutionist view until the very end of his life. On the rudimentary organs, he wrote 'The organ remains, not for the performance of a function, but with reference to a plan' (Agassiz 1859, p. 11). He considered *A. spelaea* to be an 'aberrant cyprinodont (...) created under the circumstances in which they now live' (Agassiz 1851, p. 256). And he never changed his mind: 'Have fishes descended from a primitive type? So far am I from thinking this possible, that I do not believe there is a single specimen of fossil or living fish, whether marine or fresh-water, that has not been created with reference to a special intention and a definite aim' (Agassiz 1885, 1, pp. 392–393).

The next scientist who would make significant contributions to the study of hypogean fishes was Felipe Poey (b. La Habana, Cuba, 26 May 1799; d. La Habana, 28 January 1891). He was a lawyer in Cuba, where tales of troglomorphic fishes were around since at least 1831. Poey secured the specimens and described two new species: *Lucifuga subterraneus* and *Lucifuga dentatus* (= *Stygicola dentatus*) (Poey 1858, 2, p. 100) (Figure 6). The descriptions of these two species were extraordinary on several accounts. Not only were they very detailed and precise from the external and internal anatomical viewpoints, but there was also a wealth of information on their behavior, habitat and history. Further, based on specimens he received from the United States, he described *A. spelaea*, for comparative purposes. The descriptions of the Cuban species were far superior to those published on *A. spelaea* by his American and European counterparts. Poey also demonstrated full familiarity with the published literature on cave fishes up to that time.

It would be easy to label Poey as an 'isolated genius' (sensu Beddall 1983) from social, scientific, and geographic viewpoints. He lived in a country which, at that time, was still a colony of Spain and lacked strong academic institutions. Poey himself was the founder of the first Cuban natural history museum and the first who taught zoology at the University of Havana. Yet he was not isolated from the scientific community. Poey was in contact with the most prominent contemporary researchers of cave fishes including Girard, Gill, Packard, and Putnam (see below). The latter visited Poey in Cuba in 1886 (Cockerell 1920). He received specimens of *A. spelaea* which he examined for his comparative studies adding information and making corrections to previous observations (Poey 1858, 2, pp. 104–106). He also provided American institutions such as the Museum of Comparative Zoology with specimens of the Cuban hypogean fish (Putnam 1872).

The next species was described by Charles Frédéric Girard (b. Mulhouse, France, 8 March 1822; d. Neulilly-sur-Seine, France, 29 March 1895). He had been brought to the United States by Louis Agassiz in 1847 and later worked at the Smithsonian Institution until 1860. While there, he was given some specimens

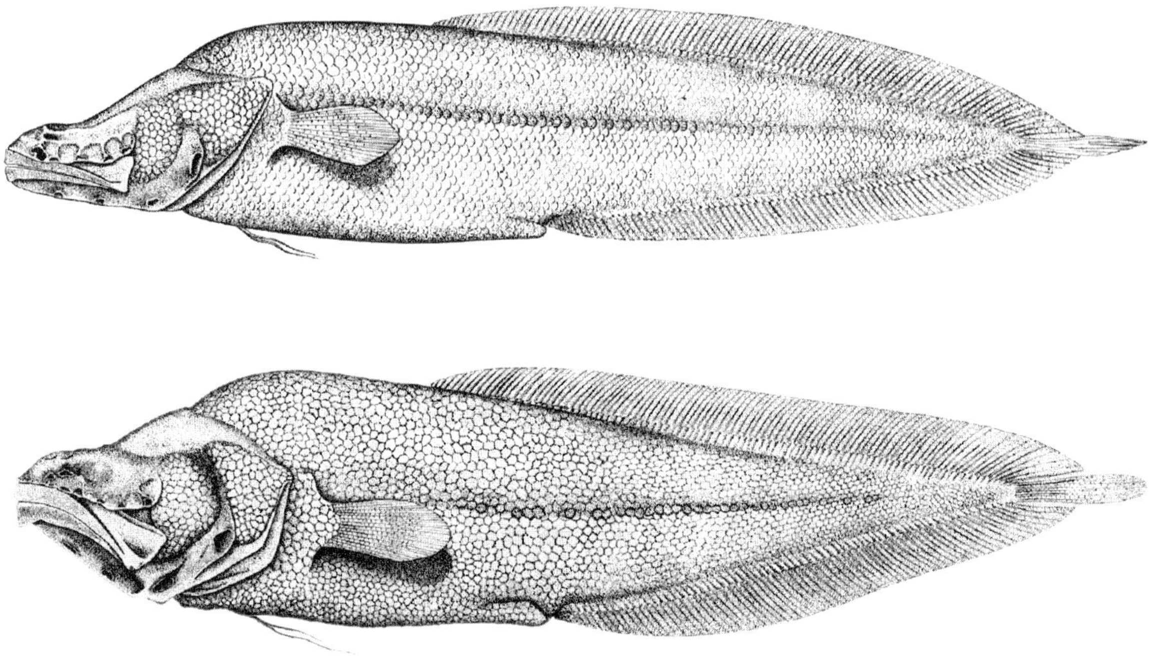

Figure 6. A 1902 illustration of *Lucifuga subterraneus* (top) and *Lucifuga dentatus* by C.H. Kennedy (courtesy of the Smithsonian Institution, NMNH, Division of Fishes).

collected by a J. E. Younglove 'from a well near Bowling Green, Ky'. He gave those specimens a new species status, *Typhlichthys subterraneus*, which he included in the family Amblyopsidae (Girard 1859). This new species seemed to have 'characters apparently transitory' between *A. spelaea* and the other species of the amblyopsid family known at that time: *Chologaster cornutus*, an epigean species. *A. spelaea* lacked eyes but had ventral fins; *T. subterraneus* lacked both eyes and ventral fins, while *C. cornuta* had eyes and lacked ventral fins.

The Civil War broke out while Girard was in Paris where he stayed and supported the Confederate cause by sending drugs, medical supplies, and arms. After the war he stayed in France where he practiced medicine (Jackson & Kimler 1999). Apparently he maintained some interest on cave fishes, because later in life he published a number of popular articles on the topic (e.g., Girard 1888).

With these species at hand, blindness and depigmentation among fishes had become synonymous with cave life which, in turn, led to inaccurate assumptions and misleading information. That was the case with a new 'species' of cave fish described by Edward Drinker Cope (b. Philadelphia, Pennsylvania, 28 July 1840; d. Philadelphia 12 April 1897). He published a paper on what he thought to be a new species and genus of troglomorphic fish, '*Gronias nigrilabris*' (Figure 7), from Pennsylvania (Cope 1864, p. 231). Although he did not present any evidence that such fish had been captured in the hypogean environment, he was quick to suggest that such fish 'is supposed to issue from a subterranean stream, said to traverse the Silurian limestone in that part of the Lancaster county, and discharge into the Conestoga'. Cope was known for his hasty conclusions and the superficiality of some of his work (Romero & Romero 1999). Further studies have shown that the specimens on which he based this description were specimens of *Ictalurus nebulosus* that had eyes present which were asymmetrically developed – probably as a result of a teratological condition. Unfortunately his assertion on this fish continued to be repeated in the literature until recently (see Romero 1999b for full history of this misconception).

Thus, by the end of this period what we have are the descriptions of four species of troglomorphic fishes, two from the United States and two from Cuba, as well as a number of morphological studies about them. Absent from this period were evolutionary discussions. That would soon change.

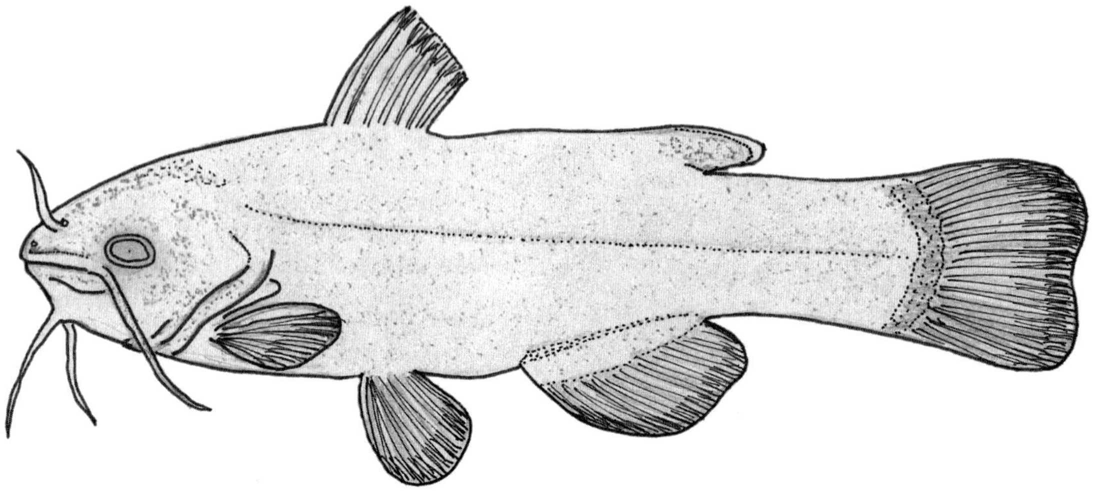

Figure 7. '*Gronias nigrilabris*', which, according to Cope, was a 'blind' cave silurid. Drawing by Olga Mayayo.

American neo-Lamarckism (1868–1919)

The divide between this and the previous era was marked by two major historical events. The first was the American Civil War (1861–1865), which paralyzed scientific field activity in those areas where hypogean fish had been found. The second was the publication of *On the Origin of Species* (Darwin 1859). This period was also marked by the dominance, for the first time, of a generation of U.S.-born ichthyologists. Together with a few European expatriates, they began cataloguing the fish fauna of North America at the same time that the United States began the systematic exploration of the West through efforts such as the Mexico Boundary and Pacific Railroad Surveys (Hubbs 1964). Although some new species were described during this time, the intellectual era of American neo-Lamarckism was characterized by the usage of cave fauna in general, and troglomorphic fishes in particular, to advance this view of evolution.

Charles Darwin (b. Shrewsbury, England, 12 February 1809; d. Downe, Kent, England, 19 April 1882) wrote on cave fauna based on the observations made of European cave animals by the Danish naturalist Jorgen C. Schiodte (1815–1884) and the observations of American cave fauna of James Dwight Dana (1813–1895) as explicit support for evolution. He noted that cave fauna was more closely related to the fauna of the surrounding regions than elsewhere, as is the case for fauna of other isolated habitats. Thus, he argued that the cave fauna descended from the fauna of the surrounding region, 'the colonists having been subsequently modified and better fitted to their new homes' (Darwin 1859, p. 403). At first Darwin considered the mechanisms of both natural selection and disuse to explain troglomorphic features, i.e., enlargement of some sensory systems and appendages for the former; blindness and depigmentation for the latter. To Darwin this suggested a 'contest . . . between selection enlarging and disuse alone reducing these organs' (op. cit., p. 296). However, in the third edition of the *Origin* (Darwin 1861) he de-emphasized the importance of natural selection, eliminating the speculation of a 'contest' between selection and disuse. This is important to keep in mind because, as we will see below, the members of the American neo-Lamarckian school closely followed Darwin in their interpretation of the mechanisms producing troglomorphic characters.

Darwin went back to much earlier explanations of evolutionary mechanisms and shielded the understanding of the evolution of cave fauna from more modern interpretations. This gave researchers of cave organisms arguments to support the 'use and disuse explanation'. In many ways, Darwin maintained a modified version of the Great Chain of Being (Bowler 1984, pp. 55–59), which was championed by the Swiss naturalist Charles Bonnet (1720–1793) and the French Philosopher Jean-Baptiste Robinet (1735–1820). They were followed by Jean Baptiste Chevalier de Lamarck (b. Bazentin-le-Petit, Picardy, France, 1 August 1744; d. Paris, France, 28 December 1829), again firmly planted into the Chain of Being tradition. Lamarck argued that organisms experience 'needs' (*besoins*) which were brought about by the environment and triggered fluids (including electricity) which, when circulated in the body, enlarged or developed the appropriate organ. In 'higher' animals, a crucial causal factor was the 'inner consciousness' (*sentiment interieur*), which makes parts respond and develop. This resulted in the inheritance of derived characters (Ruse 1996, p. 48). As we will see, even long after the development of the Modern Synthesis, this philosophy and its jargon (e.g., 'regressive evolution') would continue to be strongly imbedded in the research of hypogean fishes and in cave fauna in general.

The neo-Lamarckian period of hypogean fish research was dominated by three figures: Packard, Putnam, and Eigenmann. Alpheus Spring Packard, Jr. (b. Brunswick, Maine, 19 February 1839; d. Providence, Rhode Island, 14 February 1905) (Figure 8) studied under Agassiz and was his assistant from 1862 to 1864 at the Museum of Comparative Zoology. He received his M.D. from the Maine Medical School at Bowdoin College in 1864. After breaking with Agassiz in the so-called 'Salem Secession' he went on to become a leading figure of American neo-Lamarckism which he championed from his positions at the Boston Society of Natural History, the Peabody Academy of Sciences at Salem, Massachusetts, and at Brown University (Dexter 1965, Bocking 1988). It was Packard who coined the term 'neo-Lamarckism' and called Lamarck 'the real founder of organic evolution' (Packard 1901, p. v).

Figure 8. Alpheus Spring Packard, Jr. (courtesy of the Ernst Mayr Library of the Museum of Comparative Zoology, Harvard University ©President and Fellows of Harvard College).

In 1867, together with Edward Sylvester Morse and Alpheus Hyatt – also former Agassiz students – Packard founded *The American Naturalist*, the journal that published the most articles on cave fauna during the nineteenth century. He first examined Mammoth Cave specimens after the Indianapolis meeting of the American Association for the Advancement of Science when many of the participants visited that cave. He published an account of its fauna the same year. The Mammoth Cave fauna, including its fish, convinced him of their usefulness as a demonstration of evolution. 'We trust naturalists the world over will be led to explore caves with new zeal.' For him, the highest interest was the impact that these studies could have on evolution (Packard 1871, p. 761). In 1874 he was associated with the Kentucky Geological Survey, which intensified his interest in the fauna of Mammoth Cave and other caverns in the Midwest regarding which he produced several publications (e.g. Packard 1888).

Packard, like Cope, thought that cave fauna was of very recent origin and that the loss of certain organs was compensated by the hypertrophy of others. Packard's views were not in opposition to those of Darwin himself, but to the neo-Darwinians like August Weismann (1834–1914) and Edward Ray Lankester (b. London, England, 15 May 1847; d. London, 15 August 1929). Lankester, influenced by German biologist Anton Dohrn (1840–1909), wrote that blindness among cave animals was due to a special kind of natural selection. He began with the assumption that some animals are, by chance, born with defective eyes. Occasionally a few animals, some of which have normal eyes and some defective eyes, fall or are swept into caves. In each generation, those that have good eyes were able to see the light and escape, and eventually only those that are blind remained in the cave (Lankester 1893). Lankester also believed that one can find organisms degenerating ontogenetically and phylogenetically. He defined 'degeneration' as 'a loss of organization making the descendent far simpler or lower in structure than its ancestor', a phenomenon that he found widespread.

Figure 9. Frederick Ward Putnam (courtesy of the Ernst Mayr Library of the Museum of Comparative Zoology, Harvard University ©President and Fellows of Harvard College).

'Any new set of conditions occurring to an animal which render its food and safety very easily attained, seem to lead as a Rule of Degeneration' (Lankester 1880, p. 33).

The other leading figure of this time was Frederic Ward Putnam (b. Salem, Massachusetts, 16 April 1839; d Cambridge, Massachusetts, 14 August 1915) (Figure 9). Like Packard, Putnam studied under Agassiz and was his assistant until a split from him during the 'Salem Secession'. He was much more of an ichthyologist than Packard was, occupying such positions as curator of ichthyology and/or vertebrates in many institutions including the Boston Society of Natural History, the Essex Institute, the Peabody Academy of Science, and Harvard's Museum of Comparative Zoology. Of all of these experiences, it was his position in 1874 as Assistant of the Kentucky Geological Survey that brought him in direct contact with hypogean fishes.

Like Packard, Putnam first visited Mammoth Cave to collect fishes in 1871 after the adjournment of the meeting of the American Association for the Advancement of Science. He returned in 1874 following an invitation from Nathaniel S. Schaler, another of Agassiz's students who, as director of the Kentucky State Geological Survey, appointed Putnam as special assistant to the Survey that year. He made the first large collection of Mammoth Cave fishes from Kentucky (22 in total) in October 1874, some of which were displayed alive at the Essex Institute. Interestingly, the first public exhibition of a live amblyopsid took place not in the United States but in the Dublin Zoological Gardens, Ireland, sometime in 1870 (Baird 1872, Putnam 1874).

Although less known than Packard because he was not very much inclined to provide sweeping generalizations or engage in theoretical discussions, Putnam expressed a high degree of critical thinking. For example, he criticized Cope's interpretation that *A. spelaea* was able to survive in those waters because it had 'the projecting under jaw and upward direction of the mouth renders it easy for the fish to feed at the surface of the water (...) This structure also probably explains the fact of its being the sole representative of the fishes of subterranean waters. No doubt many other forms were carried into the caverns since the waters first found their way there, but most of them were like those of our present rivers, deep waters or bottom feeders. Such fishes would starve in a cave river, where much of the food in carried to them on the surface of the stream...'. He asked then: where are the surface forms of the 'surface feeders'? How come other surface feeders are not found in caves? He did not understand that Cope made this statement when he himself had described an alleged 'cave' fish ('*Gronias nigrilabris*') from Pennsylvania which was a bottom feeder and that the blind cave fishes from Cuba were bottom feeders as well. He also said that studies of stomach contents in *A. spelaea* had shown that they eat essentially crayfish and other fishes. He asked that if blindness was the direct result of darkness, as some contended, 'how is it that *Chologaster* from the well in Tennessee, or the 'mud fish' at Mammoth Cave are found with eyes?' (Putnam 1872, p. 24).

Putnam (1872, p. 6) also stated that 'The blind fish of the Mammoth Cave has from its discovery been regarded with curiosity by all who have heard of its existence, while anatomists and physiologists have considered it as one of those singular animals whose special anatomy must be studied in order to understand correctly facts that have been demonstrated from other sources; and, in these days of the Darwinian and development theories, the little blind fish is called forth to give its testimony, pro or con.' He viewed the amblyopsids as former marine and saltwater estuary fishes that were slowly trapped in that geographical area. He substantiated this hypothesis by pointing out that the eyed amblyopsid *C. cornuta*, was 'now living in the ditches of the rice fields of South Carolina, under very similar conditions to those under which others of the family may have lived in long preceding geological time; and to prove that the development of the family was not brought about by the subterranean conditions under which some of the species now live, we have the ones with eyes living with the one without, and the South Carolina species to show that a subterranean life is not essential to the development of the singular characters which the family possess'. He further supported this hypothesis by mentioning that the Cuban blind cave fishes belonged to the genera 'with their nearest representative in the family a marine form, and with the whole family of cods and their allies, to which group they belong, essentially marine'. Putnam (op. cit., pp. 22–23) also described a new species of amblyopsid, *Chologaster agassizi*, as the fifth troglomorphic fish. By the year of his last publication on hypogean fishes (Putnam 1874), he switched careers from ichthyology to archaeology after he was appointed as the Curator of the Peabody Museum of American Archaeology and Ethnology at Harvard University.

All discussions on the evolution of cave fauna in general and hypogean fish in particular, at the time of Packard and Putnam, took place under the influence of the so-called 'Hyatt-Cope' position or school which was based on drawing parallels between embryology and phylogeny. Cope had little time for natural selection (e.g., Cope 1864) and seeing adaptation as part of the picture, plumped for Lamarckism. Central to Cope's vision of the biological world was the way in which organisms fall in trends. 'The method of evolution has apparently been one of successional increment or decrement of parts along definite lines' (Cope 1896, p. 24). This is what was later called orthogenesis, the view that evolution has a momentum of its own that carries organisms along certain tracks. Cope also proposed the 'Law of the Unspecialized': when an organism has gone too far down a particular phylogenetic path (as cave species do, according to him), especially a particular adaptively specialized path, it can never pull itself out and evolve into an altogether new and fruitful form (Cope 1896, pp. 172–174).

Aplheus Hyatt (b. Washington, D.C., 5 April 1838; d. Cambridge, Massachusetts, 15 January 1902) (Figure 10), whose more prominent position was that of director of the Boston Museum of Natural History,

Figure 10. Alpheus Hyatt (courtesy of the Ernst Mayr Library of the Museum of Comparative Zoology, Harvard University ©President and Fellows of Harvard College).

visited Mammoth Cave in September 1859, much earlier that his contemporary colleagues, and collected seven specimens of *T. subterraneus* that Putnam later studied (Bocking 1988). Hyatt's ideas can be summarized as follows: (1) just as the individual eventually declines into old age and senescence, so also the group declines into old age and senescence; (2) before extinction, there is degeneration of species; (3) change is a function of the speeding ('acceleration') or slowing ('retardation') of development which, in turn, is propelled by use and disuse (Hyatt's 'Law of Acceleration', is a direct descendant of Haeckel's 'Principle of Recapitulation'); (4) degeneration is therefore a virtually inevitable outcome of the evolutionary process, since any organism tends to collapse into old age, and it is a matter of time before this decay is delayed in development until adulthood (for a summary of Hyatt's ideas see Brooks 1909). Thus, Hyatt worked within a progressionist framework, a kind of Americanized *Naturphilosophie*. After all, the kind of evolutionism expressed by Agassiz's students and those under their influence (which was quite vast) had more to do with the transcendentalism of the German *Naturphilosophie* than to the natural theology of essential Darwinism. This is an idea that, as we will see, would be largely embraced by the European neo-Lamarckians of the twentieth century.

By the late nineteenth century, discussions on evolution in general languished primarily because it was obvious that no progress was being made, and 'Morphology having been explored in its minutest corners, we turned elsewhere' (Bateson 1922, p. 1412). Thus, evolutionary morphology stalled and fell to the status of a second-rate science. Many morphologists went into the kind of science practiced in museums. Lankester, for example, moved from being a professor at Oxford to Director of the British Museum of Natural History (Ruse 1996). This era of cave fauna research in the United States closed in the 1870s, within a conceptual framework of Bauplane, homologies, and parallelisms between embryonic and phylogenetic development while natural selection was largely ignored.

Thus, cave fauna, with their bizarre adaptations to an extreme environment, were interpreted by neo-Lamarckians as providing excellent evidence for the effect of the environment upon the evolution of organism. Cave fauna could demonstrate the power of the physical environment as an evolutionary factor, capable of maintaining an organism's adaptation to its changing surroundings. Packard attributed more evolutionary importance to the direct effect of the environment than to the effect of changes in habits. Cave research therefore helped to establish that the changing environment could be responsible for the underlying progressive trends in evolution that Packard saw in his work in embryology.

By the 1870's Wyman, Packard, Putnam, and the rest stopped publishing on hypogean fishes. The topic seemed to have been conceptually exhausted. Nearly 20 years were to pass before anyone else would show an interest in the topic.

The next relevant figure in the history of hypogean fish research was Carl H. Eigenmann (b. Flehingen, Beden, Germany, 9 March 1863; d. Chula Vista, California, 24 April 1927) (Figure 11). Influenced by David Starr Jordan (a student of Louis Agassiz's son,

Figure 11. A little known picture of Carl. H. Eigenmann. The effect of this picture was probably achieved by making five different exposures. This picture is dated September 1914 (courtesy of the Smithsonian Institution, NMNH, Division of Fishes).

Alexander), he became a biologist particularly interested in fishes. His first experience with troglomorphic fishes took place in 1886 while at Indiana University when he received a living blind fish taken from a well in Corydon, Indiana. The next year he married Rosa Smith (1858–1947), the first woman president of the scientific society Sigma-Xi. In 1889 they established residence in California where he was named Curator of the San Diego Natural History Society. There, his wife introduced him to the blind goby found among the rocks of the California coast, *Typhlogobius californiensis* (see Eigenmann 1890 for a historical account of this encounter and how much it impressed him). In 1891 he was appointed Professor of Zoology at Indiana University, a perfect location to study the blind vertebrates of the caves in the nearby areas. This motivated him to devote a substantial part of his scientific career to the study of blind vertebrates, most of them in caves (Romero 1986b).

Between 1887 and 1909, much of his work was devoted to comprehending the process of the loss of visual structures in cave vertebrates. In May 1896 he visited Dalton's Spring (actually a cave-stream) where he secured 20 specimens of *A. spelaea*. This became his favorite collecting locality. In 1898, Eigenmann published the description of a new species of cave fish with rudimentary eyes from south-western Missouri, *Amblyopsis rosae* (= *Typhlichthys rosae*), which he named after his wife.

Eigenmann extensively visited the caves of Indiana, Kentucky, Texas, and Missouri in search of specimens for his work and in March 1902, he visited Cuba for the first time to secure cave specimens for his comparative studies. He had been working on fish reproduction in the past and quickly recognized that these two Cuban hypogean fish species were viviparous.

Contrary to Mammoth Cave, Eigenmann found the localities for the Cuban blind fish 'monotonous' (Eigenmann 1903). From 1906 to 1907 he did many laboratory studies in Europe, mostly in Germany, with the Cuban specimens he had collected. During this period he made plans to visit the Yucatán peninsula in Mexico because of persistent reports of a varied cave fauna in that part of the world. From 1898 to 1905 Eigenmann published at least 39 papers and abstracts on cave vertebrates, dealing mostly with developmental and anatomical aspects of vision loss in fishes, salamanders, lizards, and mammals as an attempt to understand the underlying process of blindness among hypogean animals. All this research was summarized in his *Cave Vertebrates of North America* (Eigenmann 1909), a 341-page, 30-plate monograph.

Although a taxonomist by training, Eigenmann quickly sought answers to the issues of the origin and evolution of the cave fauna. Originally a neo-Lamarckian, Eigenmann thought that the reduction or disappearance of organs among cave animals was a case of convergent evolution, i.e., well-defined conditions of the subterranean environment facilitate the evolutionary changes leading to blindness and depigmentation in a variety of vertebrate and invertebrate organisms which come to inhabit them. He pointed out that lack of pigmentation had to be understood as the combination of genetically fixed and epigenetically (environmentally-influenced) determined characters; in other words, that although a character was genetically determined, its degree of development may vary under certain light conditions. For Eigenmann cave evolution was essentially 'degenerative' and all successful cave-invaders were somehow pre-adapted to that medium. The origin of caves and of blind fauna were two distinct questions because of his experience with the blind fish found among the rocks of California's seacoast. He insisted on a strong link between ontogeny and phylogeny and his constant use of terms such as 'phyletic degeneration' indicates that he held orthogenetic views. He followed Herbert Spencer's (1820–1903) idea that cave faunas were not the result of 'accidents' but rather the product of an active process of colonization (Eigenmann 1909).

His last contributions in the field were the descriptions of a new species of blind fish, *Trogloglanis pattersoni*, from the artesian waters of San Antonio, Texas (Eigenmann 1919). This description was based on a specimen collected by G.W. Brackenridge of San Antonio, who gave the fish to J.T. Patterson, a professor at the University of Texas, who in turn sent it to Eigenmann at Indiana University.

Two more hypogean species were discovered during this period: *Phreatobius cisternarum* by Goeldi (1905) and *Pimelodella kronei* (= *Typhlobagrus kronei*) by Miranda-Ribeiro (1907), both from Brazil. Probably due to geography, the discovery of these species never made any important impact on general discussions of biological concepts.

Thus, from 1866 on, the neo-Lamarckian school flourished and found its strongest supporters among American naturalists interested in cave fauna (Figures 12, 13). Although Barr (1966) called the beginning of this era the 'Neolamarckian revolt against

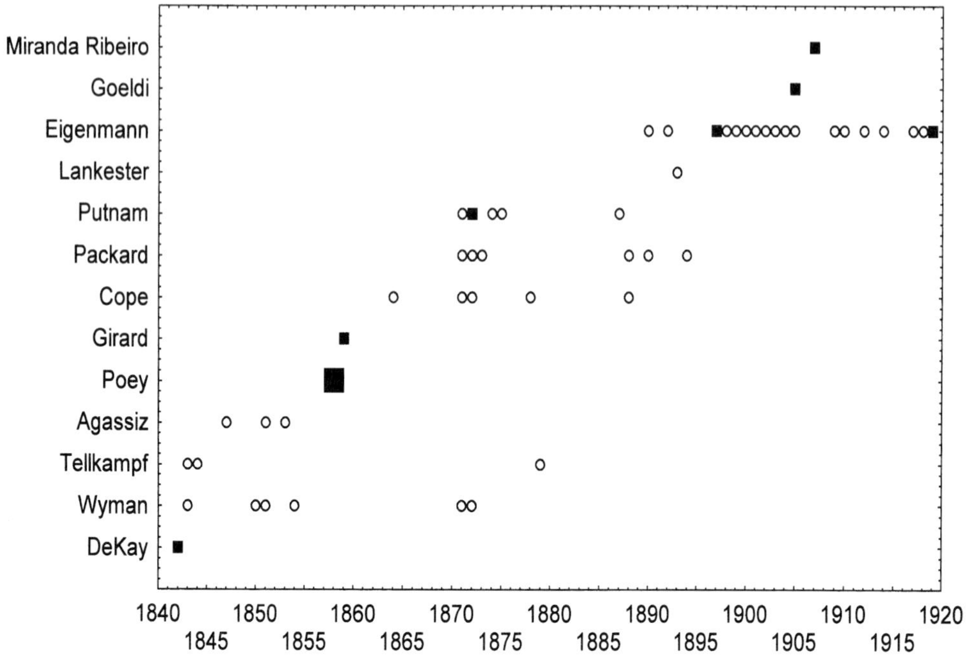

Figure 12. Time-line graph of major papers on hypogean fishes published between the first scientific description of a troglomorphic one, *Amblyopsis spelaea* in 1842, to right before the beginning of the typological era. Squares represent papers describing new species. The larger square for Poey represents a single paper describing two species. Open circles are papers on anatomy and evolution. Notice the divide created by the American Civil War (1861–1865) which interrupted virtually all research efforts.

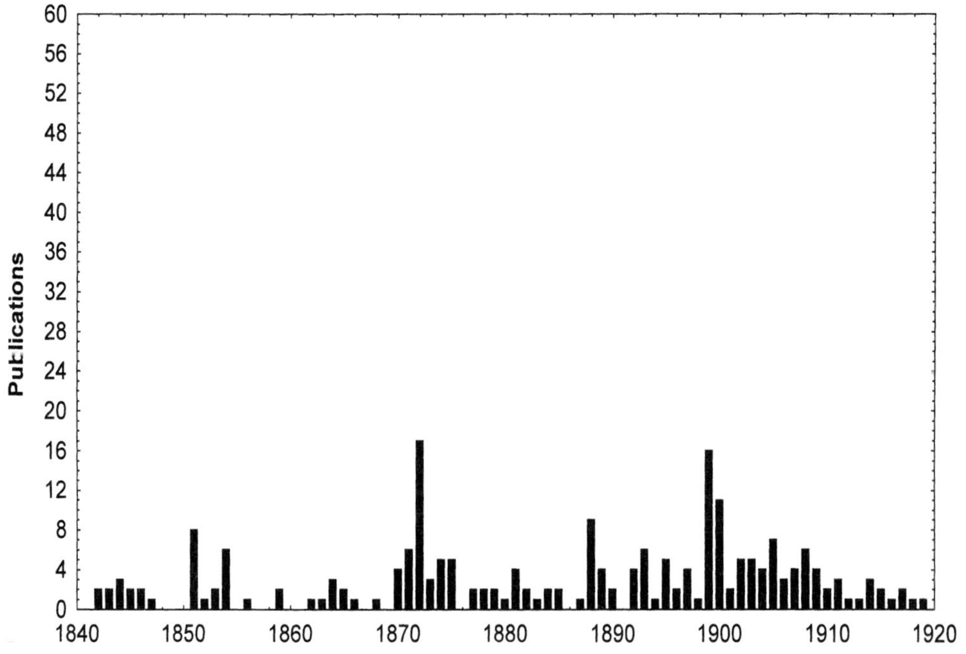

Figure 13. Number of publications on hypogean fishes between 1842 and 1919. The 'pulse' for the 1870's represent the high point of the neo-Lamarckian school's interest in these animals. The one for the early 1900's is almost entirely due to the intense publishing activity by Carl H. Eigenmann.

Darwin's theory of evolution by means of natural selection', the fact of the matter is that those neo-Lamarckians were very much in line with what Darwin himself said, particularly in later editions of his *Origin of Species*.

Dominance of typological thinking (1921–1940)

This period is characterized by three trends: (1) the incremental discovery of new species, all of them outside of the United States, which gave a broader perspective not only geographically but also taxonomically to the diversity of hypogean fishes, since most of these new species belonged to fish families with not previously reported hypogean representative; (2) the absence of researchers who devoted considerable part of their research to this topic, as Wyman, Packard, Putnam and Eigenmann had done before. During this period all authors of hypogean fish papers spent very little time on the subject, mostly because they were studying the fish fauna of a particular area where troglomorphic fishes were found. The only possible exceptions to this were Jacques Pellegrin (1873–1944) and Luisa Gianferrari (1890–1977) who published a handful of papers on the subject. Yet, those papers were the product of their interest on the fish fauna of a particular geographical region more than the study of hypogean fishes per se; (3) most of the new species/populations discovered were given a new generic status based almost exclusively on the fact that they lacked eyes and pigmentation.

Of the 12 new species/populations described during this period (Table 1, Figure 14), six were from Africa, three from Mexico, two from Japan and one from Trinidad, W.I. The new discoveries in Africa were propelled by the presence of European colonial powers in that part of the world and the specimens were collected by local colonial authorities. The ones in Mexico were carried out by American-born researchers venturing for the first time outside the United States. The ones from Japan and Trinidad were described by British researchers residing in the United Kingdom. The two new species from Japan were re-descriptions based on specimens collected by a Japanese researcher while the one from Trinidad was collected by a local natural historian.

The first description of a new species of troglomorphic fish for this period epitomizes the trend set by European researchers when dealing with African hypogean species. When King Leopold II of Belgium

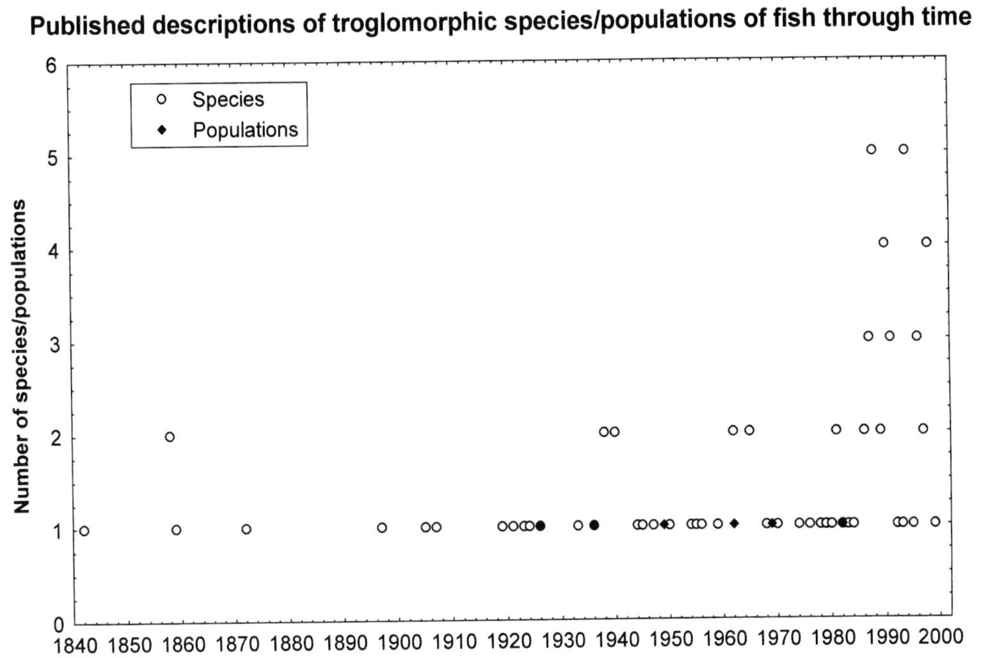

Figure 14. Published scientific descriptions of troglomorphic species/populations of hypogean fishes. There has been a steady number of new descriptions since the 1920's when the typological era started with description of new species mostly from Africa. The increased number in the late 1980's and throughout the 1990's is due to new discoveries in Asia, particularly in China.

foresaw the importance of scientific research in Africa, he instructed voyagers and explorers to collect natural history information about the Congo. Not surprisingly, many Belgian officials collected and brought to Europe unknown zoological specimens that had never been scientifically studied. One of those specimens fell in the hands of George Albert Boulenger (b. Brussels, Belgium, 19 October 1858; d. Saint Malo, France, 23 November 1937) (Figure 15). He worked from 1880 until his retirement in 1920 at the British Museum. By then, Boulenger had published 877 papers totaling more than 5000 pages, as well as 19 monographs on fishes, amphibians, and reptiles in which he described 1096 species of fish, 556 species of amphibians, and 872 species of reptiles. He was considered the world's expert on African freshwater fishes despite the fact he had never been to Africa. Boulenger's only previous stint as researcher of cave animals had been a short note on the eye development of the cave salamander *Proteus* (Boulenger 1893).

In 1921, Boulenger received a blind and depigmented fish specimen from the Congo. He was quick to

Figure 15. A young Georges Albert Boulenger about the time he started working at the British Museum of Natural History (courtesy of the Smithsonian Institution, NMNH, Division of Fishes).

recognize its uniqueness since it did not seem closely related to any extant African epigean species. He named it *Caecobarbus geertsii*, from *caeco* = blind, *barbus* = barb, and *geertsii*, honoring a mysterious person, M. Geerts, who provided him with the specimen (Boulenger 1922). This species of fish may have been originally collected in 1915 by M. Delporte, probably the first European to see this fish. Soon this species became a celebrity in the world of cave fish research. Being easy to transport alive, many European scientists started to work on it. It became so famous that in May 1951 it was exhibited at the New York Aquarium (Romero & Benz 2000).

The same can be said of other African hypogean fishes. Three were discovered in Somalia, collected by local colonial Italian officials, but then described by ichthyologists in museums of natural history of Italy (*Uegitglanis zammaranoi* Gianferrari, 1923, *Phreatichthys andruzzii* Vinciguerra, 1924, *Barbopsis devecchii* Di Caporiacco, 1926). Another was collected by a French explorer in Madagascar and sent to France for identification (*Typhleotris madagascariensis* Petit, 1933) and the last from Africa was collected in South Africa by a local physician and sent to the British Museum of Natural History (*Clarias cavernicola* Trewavas, 1936).

The case of the hypogean fish from Trinidad had a different twist. It was collected by a local and well-known naturalist, Friederick William Urich (1872–1936), who sent the specimens to London for identification. When received by the British Museum of Natural History in July 1924, they were given to John Roxborough Norman (1899–1944), who had published a few papers on the fishes of the nearby island of Tobago. Besides certain reduction in the eye size, this fish was extremely similar to *Rhamdia quelen*, a well-known epigean fish common to northeastern South America and Trinidad. Fearing that the specimen could represent an accident of nature rather than a true new fish species, Norman requested two more specimens and Urich complied. He later published its report and named the fish *Caecorhamdia urichi* (*caeco* = blind; *rhamdia* = the genus of a catfish to which this cave fish seemed most related to; *urichi* = honoring Urich, the collector) (Norman 1926). Since then, this fish species has consistently appeared in the lists of blind cave fishes of the world although today they are considered individuals of the common catfish *R. quelen*, a species of nocturnal habits showing different degrees of (but never complete) blindness and depigmentation (Romero &

Creswell 2000). This would be the first report of a cave fish population of an epigean species with certain degrees of troglomorphic features.

As previous examples demonstrate, during this period almost every cave fish population that was discovered, regardless of how close it looked to its epigean ancestor, had a new genus and species erected for it based on its taxonomic characters of blindness and depigmentation. This was the typological approach prevalent at that time, a derivative of the essentialist philosophy that has dominated Western thinking and was not replaced until after the Modern (Evolutionary) Synthesis of the late 1930s and early 1940s. As we will see later, such thinking has been retained to certain extent in some quarters of hypogean fish research.

Besides the re-descriptions by Charles Tate Regan (1878–1943) of two Japanese hypogean fishes (*Luciogobius pallidus* and *Luciogobius albus* Regan, 1940), all the other discoveries during this period took place in Mexico by American born researchers. Since they operated intellectually and geographically independently from their European counterparts, and because their discoveries opened new doors to hypogean fish research, they deserve to be treated separately despite an overlap of a few years with the efforts of their colleagues elsewhere.

American renaissance (1936–1960)

The resurgence of hypogean fish research began around the time the typological period started to wind down. Although the discovery of new species continued in Asia, Africa, and Australia, it was the work of American-born scientists doing field studies outside the United States and experimental work in American institutions that introduced a renewed interest in the subject, particularly because aspects other than taxonomy and morphology were investigated. This period is characterized by more comprehensive studies which included ecology, physiology, and behavior. This was, in part, due to the fact that the species most suitable for these kinds of studies was discovered at the very beginning of this era.

In 1936, Salvador Coronado, a young Mexican in charge of the Fish Culture Station at Almoloya del Rio near Mexico City, sent 75 fish collected at the La Cueva Chica (the little cave) to C. Basil Jordan, a fish dealer from Dallas, Texas. Coronado thought those animals were particularly interesting: blind and pinkish. Jordan was impressed because all of the fish arrived to Texas alive, something particularly interesting for someone whose business largely depended upon the ability for a fish to survive transportation.

Not being able to determine the species, Jordan sent (at least some of) the fish to William Thornton Innes (1874–1969), a well-known aquarist and aquarium writer. Strongly suspecting that he had a new species in his hands, Innes remitted on November 1936 the specimens together with Jordan's notes to Carl Leavitt Hubbs (b. Williams, Arizona, 18 October 1894; d. La Jolla, California, 30 June 1979) (Figure 16). Hubbs was, by that time, one of the most respected American ichthyologists who had been doing field work in Mesoamerica and whose knowledge on freshwater fish taxonomy was considered to be virtually encyclopedic.

It took Hubbs only a few weeks upon receiving the material to publish a joint article with Innes describing this fish from the characid family as a new genus and species: *Anoptichthys* (eyeless fish) *jordani* (honoring Jordan). As Hubbs himself put it, this was the 'most surprising, by far subterranean fish belonging to the family Characidae, of which no blind representative has ever been seen before' (Hubbs & Innes 1936, p. 1).

Hubbs' surprise was manifold. First, it was most unusual to capture so many individuals of a cave fish species in a single locality; the amblyopsids, by that time the best known cave fish family, were not so abundant. Second, the fact that all 75 individuals had arrived in the U.S. alive and were easily kept in captivity said something about the potential of this species as a research subject (Innes 1937). Third, this cave characin did not grossly display the hyperdeveloped sensory organs quite common among other cave animals. Basically, it only differed from its likely ancestor, *Astyanax fasciatus*, in lacking eyes and pigmentation. Yet, in line with the typological thinking of the times, it was given a new generic and specific status.

This Mexican cave fish was so amusing that in both Mexico and the United States a great deal of interest arose. A group of the Mexican 'Escuela Nacional de Ciencias Biologicas' that included, among others José Alvarez (1903–1986) and Bibiano F. Osorio Tafall, began the exploration of the whole cave system for the area which now has about 30 localities containing this fish (Mitchell et al. 1977). As a matter of fact, two new populations, in addition to the one in La Cueva Chica, were given new specific names: *Anoptichthys antrobius* for the La Cueva El Pachón population (Alvarez 1946) and *A. hubbsi* for the La Cueva de los Sabinos

Figure 16. Carl Leavitt Hubbs in his laboratory at the Scripps Institution of Oceanography. This picture was taken in 1945, shortly after he had made his major contributions to hypogean fish research (courtesy of the Scripps Institution of Oceanography Library).

population (Alvarez 1947). As more cave populations were discovered, it became evident that this flow of specific names was leading nowhere and that all the troglomorphic fish should be considered as a single species.

Hubbs' hypogean cave research also comprised the Yucatán peninsula. Although there had been rumors of blind fishes from the cenotes (sink holes) of Yucatan (Girard 1888, Eigenmann 1909), it was not until the 1930s that organized scientific expeditions were carried out to explore this unique subterranean habitat. There, Hubbs found a number of epigean fish living in the cenotes, some of which showed certain degree of eye reduction and depigmentation and erected new subspecies status for them. But, more importantly, he also described two new troglomorphic species: *Ogilbia pearsei* (= *Typhlias pearsei*) and *Ophisternon infernale* (= *Pluto infernalis*) in a monographic publication in which he also reviewed all known information on cave fishes to that date (Hubbs 1938).

But it was his discovery of the cave characid that revitalized hypogean fish research. Hubbs & Innes (1936) paper immediately attracted the attention of non-taxonomists. One of them was Myron Gordon (1899–1959) an internationally known fish geneticist who had been associated with the New York Aquarium since 1938. Since the late 1920s he had a particular interest on the inheritance of normal and abnormal pigment cell growth, especially melanoma tissues, in fishes. Thus, in 1939 he visited the cave in which the fish had been discovered in Mexico. There he collected more individuals which he took to New York. This fish elicited the interest of his New York colleagues, among them Edward Bellamy Gresser (1898–1951), a practicing physician and a professor of ophthalmology at New York University. He used the laboratories of the New York Aquarium to study the histology of the cave *Astyanax* and published an article about it (Gresser & Breder 1940). However, it was the co-author of that article who actually showed an intense interest on that species.

Charles Marcus Breder Jr. (b. Jersey City, New Jersey, 25 June 1897; d. Englewood, Florida, 28 October 1983), was the director of the New York Aquarium in 1939. After learning about the newly described fish and hearing Gordon's account of his recent trip to Mexico, Breder took the initiative of

Figure 17. Picture of Charles Marcus Breder (right) and Salvador Coronado during their first expedition to 'La Cueva Chica' in Mexico in 1940. This will be the beginning of an extraordinary research program by Breder and his associates (photograph by Mary Edwards Braden, courtesy of the Wildlife Conservation Society).

organizing and leading an expedition to Mexico to perform field studies, obtain enough ecological information for a cave habitat display for the Aquarium, to shoot a documentary to be presented at the 1941 annual meeting of the New York Zoological Society, and, most importantly, to bring back enough fish to conduct extensive laboratory research (Breder 1942, Romero 1984, 1986a).

In January 1940, Breder held a meeting with other scientists where the whole expedition, known as 'The Aquarium Cave Expedition to Mexico,' was organized. By 11 March 1940, the group was already in Ciudad Vallés, near La Cueva Chica, the fish type locality (Figure 17). Despite the fact that most members of the expedition suffered from 'tropical fevers' (possibly histoplasmosis) after the trip, the expedition was a complete success (Bridges 1954).

Unlike many hypogean fish researchers before him, Breder was not particularly interested in systematics. Yet, in 1944 he was offered the chair of the Department of Fishes of the American Museum of Natural History and retired in 1965 (Atz 1986). Between 1940 and 1954 Breder (sometimes coauthoring with Gresser or Priscilla Rasquin, an ichthyologist at the American Museum of Natural History) published 17 papers (168 printed pages of information) on which *Astyanax* was the major subject of research. Most of the research concentrated on behavior, particularly responses to light and chemicals. But he also published on the fish sensory organs, metabolism, genetics, ecology, and evolution. Because the cave and the surface forms interbreed to produce fertile hybrids, he was the first who strongly suspected that the blind depigmented form was nothing but a remarkably locally-adapted population of the surface species *Astyanax fasciatus* well before advanced genetic techniques proved it. Many of Breder's contributions are still cited and several of his associates and students embarked in researching this cave fish. There is no question that Breder was the dominant figure in hypogean fish research in the 1940s and 50s (Figure 18).

In summary, this period was first characterized by the non-taxonomical research of Breder and his associates first at the New York Aquarium, and later at the American Museum of Natural History. Yet, not until the late 1950s did a number of European researchers also become interested in the genetics and evolution of these fishes which would, ultimately, signal the rebirth of neo-Lamarckian ideas about the evolution of cave fauna in general and hypogean fishes in particular.

Times of philosophical conflict (1960–1990)

From 1960 on, not only the number of papers on hypogean fishes increased (Figure 19), but also two major schools of thought started to develop very strongly. One was the school of orthogenesis which, essentially, utilized cave organisms as prime examples for 'regressive evolution', and the neo-Darwinian school, which tried to use the new concepts and approaches provided by the Modern Synthesis in order to explain the evolution of those very same organisms.

Let us begin with the orthogentic school. The notion of progress has been deeply embedded in evolutionary thought since pre-Darwinian times, and it was usually referred to as a matter of levels of complexity (Nisbet 1979, Ruse 1996). For Karl Ernst von Baer (1792–1876), for example, 'The more homogeneous the whole mass of the body is, so much the lower is the grade of its development. The grade is higher when nerves and muscles, blood and cell-substance, are sharply distinguished' (von Baer 1828, p. 207). We can see the concept of 'progress' also being used in the social sciences, particularly by those who held Marxist ideas (e.g., Demoor et al. 1897, Guillaume 1971). How was this idea embraced by students of cave faunas? For an answer, we need to look back to the American neo-Lamarckian school.

As we mentioned earlier, Cope introduced the idea that evolution was directed by trends and that 'The method of evolution has apparently been one of successional increment or decrement of parts along definite lines' (Cope 1896, p. 143). He, like other neo-Lamarckians, had little regard for natural selection; thus, these 'natural' trends explained how evolution worked overall. That was the foundation of 'orthogenesis', i.e., the idea that there is a directionality in evolution. When an organism went too far in a

Figure 18. Charles Breder and his wife Priscilla next to their retirement home at Englewood on the 29 April 1979 (photograph by Eugene K. Balon).

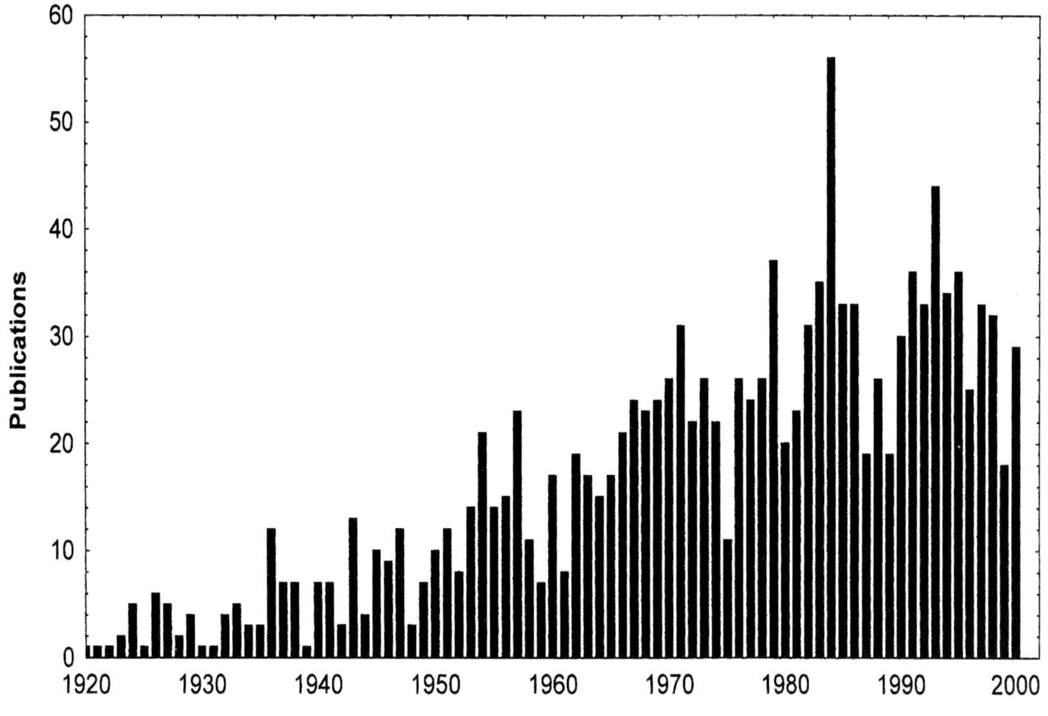

Figure 19. Number of publications on hypogean fishes from 1920 until 2000. From the beginning of the typological era in 1920, there has been a steady stream of papers published on diverse aspects of hypogean fishes. The robustness of publication numbers from the late 1960's on is largely due to the members of the Hamburg School, whose interest was prompted by the pioneering research of Curt Kosswig.

particular direction or adaptively specialized path, like cave organisms, it could not reverse into a new form (Cope 1896, pp. 172–174).

With the re-discovery of Mendel's laws and under the influence of the Modern Synthesis, orthogenetic ideas advocated by neo-Lamarckians were temporarily dismissed, particularly among researchers in English-speaking countries. The resurgence of these ideas, however, started with the work of the Romanian biologist Emil G. Racovitza (1868–1947), considered by many as the founder of biospeleology and later continued in the 1960's by the French speleologist Albert Vandel (1894–1980). For Racovitza all cave organisms were 'preadapted' to the cave environment where 'the function creates the organ'. He was also a rabid anti-neo-Darwinian (Motas 1962). For Vandel, the explanation of the evolution of cavernicoles was neither neo-Lamarckism (for which no empirical evidence could be found) nor neo-Darwinism (for which French researchers had little sympathy at that time). The explanation was 'organicism' which basically meant that 'all phyletic lines pass through several successive stages: the stage of the creation (. . .) [t]he stage of expansion and diversification, and finally the stage of specialization and senescence (. . .) The first stage of this cycle is a manifestation of progressive evolution and the last of regressive or gerontocratic evolution. The evolution of cavernicoles provides a particularly good example of regressive evolution which can occur only in the final stages of the evolution of animal lines' (Vandel 1964, p. 463). The application of these ideas to hypogean fishes can be clearly seen in Thinès (1969) book on cave and abyssal fishes. Curiously, these authors never fully acknowledged that orthogenesis was neo-Lamarckian in nature and was fully developed in the United States, but rather preferred to trace themselves back to French authors such as Henri Bergson (1859–1941), René Gabriel Jeannel (1879–1965), or Pierre Teilhard de Chardin (1881–1955).

It was a German researcher who popularized the whole notion of regressive evolution among cave animals in general and cave fishes in particular – Curt Koswigg (b. Berlin, Germany, 30 October 1903; d. Hamburg, Germany, 29 March 1982). Since the 1930's he had been working on cave fishes and showed an interest in evolutionary issues, but it was a paper he wrote while in exile in Turkey that set the basis for his studies on 'regressive' evolution (Kosswig 1949; for a more updated and accessible version see Kosswig 1960). Basically, he used the results of his own Mendelian studies to explain the rudimentation or loss of structures among cave animals. Without being a neo-Lamarckian, the fact that he used modern (at that time) genetic techniques to make his point gave respectability to his work. On the other hand, there is an underlying notion in his work about directionality in evolution which also sets him apart from contemporary neo-Darwinian thinking. As a matter of fact, the notion of 'regressive evolution' is still the central theme among members of the most productive group of hypogean fish researchers at the Zoological Museum and Zoological Institute at the University of Hamburg. This idea of 'regressive evolution' has been the source of monographic texts and, as Thinès (1969) book title, the central idea for describing (although not necessarily explaining) the common features among troglomorphic organisms.

The major problem with 'regressive evolution' is that it gives the impression that the evolution of cave organisms is not the result of ~~non-~~neo-Darwinian mechanisms and, thus, set apart from mainstream evolutionary thought as evidenced by the attempts to explain phenotypic simplification by means of neutral mutations (e.g., Wilkens 1988). However, this view of cave life had three difficulties: (1) shared characteristics among troglomorphic organisms that are phylogenetically unrelated seem to be an excellent example of convergent evolution and are therefore easily explained by using neo-Darwinian mechanisms; (2) the fact that the gain and loss of characters in evolution is a common phenomenon; for example, nobody has had the anti-anthropocentric temerity to label *Homo sapiens* the result of 'regressive evolution' because that species lost the tail, most of the pelage, and the ability to graciously jump from one tree to another from its ancestors; (3) if this evolution is regressive, then the question of 'regressive to where?' remains unanswered (for further discussion see Romero 1985).

From the neo-Darwinian side, this period begins with the work of Thomas Layman Poulson (1934–) who, in the early 1960s started to apply new ecological concepts to the study of cave adaptation for amblyopsids, a fish family whose study had largely been neglected since the times of Eigenmann. Poulson took a closer look at some of the adaptations to this environment such as metabolic rate, life history, sensory morphology and foraging behavior, and tried to draw a step-by-step description of that process from a holistic viewpoint by comparing different species within the same family. This approach was in tune with the neo-Darwinian view of evolution in which adaptationism and gradualism play a major descriptive role (Poulson 1963, 1964). However, he would later take a view that combined both neutralism and selectionism when trying to explain the evolution of troglomorphic features (Poulson 1985). He viewed neutralism and selection as complementary and not mutually exclusive for traits that show evolutionary reduction. His approach was strongly comparative where different species of amblyopsids, with different relative evolutionary time in caves, constitute natural experiments.

Another influential American biospeleologist of this period is David Clair Culver (1944–). Although he worked on cave crustaceans and not on cave fishes, his ideas were, and continue to be, influential in the field. He also took a look at the phenomenon of troglomorphy from an ecological viewpoint and applied island biogeography theory to a system that, ecologically speaking, was largely isolated (Culver 1976). Although he initially took a neutralist stance (Culver & Fong 1986), his more recent studies convinced him that selection does play a major role in the convergence of troglomorphic features (e.g., Culver et al. 1995).

One of the most influential papers during that, this time of philosophical conflict, did not come from hypogean fish specialists but rather from people trying to apply molecular techniques to the understanding of genetic variation. That was the case in the study of evolutionary genetics of the cave *Astyanax fasciatus* (Avise & Selander 1972). Largely based on the Master's thesis by John Charles Avise (1948–) directed by his adviser Robert Keith Selander (1927–), this work demonstrated that highly divergent morphologies (epigean vs. troglomorphic) can be achieved with very little genetic differentiation. This, in turn, questioned the validity of the typological thinking of assigning new species, and specially new generic status, to any troglomorphic population based only on blindness and depigmentation.

This new genetic information signaled the possibility that major phenotypic modifications could be achieved by relatively minor genetic changes, as long as those changes affected developmental genes. This notion has been supported by recent studies and is invigorating the study of cave fishes as models in developmental evolutionary biology (e.g., Yamamoto & Jeffery 2000).

Conclusions

The history of hypogean fish research has been marked by rather precise intellectual contexts. Each of the periods described above has been heavily influenced by particular lines of thoughts that were very popular at that time. It seems that most researchers have used hypogean fishes as examples to justify their own philosophical stances rather than to use these subjects for the sake of scientific discovery. The use of these fishes, first to support a neo-Lamarckian view of evolution, then to practice a typological view of classification, and third to demonstrate that they must represent some sort of stage in alleged evolutionary trends, have marred what may have been an excellent topic for more mainstream evolutionary studies. This latter view has, probably, been the most perversive of all given the jargon used by most researchers ('regressive', 'degenerative', etc.) to describe their evolutionary history.

But we should not blame them for that. After all, the idea of progress has been the backdrop to evolutionary thinking and not until recently have we started to realize that the notion of evolutionary progress is a fallacy (Maynard Smith & Szathmary 1995, 1999). In my view, it is a cultural construct that fits perfectly with anthropocentric ideas that have dominated human notions of ourselves since the dawn of history. This is an idea that has been central in the battles science has fought to enlighten humankind, from geocentrism to creationism. The idea that, somehow, humans represent the zenith of evolution and that anything else is inferior – anything not aiming there is regressive or degenerative.

However, if we see troglomorphic fauna in general and troglomorphic fish in particular, from a less philosophically charged perspective, we will find that they are part of nature and that the characters that amaze us, mostly their blindness and depigmentation, are part of common biological phenomena found everywhere in nature, such as loss of eyes and pigmentation among deep-sea fishes, loss of wings among birds and insects living on islands and caves, or loss of limbs among cetaceans. Further, that the loss of certain phenotypic characters can take place thanks to relatively minor genetic re-arrangements.

Also, the increasing interest on non-troglomorphic hypogean fishes should shed much light on our understanding of evolution as a dynamic, non-stopping process. The finding of genetically undistinguishable epigean and hypogean populations as well as hypogean species that seem to be in 'status nascendi' is also important (e.g., Borowsky & Mertz 2001, Burr et al. 2001). It tells us that the separation between both epigean and troglomorphic populations is in many cases very tenuous. To apply typological thinking by creating artificial divides to create species just for the sake of it will mask the reality of the fluidity in nature, that nothing stands stills as originally understood by Heraclitus (ca. 540–ca. 480 B.C.) (Wheelwright 1959, p. 29).

When we see troglomorphic fishes we should see them as life trying to occupy every possible space; after all, evolution is opportunistic (Dobzhansky 1970). This is a view more in tune with the Red Queen Hypothesis (Van Valen 1973), according to which organisms will keep changing their phenotype while attempting to survive both as individuals and/or through their genes. Troglomorphic fishes, by evolving from their epigean ancestors, are just trying to make the best of an environment lacking visual information.

And that brings us to the last point. One of the reasons why the notion of evolution by means of natural selection has taken so long to be accepted, and is still debated today, is because of misunderstandings at different levels of how natural selection works. Students of the hypogean fauna have historically missed the point that it is precisely this fauna that represents one of the prime examples of convergent evolution. This is understandable given that the study of convergence and the constraints of form have never been the subject of a single synthesis (Conway Morris 1998). The study of hypogean fishes has the potential to help us better understand how convergence works.

Acknowledgements

Andrew Miller helped to compiled the bibliography in the data base. Jerry Reedy of Macalester College revised my translation of Kircher's passages on cave fishes. Herman W. de Swart secured the Dutch translation of Kircher's Mundus Subterraneus at the library of the National Museum for the History of Science in Boerhaave, Leiden, and sent translations of some its portions. Janis Langins of the University of Toronto provided a copy of Montalembert's portrait. The following people provided me with information or hints where to find information about Theodore Tellkampf: Toby A. Appel of Yale University, Frederick B. Churchill of

Indiana University, Uwe Kunert (a Tellkampf relative) from Germany, Edward T. Morman of the New York Academy of Medicine, Lynn K. Nyhart of the University of Wisconsin-Madison, Jocelyn K. Wilk of the Columbia University Archives & Columbiana Library, Mary P. Winsor of the University of Torornto and Robert Young of the Museum of Comparative Zoology at Harvard University. William G. Saul of the Academy of Natural Sciences provided useful information on the specimens originally classified as '*Gronias nigrilabris*' that are deposited at the Academy's collection. Mark Sabaj of Department of Ichthyology of The Academy of Natural Sciences of Philadelphia provided information about the specimens of *A. spelaea* at that institution. G. Teugels, of the Musée Royal de l'Afrique Centrale, in Tervuren, Belgium, provided useful information on Georges Boulenger. Steven Johnson Librarian and archivist of the Wildlife Conservation Society provided useful information on Myron Gordon. James W. Atz read the section on C.M. Breder and made valuable suggestions. Olga Mayayo did the drawing of '*Gronias nigrilabris*'. Jessica Romero secured a copy of the entire collection of the Memoirs of the Wernerian Society in search of an alleged first printed reference to a Kentucky 'eyeless' fish. Lisa Palmer of the Smithsonian Institution, Carolyn Rainey of the Scripps Institution of Oceanography, Mary Messing of the Wildlife Conservation Society, and Dana A. Fisher provided some of the illustrations and permission to use them for this article. Thomas Poulson and David Culver read the portions where their work was described and contributed with suggestions. Jakob Parzefal read the section on Curt Kosswig and also provided useful information. Karen Lucatelli and Kelly M. Paulson read the entire manuscript and made valuable suggestions.

References cited

Agassiz, E. (ed.). 1885. Louis Agassiz: his life and correspondence, Vol. 1. Macmillan and Company, London. 794 pp.

Agassiz, L. 1847 [1848]. [Plan for an investigation of the embryology, anatomy and effect of light on the blind-fish of the Mammoth Cave, *Amblyopsis spelaeus*]. Proc. Amer. Acad. Arts Sci. 1: 1–180.

Agassiz, L. 1851. Observations on the blind fish of the Mammoth Cave. Amer. J. Sci. 11: 127–128.

Agassiz, L. 1859. An essay on classification. Longman, Brown, Green, Longmans, Roberts, London. 381 pp.

Alvarez, J. 1946. Revisión del género *Anoptichthys* con descripción de una especie nueva (Pisc., Characidae). An. Esc. Nac. Cienc. Biol. Mex. 4: 263–282.

Alvarez, J. 1947. Descripción de *Anoptichthys hubbsi* caracinido ciego de la cueva de los Sabinos, S.L.P. Rev. Soc. Mex. Hist. Nat. 8: 215–219.

Anonymous. 1842. [Mammoth Cave blind crayfish and fish]. Proc. Acad. Nat. Sci. Phil. 1: 175.

Appel, T.A. 1988. Jeffries Wyman, philosophical anatomy, and the scientific reception of Darwin in America. J. Hist. Biol. 21: 69–94.

Atz, J.W. 1986. C.M. Breder, Jr. 1897–1983. Copeia 1986: 853–856.

Avise, J.C. & R.K. Selander. 1972. Evolutionary genetics of cave-dwelling fishes of the genus *Astyanax*. Evolution 26: 1–19.

Baird, S.F. 1872. Living eyeless fish. Ann. Rec. Sci. Indust. 1871: 266.

Barr, T.C. 1966. Evolution of cave biology in the United States, 1822–1965. Nat. Speleol. Soc. Bull. 28: 15–21.

Bateson, W. 1922. Evolutionary faith and modern doubts. Science 55: 1412.

Beddall, B.G. 1983. The isolated Spanish genius – myth or reality? Félix de Azara and the birds of Paraguay. J. Hist. Biol. 16: 225–258.

Besson, J. 1569 [1969]. L'art et science de trouver les eaux et fontaines cachees soubs terre: autrement que par les moyens vulgaires des agriculteurs et architectes. E. Gibier, Orléans. 83 pp. [facsimile reproduction by Editions Coral, Columbus].

Blanc, M., J.-L. Gaudet, P. Bănărescu & J.-C. Hureau. 1971. European inland fish: a multilingual catalogue. Fishing News (Books) Ltd., London. 149 pp.

Bocking, S. 1988. Alpheus Spring Packard and cave fauna in the evolution debate. J. Hist. Biol. 21: 425–456.

Borowsky, R.B. & L. Mertz. 2001. Genetic differentiation among populations of the cave fish *Schistura oedipus* (Cypriniformes: Balitoridae). Env. Biol. Fish. 62: 225–231 (this volume).

Boulenger, G. 1893. Blind animals in caves. Nature 47: 608.

Boulenger, G. 1922. Description d'un poisson aveugle decouvert par M.G. Geerts dans la grotte de Thysville (Bas-Congo). Rev. Zool. Afr. 9: 252–253.

Bowler, P.J. 1984. Evolution. The history of an idea. University of California Press, Berkeley. 412 pp.

Breder, C. M. 1942. Descriptive ecology of La Cueva Chica, with especial reference to the blind fish, *Anoptichthys*. Zoologica 27: 7–15.

Bridges, W. 1954. Zoo expeditions. Curator Publications, New York. 191 pp.

Brooks, W.K. 1909. Biographical memoir of Alpheus Hyatt (1838–1902). Biogr. Mem. Nat. Acad. Sci. USA 6: 311–325.

Burgess, W.E. 1989. An atlas of freshwater and marine catfishes: a preliminary survey of the Siluriformes. T.F.H. Publications, Neptune City. 784 pp.

Burr, B.M., G.L. Adams, J.K. Krejca, R.J. Paul & M.L. Warren, Jr. 2001. Troglomorphic sculpins of the *Cottus carolinae* species group in Perry County, Missouri: distribution, external morphology, and conservation status. Env. Biol. Fish. 62: 279–296 (this volume).

Chen, Y.-R., J.-X. Yang & Z.-G. Zhu. 1994. A new fish of the genus *Sinocyclocheilus* from Yunnan with comments on its characteristic adaptation (Cypriniformes: Cyprinidae). Acta Zootax. Sinica 19: 246–253 (in Chinese).

Cockerell, T.D.A. 1920. Biographical memoir of Alpheus Spring Packard 1839–1905. Biogr. Mem. Nat. Acad. Sci. USA 9: 181–236.

Coe, W.R. 1918. A century of zoology in America. pp. 391–438. *In*: E.S. Dana (ed.) A Century of Science in America, Yale University Press, New Haven.

Conway Morris, S. 1998. The crucible of creation. The Burgess Shale and the rise of animals. Oxford University Press, Oxford. 242 pp.

Cope, E.D. 1864. On a blind silurid from Pennsylvania. Proc. Acad. Nat. Sci. Phil. 1864: 231–233.

Cope, E.D. 1896. The primary factors of organic evolution. Open Court, Chicago. 547 pp.

Culver, D.C. 1976. The evolution of aquatic cave communities. Amer. Nat. 110: 955–957.

Culver, D.C. & D.W. Fong. 1986. Why all cave animals look alike. Stygicola 2: 208–216.

Culver, D.C., T.C. Kane & D.W. Fong. 1995. Adaptation and natural selection in caves. Harvard University Press, Cambridge. 223 pp.

Cuvier, G. 1828 [1995]. Historical portrait of the progress of ichthyology: from its origins to our own time. Johns Hopkins University Press, Baltimore. 366 pp.

Darwin, C. 1859. On the origin of the species by means of natural selection. J. Murray, London. 502 pp.

Darwin, C. 1861. On the origin of the species by means of natural selection. J. Murray, London. 538 pp.

Davidson, R. 1840. An excursion to the Mammoth Cave, and the barrens of Kentucky. With some notices of the early settlement of the state. A.T. Skillman & Son, Lexington. 148 pp.

Dean, B. 1916–1923. A bibliography of fishes. American Museum of Natural History, New York, Vol. 1, 718 pp., Vol. 2. 702 pp., Vol. 3, 707 pp.

DeKay, J.E. 1842. Zoology of New York or the New-York fauna, Part IV, fishes. W. & A. White & J. Visscher, Albany. 566 pp.

Demoor, J., J. Massart & É. Vandervelde. 1897. L'évolution régressive en biologie et en sociologie. Félix Alacn, Paris. 324 pp.

Dexter, R.W. 1965. The "Salem secession" of Agassiz zoologists. Essex Inst. Hist. Coll. 101: 27–39.

Di Caporiacco, L. 1926. Un nuovo genere di ciprinide somalo delle acque di pozzo (One new cyprinid genus from a well in Somalia). Monit. Zool. Ital. 37: 23–25.

Dobzhansky, T. 1970. Genetics and the evolutionary process. Columbia University Press, New York. 505 pp.

Eigenmann, C.H. 1890. The Point Loma blind fish and its relatives. Zoe 1: 65–96.

Eigenmann, C.H. 1903. In search of blind fishes in Cuba. World Today 5: 1131–1136.

Eigenmann, C.H. 1909. Cave vertebrates of America. A study in degenerative evolution. Carnegie Institution of Washington, Washington, D.C. 241 pp.

Eigenmann, C.H. 1919. *Trogloglanis pettersoni* a new blind fish from San Antonio, Texas. Proc. Amer. Phil. Soc. 58: 397–400.

Flint, J. 1822. Letters from America, containing observations on the climate and agriculture of the western states, the manners of the people, the prospects of emigrants, & c., & c. [sic] W. & C. Tait, Edinburgh. 330 pp.

Gianferrari, L. 1923. *Uegitglanis zammaranoi* un nuovo siluride cieco africano (*Uegitglanis zammaranoi*, a new African cave silurid). Soc. Ital. Sci. Nat. Milan. 62: 1–3.

Gifford, G.E. 1967. An American in Paris, 1841–1842: four letters from Jeffries Wyman. J. Hist. Med. Allied Sci. 22: 274–285.

Girard, C.F. 1859. Ichthyological notes. Proc. Acad. Nat. Sci. Phila. 1859: 63–64.

Girard, C.F. 1888. Les poissons souterrains du nouveau monde. Le Naturaliste 10: 222.

Gresser, E.B. & C.M. Breder. 1940. The histology of the eye of the cave characin, *Anoptichthys jordani*. Zoologica 25: 113–116.

Goeldi, E.A. 1905. Nova zoologica aus der Amazonas-Region. Compt. Rend. 6th Congr. Inter. Zool. 1905: 542–549.

Guillaume, J.-L. 1971. La marche au socialism. Évolution progressive ou régressive? Centre d'Études Politiques et Civiques, Paris. 52 pp.

Günther, A.C.L.G. 1880. An introduction to the study of fishes. A. and C. Black, Edinburgh. 720 pp.

Gurnee, R. 1992. A brief history of cave studies in the United States before 1887 (16th to the 19th Century). J. Spelean Hist. 26: 11–20.

Horder, T.J. 1998. Why do scientists need to be historians? Q. Rev. Biol. 73: 175–187.

Hubbs, C.L. 1938. Fishes from the caves of Yucatan. Carnegie Inst. Wash. Publ. (491): 261–295.

Hubbs, C.L. 1964. History of ichthyology in the United States after 1850. Copeia 1964: 42–60.

Hubbs, C.L. & W. T. Innes. 1936. The first known blind fish of the family Characidae: a new genus from Mexico. Occ. Pap. Mus. Zool. (342): 1–7.

Innes, W.T. 1937. A cavern characin *Anoptichthys jordani*, Hubbs and Innes. Aquarium 5: 200–202.

Jackson, J.R. & W.C. Kimler. 1999. Taxonomy and the personal equation: the historical fates of Charles Girard and Louis Agassiz. J. Hist. Biol. 32: 509–555.

Jordan, D.S. 1905. The history of ichthyology, an address by David Starr Jordan. Proc. Amer. Assoc. Adv. Sci. 51: 427–456.

Kircher, A. 1665. Mundus subterraneus, in XII libros digestus; quo divinum subterrestris mundi opificium, mira ergasteriorum naturæ in eo distributio, verbo pantámorphou Protei regnum, universæ denique naturæ majestas & divitiæ summa rerum varietate exponuntur, 2 volumes. J. Janssonium & E. Weyerstraten, Amsterdam. 938 pp.

Kosswig, C. 1949. Phänomene der regressiven Evolution im Lichte des Genetik. Communs. Fac. Sci. Univ. Ankara 2: 110–150.

Kosswig, C. 1960. Darwin und die degenerative evolution. Abhandl. Verh. Naturw. Ver. Hamburg 4: 21–42.

Lankester, E.R. 1880. Degeneration: a chapter in Darwinism. Macmillan, London. 75 pp.

Lankester, E.R. 1893. Blind animals in caves. Nature 47: 389.

Maynard Smith, J. & E. Szathmáry. 1995. The major transitions in evolution. Oxford University Press, New York. 346 pp.

Maynard Smith, J. & E. Szathmáry. 1999. The origins of life. Oxford University Press, Oxford. 180 pp.

Miranda-Ribeiro, A. 1907. Una novedade ichthyologica. Kosmos 4: 21–22.

Mitchell, R.W., W.H. Russel & W.R. Elliot. 1977. Mexican eyeless characin fishes, genus *Astyanax*: environment, distribution, and evolution. Special Publications No. 12, Texas University Press, Lubbock. 89 pp.
Montalembert, M.-R. 1748. Observations de physique générale. Hist. Acad. Roy. Sci. 1748: 27–28.
Motas, C. 1962. Emil G. Racovitza: founder of biospeleology. Nat. Speleol. Soc. Bull. 24: 3–8.
Myers, G.S. 1964. A brief sketch of the history of ichthyology in America to the year 1850. Copeia 1964: 33–41.
Nisbet, R. 1979. History of the idea of progress. Basic Books, Inc., New York. 370 pp.
Norman, J.R. 1926. A new blind catfish from Trinidad, with a list of the blind cave-fishes. Ann. Mag. Nat. Hist. 18: 324–331.
Packard, A. 1871. The Mammoth Cave and its inhabitants. Amer. Nat. 5: 739–761.
Packard, A. 1888. The cave fauna of North America with remarks on the anatomy of the brain and origin of the blind species. Mem. Nat. Acad. Sci. 4: 1–156.
Packard, A. 1901. Lamarck, the founder of evolution, his life and work. Longmans, Green, and Co., New York. 451 pp.
Petit, G. 1933. Un poisson cavernicole aveugle des eaux douces de Madagascar. Compt. Rend. Hebd. Séanc. Acad. Scienc. 4: 347–348.
Pietsch, T.W. & W.D. Anderson, Jr. (ed.) 1997. Collection building in ichthyology and herpetology. American Society of Ichthyologists and Herpetologists, Lawrence. 593 pp.
Poey, F. 1858. Memorias sobre la historia natural de la isla de Cuba, 2 volumes. Barcina, Habana. 439 pp.
Poly, W.J. 2001. Nontroglobitic fishes in Bruffey-Hills Creek Cave, West Virginia, and other caves worldwide. Env. Biol. Fish. 62: 73–83 (this volume).
Poulson, T.L. 1963. Cave adaptation in amblyopsid fishes. Amer. Midl. Nat. 70: 257–290.
Poulson, T.L. 1964. Animals in aquatic environments: animals in caves. pp. 749–771. *In*: D.B. Dill, E.F. Adolph & C.G. Wilber (ed.) Handbook of Physiology, Section 4: Adaptation to the Environment, American Physiological Society, Washington, D.C.
Poulson, T.L. 1985. Evolutionary reduction by neutral mutations: plausibility arguments and data from amblyopsid fishes and linyphiid spiders. Nat. Speleol. Soc. Bull. 47: 109–117.
Putnam, F.W. 1872. The blind fishes of the Mammoth Cave and their allies. Amer. Nat. 6: 6–30.
Putnam, F.W. 1874. [The blind fish and some of the associated species of the Mammoth Cave, Kentucky, probably of marine origin]. Bull Essex Inst. 6: 191–200.
Regan, C.T. 1940. The fishes of the Gobiid genus *Luciogobius* Gill. Ann. Mag. Nat. Hist. 5: 462–465.
Romero, A. 1984. Charles Marcus Breder, Jr. 1897–1983. Nat. Speleol. Soc. News 42: 8.
Romero, A. 1985. Can evolution regress? Nat. Speleol. Soc. Bull. 47: 86–88.
Romero, A. 1986a. Charles Breder and the Mexican blind characid. Nat. Speleol. Soc. News 44: 16–18.
Romero, A. 1986b. He wanted to know them all: Eigenmann and his blind vertebrates. Nat. Speleol. Soc. News 44: 379–381.
Romero, A. 1999a. The blind cave fish that never was. Nat. Speleol. Soc. News 57: 180–181.
Romero, A. 1999b. Myth and reality of the alleged blind cave fish from Pennsylvania. J. Spelean Hist. 33: 67–75.
Romero, A. 2000. The speleologist who wrote too much. Nat. Speleol. Soc. News 58: 4–5.
Romero, A. 2001a. It's a wonderful hypogean life: a guide to the troglomorphic fishes of the world. Env. Biol. Fish. 62: 13–41 (this volume).
Romero, A. 2001b. The life and work of a little known biospeleologist: Theodore Tellkampf. J. Spelean Hist. (in press).
Romero, A. & K. Benz. 2000. The unsung heroes of speleology. Nat. Speleol. Soc. News 58: 106, 126.
Romero, A. & J.E. Creswell. 2000. In search of the elusive "eyeless"cave fish Trinidad, W.I. Nat. Speleol. Soc. News 58: 282–283.
Romero, A. & Z. Lomax. 2000. Jacques Besson, cave eels, and other alleged European cave fishes. J. Spelean History (in press).
Romero, A. & K.M. Paulson. 2001. Humboldt's alleged cave fish from Ecuador. J. Spelean Hist. (in press).
Romero, A. & A. Romero. 1999. On Cope, caves, and skeletons in the closet. Nat. Speleol. Soc. News 57: 341–343.
Ruse, M. 1996. Monad to man. The concept of progress in evolutionary biology. Harvard University Press, Cambridge. 629 pp.
Shaw, T.R. 1992. History of cave science. The exploration and study of limestone caves, to 1900. Sydney Speleological Society, Broadway. 338 pp.
Smallwood, W.M. 1941. Natural history and the American mind. Columbia University Press, New York. 445 pp.
Soule, G.K. 1982. A Mammoth cave chronology. J. Spelean Hist. 16: 3–9.
Tellkampf, T. 1844a. Beschreibung einiger neuer in der Mammuth-Höhle in Kentucky aufgefundener Gattungen von Gliederthieren. Arch. Vereins Freund Natur. Mecklenburg 10: 318–322.
Tellkampf, T. 1844b. Über den blinden Fisch der Mammuthhöhle in Kentucky. Muller's Arch. Anat. Phys. 1844: 381–395.
Tellkampf, T. 1870. Note respecting the eyes of *Amblyopsis spelaeus*. Ann. Lyc. Nat. Hist. N.Y. 9: 150–152
Thinès, G. 1969. L'evolution regressive des poissons cavernicoles et abyssaux. Masson et Cie, Paris. 394 pp.
Thompson, W. 1844. Notice of the blind fish, cray-fish, and insects from the Mammoth Cave, Kentucky. Ann. Mag. Nat. Hist. 13: 111–113.
Trewavas, E. 1936. Dr. Karl Jordan's expedition to South-West Africa and Angola. The fresh water fishes. Novit. Zool. 40: 63–76.
Van Valen, L. 1973. A new evolutionary law. Evol. Theor. 1: 1–30.
Vandel, A. 1965. Biospeleology. The biology of cavernicolous animals. Pergamon Press, Oxford. 524 pp.
Vinciguerra, D. 1924. Descrizione di un ciprinide cieco proveniente dalla Somalia Italiana (Description of a blind cyprinid from the Italian Somalia). Ann. Mus. Civ. Stor. Nat. Genoa 51: 239–243.
von Baer, K.E. 1828. Über Entwickelungsgeschichte der Thiere. Beobachtung und Reflexion. Bei Den Gebtdern Borntrager, Königsburg. 315 pp.
von Humboldt, A. 1805. Quatrième mémoire, sur une nouvelle espèce de pimelode, jetée par les volcans du royaume de Quito. pp. 40–48. *In*: Voyage de Humboldt et Bonpland, Deuxième

partie, Observations de Zoologie et d'Anatomie comparée, F. Schoell, Paris.

Wheelwright, P. 1959. Heraclitus. Princeton University Press, Princeton. 181 pp.

Wilkens, H. 1988. Evolution and genetics of epigean and cave *Astyanax fasciatus* (Characidae, Pisces). Support for the neutral mutation theory. Evol. Biol. 23: 271–367.

Wyman, J. 1843. Description of a 'Blind Fish,' from a cave in Kentucky. Amer. J. Sci. 45: 94–96.

Wyman, J. 1851. [Account of dissections of the blind fishes (*Amblyopsis spelaeus*) from the Mammoth Cave, Kentucky]. Proc. Boston Soc. Nat. Hist. 3: 349, 375.

Wyman, J. 1854a. The eyes and organs of hearing in *Amblyopsis spelaeus*. Proc. Boston Soc. Nat. Hist. 4: 149–151.

Wyman, J. 1854b. On the eye and the organ of hearing in the blind fishes (*Amblyopsis spelaeus* DeKay) of the Mammoth Cave. Proc. Boston Soc. Nat. Hist. 4: 395–396.

Wyman, J. 1872. Notes and drawings of the rudimentary eye, brain, and tactile organs of *Amblyopsis spelaeus*. Amer. Nat. 6: 16–20.

Yamamoto, Y. & W.R. Jeffery. 2000. Central role for the lens in cave fish eye degeneration. Science 289: 631–633.

Environmental Biology of Fishes **62:** 73–83, 2001.
© 2001 *Kluwer Academic Publishers. Printed in the Netherlands.*

Nontroglobitic fishes in Bruffey-Hills Creek Cave, West Virginia, and other caves worldwide

William J. Poly
Department of Zoology, Southern Illinois University, Carbondale, IL 62901-6501, U.S.A.
(e-mail: argulus5@siu.edu)

Received 29 February 2000 Accepted 17 December 2000

Key words: neoteny, pigment, eye

Synopsis

Six species of fishes were tagged and released in September and November 1995 and on five dates between November 1998 and October 1999 inside Bruffey-Hills Creek Cave. Most of the tagged fishes were creek chub, *Semotilus atromaculatus*, and green sunfish, *Lepomis cyanellus*. The overall recapture rate was 2.6% as only three of 117 fishes were recaptured. Forty-nine days was the longest time of residence by two *L. cyanellus* and one *S. atromaculatus*, and movement of 83.4 m was observed only for the creek chub. Tag loss was confirmed as one factor in the low recapture rate. Nine species of fishes were collected in the cave, including *Phoxinus oreas* and *Pimephales promelas*, two species never collected in a cave. Most of the fishes were pigmented normally, but many creek chubs were depigmented and appeared white or pigmentless when first observed in the cave stream. Four species, *Phoxinus oreas*, *Pimephales promelas*, *Ameiurus nebulosus*, and *Lepomis macrochirus*, were collected for the first time in Bruffey-Hills Creek Cave and in the Bruffey and Hills creeks drainage. Although data from this study did not shed light on residence time of fishes in the cave, the consistent occurrence of epigean fishes in this cave was shown. A list of epigean fishes from caves worldwide was included along with a discussion of aspects of the ecology of epigean fishes in caves and of evolution of troglobitic fishes.

Introduction

Epigean fishes inhabiting caves have been given little attention in the past by speleologists and cave biologists (Proudlove 1979, 1982, Chapman 1993). Poly & Boucher (1996) reviewed the occurrence of epigean fishes in caves and discussed the relevance of these fishes to cave ecosystems and to understanding evolution of troglobitic organisms. Information on the ecology of nontroglobitic fishes in caves remains scarce, and the present study was undertaken to help fill this void. Tagging and periodic monitoring of fishes would provide useful biological information on epigean fishes in caves. Tagging and recapturing of individual organisms in caves has not been attempted often and most previous studies have dealt with troglobitic organisms. Cooper (1975) used ink to mark three species of troglobitic crayfish in Shelta Cave, Alabama, then made periodic observations and noted movements of individual crayfish within the cave (Cooper & Cooper[1]). Hobbs (1978) performed a mark-recapture study of the crayfish, *Orconectes inermis inermis*, in Pless Cave, Indiana. Means & Johnson (1995) marked Ozark cavefish, *Amblyopsis rosae*, with alphanumeric tags to estimate population size and to document movement of fishes within a cave, and Trajano (1991, 1997) performed similar studies on *Pimelodella kronei*,

[1] Cooper, J.E. & M.R. Cooper. 1971. Studies of the aquatic ecology of Shelta Cave, Huntsville, Alabama. ASB Bull. 18: 30 (abstract). Cooper, J.E. & M.R. Cooper. 1976a. Marking crayfish for long-term ecological studies. ASB Bull. 23: 52 (abstract). Cooper, M.R. & J.E. Cooper. 1976b. Growth and longevity in cave crayfishes. ASB Bull. 23: 52 (abstract).

P. transitoria (an epigean species), and *Trichomycterus itacarambiensis* in Brazil. Bruffey-Hills Creek Cave was chosen as a study site because several species of fishes occur in the cave in fair abundance. Information is available on the vertebrate fauna of Bruffey-Hills Creek Cave (Garton et al. 1993, Poly & Boucher 1996), but the invertebrate fauna has not been examined (Holsinger et al. 1976). The goals of the study were to gather information on: (1) residence time of individual fishes in the cave, (2) movement of individuals throughout the cave, (3) growth of individuals living in the cave environment, (4) reproduction within the cave, and (5) temporal changes in pigmentation of individual fishes.

That a relationship exists between pigmentation and length of residence in a cave is suggested by the study of Rasquin (1947) and reports of depigmented fishes in caves (Hora 1924, Kosswig 1937, Proudlove 1979). Some fishes that were depigmented when captured in a cave were able to regain some pigment after several months of exposure to light (presumably a normal photoperiod, Banister 1984), whereas a 'white' fish that appeared depigmented regained a more normal coloration in 24 h (Proudlove 1979). Also, fishes held in aquaria often exhibit temporary pigment changes, appearing pallid after exposure to darkness (overnight), then appearing normal following exposure to light (personal observation). Such a temporary state of depigmentation was observed in some epigean fishes in a cave (Poly & Boucher 1996).

Methods

Fish surveys were conducted on 22 October 1994, 30 September 1995, 18 November 1995, 14 November 1998, 6 February 1999, 27 March 1999, 27 July 1999, 2 October 1999, and 29 December 1999 in Bruffey-Hills Creek Cave, Pocahontas County, West Virginia. Fishes were tagged beginning 30 September 1995 and ending 2 October 1999. The cave is approximately 4 km north of Droop, WV, and approximately 2 km southeast of Lobelia, WV, Little Levels Township, 38°07′21″ N, 80°16′42″ W. The spring indicated on the Droop WV USGS 7.5′ quadrangle map is actually the inflow and cave entrance; Bruffey-Hills Creek Cave contains 2156 m of surveyed passages (Storrick 1992). Fishes were collected with a battery-powered backpack electrofishing unit, except on 22 October 1994 when seines and dipnets were used. Distance sampled and distance of capture or observation were measured to 0.1 m with a hip chain. Distance sampled (m) on each date was (in chronological order): 479.0, 264.4, 305.2, 132.6, 136.0, 138.0, 129.2, 127.6, and 123.4 m. Because of an impassable logjam, less than 140 m of the cave could be sampled between November 1998 and December 1999. Each tagged fish was weighed to the nearest 0.1 g on an Acculab Pocket Pro 250-B portable electronic balance (a spring scale was used during 1995 – weights recorded to the nearest 1 g), then measured to the nearest 1 mm standard length (SL) and total length (TL) with dividers and a stainless steel ruler. Initial pigmentation when captured (pallid or 'normal') as well as any changes from pallid (light) to 'normal' pigmentation after exposure to light for several minutes were noted. Then fishes were tagged with visible implant tags utilizing a syringe style injector and precut, alphanumeric tags (Northwest Marine Technology, Inc.). Tags were inserted on the top of the head of cyprinids, on the lower jaw of centrarchids, and on the adipose eyelid of salmonids. Voucher specimens were preserved in 10% formalin and deposited in the Ohio State University Museum of Biological Diversity Fish Collection (OSUM), Southern Illinois University at Carbondale Fluid Vertebrate Collection (SIUC), and University of Tennessee, Knoxville (UT). Fish nomenclature follows Robins et al. (1991) and Eschmeyer (1998).

Following completion of the field work, 55 creek chubs obtained from Big and Little creeks, Union County, Illinois (16 January 2000), were tagged on 17 January 2000 and checked weekly for the presence of tags. Creek chubs were held in circular fiberglass tanks in a greenhouse and fed every other day with commercial pellets; small minnows were kept with the large creek chubs as additional forage. Smaller fishes ($n = 50$, SL = 41–82 mm, $n = 48$ measured) were kept in a separate tank from the larger chubs ($n = 5$, SL = 114–134 mm, $n = 4$ measured) to reduce loss of fishes through predation. The temperature varied from 1.5 to 19.7°C between 17 January and 5 March.

Results

One hundred and ninety-six individuals of nine fish species were collected between 1994 and 1999, and most of the fishes were creek chub and green sunfish (Table 1). One hundred and seventeen individuals of six species were tagged and released inside the cave: 90 *Semotilus atromaculatus*, four *Rhinichthys atratulus*, 20 *Lepomis cyanellus*, one *Lepomis macrochirus*, one *Ambloplites rupestris*, and one *Oncorhynchus mykiss*.

Table 1. Fish species, number collected, and distance of capture in Bruffey-Hills Creek Cave, West Virginia.

Date	Species (size range, catalog number)	Number	Distance in cave[1]
22 Oct 1994	*Semotilus atromaculatus* (26–146 mm, $n = 11$, OSUM 77874)	46[2]	4.8-476.0
	Lepomis cyanellus (61 & 69 mm, OSUM 77875)	4[2]	45.8–476.0
30 Sep 1995	*Semotilus atromaculatus* (39–167 mm, $n = 18$)	19	2.7–264.4
	Lepomis cyanellus (54–115 mm, $n = 14$)	14	2.7–198.5
	Ambloplites rupestris (68 & 104.6 mm, SIUC 24889)	2	2.7–187.6
	Rhinichthys atratulus (73 mm)	1	35.8
	Oncorhynchus mykiss (275 mm)	1	35.8
18 Nov 1995	*Semotilus atromaculatus* (53–159 mm, $n = 7$)	7	2.5–305.2
	Lepomis cyanellus (63–110 mm, $n = 6$)	6	34.7–84.8
14 Nov 1998	*Semotilus atromaculatus* (45–185 mm, $n = 27$)	27	2.9–132.6
	Lepomis cyanellus (141 mm)	1	110.5
	Lepomis macrochirus (85 mm)	1	38.8
	Rhinichthys atratulus (40 & 42 mm)	2	14.8
	Phoxinus oreas (42–53 mm, $n = 3$, SIUC 33140)	3	2.9
6 Feb 1999	*Semotilus atromaculatus* (43–95 mm, $n = 3$)	3	23.2–29.0
	Rhinichthys atratulus (51 mm)	1	50.0
	Phoxinus oreas (53 mm, UT 44.8611)	1	23.2
27 Mar 1999	*Semotilus atromaculatus* (32–49 mm, $n = 6$)	6	19.5–37.0
27 Jul 1999	*Semotilus atromaculatus* (38–150 mm, $n = 18$)	18	2.6–100.4
	Phoxinus oreas (34 & 58 mm, UT 44.8611)	2	2.9–29.4
	Ameiurus nebulosus (210 mm, UT 48.1006)	1	2.6
	Ambloplites rupestris (110 mm, UT 90.2636)	1	2.6
2 Oct 1999	*Semotilus atromaculatus* (44–96 mm, $n = 12$)	12[3]	2.4–91.7
	Lepomis cyanellus (25 & 75 mm)	2	2.4–69.8
	Pimephales promelas (53 mm, UT 44.8609)	1	29.0
	Rhinichthys atratulus (size not recorded)	1	15.1
29 Dec 1999	*Semotilus atromaculatus* (36–152 mm, $n = 6$, UT 44.8612)	6[4]	2.4–106.2
	Lepomis cyanellus (20–69 mm, $n = 7$, UT 90.2628)	7[4]	2.4–46.0

[1]A range of capture distances is given when numerous individuals were captured. Distance in meters from Bruffey Creek entrance; distance was measured at the center of a distinct pool or stream segment.
[2]Distance not recorded for all fishes.
[3]Two additional creek chubs were observed but could not be captured, and another large creek chub was found dead on the stream bank.
[4]Four other creek chubs and one centrarchid were observed but could not be captured.
Notes: Stream flow was elevated on 18 November 1995, 6 February 1999, and 27 March 1999. Sampling distance was greater on the first three dates, then was <140 m on all other dates due to the presence of a logjam. *Phoxinus oreas* from 6 February and 27 July were combined in one lot (UT 44.8611).

Two green sunfish and one creek chub were recaptured and inhabited the cave for at least 49 d (30 September to 18 November 1995). These fishes exhibited little or negative growth as determined from length and weight data (only minor differences observed). Movement was not observed for the two green sunfish as both were recaptured in the same pool (35.8 m). However, the creek chub was 83.4 m further inside the cave (180.9 m) from the initial point of capture (97.5 m).

The recaptured creek chub was initially recorded as lightly pigmented, reverting to a normal pigment pattern after exposure to light, whereas the same fish on 18 November 1995 was lightly pigmented but did not revert to a normal pigment pattern. Lightly pigmented creek chubs were captured in all years; many of these fishes had contracted melanophores because most became darker after exposure to light for several minutes. However, a few creek chubs remained lightly pigmented. Overall, of 85 creek chubs, 60 were pigmented normally when captured, whereas 10 were listed as lightly pigmented initially but returning to normal, and 15 remained lightly pigmented even after exposure to light.

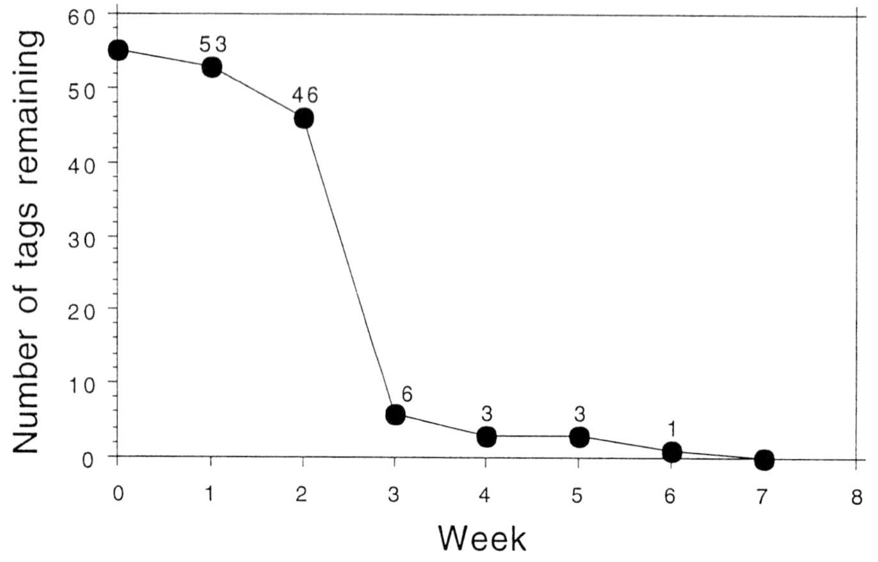

Figure 1. Number of creek chubs that retained alphanumeric implant tags at each weekly sample time (n = 55 creek chubs were tagged).

In the follow-up tagging study, most captive creek chubs lost tags within one month; the highest loss of tags occurred between the second and third weeks (Figure 1). Two of 55 creek chubs died but retained the tags, and a third moribund chub lost the tag; all were included as fishes that lost their tags. Moribund fishes were found between the second and third weeks and the fifth and sixth weeks. The longest retention time for a tag was between six and seven weeks, which is close to the 49 d residence time recorded for a creek chub and two green sunfishes.

Discussion

Phoxinus oreas and *Pimephales promelas* were captured in a cave and in the Bruffey-Hills Creek drainage for the first time (Hocutt et al. 1978, Stauffer et al. 1995, Poly & Boucher 1996). *Lepomis macrochirus* and *Ameiurus nebulosus* were not reported in this cave or in the basin previously. Brown bullheads have been collected rarely in the New River drainage of West Virginia and Virginia (Burkhead et al. 1980, Jenkins & Burkhead 1994, Stauffer et al. 1995), and Easton & Orth (1994) did not capture any brown bullheads during their survey. The only two records of brown bullhead in the Greenbrier River drainage have come from collections in caves (Poly & Boucher 1996, present study).

The low recapture rate of fishes during this study could be due to a number of factors. First, the epigean fishes might occur in the cave only for brief periods. High water during sampling and at other times between samples most likely affected collecting. The lowest abundance and diversity of fishes was obtained on dates when water levels were elevated (Table 1). A severe flood event had occurred (at least one) between September and November 1995 as evidenced by the scoured stream substrate and debris clinging to the cave walls nearly to the ceiling. The flood had not influenced the distribution of the two recaptured green sunfishes but might have contributed to the lack of recaptures. Another factor contributing to the low recapture rate was the logjam that limited the sampling area considerably. Tag loss was identified as a factor in lack of recaptures of tagged fishes, and an alternate method will be required for marking creek chubs and other cyprinids. Means & Johnson (1995) used *Phoxinus erythrogaster* (a cyprinid) and *Fundulus olivaceus* (a fundulid) as surrogates for testing visual implant tags; they found best tag retention in the caudal peduncle region, but they did not indicate results from placing tags elsewhere on the two surrogate species. No mention was made as to whether the tags could be seen well in the surrogate species when tags were in the caudal peduncle region, which has a cover of scales and pigmented epithelium in the surrogates. In this study, creek chubs were tagged on the top of the head because that region appeared to be the only position on the fish where tags could be read.

Limited information was gained on changes in pigmentation of fishes in Bruffey-Hills Creek Cave. The single creek chub that was recaptured initially became darker after exposure to light, whereas 49 d later the fish remained lightly pigmented; therefore, either the ability of this fish to rapidly expand melanophores had diminished, or the melanin content or number of melanophores might have decreased (Rasquin 1947). Rasquin (op. cit.) examined pigmentary changes of *Astyanax fasciatus* held in darkness, and a brief account of pigment loss in a crayfish was given in Kusek & Parker (1970).

No evidence of reproduction in the cave was found during the present study; tuberculate males of creek chubs have not been seen in this cave at any time. There is one report of a banded sculpin, *Cottus carolinae*, nesting just inside a cave (Craddock 1965), of a fully tuberculate male bluntnose minnow, *Pimephales notatus*, well inside a cave (Poly & Boucher 1996), and of a tuberculate male redeye chub, *Notropis harperi*, inside a cave (Brockman & Bortone 1977). Long-term residence of fishes in Bruffey-Hills Creek Cave was not demonstrated, although fishes were encountered regularly in the cave, and little information was obtained on movement of fishes within the cave.

Epigean fishes should be given greater consideration concerning their role(s) in the community of organisms within a cave, especially in cases that involve rare cave organisms. Even if epigean fishes cannot survive for extended periods within a cave, they might be capable of predation on troglophiles or troglobites during their short time of residence. Sculpins are presumed to be a likely potential predator of troglobites, but no case of such predation on troglobitic fishes or shrimps has been documented (Woods 1956, Lisowski 1983, Brown et al. 1994, Poly & Boucher 1996), although cases of predation on troglobitic and troglophilic crayfishes were noted by Brown et al. (1994). In addition, Relyea & Sutton (1974) found a troglobitic crayfish, *Procambarus lucifugus*, in the stomach of a yellow bullhead, *Ameiurus natalis*, in a Florida cave. Lee (1969) described instances of predation on cave-dwelling animals by an epigean species of frog, the bullfrog, *Rana catesbeiana*. Animals consumed by several bullfrogs included a cave salamander, *Gyrinophilus palleucus*, a troglobitic crayfish, *P. lucifugus*, and a bat, *Myotis austroriparius*. The frogs were well into the aphotic zones of the caves. Creek chubs could be potential predators in caves because even though they are active during the day, adults are known to feed and engage in various reproductive activities at night (Magnan & FitzGerald 1984, Maurakis et al. 1995); however, more light is available to the epigean fishes, even at night. Brown trout, *Salmo trutta*, living in British caves have been known to feed by 'grubbing' in the substrate for amphipods and insects (Proudlove 1979, 1982, Chapman 1993).

Occurrences of nontroglobitic fishes in caves are of interest also in that they represent various stages of evolution from accidental residents to occasional or seasonal residents (trogloxenes or troglophiles, e.g., Zhang 1986), which might lead to obligate troglobitic species. Studies on organisms in the early or intermediate intervals of cave-adapted existence were considered important for a better understanding of evolution of blind, depigmented troglobites (Dearolf 1956, Poly & Boucher 1996, Green & Romero 1997).

There are numerous reports of epigean species in caves (Appendix). Epigean fish species found in caves (or cave-like habitats) sometimes exhibit loss of pigment (Hora 1924, Kosswig 1937, Greenfield et al. 1982, Wüstemann 1990, Poly & Boucher 1996), reduced eye size (Burr et al.[2], Greenfield et al. 1982), and other structural differences (Paul et al.[3], Relyea & Sutton 1974) – convergent characters of hypogean species (Romero 1985b). A few examples of troglomorphic evolution in progress are found in some populations of *Astyanax fasciatus*, *Poecilia mexicana* (or *P. sphenops*), *Garra barreimiae*, and *Rhamdia* spp. (Norman 1926, Hubbs 1938, Gordon & Rosen 1962, Walters & Walters 1965, Greenfield et al. 1982, Banister 1984, Silfvergrip 1996, Weber & Wilkens 1998, Romero & Creswell 2000).

Neoteny (a form of paedomorphosis) appears to be a possible mechanism for troglomorphisms in fishes as discussed by Banister (1984), and he suggested that more experimental studies must be performed. The neotenic condition has been reported for a number of cave salamanders (Brandon 1971, references in Banister op. cit.), and Banister (op. cit.) provided examples of neoteny from fishes and a decapod crustacean. In a hypogean population of *Garra barreimiae* and

[2] Burr, B.M., M.L. Warren, Jr. & R.J. Paul. 1992. Cavernicolous sculpins of the *Cottus carolinae* species group from Mystery Cave, Missouri. 72nd Ann. Meeting Amer. Soc. Ich. Herp.,4–10 June 1992 (abstract).

[3] Paul, R.J., B.M. Burr & M.L. Warren, Jr. 1993. Cavernicolous sculpins of the *Cottus carolinae* species group from Perry County, Missouri: cave-life phenomena or speciation? 4th Ann. Argonne Symp. for Undergraduates in Science, Engineering and Mathematics, Argonne Natl Lab., 5–6 November 1993 (abstract).

in *Ancistrus cryptophthalmus*, eyes are visible externally in small individuals, then become covered in older fishes (Banister 1984, Reis 1987). However, in the salamander, *Typhlotriton spelaeus*, the eyes develop normally and are functional in the larvae, which inhabit springs and spring streams, but the cave-dwelling adults are blind due to tissue degeneration during and after metamorphosis (Stone 1964 cited in Brandon 1971). There could be many factors and processes involved in evolution of troglomorphisms (reviewed in Sket 1986). Experiments involving surface fishes deprived of light at early developmental stages and the converse, troglophilic or troglobitic species raised with light exposure, would provide information on the effects of light on development of eyes and other structures as well as pigment levels. Zeutzius & Rahmann (1980) found significantly fewer synapses per unit area in the cerebellum and optic tectum and lower density of synaptic vesicles in the optic tectum of rainbow trout raised without light from the larval period to 25 d. Similar studies carried out for longer time periods would help determine whether substantial troglomorphisms can develop within the lifespan of individual epigean fishes.

Acknowledgements

Field work could not have been accomplished without the assistance of Charles Boucher, Fumiko Kanekawa, and Robert Noggle. The property owners, Adriann and Ron White, allowed access to Bruffey-Hills Creek Cave. The West Virginia Association for Cave Studies (WVACS) provided lodging at their field station. Funding was provided by grants from Cave Conservancy of the Virginias (CCV) (to WJP & C. Boucher) and National Speleological Society (to WJP); the CCV grant was administered through WVACS by Verl Hickman and Chuck Frostick. Graham Proudlove and Phil Cochran provided several useful references. David Etnier cataloged fishes in the UT collection, John Baxter provided UT catalog information, and the record from CU was supplied by John Friel. John Stahl assisted by translating several papers. Helpful suggestions were made by Aldemaro Romero and two anonymous referees. This work was conducted under Scientific Collecting Permits #61-1994, #26-1995, #109-1998, and #02-1999 from the West Virginia Department of Natural Resources, coordinated by Barbara Sargent. See other acknowledgments in Poly & Boucher (1996).

References cited

Armstrong, J.G. & J.D. Williams. 1971. Cave and spring fishes of the southern bend of the Tennessee River. J. Tenn. Acad. Sci. 46: 107–115.

Bailey, V. 1933. Cave life of Kentucky, mainly in the Mammoth Cave region. Amer. Mid. Nat. 14: 385–634.

Banister, K.E. 1984. A subterranean population of *Garra barreimiae* (Teleostei: Cyprinidae) from Oman, with comments on the concept of regressive evolution. J. Nat. Hist. 18: 927–938.

Beck, B.F., M. Fram & J.R. Carvajal. 1976. The Aguas Buenas Caves, Puerto Rico: geology, hydrology, and ecology with special reference to the histoplasmosis fungus. Nat. Speleol. Soc. Bull. 33: 1–16.

Brandon, R.A. 1971. North American troglobitic salamanders: some aspects of modification in cave habitats, with special reference to *Gyrinophilus palleucus*. Nat. Speleol. Soc. Bull. 33: 1–21.

Brockman, K.A. & S.A. Bortone. 1977. Cave dwelling fishes in panhandle Florida. Florida Sci. 40: 406–407.

Brown, A.V. & C.S. Todd. 1987. Status review of the threatened Ozark cavefish (*Amblyopsis rosae*). Proc. Ark. Acad. Sci. 41: 99–100.

Brown, A.V., W.K. Pierson & K.B. Brown. 1994. Organic carbon resources and the payoff-risk relationship in cave ecosystems. pp. 67–76. *In:* J.A. Stanford & H.M. Valett (ed.) Proceedings of the Second International Conference on Ground Water Ecology, American Water Resources Association, Herndon.

Bryant, W., R. Erisman & J. Hockersmith. 1972. Sculpin Cave – an unusual blue grass cavern. Ky. Happy Hunt. Ground 28: 9–12.

Burkhead, N.M., R.E. Jenkins & E.G. Maurakis. 1980. New records, distribution and diagnostic characters of Virginia ictalurid catfishes with an adnexed adipose fin. Brimleyana 4: 75–93.

Chapman, P. 1985. Cave-frequenting vertebrates in the Gunung Mulu National Park, Sarawak. Sarawak Mus. J. 34: 101–113.

Chapman, P. 1993. Caves and cave life. Harper Collins, London. 224 pp.

Cliburn, J.W. & A.L. Middleton, Jr. 1983. The vertebrate fauna of Mississippi caves. Nat. Speleol. Soc. Bull. 45: 45–48.

Cooper, J.E. 1975. Ecological and behavioral studies in Shelta Cave, Alabama, with emphasis on decapod crustaceans. Doct. Diss., University of Kentucky, Lexington. 364 pp.

Cooper, J.E. & A. Iles. 1971. The southern cavefish, *Typhlichthys subterraneus*, at the southeastern periphery of its range. Nat. Speleol. Soc. Bull. 33: 45–49.

Cope, E.D. & A.S. Packard, Jr. 1881. The fauna of Nickajack Cave. Amer. Nat. 15: 877–882.

Craddock, J.E. 1965. Some aspects of the life history of the banded sculpin, *Cottus carolinae carolinae*, in Doe Run, Meade County, Kentucky. Doct. Diss., University of Louisville, Louisville. 157 pp.

Dearolf, K. 1956. Survey of North American cave vertebrates. Proc. Penn. Acad. Sci. 30: 201–210.

Delamare-Deboutteville, C. & C. Juberthie. 1975. Mission en République de Saint-Domingue et au Guatemala (8 Avril au 29 Avril 1975). Ann. Spéléol. 30: 767–771.

Easton, R.S. & D.J. Orth. 1994. Fishes of the main channel New River, West Virginia. Virg. J. Sci. 45: 265–277.

Eschmeyer, W.N. (ed.). 1998. Catalog of fishes. California Academy of Sciences, Spec. Publ. 1 (3 vol.). 2905 pp.

Evermann, B.W. & S.F. Hildebrand. 1914 (1916). Notes on the fishes of East Tennessee. Bull. U.S. Bur. Fish. 34: 433–451.

Franz, R., J. Bauer & T. Morris. 1994. Review of biologically significant caves and their faunas in Florida and South Georgia. Brimleyana 20: 1–109.

Garton, E.R., F. Grady & S.D. Carey. 1993. The vertebrate fauna of West Virginia caves. West V. Speleol. Surv. Bull. 11: 1–107.

Gordon, M.S. & D.E. Rosen. 1962. A cavernicolous form of the poeciliid fish *Poecilia sphenops* from Tabasco, Mexico. Copeia 1962: 360–368.

Green, S.M. & A. Romero. 1997. Response to light in two blind cave fishes (*Amblyopsis spelaea* and *Typhlichthys subterraneus*) (Pisces: Amblyopsidae). Env. Biol. Fish. 50: 167–174.

Greenfield, D.W., T.A. Greenfield & R.L. Woods. 1982. A new subspecies of cave-dwelling pimelodid catfish, *Rhamdia laticauda typhla* from Belize, Central America. Brenesia 19/20: 563–576.

Hale, J.A. & W.J. Streever. 1994. Cave fauna distribution within fully-flooded cave systems in Florida. J. Fresh. Ecol. 9: 171–174.

Hay, W.P. 1902. Observations on the crustacean fauna of Nickajack Cave, Tennessee, and vicinity. Proc. U.S. Nat. Mus. 25: 417–439.

Hobbs, H.H., III. 1978. Studies on the cave crayfish, *Orconectes inermis inermis* Cope (Decapoda, Cambaridae). Part IV: Mark-recapture procedures for estimating population size and movements of individuals. Int. J. Speleol. 10: 303–322.

Hobbs, H.H., III. 1991. A new pseudothelphusid crab from a cave in southern Costa Rica (Decapoda: Brachyura). Proc. Biol. Soc. Wash. 104: 295–298.

Hobbs, H.H., III. 1993. *Cambarus* (*Jugicambarus*) *subterraneus*, a new cave crayfish (Decapoda: Cambaridae) from Northeastern Oklahoma, with a key to the troglobitic members of the subgenus *Jugicambarus*. Proc. Biol. Soc. Wash. 106: 719–727.

Hocutt, C.H., R.F. Denoncourt & J.R. Stauffer, Jr. 1978. Fishes of the Greenbrier River, West Virginia, with drainage history of the central Appalachians. J. Biogeog. 5: 59–80.

Holsinger, J.R., R.A. Baroody & D.C. Culver. 1976. The invertebrate cave fauna of West Virginia. West V. Speleol. Surv. Bull. 7: 1–82.

Hora, S.L. 1924. Fish of the Siju Cave, Garo Hills, Assam. Rec. Indian Mus. 26: 27–31.

Hubbs, C.L. 1938. Fishes from the caves of Yucatan. Carnegie Inst. Washington Publ. 491: 261–295.

Hubbs, C.L. & M.B. Trautman. 1937. A revision of the lamprey genus *Ichthyomyzon*. Misc. Publ. Mus. Zool. Univ. Mich. 35: 1–109.

Jenkins, R.E. & N.M. Burkhead. 1994. Freshwater fishes of Virginia. American Fisheries Society, Bethesda. 1079 pp.

Jones, R.S. & W.F. Hettler. 1959. Bat feeding by green sunfish. Texas J. Sci. 11: 48.

Kosswig, C. 1937. Über pigmentverlust während des Höhlenlebens. Beobachtungen an *Squalius cephalus* L. Zool. Anz. 117: 37–43.

Kuehne, R.A. 1966. Depauperate fish faunas of sinking creeks near Mammoth Cave, Kentucky. Copeia 1966: 306–311.

Kusek, B.A. & D.J. Parker. 1970. Preliminary survey of the organisms in a portion of the Disappearing River within Mystery Cave. Thesis, Saint Mary's College, Winona. 31 pp.

Lawhon, W.T., Jr. 1969. Ecology of Alexander Cave, Perry County, Tennessee. M.S. Thesis, Memphis State University, Memphis. 33 pp.

Lee, D.S. 1969. Notes on the feeding behavior of cave-dwelling bullfrogs. Herpetologica 25: 211–212.

Lisowski, E.A. 1983. Distribution, habitat, and behavior of the Kentucky cave shrimp *Palaemonias ganteri* Hay. J. Crust. Biol. 3: 88–92.

McDaniel, V.R. & J.E. Gardner. 1977. Cave fauna of Arkansas: vertebrate taxa. Proc. Ark. Acad. Sci. 31: 68–71.

McDonald, E.F., Jr. & W.L. Pflieger. 1979. The spring cavefish, *Chologaster agassizi* (Pisces: Amblyopsidae), in southeastern Missouri. Amer. Mid. Nat. 102: 194–196.

Magnan, P. & G.J. FitzGerald. 1984. Ontogenetic changes in diel activity, food habits and spatial distribution of juvenile and adult creek chub, *Semotilus atromaculatus*. Env. Biol. Fish. 11: 301–307.

Marshall, N. 1947. The spring run and cave habitats of *Erimystax harperi* (Fowler). Ecology 28: 68–75.

Maurakis, E.G., W.S. Woolcott & W.R. McGuire. 1995. Nocturnal reproductive behavior in *Semotilus atromaculatus* (Pisces, Cyprinidae). S. E. Fish. Council Proc. 31: 1–3.

Means, M.L. & J.E. Johnson. 1995. Movement of threatened Ozark cavefish in Logan Cave National Wildlife Refuge, Arkansas. Southwest. Natur. 40: 308–313.

Menon, A.G.K. 1987. The fauna of India and the adjacent countries. Pisces. Vol. IV. Teleostei – Cobitoidea. Part 1. Homalopteridae. Zoological Survey of India, New Delhi. 259 pp.

Mitchell, R.W., W.H. Russell & W.R. Elliott. 1977. Mexican eyeless characin fishes, genus *Astyanax*: environment, distribution, and evolution. Spec. Publ. Mus. Texas Tech. Univ. 12. 89 pp.

Ng, H.H. & M. Kottelat. 1998. *Pterocryptis buccata*, a new species of catfish from western Thailand (Teleostei: Siluridae) with epigean and hypogean populations. Ichthyol. Res. 45: 393–399.

Norman, J.R. 1926. A new blind catfish from Trinidad, with a list of the blind cave-fishes. Ann. Mag. Nat. Hist. (9th Ser.) 18: 324–331.

Pembleton, E.F. & E.L. Bake. 1967. The fauna of Maze Cave. Mo. Speleol. 9: 8–13.

Perez, J.E. & G.E.E. Moodie. 1993. Genetic variation in a cave-dwelling Venezuelan catfish. Acta Cient. Venez. 44: 28–31.

Poly, W.J. & C.E. Boucher. 1996. Nontroglobitic fishes in caves: their abnormalities, ecological classification and importance. Amer. Mid. Nat. 136: 187–198.

Pradhan, R.K. & J. Biswas. 1994. Towards regressive evolution: the periodic colour change behaviour of a troglophilic fish *Nemacheilus evezardi* (Day). Int. J. Speleol. 23: 191–201.

Proudlove, G.S. 1979. Fishes in British caves: an interim report. J. Craven Pothole Club 6: 10–12.

Proudlove, G.S. 1982. Cave life. Part 2: cave fish. Caves and Caving 15: 6–7.

Pruitt, B.C., Jr. 1995. New records of fishes within Florida caves. Nat. Speleol. Soc. Bull. 57: 20–23.

Pylka, J.M. & R.D. Warren. 1958. A population of *Haideotriton* in Florida. Copeia 1958: 334–336.

Rasquin, P. 1947. Progressive pigmentary regression in fishes associated with cave environments. Zoologica 32: 35–42.

Reddell, J.R. 1967. A checklist of the cave fauna of Texas. III. Vertebrata. Texas J. Sci. 19: 184–226.

Reese, A.M. 1934. The fauna of West Virginia caves. Proc. West V. Acad. Sci. 7: 39–53.

Reis, R.E. 1987. *Ancistrus cryptophthalmus* sp. n., a blind mailed catfish from the Tocantins river basin, Brazil (Pisces, Siluriformes, Loricariidae). Rev. fr. Aquariol. 14: 81–84.

Relyea, K. & B. Sutton. 1973 (1974). Cave dwelling yellow bullheads in Florida. Florida Sci. 36: 31–34.

Robins, C.R., R.M. Bailey, C.E. Bond, J.R. Brooker, E.A. Lachner, R.N. Lea & W.B. Scott. 1991. Common and scientific names of fishes from the United States and Canada, fifth edition. Amer. Fish. Soc. Spec. Publ. (20): 1–183.

Romero, A. 1984. Behavior in an 'intermediate' population of the subterranean-dwelling characid *Astyanax fasciatus*. Env. Biol. Fish. 10: 203–207.

Romero, A. 1985a. Cave colonization by fish: role of bat predation. Amer. Mid. Nat. 113: 7–12.

Romero, A. 1985b. Can evolution regress? Nat. Speleol. Soc. Bull. 47: 86–88.

Romero, A. 1987. Observaciones preliminares sobre la conducta en individuos cavernícolas de *Trichomycterus conradi* (Pisces: Trichomycteridae). Bol. Soc. Venez. Cien. Nat. 41: 205–216.

Romero, A. 1999 (2000). Myth and reality of the alleged blind cave fish from Pennsylvania. J. Spelean Hist. 33: 67–75.

Romero, A. & J.E. Creswell. 2000. In search of the elusive 'eyeless' cave fish of Trinidad, W.I. Nat. Speleol. Soc. News. 58: 282–283.

Schmidt, K.P. 1994. The Mystery Cave expeditions. Amer. Currents (publ. of the N. Amer. Native Fish. Assoc.) 1994 (Spring): 26–31.

Silfvergrip, A.M.C. 1996. A systematic revision of the Neotropical catfish genus *Rhamdia* (Teleostei, Pimelodidae). Swedish Museum of Natural History, Stockholm. 156 pp.

Sket, B. 1985 (1986). Why all cave animals do not look alike – a discussion on adaptive value of reduction processes. Nat. Speleol. Soc. Bull. 47: 78–85.

Smith, P.W. 1948. Food habits of cave dwelling amphibians. Herpetologica 4: 205–208.

Stauffer, J.R., Jr, J.M. Boltz & L.R. White. 1995. The fishes of West Virginia. Proc. Acad. Nat. Sci. Phil. 146: 1–389.

Storrick, G.D. 1992. Caves and karst hydrology of southern Pocahontas County and the upper Spring Creek valley. West V. Speleol. Surv. Bull. 10: 1–214.

Streever, W.J. 1995. Recovery of the cave crayfish (Decapoda: Cambridae [*sic*]) population in Peacock Springs, Florida? Brimleyana 22: 61–65.

Tehsin, R., V.S. Durve & M. Kulshreshtha. 1988. Occurrence of a schizothoracine fish (snow trout) in a subterranean cave near Udaipur, Rajasthan. J. Bombay Nat. Hist. Soc. 85: 211–212.

Tellkampf, T.A. 1845. Memoirs on the blind fishes and some other animals living in the Mammoth Cave in Kentucky. N.Y. J. Med. Collat. Sci. 5: 84–93.

Trajano, E. 1991. Population ecology of *Pimelodella kronei*, troglobitic catfish from Southeastern Brazil (Siluriformes, Pimelodidae). Env. Biol. Fish. 30: 407–421.

Trajano, E. 1997. Population ecology of *Trichomycterus itacarambiensis*, a cave catfish from eastern Brazil (Siluriformes, Trichomycteridae). Env. Biol. Fish. 50: 357–369.

Viloria, A., F. Herrera & C. Galan. 1992. Resultados preliminares del estudio del material biológico colectado en Mesa Turik y Cuenca del Río Socuy. Bol. Soc. Venezolana Espel. 26: 7–9.

Walters, L.H. & V. Walters. 1965. Laboratory observations on a cavernicolous poeciliid from Tabasco, Mexico. Copeia 1965: 214–223.

Weber, A. & H. Wilkens. 1998. *Rhamdia macuspanensis*: a new species of troglobitic pimelodid catfish (Siluriformes; Pimelodidae) from a cave in Tabasco, Mexico. Copeia 1998: 998–1004.

Weise, J.G. 1957. The spring cave-fish, *Chologaster papilliferus*, in Illinois. Ecology 38: 195–204.

Williams, J.D. & W.M. Howell. 1979. An albino sculpin from a cave in the New River Drainage of West Virginia (Pisces: Cottidae). Brimleyana 1: 141–146.

Woods, L.P. 1956. Blind fishes found in cave pools and streams. Nat. Speleol. Soc. Bull. 18: 24–29.

Wüstemann, O. 1990. Bemerkenswerte anpassung einer Population der Westgroppe (*Cottus gobio*) an ein Höhlenbiotop in Harz (Osteichthyes: Cottidae). Faun. Abh. Mus. Tierkd. Dresden 17: 190–191.

Yager, J. & D. Williams. 1988. Predation by gray snapper on cave bats in the Bahamas. Bull. Mar. Sci. 43: 102–103.

Zeutzius, I. & H. Rahmann. 1980. Quantitative ultrastructural investigations on synaptogenesis in the cerebellum and the optic tectum of light-reared and dark-reared rainbow trout (*Salmo gairdneri* Rich.). Differentiation 17: 181–186.

Zhang, C. 1986. On the ecological adaptation and geographical distribution of the barbine fish *Varicorhinus* (*Scaphesthes*) *macrolepis* (Bleeker). Acta Zool. Sin. 32: 266–272 (in Chinese).

Appendix. Nontroglobitic fish species reported from caves worldwide.

Species	States or countries
Order Petromyzontiformes	
Family Petromyzontidae	
Lampetra Bonnaterre sp.	AL[12]
Lampetra appendix (DeKay)	MO[57]
Order Amiiformes	
Family Amiidae	
Amia calva Linné	FL[62]
Order Anguilliformes	
Family Anguillidae	
Anguilla anguilla (Linné)	UK[40,54]
Anguilla rostrata (Lesueur)	FL[21,29,41,62]
Order Cypriniformes	
Family Cyprinidae	
Barbus binotatus Valenciennes	Sarawak[53]
Barbus hexastichus McClelland	India[30]

Appendix continued.

Species	States or countries
Barilius barna (Hamilton)	India[30]
Barilius bendelisis (Hamilton)	India[30]
Campostoma anomalum (Rafinesque)[(1)]	MN[36], MO[26], WV[7]
Clinostomus funduloides Girard	VA[8]
Cyprinella spiloptera (Cope)	KY[10], TN[11]
Cyprinus carpio Linné	FL[62], MO[24]
Danio aequipinnatus (McClelland)	India[30]
Garra barreimiae Fowler & Steinitz	Oman[56]
Hybognathus placitus Girard	TX[38]
Leuciscus cephalus Linné	Slovenia[47]
Luxilus cornutus (Mitchill)	MN[45]
Margariscus margarita Cope	WV[7]
Nematabramis everetti Boulenger	Sarawak[53]
Nocomis micropogon (Cope)	WV[7]
Notemigonus crysoleucas (Mitchill)	FL[62], MS[16]
Notropis atherinoides Rafinesque	KY[4,8,10]
Notropis harperi Fowler	FL[13,61,62,63]
Notropis photogenis (Cope)	WV[25]
Notropis volucellus (Cope)	WV[25]
Phoxinus erythrogaster (Rafinesque)[(1)]	MO[26]
Phoxinus oreas (Cope)	WV[72]
Pimephales notatus (Rafinesque)	IN[22], WV[25], unspecified[6]
Pimephales promelas Rafinesque	WV[72]
Psilorhynchus sucatio (Hamilton)	India[30]
Rasbora volzi Popta	Sarawak[53]
Rhinichthys atratulus (Hermann)	MN[36], TN[23], WV[5,8,25,72]
Rhinichthys cataractae (Valenciennes)	MN[36], WV[5,7,8]
R. cataractae × *N. micropogon*	WV[7]
Semotilus atromaculatus (Mitchill)	IL[24], KY[10], MN[36], MO[24], MS[16], TN[8], WV[7,25,72]
Varicorhinus macrolepis (Bleeker)	China[32]
Schizothoracinae	India[34]
Family Balitoridae	
Indoreonectes evezardi (Day)	India[73]
Nemacheilus sijuensis Menon	India[30,77]
Family Catostomidae	
Catostomus commersoni (Lacépède)	KY[10], MN[36], TN[2,8], WV[25]
Hypentelium nigricans (Lesueur)	TN[8], WV[7]
Ictiobus bubalus (Rafinesque)	AL[11]
Order Characiformes	
Family Lebiasinidae	
Piabucina erythrinoides Valenciennes	Venezuela[74]

Appendix continued.

Species	States or countries
Family Characidae	
Astyanax fasciatus (Cuvier)	México[66], Costa Rica[35]
Astyanax Baird & Girard sp.	Brazil[46]
Order Siluriformes	
Family Ictaluridae	
Ameiurus catus (Linné)	FL[62]
Ameiurus melas (Rafinesque)	MO[27], TX[37,38]
Ameiurus natalis (Lesueur)	AL[11], FL[13,21,29,61,62,65], MS[16]
Ameiurus nebulosus (Lesueur)	FL[21,62,65], WV[25,72]
Ameiurus serracanthus (Yerger & Relyea)	FL[62]
Ameiurus Rafinesque sp.	IL[24], KY[10], MO[24], OK[49]
Ictalurus lupus (Girard)	TX[23]
Ictalurus punctatus (Rafinesque)	FL[62,65], TX[9,37,38]
Noturus Rafinesque sp.	Unspecified[8]
Pylodictis olivaris (Rafinesque)	FL[62]
Family Bagridae	
Leiocassis poecilopterus (Valenciennes)[(6)]	Sarawak[53]
Family Siluridae	
Pterocryptis buccata Ng & Kottelat	Thailand[76]
Family Sisoridae	
Glyptothorax platypogon (Valenciennes)	Sarawak[53]
Family Clariidae	
Clarias Scopoli sp.	Sarawak[53]
Family Pimelodidae	
Imparfinis piperatus Eigenmann & Norris	Brazil[46]
Pimellodella chagresi odynea Schultz	Venezuela[74]
Pimellodella transitoria Miranda-Ribeiro	Brazil[46]
Rhamdia laticauda (Kner)[(5)]	México[59]
Rhamdia laticauda typhla Greenfield, Greenfield & Woods	Belize[69]
Rhamdia quelen (Quoy & Gaimard)	Brazil[46], México[6,59], Trinidad[58,59,78]
Family Trichomycteridae	
Trichomycterus banneaui maracaiboensis (Schultz)	Venezuela[74]
Trichomycterus emanueli (Schultz)	Venezuela[74]
Trichomycterus conradi (Eigenmann)	Venezuela[1,60]
Family Loricariidae	
Lasiancistrus maracaiboensis Schultz	Venezuela[74]
Loricariidae	Brazil[46]

Appendix continued.

Species	States or countries
Order Esociformes	
Family Esocidae	
Esox niger Lesueur	FL[62]
Order Salmoniformes	
Family Salmonidae	
Oncorhynchus mykiss (Walbaum)	WV[19,25,72], UK[40,54]
Salmo salar Linné	UK[54]
Salmo trutta Linné	Unspecified[47], UK[40,54]
Salvelinus fontinalis (Mitchill)	WV[19,25]
Order Percopsiformes	
Family Aphredoderidae	
Aphredoderus sayanus (Gilliams)	MS[16], FL[21,62,63]
Family Amblyopsidae	
Chologaster agassizi Putnam[(2)]	IL[71], KY[10], MO[70], TN[8]
Order Cyprinodontiformes	
Family Fundulidae	
Fundulus zebrinus Jordan & Gilbert[(3)]	TX[38]
Family Cyprinodontidae	
Cyprinodon rubrofluviatilis Fowler	TX[38]
Family Poeciliidae	
Brachyraphis rhabdophora (Regan)	Costa Rica[35]
Gambusia holbrooki Girard	FL[21,61]
Poecilia reticulata Peters	PR[55]
Poecilia sphenops (Cuvier & Valenciennes)[(4)]	México[50,64]
Xiphophorus Heckel sp.	Guatemela[52]
Family Hemirhamphidae	
Hemirhamphodon pogonognathus (Bleeker)	Sarawak[53]
Order Gasterosteiformes	
Family Gasterosteidae	
Gasterosteus aculeatus Linné	UK[54]
Order Synbranchiformes	
Family Mastacembelidae	
Mastacembelus maculatus Cuvier	Sarawak[53]
Order Scorpaeniformes	
Family Cottidae	
Cottus bairdi Girard	PA[8], WV[5,7,8,25]
Cottus carolinae (Gill)[(7)]	AL[23], AR[8,20], IL[24], IN[6,8,22], KY[4,23,43,75], MO[6,8,17,18,22,24], TN[2,3,22,23,44], VA[8,22,23,28], WV[15,25]
Cottus gobio Linné[(8)]	Germany[33], UK[39,40,54]
Cottus Linné sp.	TN[51], unspecified[67]
Order Perciformes	
Family Ambassidae	
Chanda nama Hamilton	India[30]
Family Moronidae	
Morone saxatilis (Walbaum)	FL[21,62]
Family Centrarchidae	
Ambloplites rupestris (Rafinesque)	MO[24], WV[25,72]
Centrarchus macropterus (Lacépède)	FL[62]
Lepomis auritus (Linné)	FL[62]
Lepomis cyanellus Rafinesque	AR[14], MO[24], TX[9,37,38], WV[25,72]
Lepomis gulosus (Cuvier)	MO[27], FL[62], TX[37,38]
Lepomis macrochirus Rafinesque	FL[21,62], MO[24,79], WV[25,72]
Lepomis marginatus (Holbrook)	MS[16]
Lepomis microlophus (Günther)	FL[62]
Lepomis punctatus (Valenciennes)	FL[62]
Lepomis Rafinesque sp.	KY[10]
Micropterus salmoides (Lacépède)	FL[62]
Pomoxis nigromaculatus (Lesueur)	FL[21,62]
Family Percidae	
Etheostoma blennioides Rafinesque	WV[7]
Etheostoma flabellare Rafinesque	MN[36]
Etheostoma nigrum Rafinesque	MN[36]
Etheostoma parvipinne Gilbert & Swain	MS[16]
Etheostoma zonale (Cope)	IN[22]
Percina nigrofasciata (Agassiz)	FL[62]
Family Sciaenidae	
Micropogonias undulatus (Linné)	FL[62]
Family Cichlidae	
Cichlidae	FL[62]
Family Eleotridae	
Eleotris pisonis (Gmelin)	Bahamas[68]
Family Gobiidae	
Gobiosoma bosc (Lacépède)	FL[62]
Family Channidae	
Channa orientalis Bloch & Schneider[(9)]	India[30]
Order Pleuronectiformes	
Family Achiridae	
Trinectes maculatus (Bloch & Schneider)	FL[62]
Unidentified fishes	KY[31,52,48], TN[51]

Authors of genera follow Eschmeyer (1998). [(1)]Reported as extremely emaciated and with empty intestinal tracts (Smith 1948). [(2)]Now called *Forbesichthys agassizi* by some authors. [(3)]Given as *Fundulus kansae* in source. [(4)]*Poecilia sphenops* or *P. mexicana*; taxonomy of *Poecilia* requires further study. [(5)]Includes *R. reddelli* Miller and *R. zongolicensis* Wilkens according to Silfvergrip (1996). [(6)]Given as *Leiocassis micropogon* in source. [(7)]Includes subspecies. [(8)]The record in Germany was from a cave-like habitat. [(9)]Given as *Ophiocephalus gachua* in source. Data from [1]Romero 1987, [2]Cope & Packard 1881, [3]Evermann & Hildebrand 1916, [4]Bailey 1933, [5]Reese 1934, [6]Hubbs 1938, [7]Frank

J. Schwartz unpublished, data collected in 1946 from Sinks of Gandy, [8]Dearolf 1956 (record of *Erimystax* sp. probably = *N. harperi*), [9]Jones & Hettler 1959, [10]Kuehne 1966, [11]Armstrong & Williams 1971, [12]Cooper & Iles 1971, [13]Relyea & Sutton 1974, [14]McDaniel & Gardner 1977, [15]Williams & Howell 1979, [16]Cliburn & Middleton 1983, [17]Pflieger[4], [18]Burr et al.[5], [19]Garton et al. 1993, [20]Brown et al. 1994, [21]Franz et al. 1994, [22]University of Michigan Museum of Zoology (UMMZ) records, [23]National Museum of Natural History (USNM) records, [24]Southern Illinois University at Carbondale (SIUC) records, [25]Poly & Boucher 1996, [26]Smith 1948, [27]Pembleton & Bake 1967, [28]Cornell University Ichthyology Collection (CU) records, [29]Streever 1995, [30]Hora 1924, [31]Tellkampf 1845, [32]Zhang 1986, [33]Wüstemann 1990, [34]Tehsin et al. 1988, [35]Romero 1984, 1985a, [36]Schmidt 1994, [37]Reddell 1967, [38]Reddell & Mitchell[6], [39]Proudlove 1979, [40]Proudlove 1982, [41]Pylka & Warren 1958, [42]Paul et al.[7], [43]Lisowski 1983, [44]Lawhon 1969, [45]Kusek & Parker 1970, [46]Trajano 1991, [47]Kosswig 1937, [48]Hobbs 1991, [49]Hobbs 1993 (bullhead catfish, presumably *Ameiurus* sp.), [50]Gordon & Rosen 1962, [51]Hay 1902, [52]Delamare-Deboutteville & Juberthie 1975, [53]Chapman 1985, [54]Chapman 1993, [55]Beck et al. 1976, [56]Banister 1984, [57]Hubbs & Trautman 1937, [58]Norman 1926, [59]Silfvergrip 1996, [60]Perez & Moodie 1993, [61]Marshall 1947, [62]Pruitt 1995, [63]Brockman & Bortone 1977, [64]Walters & Walters 1965, [65]Hale & Streever 1994, [66]Mitchell et al. 1977, [67]Woods 1956, [68]Yager & Williams 1988, [69]Greenfield et al. 1982, [70]McDonald & Pflieger 1979, [71]Weise 1957, [72]present study, [73]Pradhan & Biswas 1994, [74]Viloria et al. 1992, [75]Bryant et al. 1972, [76]Ng & Kottelat 1998, [77]Menon 1987, [78]Romero & Creswell 2000, [79]Brown & Todd 1987 (reported *L. macrochirus*, but all other sunfishes collected in this cave have been *L. cyanellus* (personal observation, K. Lister personal communication)). The record of *Ameiurus nebulosus* from a cave in Pennsylvania (based on Cope's '*Gronias nigrilabris*') was not included because the bullheads were collected from a surface stream rather than a cave (see Romero 2000).

[4] Pflieger, W.L. 1989. Aquatic community classification system for Missouri. Missouri Department of Conservation, Aquatic Series, No. 19. 70 pp.

[5] Burr, B.M., M.L. Warren, Jr & R.J. Paul. 1992. Cavernicolous sculpins of the *Cottus carolinae* species group from Mystery Cave, Missouri. 72nd Ann. Meeting Amer. Soc. Ich. Herp., 4–10 June 1992 (abstract).

[6] Reddell, J.R. & R.W. Mitchell. 1969. A checklist and annotated bibliography of the subterranean aquatic fauna of Texas. Texas Technological College, Water Resources Center, Lubbock, Special Report 24. 48 pp.

[7] Paul, R.J., B.M. Burr & M.L. Warren, Jr. 1993. Cavernicolous sculpins of the *Cottus carolinae* species group from Perry County, Missouri: cave-life phenomena or speciation? 4th Ann. Argonne Symp. for Undergraduates in Science, Engineering and Mathematics, Argonne Natl Lab., 5–6 November 1993 (abstract).

Part 1. Anatomy, physiology and behavior

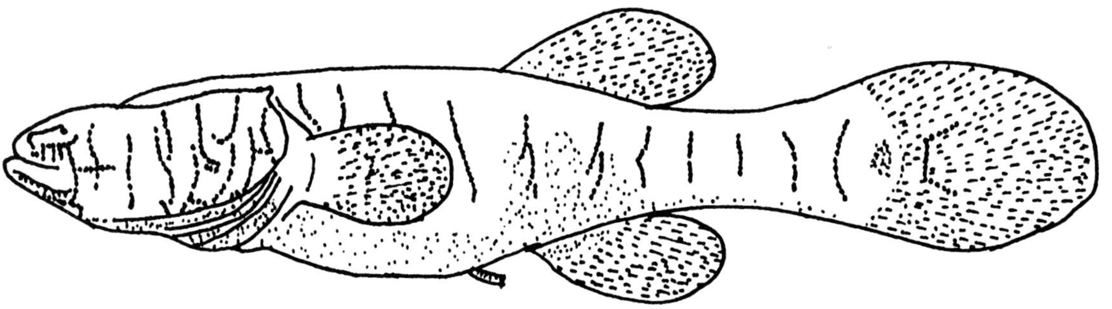

Environmental Biology of Fishes **62:** 87–96, 2001.
© 2001 *Kluwer Academic Publishers. Printed in the Netherlands.*

The mechanosensory lateral line system of the hypogean form of *Astyanax fasciatus*

John C. Montgomery[a], Sheryl Coombs[b] & Cindy F. Baker[a]
[a]*School of Biological Sciences, University of Auckland, Private Bag 92019, Auckland, New Zealand (e-mail: j.montgomery@auckland.ac.nz)*
[b]*Department of Biology, Loyola University of Chicago, 6525 North Sheridan Road, Chicago, IL 60626, U.S.A.*

Received 7 March 2000 Accepted 16 December 2000

Key words: superficial neuromast, canal organ, rheotaxis, olfactory search, prey detection, hydrodynamic imaging

Synopsis

The mechanosensory lateral line is a distributed, hair-cell based system which detects the water flow regime at the surface of the fish. Superficial neuromasts densely scattered over the surface of some cave fish detect the pattern of flow over the surface of the body and are important in rheotactic behaviors and perhaps in the localization of small vibrating sources. Canal neuromasts are very likely also involved in the detection of small planktonic prey, but seem also to play an essential role in replacing vision as the major sense by which blind cave-fish perceive their surroundings. The flow-field that exists around a gliding fish is perturbed by objects in the immediate vicinity, these perturbations are detected by the lateral line system. In this way the fish can build up a 'picture' of its environment, a process that has been called active hydrodynamic imaging. None of the lateral line behaviors exhibited by blind cave fish are necessarily exclusive to these species, but there is some evidence that their lateral line capabilities are enhanced with respect to their sighted relatives.

Introduction

One of the distinctive features of many hypogean fish species is their reduced eyes. The absence of light in caves clearly precludes the use of vision as an effective sensory modality and visual systems become correspondingly reduced, to the point of being incapable of image formation in the 'blind' cave fishes. It follows that other systems become relatively more important. In line with this expectation it has been demonstrated that non-visual senses have developed to compensate for the lack of vision.

Chemosensory, tactile, equilibrium and auditory systems no doubt make their contribution. However, in the absence of vision, the only sense most fish possess that is capable of providing detailed information of the immediate surroundings is the mechanosensory lateral line system (Hensel 1978, 1986, Coombs & Montgomery 1999). This system is found in all fishes, but it is in the blind hypogean fishes that we best see the functional potentialities of this sensory modality. Not only is the mechanosensory lateral line system important to hypogean fishes, but these fishes have made an important contribution to understanding lateral line systems. Almost all the work on lateral lines of blind cave fishes has been done on the blind Mexican species *Astyanax fasciatus*[1] (Characidae). This article concentrates particularly on this species. Unless otherwise indicated, when referring to 'blind cave fish' we mean the hypogean form of *Astyanax fasciatus*.

[1] *Astyanax fasciatus* is the currently accepted name for the blind Mexican cave fish. Previously, sighted *Astyanax* and blind *Astyanax* were considered to be separate species *A. mexicanus* and *A. hubbsi* respectively. However they have been synonymysed (see Romero 1984). A further potential source of confusion is that *A. hubbsi* was previously known by the name *Anoptichtys jordani*.

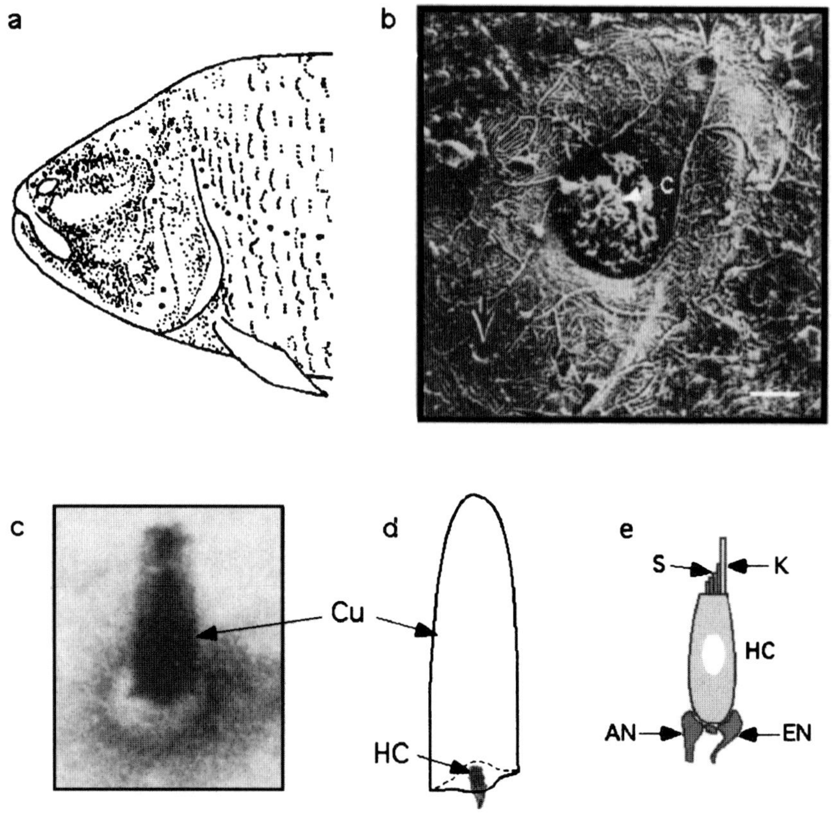

Figure 1. Superficial neuromasts of the blind cave-fish *Astyanax fasciatus*: a – Lateral view of the left anterior half of *Astyanax fasciatus*. Large dots are pore entrances to the lateral line canals. Small dots are superficial neuromasts. b – Scanning electronmicrograph of a superficial neuromast (from Teyke 1990); note the central sensory epithelium and the flanges and peg-holes to which the cupula is normally attached. c – Photomicrograph of the cupula in situ (from Teyke 1990). d – Drawing of the neuromast with the cupula drawn to scale. e – Diagram of a single hair cell (S = steriocilia; K = kinocilium; AN = afferent neuron; EN = efferent neuron; HC = hair cell; Cu = cupula; C = cilia of the sensory epithelium).

We will address the following questions: What is their mechanosensory lateral line system and how does it work? What are its capabilities in cave fish, and how do these compare with the lateral line capabilities found in other fishes?

What is the mechanosensory lateral line system and how does it work?

The functional unit of the lateral line is the neuromast. Each neuromast consists of a patch of hair cells and associated support cells that secrete a gelatinous cupula (Figure 1). The cupula forms the mechanical connection between the surrounding water and the cilia of the hair cells. Water movement displaces the cupula, which in turn deflects the cilia of the hair cells. This process opens stretch-sensitive ion channels altering the internal electrical potential of the hair cell. Variations in the resting potential modulate synaptic release at the base of the hair cell, which in turn affects the flow of action potentials in the afferent nerve fibers that connect the sensory periphery to the brain. Understanding the lateral line periphery requires a functional understanding of the nature of hydrodynamic signals in the immediate environment of the fish, how these signals are modified/processed by the ancillary structures associated with the neuromasts, the biomechanics of hair cell stimulation, and the chemical/electrical transduction of these signals by the hair cell and hair cell afferent synapse. Fortunately there appears to be a relatively simple relationship between bending of the hair

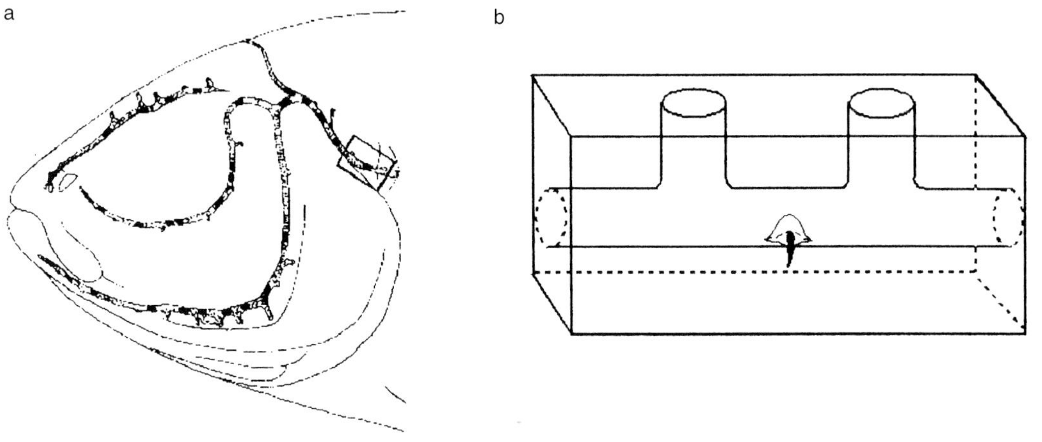

Figure 2. Canal lateral line system of the blind cave-fish *Astyanax fasciatus*: a – diagrammatic view of the head lateral line canals of *Astyanax fasciatus*, b – block diagram of a section of the lateral line canal (outlined in a) showing the position of the canal neuromast.

cell and afferent firing. So for practical purposes, the mechanosensory lateral line afferent firing is generally assumed to be proportional to the displacement of the cupula.

Each hair cell has a single axis of maximal sensitivity. Displacement of the cilia in one direction along this axis is excitatory, and displacement in the opposite direction is inhibitory. The sensitivity of the hair cell to an oblique stimulus follows an approximate cosine function, such that a stimulus at right angles to axis of sensitivity has no effect. The functional polarization of the hair cells is evident in their morphology. The cilia form a graded stack, increasing in height across the apex of the cell. At one side of the stack is a single long cilium called a kinocilium. Deflection of the cilia towards the kinocilium depolarizes the hair cell. All the hair cells in a neuromast are aligned along a single axis, such that approximately half the population has their kinocilium on one side, and the other half have the kinocilium on the opposite side. Afferent fibers connect hair cells of only one orientation, and exhibit a spontaneous firing rate at rest. So that within the nerve branch innervating a neuromast, some fibers will respond with an increase in firing rate to a particular deflection of the cupula, and the others will respond with a decrease in firing rate.

In most fishes, including blind cave fish, there are two lateral line submodalities: superficial neuromasts, that are isolated neuromasts on the surface of the skin and the scales, and canal neuromasts embedded in subdermal canals (Figure 2). The pattern of canal neuromasts in blind cave fish is very similar to that found in many fishes, with canals occurring across the top of the head, above and below the regressed eye, running vertically down the preopercule, and out onto the lower jaw, and finally down the lateral flank of the fish. The plethora (i.e. thousands) of superficial neuromasts found over nearly the entire body surface of blind cave fish is unusual. Most fish that have been examined have only relatively few such organs found in short, discrete lines on the head and lines often associated with the trunk lateral line system (Coombs et al. 1988). However, such an abundance of superficial neuromasts is also found in related fishes (Schemmel 1967, Puzdrowski 1989) so is not distinctively associated with the hypogean condition.

In general terms, the superficial neuromasts respond to the flow of water across the skin of the fish. In many fish species the axis of sensitivity of the hair cell population corresponds with the long axis of the neuromast, which is also aligned along the long axis of the cupula. Flow of water past the cupula creates a frictional force on the cupula generating the displacement that is the effective stimulus. In other words, many superficial neuromast are 'velocity sensitive'. In blind cave fish, Teyke (1990) reports that the long axis of the cupula is at right angles to the axis of hair cell sensitivity. The long axis of the cupula would move directly with any cross flow, making the neuromast 'position (or displacement) sensitive' rather than 'velocity sensitive'. This arrangement will tend to increase the sensitivity of the superficial neuromast. A longer cupula will also increase flow sensitivity. This is because the longer cupula protrudes through the stationary boundary layer surrounding the fish and because a larger cupula provides a greater surface area of contact with the water.

Teyke (1990) conducted a study of the morphology of the superficial neuromast system in blind cave fish *Astyanax hubbsi* and the sighted river fish *Astyanax mexicanus* (the two groups now considered to be the same species). In blind *Astyanax*, the neuromasts are about 80 × 50 μm and possess cupulae with an oval base shape. Cupulae were longest at the head (up to 300 μm) and decreased in height towards the rear of the fish (average length on the trunk 100 μm). In sighted *Astyanax* the cupulae of the neuromasts are shorter by about half. In amblyopsid fishes there is also an increase in cupula length across a series from epigean to hypogean species (Poulson 1963).

The morphology of the canal neuromasts has not been specifically studied in blind cave fish, but there is no reason to believe that they are different from the typical canal neuromasts found in other fishes. Canal neuromasts tend to have more hair cells than superficial neuromasts, the kinocilia are longer, but the cupulae are shorter. The functional implications of these differences have not been fully explored, but it is generally accepted that the canal itself functions as a high pass mechanical filter (Denton & Gray 1988, Montgomery et al. 1994). In essence this means that canal neuromasts respond less to steady currents and low frequency flows, and are better suited to encode higher frequency signals.

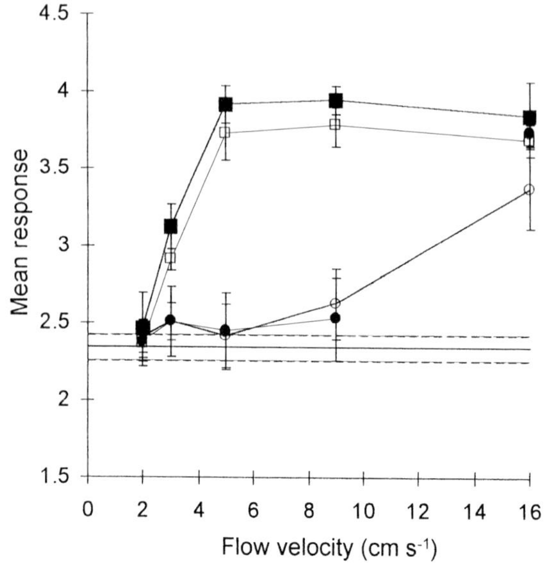

Figure 3. Mean number of *A. fasciatus* exhibiting rheotaxis under different experimental conditions (■ unmanipulated fish; □ fish treated with gentamicin sulfate; ● fish treated with cobalt chloride; ○ fish after a physical ablation of the superficial neuromasts).

What are the lateral-line mediated capabilities of cave-fish, and how do these compare with the capabilities found in other fishes?

As might be predicted from the consideration of the structural and functional properties of the lateral line end-organs, superficial neuromasts provide an important source of information on water currents in the environment of the fish. Orientation to water currents is termed rheotaxis. Rheotaxis is important for maintaining position within a flowing water system and for olfactory orientation and localization. Recent studies of the sensory basis of rheotaxis in *Astyanax* (Montgomery et al. 1997, Baker & Montgomery 1999) provide clear evidence for a role of the superficial neuromasts in rheotactic behaviour. A fish embedded within a smooth flow without any external reference frame has no basis on which to know whether it is moving or stationary. In sighted fish, movement relative to stationary objects provides an earth-fixed visual reference frame against which to judge the fish's own movements. Blind cave fish must use occasional and repeated substrate contact to maintain a reference frame. These behavioural studies show that for slow flow velocities *Astyanax* uses its superficial neuromasts in addition to tactile contact to maintain an upstream orientation (Figure 3). Disabling of the entire lateral line system elevates the rheotactic threshold to greater than 8 cm s^{-1}. Mechanical damage of the superficial system has the same effect, whereas selective blockade of the canal system has no effect on rheotactic behaviour.

For cave fish, rheotaxis will be important for station holding and maintaining position within the cave. However, rheotaxis is also an important component of olfactory search behaviors in fish. For example, there is evidence that the olfactory search behavior of the Antarctic fish *Trematomus bernacchii* includes a rheotactic component (Montgomery et al. 1999). Olfactory search will be important for food location, particularly in blind *Astyanax* that feed on bat guano and are guided to food by water movements and chemical signals released from the food (Parzefall 1993). A very good indication of the important interaction of rheotaxis and olfaction is the finding that

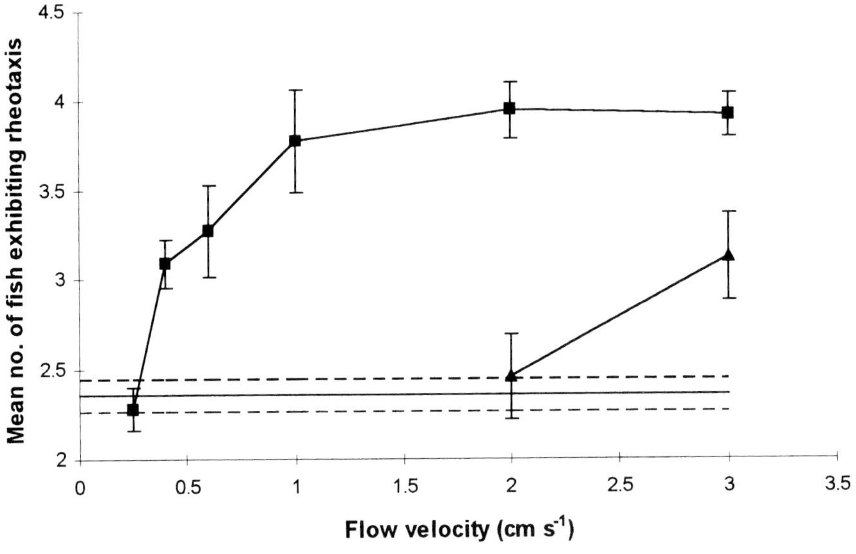

Figure 4. Mean number of *A. fasciatus* exhibiting rheotaxis in the absence of an odor stimulus (■) and in the presence of a dilute odour stimulus (♦). Note that the control condition shows a rheotaxis threshold of between 2 and 3, whereas the presence of the diffuse odor lowers the threshold to less than 0.4 cm s^{-1}.

the presentation of a diffuse food odor produces a marked reduction in the rheotactic threshold of *Astyanax* (Figure 4) (Baker & Montgomery 1999). This observation also serves as a warning that the lowest flow velocities that produce an unconditioned rheotaxis, are clearly not the lowest flow velocities that can be detected by the fish. Olfactory search is also likely to be important in reproductive interactions. Species specific pheromones have been demonstrated in *Poecilia mexicana* (Zeiske 1971, quoted in Parzefall 1993) and are likely to be found in other cave species.

Invertebrate prey are the main food source for amblyopsid cave fish (Poulson 1963). Animate prey present quite a different set of problems with respect to food localization and capture. They are often small, can be quite fast moving and capable of avoiding capture. The active swimming movements of invertebrates form a potent stimulus to the lateral line system. Lateral line feeding on planktonic prey has not been extensively studied in cave fishes, but is likely to be similar to the feeding behaviour demonstrated in number of other fish including detection of daphnia by mottled sculpin (Hoekstra & Janssen 1985) and marine zooplankton by the Antarctic fish *Pagothenia borchgrevinki* (Montgomery & Macdonald 1987, Montgomery 1989). The swimming movements of planktonic prey produce a range of higher frequency water movements that are particularly suited to detection by the lateral line canal system.

Although there is no a priori reason why superficial neuromasts could not participate in the localization and capture of small prey, there are several lines of evidence to indicate that this ability is mediated by lateral line canal organs in at least one species of fish – the Lake Michigan mottled sculpin, *Cottus bairdi*. Mottled sculpin respond to both live (e.g. *Daphnia*) and artificial (e.g. vibrating sphere) prey with an unconditioned orienting and feeding response (Hoekstra & Janssen 1985, 1986). By varying the amplitude of artificial prey vibration over a wide range of frequencies (10–200 Hz), Coombs & Janssen (1990) determined that threshold levels of detection, measured with the orienting response of the mottled sculpin, were very similar to threshold levels of spike rate increments measured extracellularly from canal neuromast fibers, but dissimilar to threshold levels from superficial neuromast fibers. Furthermore, when canal neuromasts were selectively damaged with gentamicin, the orientation of sculpin to artificial prey was reduced to spontaneous levels of response and the time course of recovery to pre-treatment levels (70–100%) followed the time course of hair cell regeneration in canal neuromasts, as determined from scanning electron microscopy (Coombs et al. 2000). Conversely, when superficial neuromasts were selectively damaged, response levels

Table 1. Reaction distances of three amblyopsid species to 2 types of invertebrate prey (data from Poulson 1963).

	Chologaster agassizi	*Typhlichthys subterraneus*	*Ambloysis splelaea*
Habitat	Troglophile	Hypogean	Hypogean
Hyalella	20–30 mm	40–50 mm	30–45 mm
Adult *Daphnia magna*	10 mm	30–40 mm	20–30 mm

were maintained at pre-treatment levels. These results were independent of whether prey vibration was within the frequency range of best sensitivity for superficial (10 Hz) or canal neuromasts (50 Hz).

Poulson (1963) gives some indication of the effective range of lateral line mediated prey detection in amblyopsid cave fishes (Table 1). Evident in this data is the increased reaction distance in the hypogean species when compared with their troglophilic relatives. Avoidance conditioning experiments also indicate an increased sensitivity of the lateral line of *Astyanax* when compared with goldfish (Weiss & Martini 1970). These findings are entirely consistent with what would be predicted from the longer cupula reported for hypogean vs. epigean species (Poulson 1963, Teyke 1990).

Perhaps the best studied and most interesting aspect of lateral line mediated behavior in cave fish is their ability to detect stationary objects, to avoid obstacles and to build up a 3-dimensional 'image' of their environment. This ability has been termed 'hydrodynamic imaging' (Hassan 1989) and it is an active use of the lateral line system comparable with, and analogous to, bat echolocation and electrolocation in weakly electric fish. When a fish glides through open water, a flow field is set up around the fish's body. Water displaced at the head flows back past the fish and is drawn back into the space vacated by the body as it moves forward (Figure 5). When the fish glides in the vicinity of an obstacle, the flow field will be altered in a characteristic way that depends on the geometry and dimensions of the obstacle. The essentials of this hypothesis were first proposed by Dijkgraaf (1963) and later applied to blind cavefish by von Campenhausen et al. (1981).

Teyke (1985) provided behavioral support for the hypothesis by showing that *Astyanax* introduced to new surroundings maintain a high swimming velocity for the first 24 h after which their swimming velocity decreases. The interpretation of this observation is that the higher swimming velocity is an exploratory behavior, where the increased velocity leads to an increase in the amplitude of the stimulus to the lateral line. One other observation in these experiments was that collisions that occurred with stationary objects were correlated with active tail movements; the suspicion being that active swimming movements may activate the efferent system and actively decrease the sensitivity of the lateral line. In other words, the lateral line is likely to be more sensitive during gliding movements, and during gliding movements the fish is more sensitive to the alterations in flow generated by surrounding objects.

Hassan (1986, 1989) further developed the hypothesis by both behavioral and theoretical analyses. He was able to show that in spatial discrimination tasks *Astyanax* was able to differentiate between girds of equidistant vertical bars when the difference between the bar spacing amounted to only about 1.5 mm. Hassan's (1992a,b) elegant mathematical analyses also clearly demonstrated the kind and quality of the information available to the lateral line for different objects, at different distances as a function of glide speed of the fish (Figure 5).

Further predictions of the 'hydrodynamic imaging' hypothesis were confirmed by Teyke (1988). He showed that individual cave fish prefer particular swimming velocities, and that the preferred velocity depends on the cross-sectional area of the fish. Fish with a smaller cross-sectional area had a higher swimming velocity. As implied above, the higher swimming velocity is important for generating the lateral line stimulus by which the fish explores its surroundings. Smaller fish must swim faster to generate an equivalent lateral line stimulus. The implication of these observations is that swimming velocity is under active control to optimize the information available to the fish.

Hassan et al. (1992) directly tested this contention by decreasing the sensitivity of the lateral line with the divalent ions Ca^{++} and Co^{++}. Partial chemical blockade of the lateral line resulted in the predicted compensatory increase in swimming velocity. It is interesting to note that the swimming behavior of blinded goldfish is similar to that of the cave fish (Teyke 1988). Blinded goldfish swim continuously whereas sighted fish generally hover, or move intermittently. Although there is no empirical evidence that blinded goldfish can form active hydrodynamic images of their environment, this change in behavior suggests that they might. However, unlike blind cavefish, blinded goldfish do not decrease their swimming velocity with increasing familiarity

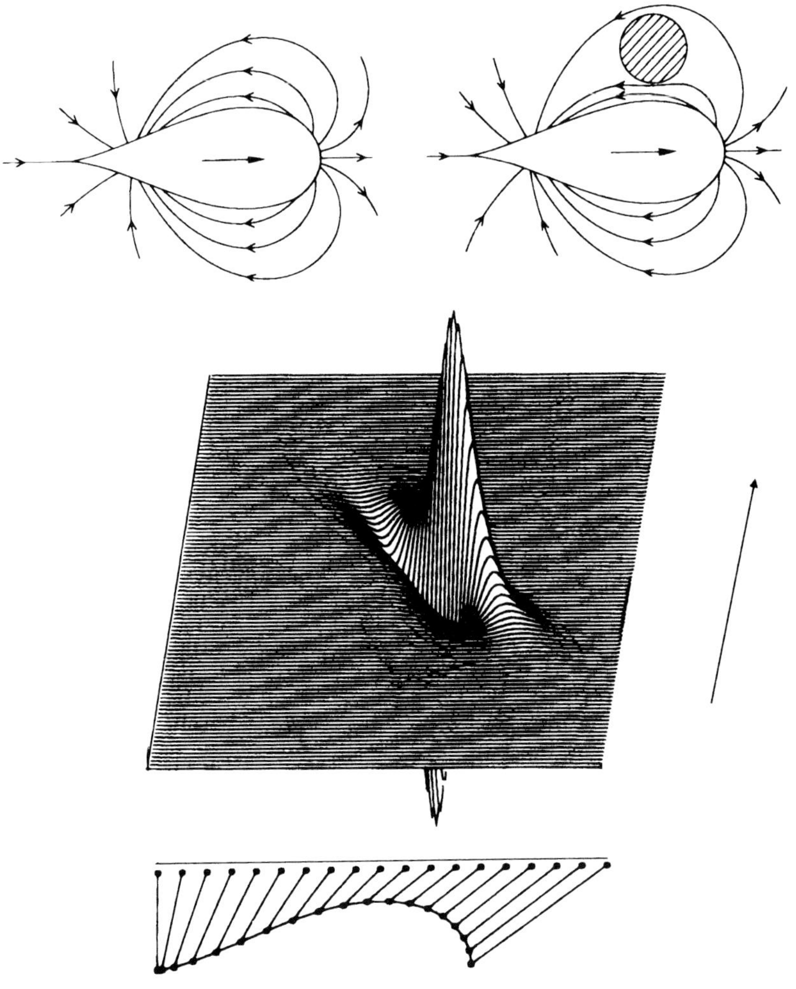

Figure 5. The physical basis of hydrodynamic imaging (from Hassan 1989). The top panel illustrates the flow field generated by a gliding fish and the distortion in that flow field generated as the fish passes a nearby object. The three dimensional graph is derived from Hassan's model of the pressure gradient pattern generated on the surface of the fish. The lateral profile of the fish is mapped onto the x-axis of the graph. Time progresses from the front to the back of the graph, such that the lateral flank of the fish passes the obstacle at approximately the middle of the time axis. The vertical height of the graph is the calculated pressure gradient at each point on lateral surface of the fish. The main purpose of the figure is to illustrate the strong pressure gradient signal that exists on the lateral flank of the fish as the fish glides past the object.

with their surroundings (Teyke 1988), suggesting that if they do form active images, they don't do it as well.

Teyke (1989) used the elevated swimming behavior of cave fish in novel environments to examine their recollection of previously experienced locations. He found that fish removed from a familiar environment for periods of up to 2 days spent less time exploring when returned than when they were initially placed in the environment. These data support the idea that during lateral line exploration, blind cave fish develop an internal (cognitive) map of their environment that contributes to subsequent spatial orientation. Although hydrodynamic imaging may not be the sole prerogative of blind cave fish, these fish appear to have evolved rather sophisticated mechanisms for processing and using the images.

In all of the early work on the flow-field hypothesis it had been assumed that the large numbers of superficial neuromasts found on *Astyanax* would be

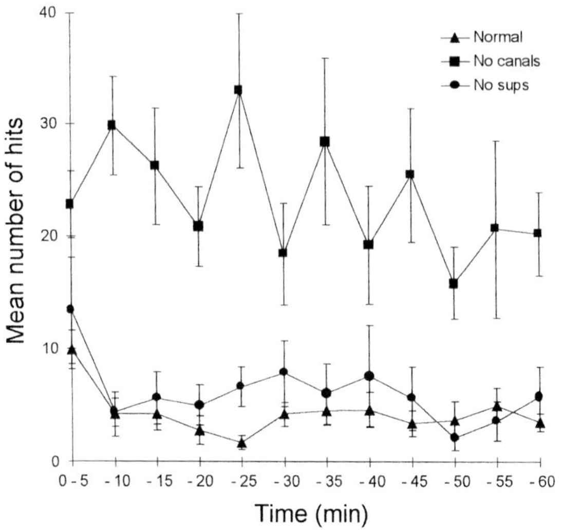

Figure 6. Mean number of hits during 5 minute intervals of a grid partition by 3 groups of 5 cavefish under different experimental treatments. Normal control fish (Normal); fish exposed to gentamicin sulfate causing block of the lateral line canal system (No canals) (procedure from Song et al. 1997); fish with physical ablation of their superficial neuromast subsystem of the lateral line (No sups) procedure from Baker & Montgomery 1999).

involved in hydrodynamic imaging. It was somewhat of a surprise to find that destruction of the canal system impaired the ability of *Astyanax* to discriminate spatial structure (Abdel-Latif et al. 1990). Without functional canal organs the fish swam more slowly, less often along the walls, and did not discriminate between a grid and no-grid pattern. This result has recently been confirmed in our own laboratory (Figure 6). Using an electronic contact detection devise it was found that intact *Astyanax* initially collided with a grid partition introduced into their tank, but after a short interval (<10 min) the number of collisions dropped to a sustained low level. *Astyanax* with physical ablation of the superficial neuromasts were not different from controls whereas gentamicin treated fish (gentamicin specifically disables canal neuromasts) had much higher initial and sustained rates of contact.

It would be nice to conclude from the above that there is a clear division of labour between the two submodalities of the lateral line system. That canal neuromasts are responsible for hydrodynamic imaging of stationary objects in the environment and for the localization of small prey items and that superficial neuromasts are more concerned with larger scale flows over the surface of the body. However, this functional and behavioral division of labor may not completely hold up. One other surprising finding of the Abdel-Latif et al. (1990) study was that the localization of a small vibrating bead was unaffected by the canal ablation. It is possible that the inner ear, rather than superficial neuromasts, could have been responsible for the blind cavefish's responses to the vibrating source. Nevertheless, it is probably too simplistic to view the superficial system and the canal system as having non-overlapping functions. From what is known in other fishes, the essential difference at the periphery is that the canals should be biased towards higher frequencies. The behavioral use of the information will depend on how the information from both canal and superficial neuromasts is preserved, integrated and analysed by the central nervous system.

Conclusion

We have to some extent answered the questions posed at the beginning of this paper. The mechanosensory lateral line is a distributed, hair-cell based system which detects the water flow regime at the surface of the fish. Superficial neuromasts densely scattered over the surface of some cave fish detect the pattern of flow over the surface of the body and are important in rheotactic behaviors and perhaps in the localization of small vibrating sources. Canal neuromasts are very likely also involved in the detection of small planktonic prey, but seem also to play an essential role in active hydrodynamic imaging. None of the lateral line behaviors exhibited by blind cave fish are necessarily exclusive to these species, but there is some evidence that their lateral line capabilities are enhanced with respect to their sighted relatives. For example, lateral line prey detection distances are increased, and there is some indirect evidence that cave-fish may have more sophisticated mechanisms for generating and processing hydrodynamic images.

There is still much to learn about lateral line systems, lateral line based capabilities and particularly the way in which mechanosensory information from the periphery is processed by the central nervous system. Most of the classical sensory neuroethology has been done on animals with extreme sensory capabilities, such as bats, barn owls and electric fishes. We can anticipate that cave fishes could have an important role to play in furthering our understanding of the mechanosensory lateral line system. It is equally the case that the lateral line is central to developing our understanding of many aspects of the biology of hypogean fishes.

Acknowledgements

The authors wish to dedicate this chapter to El-sayed Hassan.

References cited

Abdel-Latif, H., E.S. Hassan & C. von Campenhausen. 1990. Sensory performance of blind cave fish after destruction of the canal neuromasts. Naturwiss. 77: 237–239.

Baker, C.F. & J.C. Montgomery. 1999. The sensory basis of rheotaxis in the blind Mexican cave fish, *Astyanax fasciatus*. J. Comp. Physiol. 184A: 519–527.

Campenhausen, C. von, I. Riess & R. Weissert. 1981. Detection of stationary objects by the blind cave fish *Anoptichthys jordani* (Characidae). J. Comp. Physiol. 143A: 369–374.

Coombs, S., J. Janssen & J. Webb. 1988. Diversity of lateral line systems: evolutionary and functional considerations. pp. 553–593. *In*: J. Atema, R.R. Fay, A.N. Popper & W. N. Tavolga (ed.) Sensory Biology of Aquatic Animals, Springer-Verlag, NewYork.

Coombs, S. & J. Janssen. 1990. Behavioral and neurophysiological assessment of lateral line sensitivity in the mottled sculpin, *Cottus bairdi*. J. Comp. Physiol. 167A: 557–567.

Coombs, S. & J.C. Montgomery. 1999. The enigmatic lateral line system. pp. 319–362. *In*: R.R. Fay & A.N. Popper (ed.) Comparative Hearing: Fish and Amphibians, Springer Handbook of Auditory Research, Springer-Verlag, New York.

Coombs, S., C.B. Braun & B. Donovan. 2000. Orienting response of mottled sculpin is mediated by canal neuromasts. J. Exp. Biol. (in press).

Denton, E.J. & J.A.B. Gray. 1988. Mechanical factors in the excitation of the lateral line of fishes. pp. 595–617. *In*: J. Atema, R.R. Fay, A.N. Popper & W. N. Tavolga (ed.) Sensory Biology of Aquatic Animals, Springer-Verlag, NewYork.

Dijkgraaf, S. 1963. The functioning and significance of the lateral-line organs. Biol. Rev. 38: 51–105.

Hassan, E.S. 1986. On the discrimination of spatial intervals by the blind cave fish (*Anoptichthys jordani*). J. Comp. Physiol. 159A: 701–710.

Hassan, E.S. 1989. Hydrodynamic imaging of the surroundings by the lateral line of the blind cave fish *Anoptichthys jordani*. pp. 217–228. *In*: S. Coombs, P. Gorner & H. Munz (ed.) The Mechanosensory Lateral Line Neurobiology and Evolution, Springer-Verlag, New York.

Hassan, E.S. 1992a. Mathematical description of the stimulus to the lateral line system of fish derived from a three-dimensional flow field analysis. I. The cases of moving in open water and of gliding towards a plane surface. Biol. Cybern. 66: 443–452.

Hassan, E.S. 1992b. Mathematical description of the stimulus to the lateral line system of fish derived from a three-dimensional flow field analysis. II. The case of gliding alongside or above a plane surface. Biol. Cybern. 66: 453–461.

Hassan, E.S., H. Abdel-Latif & R. Biebricher. 1992. Studies on the effects of Ca^{++} and Co^{++} on the swimming behaviour of the blind Mexican cave fish. J. Comp. Physiol. 171: 413–419.

Hensel, K. 1978. Morphology of lateral-line canal system of the genera *Abramis*, *Blicca* and *Vimba* with regard to their ecology and systematic position. Acta Univ. Carolinae, Biologica 12: 105–153.

Hensel, K. 1986. Morphologie et interpré tation des canaux et canalicules sensorieles céphaliques de *Latimeria chalumnae* Smith, 1939 (Osteichthyes, Crossopterygii, Celacanthiformes). Bull. Mus. Natn. Hist. Nat. Paris 4e seŕ. A2: 379–407.

Hoekstra, D. & J. Janssen. 1985. Non-visual feeding behavior of the mottled sculpin, *Cottus bairdi*, in Lake Michigan. Env. Biol. Fish. 12: 111–117.

Hoekstra, D. & J. Janssen. 1986. Non-visual feeding behavior of the mottled sculpin (*Cottus bairdi*). Copeia 1986: 91–96.

Montgomery, J.C. 1989. Lateral line detection of planktonic prey. pp. 561–574. *In*: S. Coombs, P. Gorner & H. Munz (ed.) The Mechanosensory Lateral Line Neurobiology and Evolution, Springer-Verlag, New York.

Montgomery, J.C., C.F. Baker & A.G. Carton. 1997. The lateral line can mediate rheotaxis in fish. Nature 389: 960–963.

Montgomery, J.C., S. Coombs & J. Janssen. 1994. Aspects of structure and function in the anterior lateral line of six species of antarctic fish of the suborder Notothenioidei. Brain Behav. Evol. 44: 299–306.

Montgomery, J.C., C.E. Diebel, M.B.D. Halstead & J. Downer. 1999. Olfactory search tracks in the antarctic fish *Trematomus bernacchii*. Polar Biol. 21: 151–154.

Montgomery, J.C. & J.A. Macdonald. 1987. Sensory tuning of lateral line receptors in antarctic fish to the movements of planktonic prey. Science 235: 195–196.

Parzefall, J. 1993. Behavioural ecology of cave-dwelling fishes. pp. 433–458. *In*: T. Pitcher (ed.) Behaviour of Teleost Fishes, 2nd ed., Chapman & Hall, London.

Poulson, T.L. 1963. Cave adaptation in amblyopsid fishes. Amer. Midl. Nat. 70: 257–290.

Puzdrowski, R.L. 1989. Peripheral distribution and central projections of the lateral line nerves in goldfish, *Carassius auratus*. Brain Behav. Evol. 34: 110–131.

Romero, A. 1984. Behavior in an 'intermediate' population of the subterranean-dwelling characid *Astyanax fasciatus*. Env. Biol. Fish. 10: 203–207.

Schemmel, C. 1967. Vergleichende Untersuchungen an den Hautsinnesorganen ober- und unterirdisch lebender *Astyanax*-Formen. Z. Morph. Tiere 61: 255–316.

Song, J., H.Y. Yan & A.N. Popper. 1995. Damage and recovery of hair cells in fish canal (but not superficial) neuromasts after gentamicin exposure. Hear. Res. 91: 63–71.

Teyke, T. 1985. Collision with and avoidance of obstacles by blind cave fish *Anoptichthys jordani* (Characidae). J. Comp. Physiol. 157A: 837–843.

Teyke, T. 1988. Flow field, swimming velocity and boundary layer: parameters which affect the stimulus for the lateral line organ in blind fish. J. Comp. Physiol. 163A: 53–61.

Teyke, T. 1989. Learning and remembering the environment in the blind cave fish (*Anoptichthys jordani*). J. Comp. Physiol. 164A: 655–662.

Teyke, T. 1990. Morphological differences in neuromasts of the blind cave fish *Astyanax hubbsi* and the sighted river fish *Astyanax mexicanus*. Brain Behav. Evol. 35: 23–30.

Weiss, B.A. & J.L. Martini. 1970. Lateral-line sensitivity in the blind cavefish (*Anoptichthys jordani*). J. Comp. Physiol. Psych. 71: 34–37.

Environmental Biology of Fishes **62:** 97–105, 2001.
© 2001 *Kluwer Academic Publishers. Printed in the Netherlands.*

Metabolic rate and natural history of Ozark cavefish, *Amblyopsis rosae*, in Logan Cave, Arkansas

Ginny L. Adams[a] & James E. Johnson
Arkansas Cooperative Fish and Wildlife Research Unit, Department of Biological Sciences, University of Arkansas, Fayetteville, AR 72701, U.S.A.
[a]*Present address: Department of Zoology and Center for Systematic Biology, Southern Illinois University, Carbondale, IL 62901-6501, U.S.A. (e-mail: gadams@siu.edu)*

Received 3 April 2000 Accepted 13 December 2000

Key words: Amblyopsidae, seasonal metabolism, physiology, oxygen consumption

Synopsis

We investigated the effects of mass, season, and activity on oxygen consumption of the federally threatened Ozark cavefish, *Amblyopsis rosae*, at Logan Cave National Wildlife Refuge, Benton County, Arkansas. We used an acclimatization protocol to measure in-situ metabolic rates. There was a significant effect of mass on the metabolic rate of Ozark cavefish and this relationship differed significantly among seasons. We observed a positive relationship between mass and oxygen consumption during summer and autumn but a negative relationship for winter and spring. There was a 1°C water temperature change between summer/autumn sampling and winter/spring sampling, but it is not likely temperature alone accounted for seasonal variation in mass-oxygen consumption relationships. Activity in the respirometer did not vary significantly among seasons and was not correlated with fish mass or time of day. Seasonal relationships between mass and oxygen consumption may therefore reflect alterations in environmental conditions (i.e. food availability, ambient dissolved oxygen), condition, changes in susceptibility to handling stress, or may reflect low numbers of fish tested. Natural history observations were made throughout the study. Four females were seen in late August with ova visible in the body cavity. On 20 June 1996 we first observed five small cavefish (approximately 10 mm TL) in a pool just upstream of the sinkhole entrance that were likely less than a month old. One large adult (55 mm TL) was seen in the same pool with the young fish for six weeks, until two small fish were displaced by a storm event. Although this does not provide a clear sign of parental care, it does suggest that cannibalism is not always the rule for this species as previously suggested.

Introduction

Cave adaptation of organisms is a highly convergent event that has long fascinated biologists (Christiansen 1962, Poulson 1963). It is characterized by troglomorphy, defined as a lack of or reduction of eyes and pigmentation, increased size and numbers of sensory organs, elongated appendages, and metabolic rates lower than epigean (surface) relatives (Christiansen 1962). Cave colonization by fishes has occurred in approximately 90 species from 14 families, representing mostly freshwater fish (Helfman et al. 1997). Along with alterations in morphology and physiology (Poulson & White 1969, Caine 1978), cave-adapted fishes have exhibited variation in reproduction, behavior (mostly those associated with vision), and circadian rhythms (Wilkens 1988) when compared to epigean relatives.

Many studies on oxygen consumption of cave-adapted species have compared hypogean (cave) species to a closely related epigean relative (see Huppop 1986 for review). The basic conclusion is that hypogean species have lower metabolic rates compared to epigean species. It is hypothesized this reduction may be a result of alterations in food availability (Huppop 1986) or ambient dissolved oxygen (Caine 1978, Gannon et al. 1999). Poulson (1963) found metabolic rates (standard, routine, and active) decreased with increasing cave adaptation within Amblyopsidae.

Epigean organisms in temperate regions experience seasonal fluctuations in environmental parameters such as photoperiod, temperature, and food availability. These parameters affect metabolic rates of organisms and often serve as cues for reproduction, migration, and hibernation. Temperature has long been considered the most important abiotic factor governing physiological variation in organisms. However, physiological variation without temperature change has been well documented in fishes (Beamish & Mookherjii 1964, Fry 1971, Meakins 1975, Evans 1984, Facey & Grossman 1990) and has been related to changes in photoperiod and spawning (Wells 1935, Beamish 1964, Withey & Saunders 1973, Pandey 1976, Dickson & Kramer 1971).

To our knowledge, no studies have compared seasonal variation in metabolic rate within any troglobitic species. Poulson and White (1969) hypothesized that although caves lack or experience reduced seasonal cues, seasonal periodicity (endogenous annual cycles) exist for the reproductive cycle of some troglobitic organisms. Poulson (1963) related egg-laying in Amblyopsidae to spring floods that cause a shift in temperature and increased food availability.[1] If there are seasonal alterations in environmental conditions in cave waters, and cavefish respond to these cues to determine onset of reproduction, it is feasible to assume there will be a concurrent alteration in physiology.

Our research was conducted to obtain information on natural history, metabolism, and condition of Ozark cavefish in Logan Cave, Arkansas. We addressed the following questions: (1) Does routine metabolic rate change seasonally for Ozark cavefish? (2) Does condition (length/weight) vary seasonally? Ozark cavefish in northwest Arkansas inhabit cave streams that experience numerous potential threats within their recharge areas. Increased human population growth and associated construction may further deteriorate habitat available to this species. Due to the current protected status of the species (USFW[2]), any data pertaining to natural history should provide useful information for its future management and recovery.

Species adapted to stable cave environments (i.e., lack of photoperiod, reduced temperature change) may not be able to endure large shifts in environmental parameters (Culver 1982). Poulson and White (1969) suggested that troglobitic fishes might exhibit increased sensitivity to changes in dissolved oxygen and temperature when compared to epigean species, possibly making them more susceptible to some of the anthropogenic disturbances from agricultural runoff and related land-use practices. However, addition of nutrients (nitrates, phosphates) to subterranean streams may not induce the same response as in epigean streams. Lack of light in caves prevents photosynthesis, so nutrients may become directly deleterious to organisms rather than increasing productivity with its associated ills.

Study species

The family Amblyopsidae is comprised of five genera and six species found in the eastern United States (Woods & Inger 1957, Lee et al. 1980, Berra 1981, Robins et al. 1991). The Ozark cavefish, *Amblyopsis rosae*, was listed as a threatened species in December 1984 by U.S. Fish & Wildlife Service. Ozark cavefish inhabit several caves in northwest Arkansas, southwest Missouri and northeast Oklahoma (Willis & Brown 1985, Robison & Buchanan 1988, Pflieger 1997). Willis and Brown (1985) believed the range of the species was declining but noted that data were limited.

Poulson (1961, 1963) provided extensive biological information on amblyopsids. General family characteristics include small size (maximum length of 90 mm), absent or reduced pelvic fins, well-developed sensory organs, and a jugular anus (Cooper & Kuehne 1974, Robison & Buchanan 1988). *A. rosae* is described as one of the most specialized amblyopsid (Poulson 1963), characterized by light pink coloration, vestigial eyes, and a maximum length of 65 mm (Means & Johnson 1995). Food preference for *A. rosae* is reported as small crayfish, isopods, copepods, ostracods, salamander larvae, and young cavefish (Poulson 1961, Poulson 1963). Reproduction in Ozark cavefish

[1] Poulson, T.L. & P.M. Smith. 1969. The basis for seasonal growth and reproduction in aquatic cave organisms. Proc. 4th Int. Congr. Speleol. Lljubjana 4–5: 197–201.

[2] U.S. Fish & Wildlife Service, 1989. Ozark cavefish recovery plan. Atlanta. 30 pp.

has not been described, but Poulson (1963) speculated *A. rosae* followed the same pattern as *A. spelaea* in both reproductive output and breeding time. Poulson (1963) found only 10% of reproductively mature *A. spelaea* spawned in a given year and suggested that energy limited reproductive output of any individual to a single effort, with some members never reproducing. *Amblyopsis spelaea* are reported to carry eggs and embryos in the gill cavity before hatching (Eigenmann 1898, Poulson 1961).

Materials and methods

Study site

Logan Cave National Wildlife Refuge is located in Benton County, Arkansas (Gallaton Quadrangle T. 18N, R. 32W, Sec. 33) in the Mississippian cherty – limestone Boone Formation of the Springfield Plateau and is part of the Ozark Highlands[3]. The subterranean portion of Logan Cave Stream flows through the cave about 1300 m to the cave mouth (Means & Johnson 1995). About 365 m upstream from the cave mouth, a sinkhole provides a second entrance into Logan Cave. The sinkhole provides exogenous input of organic carbon through rainfall and leaf litter into the lower reach of the cave stream[4].

Aley & Aley (unpublished observations) described the recharge area for Logan Cave Stream as karst topography with an area of 26 km^2 and classified the system as a rapid–flow through discrete aquifer. Means (1993) noted several sinking streams in the area. The recharge area for Logan Cave supports extensive and intensive agriculture including pastures and concentrated poultry rearing, uses that affect both water quality and rate of inflow from discrete sinking streams.

Logan Cave harbors the second largest known population of Ozark cavefish (Means & Johnson 1995) and largest known concentration of endangered cave crayfish, *Cambarus aculabrum* (Hobbs & Brown 1987). Other stygobites in Logan Cave include an isopod *Caecidotea antricola* and an amphipod *Stygobromus ozarkensis*. Troglophiles that inhabit the cave include grotto salamander *Typhlotriton spelaeus*, cave salamander *Eurycea lucifuga*, and dark-sided salamander *Eurycea longicauda melanopleura*. Logan Cave supports a maternity colony of approximately 20 000 endangered gray bats, *Myotis grisescens* (M.J. Harvey personal communication); eastern pipistrel *Pipistrellis subflavus* and endangered Indiana *Myotis sodalis* bats also inhabit the cave in smaller numbers.

Water quality of Logan Cave Stream has been measured over the past several years (Means 1993, Brown 1996). Temperature remains relatively constant throughout the year with a mean of 14.7°C, fluctuating approximately 2°C (Means 1993). Rapid temperature fluctuations of up to 2°C are associated with warm rainfall events in summer and cold rain and snowmelt during winter. During warmer months, coliform bacteria and organic debris increase, due in part to presence of bats in the cave, but also related to rapid percolation of water carrying agricultural wastes (Means 1993).

Metabolism

Research was conducted during each season beginning in summer 1996. *Amblyopsis rosae* were collected by dipnet, just prior to sunset, to minimize disturbance to bats. To reduce stress and handling time, individual cavefish were placed directly into holding chambers that also served as test chambers. Holding chambers consisted of 250 ml glass jars submerged in the stream and covered with a mesh screen (1.0 mm square mesh) to provide oxygen while limiting food availability. Individuals were held a minimum of 12 h prior to testing to induce a postabsorpative state and allow habituation to the chamber.

Preliminary trials were conducted on surrogate species in the laboratory to perfect techniques and reduce impact to *A. rosae*. Initial techniques were then refined using a captive *A. rosae* held at the University of Arkansas. Trials determined the approximate time needed to detect a reliable change in dissolved oxygen concentration within the test chamber, and were considered routine respiration rate since chamber size did not restrict movement of the organism, nor was activity forced (Fry 1957, Winberg 1960). Test organisms remained fairly quiescent throughout the trials but some spontaneous activity did occur and was recorded.

After the habituation period, routine respiration rate was measured by fitting an air-tight rubber stopper to the holding chamber. Ports in the stopper allowed entry of a polarographic oxygen electrode to measure in-situ oxygen depletion and a stirrer to circulate water

[3] Haley, B.R. 1976. Geologic map of Arkansas. Arkansas Geological Commission, Little Rock, 1 sheet.

[4] Brown, A.V., W.K. Pierson & K. B. Brown. 1994. Organic carbon resources and the payoff-risk relationship in cave ecosystems. 2nd Int. Conf. Groundw. Ecol. March 1994: 67–76.

in the chamber. A mesh screen around the stirrer protected cavefish from harm. The oxygen electrode was connected to an Orion 810 oxygen meter that was in turn attached to a laptop computer. Dissolved oxygen measurements were recorded at 30-sec intervals. Individuals were tested in the dark except when activity was measured visually at five-min intervals. Activity measures were taken to provide a more accurate estimate of routine metabolic rates. Pectoral-fin-beat and tail-fin-beat counts every 30 sec were used as activity measures and gill-beat-frequency was recorded when possible. To ensure constant temperature throughout the experiment, the test chamber remained submerged in the cave stream. It was necessary to test *A. rosae* for approximately four hours to obtain a reliable accuracy.

After a trial, length and weight measurements were collected for each individual. Condition was calculated for each individual cavefish tested. Condition measurements provided seasonal comparisons of relative health of individuals within a population, which may in turn affect oxygen consumption. The equation $\log_{10} TW = a + b(\log_{10} TL)$ was used, where a is the Y intercept, b is the slope, TW is the total weight, and TL is the total length (Bogler & Connolly 1989, Trippel & Hubert 1990).

Feces from *A. rosae* often accumulated in the acclimation chamber before testing began; these were removed after testing and preserved for later analysis. During tests, a small rock was placed in the jar to provide cavefish refuge from the minute current produced by the stirrer. Volume of the rock, stirrer and probe was subtracted from volume of the chamber. After testing, *A. rosae* were returned to their site of collection. A blank of stream water was tested to determine background bacterial respiration and respiration rates of cavefish were adjusted accordingly. Also, background respiration was determined with the rock and feces (when present) after measurements were made on *A. rosae*.

All measurements were conducted during daylight hours to eliminate any diurnal effect on respiration rate. Measurements within a season were taken within a window of 40 to 60 days to reduce intra-seasonal variance in respiration rates. The tests were conducted during the following dates each season and with the following sample sizes: Summer (17 July to 27 August, n = 5), Autumn (5 October to 11 December, n = 6), Winter (16 January to 25 February, n = 5), and Spring (15 April to 11 May, n = 6).

Data analysis

Analysis of covariance (ANCOVA) was used to assess the effects of season on oxygen consumption, with mass as the covariate and used to determine effect of season on mass, with length as the covariate. Oxygen consumption, mass data and length were log-transformed. Assumptions of ANCOVA including normality, homogeneity of variance, and homogeneity of slopes were tested. Means in the text are reported as log-transformed oxygen consumption adjusted for mass or as relative respiration rates (mg $O_2\,kg^{-1}\,h^{-1}$) for purposes of comparison. Significant pairwise differences for season were detected using Tukey HSD tests (alpha = 0.05). Analysis of variance (ANOVA) was used to compare activity measures among seasons.

Results

Water temperature was constant for tests conducted within a season (Summer and Autumn 14.7°C; Winter and Spring 13.7°C) and a 1°C change was recorded between summer/autumn and winter/spring testing. There was not a significant difference in mean pelvic fin beat (PFB) (ANOVA, $F_{3,17} = 0.446$, $p = 0.723$) or caudal fin beat (CFB) (ANOVA, $F_{3,17} = 0.247$, $p = 0.862$) among seasons (Table 1). There was no significant correlation between time of day and PFB or CFB and no significant correlation between measures of activity and oxygen consumption. Although PFB and CFB were not good indicators of the amount of time spent swimming, they did provide some measure of activity. Often fish would move under the rock and remain quiescent throughout a majority of the test. Movement of fins often occurred while the fish remained in a stationary position. Gill-beat frequency could only be obtained for five fish during the study and was not used for statistical comparisons.

Table 1. Mean pelvic fin beat (PFB) and tail fin beat (TFB) for Ozark cavefish in Logan Cave, Arkansas.

Season	Mean PFB (±2SE)	Mean TFB (±2SE)
Summer	4.98 (1.75)	2.70 (1.95)
Autumn	3.63 (3.58)	1.89 (1.58)
Winter	3.30 (3.06)	2.32 (1.62)
Spring	3.38 (0.52)	2.76 (0.79)

Metabolism and condition

Mass-specific respiration rates were plotted for each season and compared to data collected by Poulson (1963) (Figure 1). We found a significant effect of mass on metabolic rate of *A. rosae* (ANCOVA, $F_{1,18} = 5.87$, $p = 0.026$). However, there was a significantly different relationship between log mass and log metabolic rate among seasons (Test of Parallelism: $F_{3,15} = 4.466$, $p = 0.019$). Because of this, ANCOVA could not be employed to compare metabolism among seasons (Figure 2). However, there was no significant difference when we compared summer and autumn metabolic rates (ANCOVA, $F_{1,8} = 0.468$, $p = 0.513$) and winter and spring rates (ANCOVA, $F_{1,9} = 0.411$, $p = 0.537$). Although sex could not be determined based on external characters, three females with visible ova were tested during the summer season and no significant difference was found between fecund females with ova and other fish tested.

Mean length and weight values and condition are reported by season in Table 2. There was a significant effect of length ($F_{1,18} = 46.97$, $p < 0.001$) and season on mass ($F_{3,18} = 4.69$, $p < 0.01$) for *A. rosae*. Condition in summer and autumn seasons was significantly higher than winter and spring (Tukey HSD, $p < 0.01$).

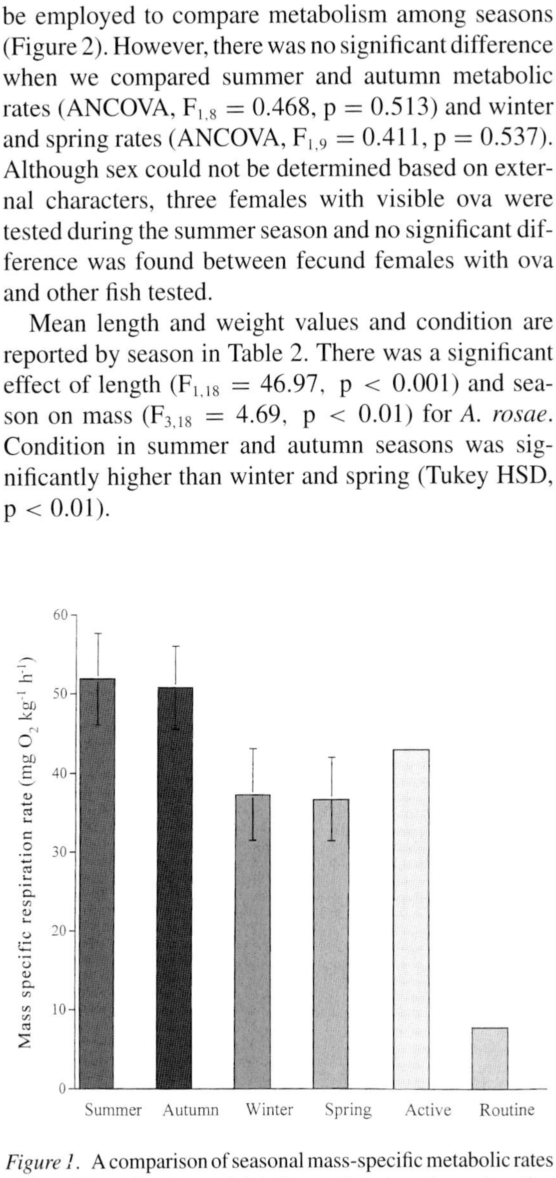

Figure 1. A comparison of seasonal mass-specific metabolic rates collected from Ozark cavefish in Logan Cave to active and routine rates collected by Poulson (1963).

Natural history

Between 22 and 25 August 1996, four female *A. rosae* were found with large, visible ova in their abdomen. The ova were observed by shining a light through the translucent body wall. Five eggs could be seen in the abdomen of one female, but none of the fecund fish was sacrificed to verify egg counts or ova size. One fecund female was located at the upstream end of the pool section of the cave (see map in Means & Johnson 1995) and three were found approximately 135 meters upstream of that location in a low-current area. All four fecund females were found over silt substrate in shallow water. Length of the fecund females ranged from 51.5 to 58.7 mm TL and all had a distinct yellow coloration (as opposed to the opaque white coloration usually reported). One female (56.8 mm TL) had an alphanumeric tag from a previous study (Brown & Johnson 2001) conducted in January 1994. In the 2.5 years since it was last captured, the total length of the fish increased 4.8 mm and it was found in the same pool section of the cave. Three of the fecund females were held overnight for physiological testing and were quiescent during physiological measurements.

On 20 July 1996, one large and five very small Ozark cavefish were found together approximately 15 m upstream of the sinkhole. These fish were found in a deep pool (> 3 m) and appeared to associate with each other. The small cavefish were approximately 10–11 mm TL. Although *A. rosae* has only vestigal eyes, two of the small cavefish approached and touched the dive light. These six cavefish remained as a group in this area for six weeks until a spate increased water levels in the cave stream. After water levels receded enough to allow entry back into the cave (4 days), two of the young cavefish had moved downstream a few meters while the rest remained at the same locality.

Brown (1996) found the majority of Ozark cavefish in Logan Cave in the large pool upstream of the sinkhole. Many of the cavefish collected during this study were found above this pool, near fast-flowing riffle areas of the cave. Cavefish were never seen actively swimming against current but rather moving laterally into and out of the current or drifting with it. Cavefish were often found in or just below riffle areas, behind rocks, maintaining station. Two cavefish were observed wedged between rocks in high-flow areas. When water levels rose in the cave, cavefish moved to the periphery and out of the main current.

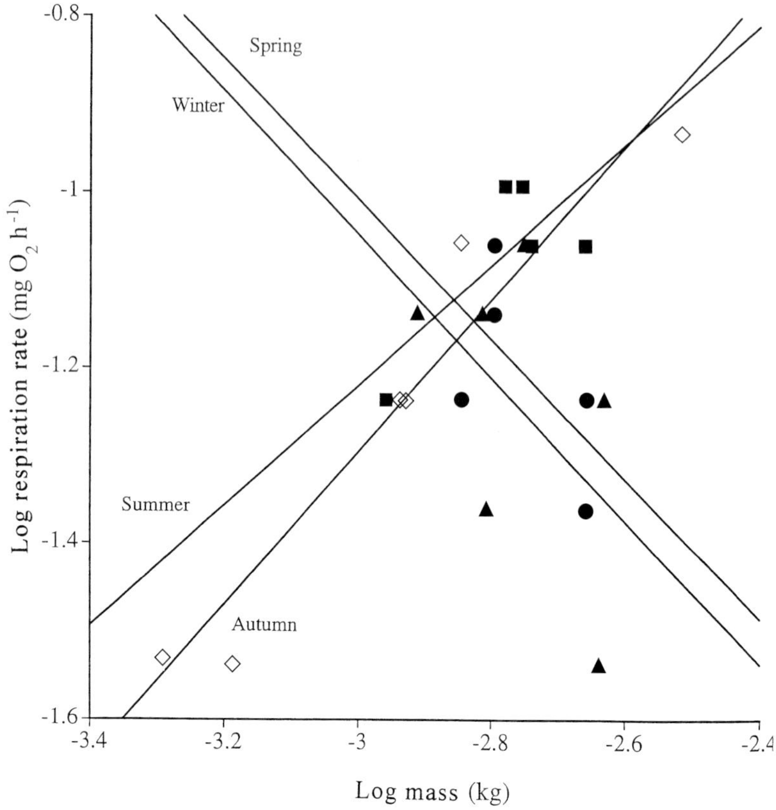

Figure 2. Seasonal relationships between log respiration rate and log mass for Ozark cavefish in Logan Cave. Equations and symbols are represented as follows: Summer (■) $y = 0.681x + 0.832$, $r^2 = 0.575$; Autumn (◇) $y = 0.876x + 1.333$, $r^2 = 0.914$; Winter (●) $y = -0.794x - 3.391$, $r^2 = 0.369$; Spring (▲) $y = -0.814x - 3.489$, $r^2 = 0.233$.

Table 2. Ozark cavefish weight, length, seasonal mass specific metabolic rates, and condition in Logan Cave, Arkansas.

Season	Sample size	Mean mass (g) (±2) SE	Mean total length (mm) (±2) SE	Mass specific metabolic rate (mg O_2 kg^{-1} h^{-1}) (±2) SE	Adjusted condition (adjusted log weight) (±2) SE
Summer	5	1.61 (0.34)	48.58 (5.78)	51.89 (11.6)	0.24 (0.05)
Autumn	6	1.70 (0.57)	49.45 (5.78)	50.81 (10.6)	0.25 (0.04)
Winter	5	1.64 (0.13)	58.34 (6.32)	37.32 (11.6)	0.15 (0.05)
Spring	6	1.61 (0.37)	56.09 (5.34)	36.75 (10.6)	0.01 (0.05)

Discussion

Metabolism and condition

Amblyopsis rosae residing in Logan Cave Stream exhibited a difference in the relationship between body mass and metabolic rate among seasons. The relationship was positive during the summer and autumn seasons and had a high R^2 value (0.575 and 0.914, respectively). The slopes of the lines for both summer and autumn are similar to those reported for other species (Withers 1992). Although, winter and spring slopes were negative, the R^2 values were low (0.369 and 0.253, respectively). If these negative slopes are not an artefact of low R^2 values and small sample sizes, they are the first reported for a fish species and may be a reflection of seasonal changes in the cave environment. Ozark cavefish lacked the important metabolic cue of photoperiod, but experienced varying seasonal fluctuations in food availability and temperature. The gray bat maternity colony present in Logan Cave from late March to early September likely increased nutrients in the cave environment. Brown and Johnson (2001) found most growth of *A. rosae* between April and October and related this to increased food availability and presence of bats. Improved condition observed during this study also correlates with presence of bats. Definite periods of growth and improved condition during the time bats were present likely affected the relationship between mass and metabolic rate.

Although temperature did not change drastically (<1°C) throughout the year, troglobitic organisms may be more sensitive to changes in temperature than epigean animals. Poulson and White (1969) found *A. rosae* exhibited a fourfold increase in metabolism in response to an acute temperature increase from 10°C to 15°C, suggesting that *A. rosae* react dynamically to rapid changes in water temperature. However, when *A. rosae* were given time to acclimate to different temperatures, there was only a 1.3 metabolic increase between 10 and 15°C (T.L. Poulson unpublished data). It seems unlikely that a 1°C increase in water temperature was solely responsible for the change in metabolism of Logan Cave cavefish, but it may have acted as a cue for change. Although water levels did not correlate directly with metabolic rate changes, they too may have acted, in conjunction with temperature, as a cue for a change in food availability. Water levels were highest during the spring season and lowest during winter.

Biases due to sex ratios may have also influenced seasonal metabolic rates since fish sex could not be determined. Other possible influences to metabolism include seasonal or size susceptibility to handling stress.

Poulson (1963) found *A. rosae* able to survive 401 days without food before spontaneous movement ceased, implying an ability of cavefish to greatly alter metabolic rate in response to food availability. Routine metabolic measurements taken during this study were higher than routine rates collected by Poulson (1963) (Figure 1) and more similar to his active rates. Differences may be attributed to variation in protocol between the two studies. Poulson's fish were held in the laboratory for many weeks and acclimated to an experimental temperature of 13 ± 1°C, while cavefish used in this study were field tested at ambient temperature. Also, Poulson held cavefish for 48 h without food before testing began as opposed to a maximum of 12 h in this study. The habituation period we used appeared to provide sufficient time for cavefish to reach postabsorpative state since feces were often collected during the habituation period but never during testing.

Natural history

Young fish (ca. 10 mm) were observed on 20 July and females with ovarian eggs in late August. Hatching of embryos in Ozark cavefish in Logan Cave is estimated to have occurred between June and August in 1996. This is earlier than cavefish reproduction reported by Poulson (1963) who found mature females in Cave Springs Cave in northwest Arkansas between August and December. Poulson (1963) estimated cavefish fecundity at a mean of 23 ova for the eight females he dissected. Cavefish in Logan Cave had a lower fecundity (five eggs that could be counted in one female), but we did not dissect fish and may have missed a few eggs blocked from view.

Eigenmann (1909) and Poulson (1963) hypothesized Ozark cavefish to be gill brooders, based on the discovery by both authors of eggs and young in the gill cavity of *A. spelaea*, but gill brooding has never been confirmed for *A. rosae*. Both Eigenmann and Poulson theorized the jugular anus found in all members of the family facilitated gill brooding behavior. Gill brooding was also reported for the closely related pirate perch, *Aphredoderus sayanus* (Boltz & Stauffer 1986) that also has a jugular anus, but other authors have reported nesting behavior for this species (Forbes & Richardson 1920, Katula 1992). The jugular location of the anus is the one character linking gill brooding behavior among species (Pflieger 1997), however movement of eggs from the anus to gill cavity has never been witnessed.

Over the past eight years of work in Logan Cave more than 200 cavefish have been handled, most of them repeatedly, without finding eggs in the gill chamber. It is possible gill brooding does not occur in *A. rosae* as previously suggested. Parental-young aggregations have not been witnessed in *A. rosae* until this study that found one large and five small fish together for more than six weeks. In fact, Poulson (1963) suggested adult cavefish may be cannibalistic on other young cavefish.

Poulson (1963) did not provide any information about habitat of adult females with mature ova; we report their use of slack water areas over silty substrate. Female cavefish with large ova appeared awkward while swimming and may have had difficulty occupying even low current habitats. During physiological testing we noted females with mature ova often appeared 'slow' and would move away from the minute current created by the stirrer. Gill-beat frequency was measured for two of these females and averaged 12 beats every 30 sec for one fish and 20 beats every 30 sec for the second. This average appears higher than the mean obtained for three other fish tested (10 beats every 30 sec) but low sample size made statistical tests unreliable.

Acknowledgements

We thank Arkansas Game and Fish Commission and the U.S. Fish and Wildlife Service for funding portions of this study and for collecting permits. We appreciate the help of the over thirty people who assisted in the collection of cave organisms. Helpful comments on the manuscript were provided by Steve Beaupre and J.V. Brahana. J.V. Brahana also shared water level data.

References cited

Beamish, F.W.H. 1964. Respiration of fishes with special emphasis on standard oxygen consumption II. Influence of weight and temperature on respiration of several species. Can. J. Zool. 42: 177–188.

Beamish, F.W.H. & P.S. Mookherjii. 1964. Respiration of fishes with special emphasis on standard oxygen consumption. Influence of weight and temperature on respiration of goldfish *Carassius auratus*. Can. J. Zool. 42: 161–175.

Berra, T.M. 1981. An atlas of distribution of the freshwater fish families of the world. University of Nebraska Press, Lincoln. 197 pp.

Bogler, T. & P.L. Connolly. 1989. The selection of suitable indices for the measurement and analysis of fish condition. J. Fish Biol. 34: 171–182.

Boltz, J.M. & J.R. Stauffer. 1986. Branchial brooding in the pirate perch, *Aphredoderus sayanus* (Gilliams). Copeia 1986: 1030–1031.

Brown, J.Z. 1996. Population dynamics and growth of Ozark cavefish in Logan Cave, Benton County, Arkansas. M.S. Thesis, University of Arkansas, Fayetteville. 87 pp.

Brown, J.Z. & J.E. Johnson. 2001. Population biology and growth of Ozark cavefish in Logan Cave National Wildlife Refuge, Arkansas. Env. Biol. Fish. 62: 161–169 (this volume).

Caine, E.A. 1978. Comparative ecology of epigean and hypogean crayfish (Crustacea : Cambaridae) from northwestern Florida. Amer. Midl. Nat. 99: 315–329.

Christiansen, K.A. 1962. Proposition pour la classification des animaux cavernicoles. Spelunca 2: 76–78.

Cooper, J.E. & R.A. Kuehne. 1974. *Speoplatyrhinus poulsoni*, a new genus and species of subterranean fish from Alabama. Copeia 1972: 486–493.

Culver, D.C. 1982. Cave life: evolution and ecology. Harvard University Press, Cambridge. 189 pp.

Dickson, I.W. & R.H. Kramer. 1971. Factors influencing scope for activity and active and standard metabolism of rainbow trout (*Salmo gairdneri*). J. Fish. Res. Board Can. 28: 587–596.

Eigenmann, C.H. 1898. A new blind fish. Proc. Indiana Acad. Sci. (1897): 231.

Eigenmann, C.H. 1909. Cave vertebrates of America. Carnegie Institution of Washington, Washington D.C. 241 pp.

Evans, D.O. 1984. Temperature independence of the annual cycle of standard metabolism in the pumpkinseed. Trans. Amer. Fish. Soc. 113: 494–512.

Facey, D.E. & G.D. Grossman. 1990. The metabolic cost of maintaining position for four North American stream fishes: effects of season and velocity. Phys. Zool. 63: 757–776.

Forbes, S.A. & R.E. Richardson. 1920. The fishes of Illinois. Illinois Natural History Survey, Urbana. 357 pp.

Fry, F.E.J. 1957. The aquatic respiration of fish. pp. 1–55. *In:* M.E. Brown (ed.) The Physiology of Fishes, Vol. 1, Academic Press, New York.

Fry, F.E.J. 1971. The effect of environmental factors on the physiology of fish. pp. 1–98. *In:* W.S. Hoar & D.J. Randall (ed.) Fish Physiology, Vol. 6, Academic Press, New York.

Gannon, A.T., V.G. Demarco, T. Morris, M. G. Wheatly & Y. Kao. 1999. Oxygen uptake, critical oxygen tension, and available oxygen for three species of cave crayfishes. J. Crust. Biol. 19: 235–243.

Helfman, G.S., B.B. Collette & D. E. Facey. 1997. The diversity of fishes. Blackwell Science, Malden. 528 pp.

Hobbs, H.H., Jr. & A.B. Brown. 1987. A new troglobitic crayfish from northwestern Arkansas (Decapoda: Cambaridae). Proc. Biol. Soc. Wash. 100: 1040–1048.

Huppop, K. 1986. The role of metabolism in the evolution of cave animals. Nat. Speleol. Soc. Bull. 47: 136–146.

Katula, R. 1992. The spawning mode of the pirate perch. Trop. Fish Hobb. August 1992: 156–159.

Lee, D.S., C.R. Gilbert, C.H. Hocutt, R.E. Jenkins, D.E. McAllister & J.R. Stauffer, Jr. 1980. Atlas of North American freshwater fishes. North Carolina State Museum of Natural History, Raleigh. 854 pp.

Means, M.L. 1993. Population dynamics and movement of Ozark Cavefish in Logan Cave NWR, Benton County, Arkansas, with additional baseline water quality information. M.S. Thesis, University of Arkansas, Fayetteville. 126 pp.

Means, M.L. & J.E. Johnson. 1995. Movement of threatened Ozark cavefish in Logan National Wildlife Refuge, Arkansas. Southw. Nat. 40: 308–313.

Pandey, B.N. 1976. Effects of gonadal steroids on oxygen consumption in *Geteropneustes fossilis* BI (Pisces). Polish Arch. Hydrobiol. 23: 327–332.

Pflieger, W.L. 1997. The fishes of Missouri. Missouri Department of Conservation, Jefferson City. 372 pp.

Poulson, T.L. 1961. Cave adaptation in amblyopsid fishes. Ph.D. Dissertation, University of Michigan, Arm Arbor. 185 pp.

Poulson, T.L. 1963. Cave adaptation in amblyopsid fishes. Amer. Midl. Nat. 70: 257–290.

Poulson, T.L. & W.B. White. 1969. The cave environment. Science 165: 971–981.

Robins, C.R., R.M. Bailey, C.E. Bond, J.R. Brooker, E.A. Lachner, R.N. Lea & W.B. Scott. 1991. Common and scientific names of fishes from the United States and Canada. Amer. Fish. Soc. Spec. Publ. 20: 1–183.

Robison, H.W. & T.M. Buchanan. 1988. The fishes of Arkansas. University of Arkansas Press, Fayetteville. 536 pp.

Trippel, E.A. & J.J. Hubert. 1990. Common statistical errors in fishery research. pp. 93–102. *In:* J. Hunter (ed.) Writing for Fishery Journals, American Fisheries Society, Bethesda.

Wells, N.A. 1935. Variations in the respiratory metabolism of the Pacific killifish *Fundulus parvipinnis* due to size, season, and continued constant temperature. Phys. Zool. 8: 318–337.

Wilkens, H. 1988. Evolution and genetics of epigean and cave *Astyanax fasciatus* (Characidae, Pisces). Support for the neutral mutation theory. Evol. Biol. 23: 271–367.

Willis, L.D. & A.V. Brown. 1985. Distribution and habitat requirements of the Ozark cavefish (*Amblyopsis rosae*). Amer. Midl. Nat. 114: 311–317.

Winberg, G.G. 1960. Rate of metabolism and food requirements of fishes. Fisheries Research Board of Canada, Translation Series 194, Nanaimo.

Withers, P.C. 1992. Comparative animal physiology. Saunders College Publishing, Orlando. 949 pp.

Withey, K.G. & J.L. Saunders. 1973. Effect of a reciprocal photoperiod regime on standard rate of oxygen consumption of post-smolt Atlantic salmon (*Salmo salar*). J. Fish. Res. Board Can. 30: 1898–1900.

Woods, L.P. & R.F. Inger. 1957. The cave, spring and swamp fishes of the family Amblyopsidae of central and eastern United States. Amer. Midl. Nat. 58: 232–256.

Environmental Biology of Fishes **62:** 107–114, 2001.
© 2001 *Kluwer Academic Publishers. Printed in the Netherlands.*

Locomotory responses of the cave cyprinid *Phreatichthys andruzzii* to chemical signals from conspecifics and related species: new findings

Roberto Berti & Luca Zorn
Dipartimento di Biologia animale e Genetica dell'Università, Via Romana 17, 50125 Firenze, Italy (e-mail: berti@dbag.unifi.it)

Received 2 March 2000   Accepted 11 January 2001

Key words: chemotaxis, chemical communication, fish behaviour, blind fish, cave adaptation

Synopsis

Previous research on chemical communication among hypogean fishes have suggested that these species are attracted, as are epigean species, by the odour of non-injured conspecifics. Considerations on the biological meaning of such responses and some doubts about the reliability of the experimental procedure previously utilised, have led us to reconsider the matter and verify the results already obtained in the cave cyprinid *Phreatichthys andruzzii*. Using a new methodological approach we ascertained not only the absence of positive chemotaxis to odour of known or unknown conspecifics, but also the presence of a strong negative chemotaxis towards the odour of unfamiliar conspecifics and of individuals belonging to the closely related species *Barbus conchonius*. The biological meaning and the adaptive value of these reactions are briefly discussed.

Introduction

Many fish are strongly attracted by the odours of non-injured conspecifics. This phenomenon has long been known and seems to be common also among phylogenetically distant species (Quinn & Tolson 1986, Quinn & Busack 1985, Wrede 1932, Keenleyside 1955, Hemmings 1966, Höglund & Åstrand 1973). Despite being very common, this response does not always appear to have the same biological meaning. In species characterised by a strong migrating phenomenon, like salmonids (Nordeng 1971, 1977, Solomon 1973) and eels (Miles 1968), conspecific odours would constitute an additional stimulus in the orientation of upstream migrating fish (for an exhaustive critical review on the role of conspecific odours in salmon homing, see Quinn 1990). In non-migrating species, chemical information from single individuals would have the effect of grouping the shoal. This would prevent the scattering of the shoal, especially during night time when visual information and consequently optical-motor coordination is not possible (Hemmings 1966).

As epigean species do, hypogean ones also show an attraction to conspecific odours. This response was found in *Caecobarbus geertsii* (Berti & Thinès 1980), in the cave forms of *Astyanax fasciatus* (Quinn 1980), and in *Phreatichthys andruzzii* (Berti et al. 1982). Moreover, the cave populations of *Astyanax fasciatus* (formerly *Anoptichthys* spp.) show that responses to these stimuli vary depending on the concentration of the stimulus (de Fraipont 1987) and, at the same concentration, depending on the number of individuals that release the stimulus (de Fraipont 1986). In *Phreatichthys andruzzii* the response seems to be species-specific (Berti et al. 1989). In both species a more intense response appears when the chemical information comes from conspecifics known to the tested fish (Berti et al. 1982, de Fraipont & Thinès 1986).

Despite the great amount of data gathered on the subject, the biological meaning of such a response in hypogean species is still unclear. The aforementioned hypothesis of Hemmings (1966), which finds biological meaning in using chemotaxis to maintain group cohesion when visual information is missing, does not seem to apply to hypogean fishes since Hemmings's hypothesis presumes the existence of shoaling behaviour, behaviour that does not seem to have a relevant adaptive value in a hypogean habitat. Foraging as a group increases food competition, and this cannot be reconciled with the scarcity of food resources that is commonly attributed to the subterranean habitat. The anti-predatory function of shoaling behaviour also seems unnecessary: in the life-histories of nearly one hundred cave-dwelling fish known to science, sympatric predators are absent. In particular, no sympatric fish are known at all for *C. geertsii, A. fasciatus* and *P. andruzzii*.

This, together with the fact that the three species mentioned above do not show shoaling behaviour, may suggest that the attraction reported previously may be the result of an experimental bias and not the natural behaviour of the fish. With the exception of the study conducted by Quinn (1980), all the tests used water extracted from the breeding aquariums as stimulus. It is quite possible that this water contained not only the 'odour' of conspecifics but also a mixture of other chemicals. They probably included the smell of food remains or chemical information derived from the final products of metabolism that could have given direct or indirect information on a possible food source.

Therefore, we decided to reconsider the matter and verify the results obtained some time ago in the cave cyprinid *Phreatichthys andruzzii* (Berti et al. 1982, 1989). On the basis of a new methodological approach, we once again investigated the locomotory response of isolated individuals when presented with the odour of both conspecifics and closely related species.

Materials and methods

For the present study a total of 320 individuals of *Phreatichthys andruzzii* were used. All these individuals had been bred in the laboratory in October 1994 from fish collected from the wells of Bud-Bud (Central Somalia) in December 1982. Four months after hatching, the fish were subdivided into eight groups, each composed of 40 individuals. Each group was kept separated from each other with no chemicals being exchanged among the different aquariums at any time.

For all groups we used aquaria of identical size (100×40×50 cm; water height: 40 cm). Each aquarium contained about 160 l of water so that animal density was one individual per 4 l. The aquarium was equipped with a mobile divider capable of moving from right to left so that the environment could be subdivided into two compartments of variable volume. This device allowed us to separate tested from non-tested individuals and, at the same time, to keep constant the ratio of fish to water volume. The size of the two complimentary compartments varied as a function of the number of tested and non-tested individuals that increased and decreased respectively through time. The divider had a grill that allowed the circulation of the water between the two compartments. Thus, the chemical characterisation of the group was maintained. In cases in which experiments tested chemical information released by fish belonging to different groups, the already tested fish were isolated in another aquarium in order to not compromise the possible chemical identity of the group of origin with traces of foreign odours.

A group of 45 individuals of *Barbus conchonius*, closely related epigean species of *P. andruzzii*, was also used but only as releaser of chemical information. These fish, bought some months before the beginning of the experiments, formed a single lot and were kept in an aquarium similar to those of *P. andruzzii*.

The breeding aquariums were equipped with an adsorbing charcoal filter and aerator. Every other month, filters were changed together with one-third of the water and twice a year the water was completely renewed. Water temperature was kept constant at 27°C and 24°C in the aquaria for *P. andruzzii* and *B. conchonius*.

Four to seven times a week the fish were fed frozen chironomid larvae, rinsed before they were given to the fish to reduce chemical contamination as much as possible. The given quantity was usually completely eaten in a few minutes. When present, food remains were removed from the aquaria within a few hours.

The experimental aquarium (Figure 1) was a long parallelepiped (size: 105 × 8 × 20 cm; water height: 8 cm; water temperature: 27°C), internally subdivided by two removable doors into three compartments of 35 cm in length. The individual to be tested was taken from its breeding tank, 'rinsed' and then placed in the central compartment of the experimental aquarium. The rinsing procedure consisted of three consecutive washes, of a few minutes each, in buckets filled with water devoid of any odours. This treatment appeared to be borne well by the fish, that never appeared more stressed than after an usual aquarium change. After about 22 h, time considered sufficient for the fish to overcome transfer stress and to adapt itself to the new environment, two samples of 300 cc of water,

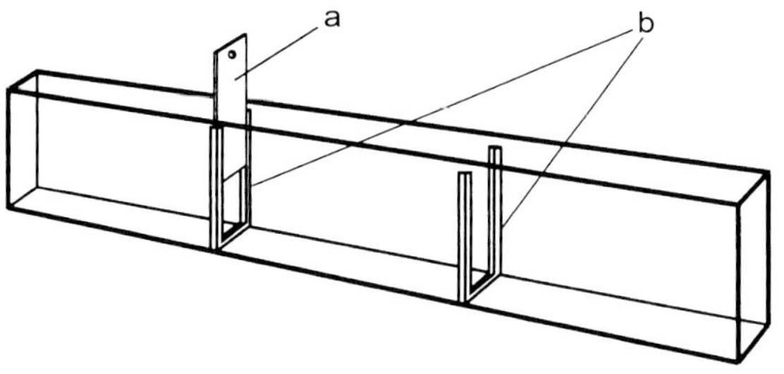

Figure 1. Schematic representation of the test aquarium. The three compartments separated by removable doors (a) are shown, as well as the sliding door guides (b). For further information see text.

one containing the chemical information and the other without any odour, were simultaneously poured at the two opposite ends of the aquarium. A third sample of 300 cc of water without odours was added in the central compartment with the fish in it. In this way, violent mixing of the water at the moment of the opening of the doors was prevented, by preserving similar water levels in all three compartments.

Five minutes after the introduction of the water samples, the necessary time for diffusion of the samples in the two extreme compartments, the doors were removed so the fish could freely move through the entire aquarium. From this moment on, every 30 s the compartment in which the fish was found was recorded. A total of 60 successive recordings were made for a period of 30 min.

For each series of experiments, the chemical information was introduced an equal number of times at the right and at the left end of the experimental aquarium, following a sequence predetermined using a random number table.

Both experimental methods and equipment employed were those already used in previous studies conducted on the same species (Berti et al. 1982, 1989). We thought it best not to introduce any innovations other than that of the nature of the chemical stimulation. This stimulus was represented by samples of water in which 'odour suppliers' had stayed for some time. Immediately after the introduction of the test fish into the experimental aquarium, the odour-supplying fish were taken out of their breeding aquarium and were transferred to an 'odour-releasing chamber' with a size of 18 × 14 × 23 cm and containing 4 l of water. As was the test fish, the odour releasers were also 'rinsed' during the transfer so as to eliminate the odour of their breeding aquarium. At the moment of the experiment, that is about 22 h after the transfer, 300 cc of water was taken from the odour-releasing chamber and used as chemical stimulation. Two different concentrations of the stimulation were used. These were obtained using either one or four fish as odour releasers. In the first case, the ratio of number of odour-releasing fish to water volume was equal to that in the breeding aquarium (one individual to 4 l of water); in the second case, this was four times greater (one individual to 1 l of water).

The odour-releasing chamber was heated in a thermostatic water bath with a second chamber containing the non-odorous water which would be added to both the central compartment of the experimental aquarium and the odourless extreme one. In this way, formation of thermal gradients in the experimental aquarium was prevented. When the odour supplying fish were *P. andruzzii*, the temperature of the thermal bath was regulated at 27°C and when *B. conchonius* were used, the temperature was kept at 24°C.

A total of 10 series of tests, two control and eight experimental ones, were conducted. Control tests were made to verify both the absence of bias in the materials and in the adopted testing procedures (C1 series: 40 tests), and the validity (C2 series: 60 tests) of the results already obtained in the same species (Berti et al. 1982). Thus the tests of the C1 series were conducted recording the fish behaviour in the absence of any chemical trace, whereas during the test of the C2 series the reactions to water samples extracted from the breeding aquarium of the test fish were investigated.

Of the eight series of experimental tests, two were conducted using chemical stimulation from familiar conspecifics (E1 series: one odour releaser, 60 tests; E2 series: four odour releasers, 60 tests), two with the odour of unfamiliar conspecifics (E3 series: one releaser, 40 tests; E4 series: four releasers, 40 tests) and two with the odour of *B. conchonius* (E5 series: one releaser, 40 tests; E6 series: four releasers, 40 tests). The other two series were conducted by simultaneously introducing the odour of unfamiliar conspecifics and that of *B. conchonius*, respectively, at the two opposite ends of the test aquarium (E7 series: one releaser, 40 tests; E8 series: four releasers, 40 tests). In this last case, the temperatures of the releasing chambers of *P. andruzzii* (27°C) and *B. conchonius* (24°C) being different, the water samples containing the odours of the two species were brought to the intermediate temperature of 25.5°C before their use. The formation of temperature gradients between the opposite ends of the test aquarium was thus avoided.

The statistical analysis of the data was conducted by classifying each test as positive, neutral, or negative according to whether the number of times the fish was registered in the odorous end was greater, equal or less compared to the number of registrations in the opposite non-odorous compartment. With regard to the last two series (E7, E8), each test was considered as positive, neutral, or negative according to whether the number of registrations in the end containing the chemical traces from conspecifics was greater, equal or less compared to the number of registrations in the opposite compartment with the *B. conchonius* odour.

For each series of tests the significance of the difference between the number of positive and negative tests was tested with the chi-square one-sample test (Siegel 1956). Probability levels were considered significant when $p < 0.05$. The evolution in time of the response was also analysed subdividing the observation period (30 min) into six sub-periods of 5 min each. The same procedures described above were applied to each sub-period.

The tests were conducted from August 1996 to February 1998. A total of 460 tests were performed using an equal number of test fish and 180 odour-releasing fish. Since only 320 individuals of *P. andruzzii* were available, some individuals were necessarily used more than once both as odour-releasing and test fish. Careful attention was given never to use the same individual more than twice as test fish and never with the same type of stimulation. Furthermore, the time interval between two tests conducted on the same individual was never less than two weeks.

During each test session, three or four individuals were simultaneously tested using as many experimental aquariums. Since this species is known to be scotophilic (i.e., photophobic) (Ercolini & Berti 1975), the room in which the tests were carried out was illuminated by a defused light of low intensity. Moreover, the light sources were placed in such a way as to avoid differences in light intensity between the end compartments of the experimental aquariums.

The fish were fed only after both the test and releasing fish had been transferred from the breeding aquariums. This meant that at the moment of the transfer these individuals were starved at least for 24 h, and at the beginning of the test their fasting was never less than 46 h.

Finally, two water sources complete the experimental equipment. These were provided by a filtering system similar to that of the breeding aquariums and supplied non-odorous water at a temperature of 24°C and 27°C (for *B. conchonius* and *P. andruzzii*, respectively) to the test aquaria, the releasing chambers, and the buckets where the fish were 'rinsed' during the transfers.

Results

Control tests

Controls showed the absence of experimental bias and confirmed the chemotactic reaction of *P. andruzzii* as previously reported. The results of the C1 series of tests indicate that the fish tested in absence of chemical information did not show any preferences for the right or left compartment of the experimental aquarium (Table 1). Those tests in which the fish stayed more often in the left or in the right compartment were named L-tests and R-tests, respectively, while those in which the fish were found in the right and left compartments the same number of times (neutral tests) were named N-tests. The statistical analysis conducted on the entire observation period (30 min, 60 recordings) showed that, out of a total of 40 experiments, the difference between the number of L-tests and R-tests is not significant. Furthermore the behaviour seems to be extremely constant in time. In fact, the difference between the number of L-tests and R-tests, relative to each of the six sub-periods (5 min long, 10 recordings) into which the test can be subdivided, oscillated constantly around the zero value.

The tests of the C2 series, conducted using a stimulation similar to that used in the previous studies (represented by the water directly taken from the breeding aquarium), showed a significant positive chemotactic response, as expected. In this case, statistical analysis (Table 2) showed that there was a clear preference for the compartment containing the chemical stimulation. Out of a total of 60 tests, the difference between the

Table 1. C1 series: 40 tests. Locomotory response of *Phreatichthys andruzzii* in odourless water. Statistical analysis of results.

Intervals of analysis (min)	Experiments			χ^2	
	L-tests	N-tests	R-tests		
0–30	17	2	21	0.421	n.s.
0–5	14	9	17	0.290	n.s.
5–10	17	9	14	0.290	n.s.
10–15	16	7	17	0.030	n.s.
15–20	18	7	15	0.273	n.s.
20–25	18	10	12	1.200	n.s.
25–30	16	4	20	0.444	n.s.

Table 2. C2 series: 60 tests. Locomotory response of *Phreatichthys andruzzii* to water from own home aquarium: chemical stimulation as in previous studies.

Intervals of analysis (min)	Experiments			χ^2	
	Positive	Neutral	Negative		
0–30	45	1	14	16.288	$p < 0.001$
0–5	31	7	22	1.528	n.s.
5–10	43	5	12	17.473	$p < 0.001$
10–15	46	2	12	19.931	$p < 0.001$
15–20	35	8	17	6.231	$p < 0.02$
20–25	29	11	20	1.653	n.s.
25–30	25	8	27	0.077	n.s.

Table 3. E1 series: 60 tests. Locomotory response of *Phreatichthys andruzzii* to the odour of familiar conspecifics: one releaser.

Intervals of analysis (min)	Experiments			χ^2	
	Positive	Neutral	Negative		
0–30	32	5	23	1.473	n.s.
0–5	25	11	24	0.020	n.s.
5–10	29	9	22	0.961	n.s.
10–15	31	6	23	1.185	n.s.
15–20	22	11	27	0.510	n.s.
20–25	28	8	24	0.308	n.s.
25–30	24	6	30	0.667	n.s.

Table 4. E2 series: 60 tests. Locomotory response of *Phreatichthys andruzzii* to the odour of familiar conspecifics: four releasers.

Intervals of analysis (min)	Experiments			χ^2	
	Positive	Neutral	Negative		
0–30	32	2	26	0.621	n.s.
0–5	32	8	20	2.769	n.s.
5–10	22	10	28	0.720	n.s.
10–15	24	9	27	0.176	n.s.
15–20	26	12	22	0.333	n.s.
20–25	28	6	26	0.074	n.s.
25–30	27	7	26	0.019	n.s.

Table 5. E3 series: 40 tests. Locomotory response of *Phreatichthys andruzzii* to the odour of unfamiliar conspecifics: one releaser.

Intervals of analysis (min)	Experiments			χ^2	
	Positive	Neutral	Negative		
0–30	7	1	32	16.026	$p < 0.001$
0–5	3	0	37	28.900	$p < 0.001$
5–10	6	1	33	18.692	$p < 0.001$
10–15	8	5	27	10.314	$p < 0.01$
15–20	12	4	24	4.000	$p < 0.05$
20–25	28	2	10	8.526	$p < 0.01$
25–30	26	2	12	5.158	$p < 0.05$

number of positive and negative tests appears highly significant. Statistical analysis conducted on the 5 min periods shows that the response increased in its intensity during the first part of the experiment with a peak around 10–15 min. The response became progressively weaker until 20 min, when non-significant results were obtained.

Experimental tests

These experiments yielded results different from the expected ones. Most striking was the complete indifference shown by the fish towards the stimulation when this was released by known conspecifics, in particular when the odour releasing fish belonged to the same group as the fish being tested. The results of the E1 series, conducted with the chemical information coming from only one releaser, are shown in Table 3. The statistical analysis conducted on the entire observation period (30 min) reveals that the difference between the number of positive and negative tests is not significant. Also the analysis of the data relative to the six subperiods of time (each of 5 min) never shows significant differences. Even in the course of the E2 series tests, when a more concentrated stimulation obtained from four releasers was used, the tested fish still showed no response to chemical stimulation (Table 4).

On the other hand, when tests were made with stimulations released by unfamiliar conspecifics, the fish displayed a strongly oriented locomotory response. Data from the E3 series (one releasing fish) show that the tested fish clearly preferred the non-odorous compartment (Table 5). Over the 30 min observation period, out of a total of 40 tests, the difference between the number of negative and positive tests is highly significant. Of particular interest is how the response varies in time. The tendency to avoid the compartment with the chemical information is very strong at the beginning, then progressively weakens until it disappears after about 20 min. During the last 10 min the response reverses and the test fish shows a significant preference for the compartment containing the chemical information. The tests of the E4 series (four releasers) seem to confirm the existence of a negative chemotaxis triggered by the stimulus released by unfamiliar conspecifics. Statistical analysis (Table 6) shows a substantial analogy with

Table 6. E4 series: 40 tests. Locomotory response of *Phreatichthys andruzzii* to the odour of unfamiliar conspecifics: four releasers.

Intervals of analysis (min)	Experiments			χ^2	
	Positive	Neutral	Negative		
0–30	4	1	35	24.641	$p < 0.001$
0–5	6	2	32	17.789	$p < 0.001$
5–10	9	2	29	10.526	$p < 0.01$
10–15	8	4	28	11.111	$p < 0.001$
15–20	9	1	30	11.308	$p < 0.001$
20–25	13	4	23	2.778	n.s.
25–30	16	8	16	0.000	n.s.

Table 7. E5 series: 40 tests. Locomotory response of *Phreatichthys andruzzii* to the odour of *B. conchonius* specimens: one releaser.

Intervals of analysis (min)	Experiments			χ^2	
	Positive	Neutral	Negative		
0–30	6	0	34	19.600	$p < 0.001$
0–5	5	4	31	18.778	$p < 0.001$
5–10	9	7	24	6.818	$p < 0.01$
10–15	11	3	26	6.081	$p < 0.02$
15–20	9	4	27	9.000	$p < 0.01$
20–25	13	3	24	3.270	n.s.
25–30	19	2	19	0.000	n.s.

Table 8. E6 series: 40 tests. Locomotory response of *Phreatichthys andruzzii* to the odour of *B. conchonius* specimens: four releasers.

Intervals of analysis (min)	Experiments			χ^2	
	Positive	Neutral	Negative		
0–30	2	0	38	32.400	$p < 0.001$
0–5	8	4	28	11.111	$p < 0.001$
5–10	7	4	29	13.444	$p < 0.001$
10–15	6	1	33	18.692	$p < 0.001$
15–20	10	3	27	7.811	$p < 0.01$
20–25	5	7	28	16.030	$p < 0.001$
25–30	10	3	27	7.811	$p < 0.01$

Table 9. E7 series: 40 tests. Locomotory response of *Phreatichthys andruzzii* to the odour of unfamiliar conspecifics and to that of *B. conchonius* specimens: one releaser.

Intervals of analysis (min)	Experiments			χ^2	
	Positive	Neutral	Negative		
0–30	33	1	6	18.692	$p < 0.001$
0–5	33	3	4	22.730	$p < 0.001$
5–10	32	3	5	19.703	$p < 0.001$
10–15	30	8	2	24.500	$p < 0.001$
15–20	25	5	10	6.429	$p < 0.02$
20–25	9	5	26	8.257	$p < 0.01$
25–30	5	5	30	17.857	$p < 0.001$

the E3 series. The difference between the number of negative and positive tests is highly significant and also the trend in time of the response appears to be similar, with χ^2 values decreasing throughout the course of the experiment. Non-significant values are reached after 20 min but, contrary to the E3 series, the response does not reverse in the last 10 min.

The response towards the odour of *B. conchonius* displayed some interesting analogies with the responses displayed towards the odour of unfamiliar conspecifics, both regarding the sign of the chemotactic response and its changes over time. The results of the tests of the E5 series, conducted with the chemical information coming from only one releasing *B. conchonius* (Table 7), show that there is a strong tendency to avoid the compartment with the chemical information. Considering the entire period of the test, the difference between the number of negative and positive tests is highly significant. In the course of the test, the intensity of the response decreases and the χ^2 values, extremely high at the beginning of the experiment, become non-significant in the last 10 min. However, no reversal of the response is recorded in this phase. The results of the E6 series of tests (four releasers) show an even stronger tendency to avoid the chemical stimulation (Table 8). Moreover, the fish response remains almost constant with time, showing high levels of significance even towards the end of the experiment.

Finally, when simultaneous stimulation coming from unfamiliar conspecifics on one side and from *B. conchonius* on the other side of the aquarium was tested, the fish showed a tendency to prefer the conspecific odour or simply avoided the compartment with the odour of *B. conchonius*. The results of the E7 series (one releasing fish for each species) are shown in Table 9. The statistical analysis conducted on the entire 30 min observation period shows the high level of significance of this preference. The response appears extremely intense during the first 15 min of the experiment, then weakens. In the last 10 min a clear change in the sign of the response takes place. Substantially similar are the results of the E8 series (four releasers for each species). The only difference is an even stronger preference for the odour of the conspecifics (Table 10). No differences were found compared to previous set of experiments, when temporal variations of the response

Table 10. E8 series: 40 tests. Locomotory response of *Phreatichthys andruzzii* to the odour of unfamiliar conspecifics and to that of *B. conchonius* specimens: four releasers.

Intervals of analysis (min)	Experiments			χ^2	
	Positive	Neutral	Negative		
0–30	39	0	1	36.100	$p < 0.001$
0–5	36	3	1	33.108	$p < 0.001$
5–10	36	2	2	30.421	$p < 0.001$
10–15	33	5	2	27.457	$p < 0.001$
15–20	28	5	7	12.600	$p < 0.001$
20–25	8	4	28	11.111	$p < 0.001$
25–30	11	8	21	3.125	n.s.

are considered. Again in this case there is a tendency to invert the response after 20 min.

Discussion and conclusions

The results of our observations are consistent with our original hypothesis, since the fish responses were different with those previously reported. Before analysing in detail the responses shown by the fish in relation to the chemical stimulation, it is important to examine the characteristics of the response itself as can be deduced by its changes over time. As a matter of fact, when the chemotactic response is present, this is clearly evident for the first 20 min, then it progressively weakens or even reverses. The causes behind this behavioural reversal are not clear. Some possible explanations could be: (a) a complete diffusion of the chemical signal through the entire experimental corridor; (b) a sensory adaptation of the fish; (c) a habituation to the stimulation; (d) a resumed exploratory activity. An indication favouring the second or third hypothesis is represented by the fact that the decrease in the response appears to be less important when four individuals are employed. In fact this presumes not only a greater concentration of the chemical information but especially a less monotonous stimulus, since it consists of signals from four individuals and of the product of their possible interactions.

Considering the responses shown by the tested fish in relation to the different chemical stimulations, it appears that the fish are basically indifferent to the odour of familiar conspecifics, but show a marked tendency to avoid the odour of both unfamiliar conspecifics and another species (*B. conchonius*). However, this behaviour should not be interpreted as the result of a general tendency to indiscriminately avoid the odour of any unfamiliar individual. *P. andruzzii* clearly chooses a conspecific odour when given the possibility of choosing between the odour of unfamiliar conspecifics and that of *B. conchonius*. This not only shows that the fish is capable of species-specific recognition, but also of confronting two opposing stimuli that evidently are characterised by repulsive charges of different intensity.

Contrary to previous studies conducted on the same species (Berti et al. 1982, 1989) and the control tests of the C2 series, in none of the experimental series of the present work did the chemical stimulation attract the fish. In previous studies, *P. andruzzii* showed a marked positive chemotaxis if the water was taken from an aquarium occupied by a group of conspecifics and the reaction was even more intense if the individuals were known to the tested fish. By contrast, indifferent responses were elicited in these previous experiments when chemical stimulations were provided by *Barbus schuberti*. The comparison between the previous studies and the present one shows a great decrease of the positive chemotactic response in this set of experiments. This phenomenon is common to all series of experimental tests and is independent of the nature of the stimulation (odour from both familiar and unfamiliar conspecifics and from *B. conchonius*). The decrease in positive chemotaxis is probably related to the use of the odour- releasing chamber. This technique may have allowed us to obtain purified chemical information, in which any 'odorous components' (food traces, catabolites) presumably found in the water of the breeding aquariums were absent.

These 'odorous components' were probably responsible for the attractive capacity of the water samples extracted from the breeding aquarium of the tested fish and, to a lesser extent, by those taken from the tank of unfamiliar conspecifics (Berti et al. 1982). In this second case, in fact, the attraction elicited by the 'odorous components' would be partially compensated for by the repulsion provoked by chemical traces released by the unfamiliar conspecifics. Finally, if the water comes from an aquarium occupied by another species, the odour of which elicits an even more intense repulsion, the compensation should be absolute. This would explain the indifference displayed towards *B. schuberti* (Berti et al. 1989).

Regarding the meaning of the responses exhibited by the fish, the tendency to avoid unfamiliar conspecifics could have the function, at the individual level, to reduce risks of predation and, at the population level, to maintain the population in a familiar area. The highly different responses shown towards familiar and unfamiliar conspecifics support the hypothesis of the existence of distinct stocks, each characterised by a 'stock-specific' odour. The perception of the odour of unfamiliar conspecific could signal the approach of an area occupied by another stock. In this way trespassing would be avoided and risks of inter-stock cannibalism would be reduced. Both laboratory and field observations (unpublished data) indicate that risks of being attacked, not only by unfamiliar but also by familiar conspecifics, are real.

Finally, the extreme repulsion induced by the odour of *B. conchonius* seems to have an even greater adaptive value. No sympatric species are known for *P. andruzzii* and the perception of an hetero-specific odour could signal to the fish the limits of the subterranean habitat. Trespassing for a hypogean fish is an event to be avoided in all cases, since risks in the epigean habitat are high. Trespassing during daytime is mainly avoided due to the marked scotophilia (Ercolini & Berti 1975), whereas the chemotactic response represents the only means to keep the species within the borders of the underground world by night, when scotophilia is obviously ineffective.

References cited

Berti, R. & G. Thinès. 1980. Influence of chemical signals on the topographic orientation of the cave fish *Caecobarbus geertsi* Boulenger (Pisces, Cyprinidae). Experientia 36: 1384–1385.

Berti, R., G. Thinès & B. Lefèvre. 1982. Effets des informations chimiques provenant d' un milieu habité par des congénères sur l'orientation topographique du poisson cavernicole *Phreatichthys andruzzii* Vinciguerra (Pisces, Cyprinidae). Int. J. Speleol. 12: 103–117.

Berti, R., R. Vezzosi & A. Ercolini. 1989. Locomotory response of *Phreatichthys andruzzii* Vinciguerra (Pisces, Cyprinidae) to chemical signals of conspecifics and of closely related species. Experientia 45: 205–207.

de Fraipont, M. 1986. La détection chimique de l'odeur des congénères chez *Astyanax mexicanus* (forme cavernicole). Réponses observées pour des concentrations égales établies à partir de groupes de densités différentes. Stygologia 2: 276–285.

de Fraipont, M. 1987. La détection chimique chez *Astyanax mexicanus* (Teleostei, Characidae) (forme cavernicole) en fonction de la densité des groupes. Ann. Soc. roy. zool. Belg. 117: 63–76.

de Fraipont, M. & G. Thinès. 1986. Responses of the cavefish *Astyanax mexicanus (Anoptichthys antrobius)* to the odor of known or unknown conspecifics. Experientia 42: 1053–1054.

Ercolini, A. & R. Berti. 1975. Light sensitivity experiments and morphology studies of the blind phreatic fish *Phreatichthys andruzzii* Vinciguerra from Somalia. Monitore zool. ital. (N.S.) Suppl. 6: 29–43.

Hemmings, C.C. 1966. Olfaction and vision in fish schooling. J. Exp. Biol. 45: 449–464.

Höglund, L.B. & M. Åstrand. 1973. Preferences among juvenile char (*Salvelinus alpinus* L.) to intraspecific odours and water currents studied with the fluviarium technique. Rep. Inst. Freshwat. Res. Drottningholm 53: 21–30.

Keenleyside, M.H.A. 1955. Some aspects of the schooling behaviour of fish. Behaviour 8: 183–248.

Miles, S.G. 1968. Rheotaxis of elvers of the American eel (*Anguilla rostrata*) in the laboratory to water from different streams in Nova Scotia. J. Fish. Res. Board Can. 25: 1591–1602.

Nordeng, H. 1971. Is the local orientation of anadromous fishes determined by pheromones? Nature 233: 411–413.

Nordeng, H. 1977. A pheromone hypothesis for homeward migration in anadromous salmonids. Oikos 28: 155–159.

Quinn, T.P. 1980. Locomotor responses of juvenile blind cave fish, *Astyanax jordani*, to the odors of conspecifics. Behav. Neur. Biol. 29: 123–127.

Quinn, T.P. 1990. Current controversies in the study of salmon homing. Ethol. Ecol. Evol. 2: 49–63.

Quinn, T.P. & C.A. Busack. 1985. Chemosensory recognition of siblings in juvenile coho salmon *(Oncorhynchus kisutch)*. Anim. Behav. 33: 51–56.

Quinn, T.P. & G.M. Tolson. 1986. Evidence of chemically mediated population recognition in coho salmon (*Oncorhynchus kisutch)*. Can. J. Zool. 64: 84–87.

Siegel, S. 1956. Nonparametric statistics for the behavioral sciences. International Student Edition, Kogakusha Co., Tokio. 312 pp.

Solomon, D.J. 1973. Evidence for pheromone-influenced homing by migrating Atlantic salmon, *Salmo salar* (L.). Nature 244: 231–232.

Wrede, W.L. 1932. Versuche über den Artduft der Elritzen. Z. vergl. Physiol. 17: 510–519.

Environmental Biology of Fishes **62:** 115–118, 2001.
© 2001 *Kluwer Academic Publishers. Printed in the Netherlands.*

Responses to light in epigean and hypogean populations of *Gambusia affinis* (Cyprinodontiformes: Poeciliidae)

Michele Maria Camassa
Istituto Sperimentale di Biologia del Sottosuolo "P. Parenzan" 91, V. Santa Margherita-I-72022 Latiano (Brindisi), Italy (e-mail: isbios@libero.it)

Received 30 March 2000 Accepted 8 January 2001

Key words: phototactic behaviour, cave adaptation, light intensity preferences

Synopsis

The responses to light among the fish *Gambusia affinis*, permanent residents of hypogean brackish water basins of the 'Spunnulate' dolinas system of Torre Castiglione, in southeastern Italy, were studied at light intensities of 180 and 430 lux. These fish showed, on the average, a negative phototactic behaviour (scotophilia) with a low coefficient of variability. On the other hand, epigean, spring-dwelling individuals of *G. affinis* from the same region showed a positive phototactic behaviour with a greater variability in the responses when compared with cave-dwelling conspecifics. Possible causes for the detachment of hypogean from epigean forms are discussed.

Introduction

Recently, a number of studies have been carried out on a particular coastal system of dolinas, locally known as 'spunnulate', of Torre Castiglione near Porto Cesareo, Lecce, Apulia, southeastern Italy (Ariani et al.[1], Camassa et al.[2]). These dolinas open in calcarenitic rocks that are overlaying a calcareous basement of Cretaceous (Caenomanian-Senonian) limestones. The calcarenite formations are quaternary, pleistocene or possibly of more recent age (Parenzan 1984). The system of the spunnulate dolinas represents a combination of epigean and hypogean brackish water habitats. The bottoms of most dolinas studied are filled with brackish water (S = 6–8‰, pH = 7.0–9.5) mostly of groundwater origin. The influence of seawater increases with decreasing distance from the coast of the Ionian Sea. These factors suggest that the water basins of the various dolinas are part of a subterranean system connected to the sea. Most dolinas include zones with low light intensities (hypophotic areas), accessible by humans only in few cases (sample sites SP-1, SP-2, SP-4, SP-13, SP-14) (Figure 1).

The fauna of the 'spunnulate' dolinas comprises marine species such as the crustacean *Tanais cavolinii* and fishes *Mugil cephalus* and *Blennius* spp., continental epigean species of marine origin such as the crustaceans *Diamysis, Gammarus, Orchestia*, and hypogean thalassoid forms such as the stygobiont crustacean *Spelaeomysis bottazzi* and certain harpacticoid copepods. The combination of the various gradients of physical and chemical parameters (mainly light, temperature, pH, salinity) in the dolinas determines both distinct faunistic and distributional patterns of aquatic animals; yet, all faunistic elements of the spunnulate could be found elsewhere in certain epigean or hypogean environments.

Gambusia affinis was introduced in the area of Torre Castiglione as a result of anti-mosquito measures

[1] Ariani, A.P., M.M. Camassa & K.J. Wittmann. 1999. Faunistic and biocenotic aspects of a semi-hypogean water system: the 'Spunnulate' of Torre Castiglione (Apulia, Southern Italy). Abst. XIVth Intern. Symp. Biospel., Makarska, 19–26 Sept. 1999: 31.

[2] Camassa M.M., A.P. Ariani & K.J. Wittmann. 1999. The dolina system of Torre Castiglione (Apulian, Southern Italy): description and environmental characteristics. Abst. XIVth Intern. Symp. Biospel., Makarska, 19–26 Sept. 1999: 36.

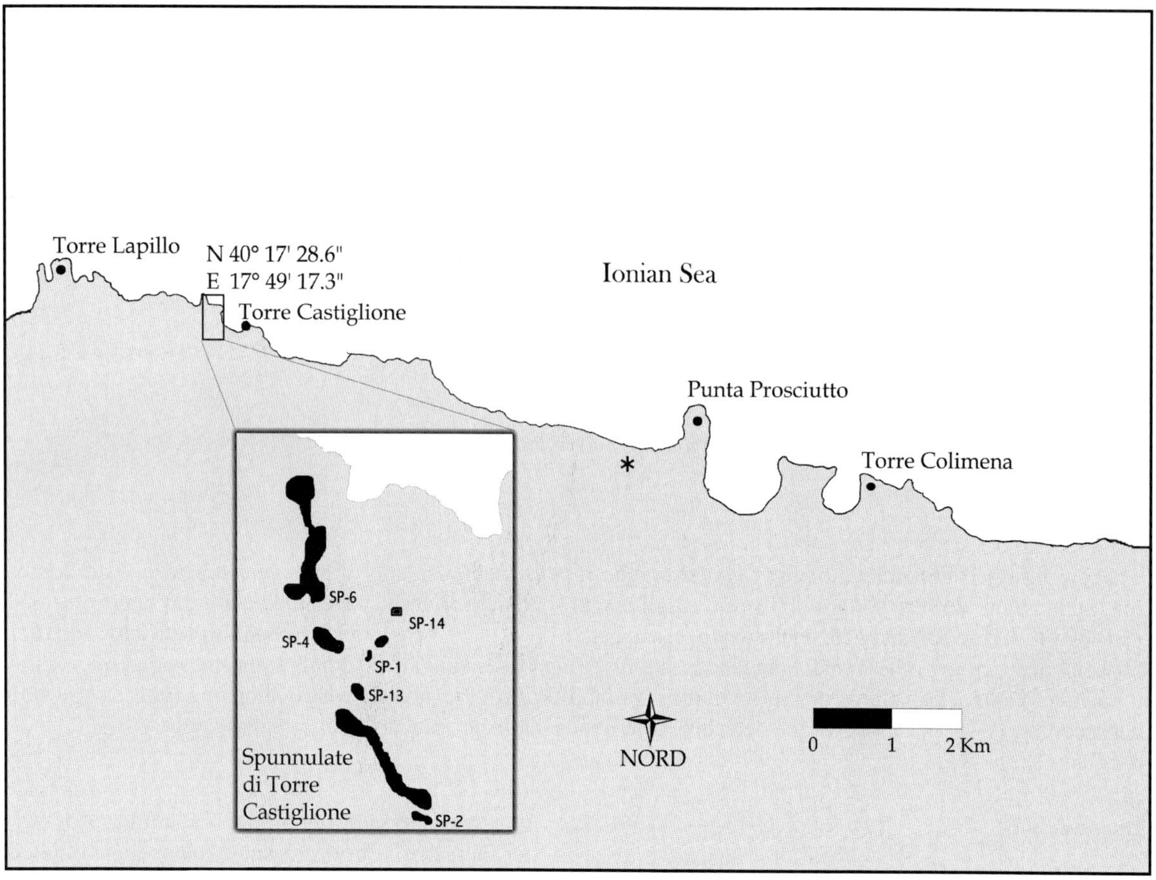

Figure 1. Map of the sites where the *Gambusia affinis* was colleted. The panel shows the dolinas, locally known as 'spunnulate' from which the samples of cave (*Cf*) originate; the asterisk points marks the site where the fish sample from the surface water (*Sf*) was obtained.

during the first half of the twentieth century. In the sites SP-1, SP-13 and SP-14, *G. affinis* was found only in the hypophotic zone and was completely absent in well illuminated habitats. This distributional preference could help this species to escape from predation by larger species of marine fishes which episodically invade the dolinas. The aim of the present study is to estimate the degree of responses to different light intensities of cave-dwelling fish and to compare these results with that of spring-dwelling epigean conspecifics.

Materials and methods

Individuals of *Gambusia affinis* were collected in June 1999 from two different sites separated by a distance of about 6 km. Cave-dwelling (*Cf*) fish (10 individuals) were collected from dolina SP-1 of the spunnulate system with a close-mesh net, and immediately placed into a 2 l transport tank. Spring-dwelling (*Sf*) fish (6 individuals) were collected and handled in the same way. In the laboratory the fish from both locations were introduced into separate aquaria of 20 l each filled with water taken from the respective sample sites and aerated using small porous ceramic cylinders. The aquaria were exposed to indirect sunlight with natural photoperiods. Here the animals were allowed to acclimatize until the beginning of the experiments in October 1999. Fish were regularly fed with Serovipan® food.

To measure the response of *Gambusia* to light, a modified version of the technique described by Vawter et al. (1987) was employed. A system of ten light-dark (L-D) chambers was prepared, each with 15 ml capacity and divided in two joined sectors (L and D). Each sector was shaped as a trunk-conic tiny bottle of transparent glass. Only the sector D was made opaque by a triple coat of black paint on the external surface. The L and D chambers were filled with water and put into a larger container filled with water maintained at a temperature of 18°C (±1.5°C). Individuals of *G. affinis* could choose between sectors L and D ($P_L = P_D = 0.5$). This system was situated in an isolated, dark section of the laboratory for 24 h. Two achromatic lamps (35 W and 50 W, corresponding to about 180 lux and 250 lux, respectively) were arranged 25 cm above this setup. These lamps could be set independently to illumination levels of approximately 180, 250 or 430 lux. The animals were acclimatized for 24 h at ambient temperature and at the selected light intensity before each experiment. For the first session of each experiment only two light intensities, 180 and 430 lux respectively, were tested. A videocamera was placed above the system for non-invasive observations from a remote monitor.

The choice made by the animals was recorded at 5 min intervals over a period of 90 min. The response was calculated as *phototactic index* (ϕ) using the formula

$$\phi_t^i = 1 - 2.\frac{n_D(i)}{n},$$

where i is the i-th interval of t minutes, n_D the number of fish present in the D-sector and n the total number of fish in the system. The index ϕ_t^i can adopt values within the interval of $[-1, 1]$ and expresses conditions of photophobia ($\phi_t^i < 0$), phototaxis ($\phi_t^i > 0$), or neutrality to light ($\phi_t^i = 0$). A total of eight experimental sessions (four at 180 lux and four at 430 lux) were performed. In each session ten *Cf* and six *Sf* specimens were tested at the same time.

Table 1. Statistical report of data recorded at 180 and 430 lux to measure the *phototactic index* (ϕ) of the mosquitofish *Gambusia affinis*. Sf = spring-dwelling fishes; Cf = cave-dwelling fishes; p is the probability to find a sample in the D sector of LD chamber [$p = \frac{1}{2}(\phi + 1)$].

	180 lux		430 lux	
	Sf	*Cf*	*Sf*	*Cf*
n	128	128	128	128
ϕ mean (p)	0.163	−0.592	−0.010	−0.579
	(0.58)	(0.20)	(0.49)	(0.21)
coeff. of var. (v)	0.428	0.190	0.333	0.162
DS	0.657	0.438	0.580	0.404

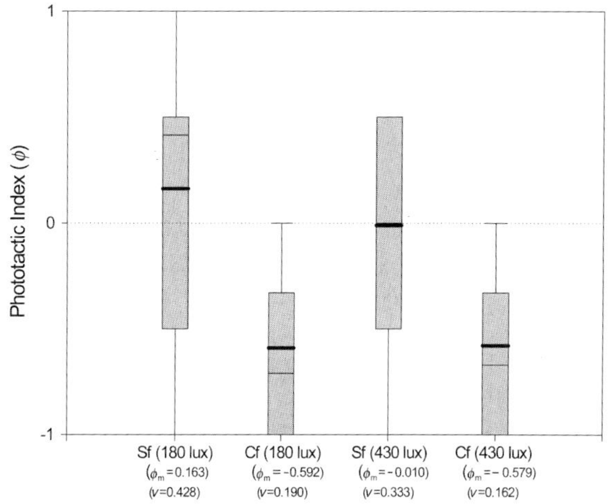

Figure 2. Phototactic index (ϕ) of mosquitofish *Gambusia affinis* recorded under 180 and 430 lux light intensity. (*Sf* = spring-dwelling form; *Cf* = cave-dwelling form; v = coefficient of variability).

Results

Individuals of *G. affinis* from *Cf* and *Sf* groups displayed different responses to light. At 180 lux, cave-dwelling (*Cf*) individuals showed a significant degree of photophobia (scotophilia) ($\phi_{mean} = -0.59$; $v = 0.19$) while on average the spring-dwelling (*Sf*) individuals did not respond to light ($\phi_{mean} = -0.16$) although the responses among individuals were variable ($v = 0.43$, Table 1, Figure 2). When subject to higher light intensity (430 lux), *Sf* individuals still did not display significant responses ($\phi_{mean} = -0.01$; $v = 0.33$) and *Cf* specimens continued to display the same photophobic responses which was very homogeneous among individuals ($\phi_{mean} = -0.58$; $v = 0.16$). Applying the relationship

$$p = \frac{1}{2}(\phi_t^i + 1)$$

the phototactic index can be converted into the probability p of finding the fish in the L-sector. The theoretical expectation (null hypothesis) for the fish would be $p = 0.50$. The deviation of the *Cf* group from the null hypothesis suggests that individuals of the cave populations of *G. affinis* do respond to light.

Discussion

Gambusia affinis was not known for preferring hypogean habitats. It is unclear on whether this preference for remaining in such environmental conditions is the result of an active (e.g., search of food, protection from predators) or passive (entrapment) explanation. Whatever the reason, there is a behavioural distinction between the hypogean (*Cf*) and epigean (*Sf*) populations. Unlike their epigean counterparts, individuals of the cave populations were photophobic. Further, there were different variation coefficients (υ) for the two groups: epigean individuals responded to light with a great deal of variation even though as a group their phototactic index was near $\phi_t^i = 0$ while the *Cf* group showed an average of $\phi_t^i < 0$ and a low coefficient of variability. This suggests that this natural variability could be the basis for the evolution of the responses to light among cave individuals. The high coefficient of variability in the *Sf* group could have resulted from the formation of a stock with high morphological variability with co-occurence of photonegative and photopositive forms. Undoubtedly scotophilia represents a condition of preadaptation to cave life. Nevertheless this would represent a sufficient condition but not a necessary one. Normally the photonegativity of animals favours the possibility of active colonization of caves and the subsequent regression of eyes and pigments (Ludwig 1942, Romero 1985, Langecker 1992). The active transition model from epigean to groundwater habitats appears to be the more probable one, representing the first step for the processes of troglomorphosis (Christiansen 1962). Accordingly, fish may be more and more evolutionarily adapted to near-hypogean conditions and in this way increase the capability to survive under truly hypogean environmental conditions. Parallel evolutionary trends may have occurred in mysid crustaceans of the genus *Diamysis*, which in part share the same habitats with *Gambusia* fish in the same near hypogean to hypogean environments of the spunnulate dolinas of Torre Castiglione (Wittmann & Ariani[3]).

[3] Wittmann K.J. & A.P. Ariani. 1999. How a mysid goes underground: examples from *Diamysis* and related genera (Crustacea: Mysidacea). Abst. XIVth Intern. Symp. Biospel., Makarska, 19–26 Sept. 1999: 75.

Acknowledgements

The author is greatly indebted to Antonio Ariani at the University of Naples and Karl Wittmann at the University of Wien for critically reading this article.

References cited

Christiansen, K. 1962. Proposition pour la classification des animaux cavernicoles. Spelunca 2: 76–78.

Langecker, T.G. 1989. Studies on the light reaction of epigean and cave populations of *Astyanax fasciatus* (Characidae, Pisces). Mém. Biospéol.16: 169–176.

Langecker, T.G. 1992. Light sensitivity of cave vertebrates behavioral and morphological aspects. pp. 295–326. *In*: A.J. Camacho (ed.) The Natural History of Biospeleology, Museo Nacional de Ciencias Naturales, Madrid.

Ludwig W. 1942. Zur evolutorischen Erklärung der Hőhlentiermerkmale durch Allelelimination. Biol. Zbl. 62: 447–482.

Parenzan P. 1983. Puglia marittima. Galatina et Lecce, Congedo. 688 pp.

Romero, A. 1985. Ontogenetic change in phototactic responses of surface and cave populations of *Astyanax fasciatus* (Pisces: Characidae). Copeia 1985: 1004–1011.

Vawter A.T., D.W. Fong & D.C. Culver. 1987. Negative phototaxis in surface and cave populations of the amphipod *Gammarus minus*. Stygologia 3: 83–88.

Environmental Biology of Fishes **62:** 119–129, 2001.
© 2001 *Kluwer Academic Publishers. Printed in the Netherlands.*

Temporal organization in locomotor activity of the hypogean loach, *Nemacheilus evezardi*, and its epigean ancestor

Atanu Kumar Pati
School of Life Sciences, Pt. Ravishankar Shukla University, Raipur-492 010, India
(e-mail: akpati19@hotmail.com)

Received 30 June 2000 Accepted 24 January 2001

Key words: circadian bimodal rhythm, endogenous oscillators

Synopsis

Locomotor activity rhythm in the hypogean population of *Nemacheilus evezardi* was recorded first under light-to-dark (LD) 12 : 12 h cycle and then DD. The results were compared with that of its epigean counterpart held under comparable regimes. In LD 12 : 12, while hypogean loach exhibited a distinct bimodality in its locomotor activity rhythm, it was altogether absent in the case of epigean population. In hypogean loach, dark-to-light transition peak in LD was observed to free-run under DD. The same was not discernible in case of epigean loach. The locomotor activity rhythm in epigean fish was noticed to free-run in DD either from the dawn peak or dusk peak in LD. It is hypothesized that the hypogean fish still possesses a functional oscillator underlying its overt circadian rhythm in locomotor activity. The ecophysiological significance of these findings is yet to be fully understood.

Introduction

Hypogean animals provide an excellent opportunity for the understanding of the evolutionary processes of adaptation to subterranean life. The extremely evolved cavernicolous animals exhibit remarkable alterations in morphology, physiology and behavior as compared with their epigean relatives. The most obvious ones among those are the visual and pigment systems (Behrens et al. 1997). Additionally, these animals slow down their metabolic rate, optimize energy expenditure (Hüppop 1986, 1989), and acquire well-developed sensory mechanisms (Riedel & Krug 1997, Teyke & Schaerer 1994, Peters et al. 1993, Adbel Latif et al. 1990).

The Kotumsar Cave in the Kanger Valley National Park, Jagdalpur, Bastar, India, has been successfully colonized by the epigean loach, *Nemacheilus evezardi*. The exact time when the cave populations of *N. evezardi* became separated from epigean populations is not known. Interestingly, both forms, as is the case for *Astyanax fasciatus* (= *mexicanus*), belong to the same species (Erckens & Martin 1982). However, unlike in *A. fasciatus*, experimental studies on *N. evezardi* are few.

Biological rhythms in hypogean animals are interesting because they live under a regime of constant darkness and at constant temperature and humidity. On the other hand, lengths of the day in an epigean environment are predictably cyclical and have been known to entrain biorhythms. Circadian rhythms in distribution (Helfman 1986), swimming or locomotor activity (Erckens & Martin 1982, Cordiner & Morgan 1991, Zafar & Morgan 1992) and melatonin synthesis in the pineal organ (Wilkens et al. 1993) have already been documented for various hypogean fishes. However, these studies have not fully answered some basic questions. While most authors have not compared their findings on biological rhythms of hypogean fishes with the epigean ones (for exceptions see Erckens & Martin 1982, Cordiner & Morgan 1991), others have not adopted adequate methodology to characterize endogenous biological rhythms.

Many in-depth studies on biological rhythms in hypogean *N. evezardi* or in its epigean population are yet to be conducted. Statistically significant 24-h rhythms in burying behavior (Biswas 1990, Biswas et al. 1990b), phototactic behavior (Pradhan et al. 1989, Biswas 1994), color change behavior (Pradhan & Biswas 1994), air-gulping behavior (Biswas et al. 1990a) and in a number of biochemical constituents of different tissues (Pradhan 1984) have been reported. However, all of these are studies conducted along either a 24-h or a 48-h timescale.

The main objective of the experiments included in this paper was to investigate further the characteristics of circadian rhythm in hypogean population of *N. evezardi* and to compare the findings with that of its epigean ancestor. The locomotor activity was selected as the test variable. The experiments were designed to evaluate the entrainment properties of locomotor activity rhythm of both hypogean and epigean populations of loach to a light-to-darkness (LD) 12 : 12 photoperiod and to examine the properties of the free-running rhythm under DD (24-h darkness) conditions.

Material and methods

Collection of animals

Individuals of adult hypogean and epigean *N. evezardi* were caught in their natural habitats using hand-held nets. The hypogean loach lives in the water pools of the subterranean Kotumsar Cave located in the Kanger Valley National Park (18°52′09″ N, 81°56′05″ E) in Jagdalpur, Bastar, India. The geophysical and chemical characteristics of the Kotumsar Cave have been described elsewhere (Biswas 1993). The epigean loach inhabits the fast flowing River Kanger in the above national park. The individuals of both hypogean and epigean loaches were transported to Raipur, India, and were acclimated to the conditions of the laboratory.

Acclimation and maintenance

The hypogean and epigean fishes were maintained separately in different stock aquaria, each measuring 60×25×25 cm, placed in two chronocubicles. The hypogean fishes were maintained under complete darkness (DD), whereas epigean fishes were maintained under LD 12 : 12 photoperiod (75 lux : 0 Lux) with lights-on at 6:00 and lights-off at 18:00 h. The lights-on and lights-off were automatically controlled by a digital clock timer (Pandeyatronics, Raipur). The water and room temperatures were relatively constant and the means and standard deviations of recorded temperatures of both cubicles during the acclimation and the entire period of experimentation were found to be 28 ± 0.6°C and 30 ± 1.8°C, respectively. The fishes were fed (Tokyo floating fish food) at least twice in a week and fresh water supplied a day after each feeding at random timings of the day. The irregular schedules of feeding and water replenishments were practiced to ensure that these regimens do not entrain the rhythm in locomotor activity. Restricted meal schedules can modulate phase of the phototactic behavior rhythm in the hypogean loach, *N. evezardi* (Pradhan et al. 1989).

Data acquisition

The locomotor activity of test fishes was monitored and recorded by using a PC-based LIP event recorder (Lakshya Infotech Pvt. Ltd., Raipur). This device employs the principle of IR beam interruptions to record locomotor activity effectively and is an improved version of the device we have been using earlier (Pati et al. 1998). The number of interruptions of the IR beam is proportional to the amount of locomotor activity in a given test fish. Each test fish was kept in a smaller aquarium (15×10×15 cm) with a water outlet at a height of 8 cm in one of the shorter flanks and a water inlet from the top cover. Each aquarium holds ca. 1.25 l of water. The final design of the aquarium was executed after a number of different designs were evaluated. A pair of IR switches (T_x and R_x) was positioned, one each on the opposite sides of the longer flanks of the aquarium. Thus, a single IR beam was produced at a height of 4 cm. The beam alignment was confirmed in the monitor of a PC. The results of IR beam interruptions were accumulated for interval of 15 min. They were then saved in two data files. One file represented a time series consisting of quantitative data with equidistant intervals and the other file contained data of each test animal as a double-plotted actogram.

Experiment 1

Seven hypogean loaches approximately of the same snout to vent length (4.0 ± 0.5 cm) were transferred from DD to LD 12 : 12 (lights-on at 6:00 h) and their locomotor activity was monitored for eight days. Thereafter, they were maintained in DD by removing the lights-on switch from the programmable digital clock timer for the next consecutive 18 days. The animals were neither fed nor water of their aquaria was replenished during the entire period of the experiment.

The door of the chronocubicle was kept completely shut and no one, including the investigator, entered inside the cubicle until after the end of the experiment.

Experiment 2

In this experiment eight epigean loaches approximately of the same snout to vent length (6 ± 0.5 cm) were used. The experimental protocol was identical with the above except that the animals were kept under LD 12 : 12 only for six days. The water temperature, room temperature and locomotor activity were monitored daily in a PC stationed in the control room.

Data presentation and analyses

The qualitative records of locomotor activity were presented fishwise as double-plotted actogram. The files containing time series data on locomotor activity were subjected to cosinor rhythmometry with $\tau = 24$ h and $\tau = 12$ h to validate statistically significant rhythms (Nelson et al. 1979). A rhythm was characterized by three parameters: the mesor (rhythm-adjusted average), the amplitude (half of the difference between the minimum and the maximum of the fitted cosine function), and the acrophase (time of the maximum in fitted cosine function). The acrophase was obtained with its 95% confidence limits, if a rhythm was detected. Two peaks (acrophases) were obtained for the time series fitted with cosine function at $\tau = 12$ h. Rhythm detection was achieved if the null amplitude hypothesis was rejected with $p < 0.05$. The acrophases were expressed in hours (h). A power spectrum method was also employed for detecting prominent periods in locomotor activity of the fishes under LD and DD (de Prins et al. 1986). This method is suitable for time series with missing data as well as for data collected with unequal time intervals (Motohashi et al. 1987).

Results

Locomotor activity in LD

Hypogean fish

The daily records of locomotor activity from two hypogean loaches (CF # 05 & CF # 07) chosen as representatives are depicted in Figure 1. All fishes, including CF # 05 and CF # 07, in LD exhibited bimodal activity. It appears from the actograms that the phase angle to lights-off in most cases was nearly zero. Visually, the time of activity onset was much more precise. The activity began almost synchronously with the onset of darkness. Further, a spurt of activity in each fish was noticed about 3 h before the onset of light and it continued till the end of the first quarter of the photophase. However, the activity cessation was not as precise as that of the activity onset. A sporadic activity was noticed during the latter three-fourth quarters of the photophase.

A statistically significant 24-h rhythm in locomotor activity was detected for all hypogean fishes except CF # 07, when all time series were subjected to cosinor analysis separately. However, the acrophase spread was very large (about 11.7 h), the earliest being at 23.2 h in CF # 07 and the latest at 10.9 h in CF # 04. When a 12-h τ was fitted to the same sets of time series, the acrophase spread was drastically reduced to 1.1 h only, the earliest being at 6.2 h in CF # 02 and the latest at 7.3 h in CF # 01 or CF # 03. Further, the amplitudes of 12-h rhythm were higher in most of the hypogean fishes, except CF # 02, as compared to that of the 24-h rhythm (Table 1).

Prominent periods resulting from power spectrum analysis are shown in Table 5. Results indicate that four out of seven hypogean fishes under LD 12 : 12 documented a 12-h τ in their locomotor activity rhythm (see Figure 3 for examples). However, CF # 02 and CF # 06 had a 24-h period whereas CF # 03 had a $\tau = 26.18$ h. The next prominent period in locomotor activity rhythm of CF # 06 was 12.0 h.

Epigean fish

The actograms (Figure 2) depict locomotor activity of epigean fishes (RF # 02 and RF # 08) maintained first under LD 12 : 12 and then DD. Most of the epigean fishes were entrained to the LD cycle and exhibited the greatest part of their activity during the dark phase (lights-on at 6:00 h and lights-off at 18:00 h). Both activity onset and cessation were equally precise. In the actograms the bimodal pattern in activity was not apparent in any one of the epigean fishes.

A statistically significant 24-h rhythm was documented for six out of eight epigean fishes (Table 2). However, a statistically significant 12-h rhythm also was found for all epigean fishes. The acrophase spread (ca. 3.0 h) was shorter when the fitted τ was 24 h as compared to that (about 5.0 h) with $\tau = 12$ h. In most of the cases the amplitude of 12-h rhythm was shorter than that of the 24-h rhythm.

In four out of eight LD-acclimated epigean fishes, the prominent period was 24 h and in another two

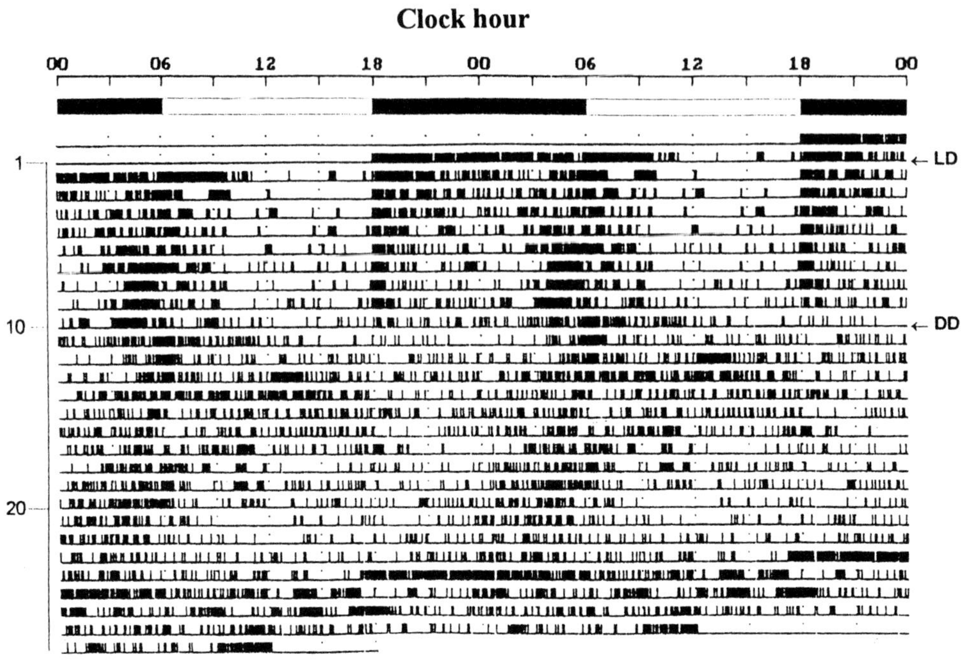

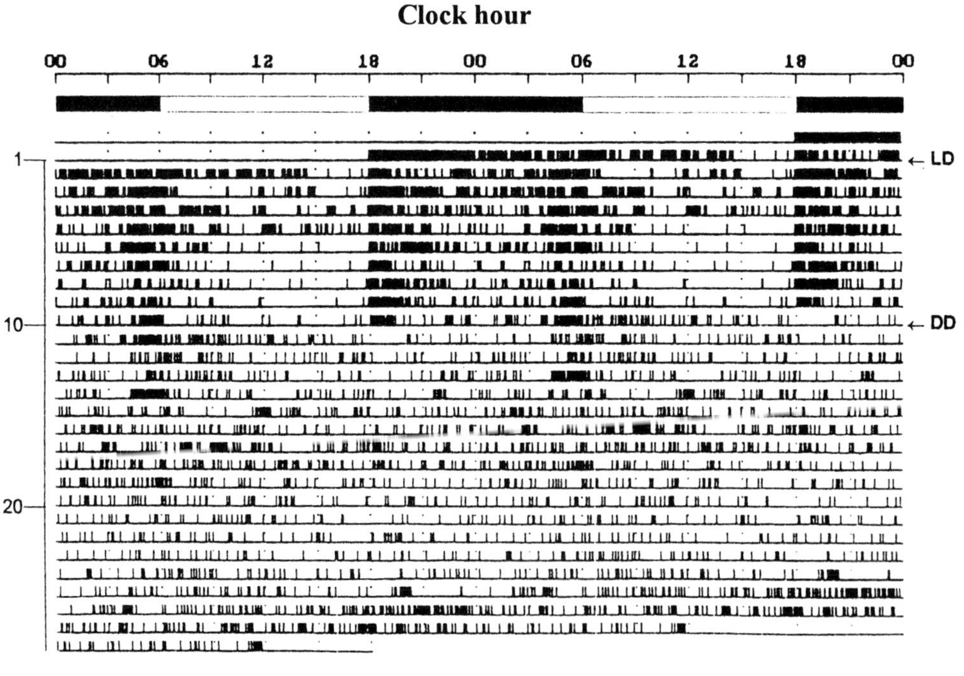

Figure 1. Actogram records in double plot showing locomotor activity rhythms of two hypogean fishes maintained firstly under LD cycles, and then DD. Upper panel = CF#05; lower panel = CF#07.

Table 1. Cosinor summary of locomotor activity rhythm in LD-acclimated hypogean loach, *Nemacheilus evezardi*.

Animal code	Data point	Fitted τ[a]	Rhythm detection (p)[b]	Rhythm-adjusted mean[c] ±1 SE	Amplitude, A[d] (95% CL)	Acrophase Ø[e] in h (95% CL)
CF # 01	867	24	<0.001	3.26 ± 0.19	1.89 (1.23, 2.56)	6.2 (04.8, 07.5)
		12	<0.001		3.04 (2.41, 3.68)	7.3 (06.9, 07.7)
						19.3 (18.9, 19.7)
CF # 02	866	24	<0.001	15.70 ± 0.82	16.48 (13.63, 19.34)	0.3 (23.3, 01.0)
		12	0.007		4.06 (0.92, 7.21)	6.2 (04.5, 07.9)
						18.2 (16.5, 19.9)
CF # 03	867	24	<0.001	4.50 ± 0.36	2.13 (0.89, 3.38)	0.5 (22.2, 02.9)
		12	<0.001		2.50 (1.26, 3.74)	7.3 (06.3, 08.3)
						19.3 (18.3, 20.3)
CF # 04	867	24	0.012	1.86 ± 0.09	0.39 (0.07, 0.70)	10.9 (07.2, 14.6)
		12	<0.001		0.90 (0.59, 1.21)	6.9 (06.2, 07.5)
						18.9 (18.2, 19.5)
CF # 05	867	24	<0.001	4.15 ± 0.23	1.54 (0.73, 2.34)	6.0 (03.9, 08.1)
		12	<0.001		2.91 (2.13, 3.69)	7.2 (06.7, 07.7)
						19.2 (18.7, 19.7)
CF # 06	867	24	<0.001	2.75 ± 0.11	1.84 (1.44, 2.23)	0.9 (00.1, 01.7)
		12	<0.001		1.73 (1.33, 2.13)	6.8 (06.3, 07.2)
						18.8 (18.3, 19.2)
CF # 07	867	24	0.599	4.35 ± 0.24	0.34	23.2
		12	<0.001		2.59 (1.79, 3.39)	7.1 (06.5, 07.7)
						19.1 (18.5, 19.7)

[a]Fitted period.
[b]From F-test of null amplitude rejection hypothesis.
[c]Rhythm-adjusted mean of best fitting cosine function ±1 standard error.
[d]Half of the difference between the maximum and the minimum of best-fitting cosine function (95% confidence interval).
[e]Time in hour of maximum in the best-fitting cosine function (95% confidence interval).

(CF # 01 and CF # 04) it was 21 h (Table 5, examples in Figure 4). In RF # 06 and RF # 07 the prominent periods were 12 and 18.66 h, respectively.

Locomotor activity in DD

Hypogean fish
A visual analysis of the DD activity records indicated that at least five hypogean individuals exhibited locomotor activity rhythms with a 24-h component. That is, although free-running rhythms of locomotor activity were evident in DD, the precise locations of activity onset, activity cessation, or activity midpoints were absent (Figure 1).

Results of cosinor analysis for CF # 01, CF # 04 and CF # 05 neither reveal statistically significant 24-h nor 12-h rhythm (Table 3). Further while CF # 03 and CF # 06 exhibited statistically significant 24-h rhythm, both 24-h and 12-h rhythms were witnessed in CF # 02 and CF # 07. The amplitudes of both rhythms declined in most of the cases.

Results of spectral analysis reveal that in five out of eight cases locomotor activity rhythm free-ran with periods very close to 24 h (Table 5, examples in Figure 3). The free-running periods in CF # 05 and CF # 07 were found to be 21.71 h and 25.33 h, respectively (Figure 3).

Epigean fish
The actograms of epigean fishes maintained under DD, much like their hypogean counterpart, did not exhibit a distinct free-running rhythm in locomotor activity (Figure 2). Most of the fishes neither exhibited statistically significant 24-h rhythm nor 12-h rhythm when data were subjected to cosinor rhythmometry (Table 4). However, interestingly the average level of locomotor activity increased dramatically in most of the cases as compared with that of the hypogean fishes maintained either in LD or DD and epigean fishes maintained under LD (Tables 1–4). Results of spectral analysis reveal that most of the epigean fishes had τ very close to 24 h (Table 5, examples in Figure 4).

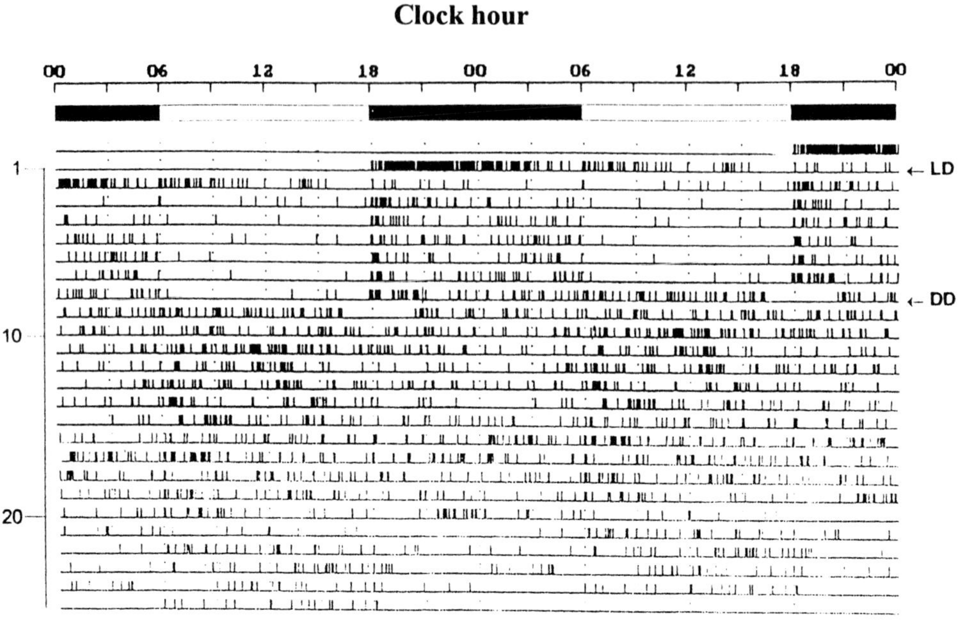

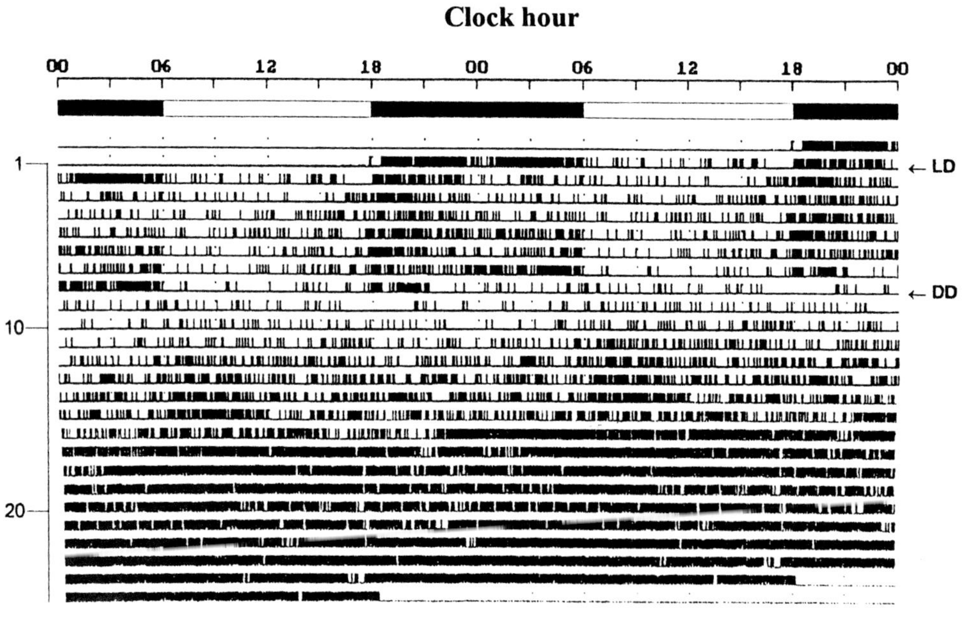

Figure 2. Actogram records in double plot showing locomotor activity rhythms of two epigean fishes maintained firstly under LD cycles, and then DD. Upper panel = RF#02; lower panel = RF#08.

Discussion

The hypogean loach, *N. evezardi*, showed a strong bimodal pattern in its locomotor activity rhythm. It was characterized by the presence of two peaks, one at the time of lights-off and the other at the lights-on. Almost all fishes anticipated dark-to-light transitions, but not light-to-dark transitions. The statistical validation of

Table 2. Cosinor summary of locomotor activity rhythm in LD-acclimated epigean loach, *Nemacheilus evezardi.*

Animal code	Data point	Fitted τ[a]	Rhythm detection (p)[b]	Rhythm-adjusted mean[c] ±1 SE	Amplitude, A[d] (95% CL)	Acrophase, Ø[e] in h (95% CL)
RF#01	672	24	<0.001	2.37 ± 0.14	1.05 (0.55, 1.55)	20.6 (18.7, 22.5)
		12	0.035		0.53 (0.03, 1.04)	7.0 (04.6, 09.4)
						19.0 (16.6, 21.4)
RF#02	672	24	<0.001	1.30 ± 0.13	1.29 (0.85, 1.73)	22.4 (21.0, 23.7)
		12	0.001		0.67 (0.22, 1.12)	8.7 (07.3, 10.1)
						20.7 (19.3, 22.1)
RF#03	672	24	<0.001	1.31 ± 0.12	1.54 (1.12, 1.95)	22.6 (21.6, 23.7)
		12	<0.001		0.77 (0.34, 1.21)	8.3 (07.1, 09.4)
						20.3 (19.1, 21.4)
RF#04	672	24	0.339	2.21 ± 0.12	0.24	21.8
		12	0.023		0.45 (0.05, 0.85)	7.7 (05.6, 09.8)
						19.7 (17.6, 21.8)
RF#05	672	24	<0.001	5.07 ± 0.48	6.96 (5.29, 8.63)	22.6 (21.7, 23.5)
		12	<0.001		2.73 (0.95, 4.51)	10.0 (08.6, 11.3)
						22.0 (20.6, 23.3)
RF#06	672	24	0.052	4.01 ± 0.18	0.61	23.4
		12	<0.001		1.78 (1.18, 2.38)	7.4 (06.8, 08.1)
						19.4 (18.8, 20.1)
RF#07	672	24	<0.001	1.38 ± 0.12	1.01 (0.59, 1.44)	21.7 (20.1, 23.4)
		12	0.011		0.53 (0.10, 0.96)	7.7 (05.9, 09.5)
						19.7 (17.9, 21.5)
RF#08	672	24	<0.001	6.39 ± 0.47	4.83 (3.18, 6.47)	23.6 (22.3, 00.9)
		12	0.019		1.95 (0.25, 3.65)	5.0 (03.0, 07.0)
						17.0 (15.0, 19.0)

For legend see Table 1.

bimodality was achieved by analyzing data with the help of cosinor rhythmometry. A statistically significant 12-h rhythm was found in all test animals. Further, results of spectral analysis complemented both visual observations of bimodality and detection of cosinor-validated 12-h rhythms. Interestingly, this bimodality in locomotor activity pattern was covert in case of LD acclimated epigean loaches. We term it covert because visual bimodality in actograms was not as clear and apparent as it was in case of hypogean loaches. Further, there was a distinct disagreement between cosinor and spectral analysis results. A 12-h τ was detected by power spectrum method only in one epigean fish, notwithstanding detection of statistically significant 12-h rhythm (by cosinor) in all epigean animals. This obvious difference between the hypogean and the epigean stocks concerning bimodality in locomotor activity rhythm is difficult to explain. There are no comparable findings for any other hypogean species. Further, the pattern of cessation of locomotor activity was another apparent difference between the epigean and hypogean fishes. The former exhibited an abrupt cessation of activity following lights-on whereas the latter continued to remain active until 3 h after light onset. At present it is difficult to explain the significance of this difference.

Both stocks of *N. evezardi*, however, seem to be night active because a major chunk of their total activity on any given day occurred in the dark phase of the LD regime. The phenomenon of entrainment of locomotor activity rhythm to LD 12 : 12 photoperiod has been documented in many other hypogean species, namely the Pachón and La Cueva Chica populations of *A. fasciatus* (Erckens & Martin 1982, Zafar & Morgan 1992, and Mitchell et al. 1977, respectively), for cave crayfish, *Orconectes pellucidus* (Brown 1961), and the cave cricket, *Hadenoecus subterraneus* (Reichle et al. 1965). Most of these cavernicolous species lack the ability to self-sustain their rhythms under DD. There has been a notable exception in case of the Pachón population of *A. fasciatus*. This population possesses an oscillator for at least surface activity rhythm with a restricted ability of self-sustenance. However, if the period of the zeitgeber is non-24 h, for example LD

Table 3. Cosinor summary of locomotor activity rhythm in DD-acclimated hypogean loach, *Nemacheilus evezardi.*

Animal code	Data point	Fitted τ^a	Rhythm detection $(P)^b$	Rhythm-adjusted mean[c] ± 1 SE	Amplitude, A[d] (95% CL)	Acrophase, Ø[e] in h (95% CL)
CF # 01	1708	24	0.796	0.82 ± 0.04	0.04	20.7
		12	0.285		0.09	6.2
						18.2
CF # 02	1708	24	<0.001	19.67 ± 0.61	13.11 (11.00, 15.22)	7.7 (07.1, 08.3)
		12	<0.001		4.97 (2.73, 7.21)	8.0 (07.1, 08.9)
						20.0 (19.1, 20.9)
CF # 03	1708	24	<0.001	3.21 ± 0.16	0.96 (0.39, 1.52)	6.0 (03.6, 08.4)
		12	0.150		0.45	8.5
						20.5
CF # 04	1708	24	0.959	0.65 ± 0.04	0.01	7.1
		12	0.342		0.08	5.7
						17.7
CF # 05	1708	24	0.067	6.20 ± 0.25	0.84	6.2
		12	0.345		0.53	0.4
						12.4
CF # 06	1708	24	0.05	5.23 ± 0.16	0.55 (0.01, 1.09)	04.1 (22.2, 10.0)
		12	0.299		0.35	7.5
						19.5
CF # 07	1708	24	<0.001	2.22 ± 0.09	0.51 (0.19, 0.82)	5.1 (02.6, 07.7)
		12	0.022		0.35 (0.04, 0.67)	7.0 (05.0, 09.1)
						19.0 (17.0, 21.1)

For legend see Table 1.

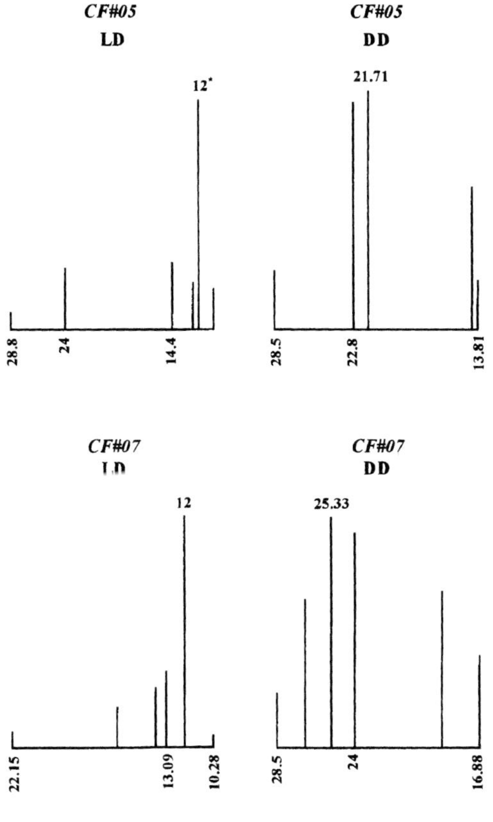

Figure 3. Examples showing power spectra for locomotor activity rhythm in the LD- and DD-acclimated hypogean fishes (CF # 05 & CF # 07). *Period (τ).

8 : 8, before transfer under DD then this fish does not exhibit free-running circadian rhythm (Erckens & Martin 1982).

The present study reveals that in the majority of hypogean loaches, the locomotor activity rhythm free-ran with individually varying periods. In some animals the period length was less than 24 h while in others it was greater than 24 h. However, in both cases, theoretical free-runs could not be revealed in the actogram. It was indeed difficult to locate precise timings of either activity onset or activity cessation, though a free-running pattern was discernible. This conclusion is supported by results of power spectrum analysis. This strongly suggests that the hypogean loach has an endogenous circadian rhythm in locomotor activity. The epigean population also behaved identically with that of the hypogean population under DD. Although visual inspection of the actograms did not reveal clear free-running rhythms in DD, power spectrum analysis indicated that six out eight epigean fishes exhibited free-running rhythms with period length close to 24 h. It has been reported that in fish, in gen-

Table 4. Cosinor summary of locomotor activity rhythm in DD-acclimated epigean loach, *Nemacheilus evezardi*.

Animal code	Data point	Fitted τ^a	Rhythm detection (p)[b]	Rhythm-adjusted mean[c] ±1 SE	Amplitude, A[d] (95% CL)	Acrophase, Ø[e] in h (95% CL)
RF#01	1708	24	0.938	5.85 ± 0.97	0.49	0.3
		12	0.810		0.89	1.3
						13.3
RF#02	1708	24	0.330	4.70 ± 1.08	2.27	2.8
		12	0.490		1.82	2.0
						14.0
RF#03	1708	24	0.021	15.76 ± 0.54	2.15 (0.26, 4.04)	5.8 (01.7, 09.9)
		12	0.331		1.14	1.4
						13.4
RF#04	1708	24	0.736	16.61 ± 1.39	1.54	8.0
		12	0.791		1.34	3.5
						15.5
RF#05	1708	24	0.053	11.15 ± 1.43	4.87	1.3
		12	0.475		2.46	3.1
						15.1
RF#06	1708	24	0.063	33.57 ± 1.48	4.91	9.5
		12	0.241		3.52	8.3
						20.3
RF#07	1708	24	0.953	15.84 ± 1.57	0.69	10.2
		12	0.949		0.72	10.8
						22.8
RF#08	1708	24	<0.001	26.23 ± 1.48	9.66 (4.51, 14.81)	7.7 (05.6, 09.8)
		12	0.630		2.03	5.7
						17.7

For legend see Table 1.

eral, a free-running locomotor activity rhythm is rare (Eriksson & van Veen 1980, Gibson 1971, Eriksson 1975, Schwassmann 1971).

Interestingly, in the hypogean loach the dark-to-light transition peak in LD was found to free-run under DD. This suggests that the dawn peak seems to be an endogenous component of circadian rhythm entrained to LD cycle. Concerning the initiation of activity the hypogean fish always exhibited a positive phase angle with respect to lights-on timings under LD. This suggests that the hypogean loach is still capable of anticipating the time of light onset. However, it is not known if they would continue to self-sustain their circadian rhythms following their transfer from any non-circadian LDs to DD. The epigean fish, however, exhibited a dual pattern. In some cases light-to-dark transition peak in LD seems to free-run while in others dark-to-light peak free-runs. The present study is not adequate to explain this duality.

Present results suggest that probably both epigean and hypogean loaches have two circadian oscillators. In the latter while one is functional the other may be dormant or in the process of being functionally reduced. The dusk (light-to-dark transition) peak in locomotor activity of the hypogean fish under LD 12 : 12 could also be a reflection of masking attributed to sudden switching off of the light. Experiments simulating natural dawn and dusk would resolve the above uncertainty.

These results strongly suggest that the hypogean loaches seem to possess at least one functional oscillator that underlies the overt circadian rhythm. Theoretically the hypogean fish should have lost its sense of time and its circadian clock. Instead it exhibits a strong circadian bimodal rhythm in LD and expresses an endogenous circadian component under DD. At present, there could be two possible explanations: (1) The hypogean fish is relatively recent occupant of the subterranean environment and is still undergoing adaptational changes before finally losing its circadian abilities. (2) There may be some subtle non-photic cues present inside Kotumsar Cave thus facilitating retention of the time-measuring mechanism. The exact age when cave population became separated from its epigean counterpart is yet to be ascertained. Furthermore, currently work is in progress to identify various cues that might be present inside the Kotumsar Cave

Table 5. Prominent periods[a] in LD- and DD-acclimated hypogean and epigean, *Nemacheilus evezardi.*

Animal code	Hypogean fish		Epigean fish		
	LD	DD	Animal code	LD	DD
CF # 01	12	8.29	RF # 01	21	21
CF # 02	24	24	RF # 02	24	22.4
CF # 03	26.18	25.33	RF # 03	24	25.84
CF # 04	12	12.32	RF # 04	21	25.84
CF # 05	12	21.71	RF # 05	24	24
CF # 06[b]	24 (12)	28.5, 22.8, 21.71	RF # 06	12	21
CF # 07	12	25.33	RF # 07	18.66	24
			RF # 08	24	25.84

[a]By power spectrum analysis (de Prins et al. 1986).
[b]The power for locomotor activity rhythm in DD for CF # 06 was the same for three different τ.

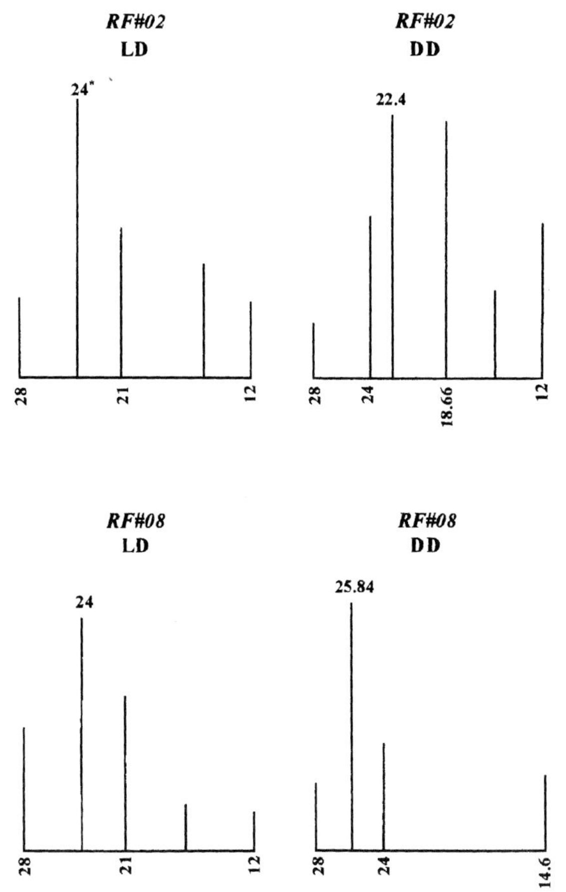

Figure 4. Examples showing power spectra for locomotor activity rhythm in the LD- and DD-acclimated epigean fishes (RF # 02 & RF # 08). *Period (τ).

and likely to provide an advantage to the hypogean population of *N. evezardi* thereby forcing it to maintain a circadian bimodal rhythm in its locomotor activity. Further studies focusing on the effects of LD with varying light intensities and LL on characteristics of locomotor activity in both population of *N. evezardi* may yield interesting results.

Acknowledgements

I thank Arvind Agrawal and Arti Chandrawanshi in helping preparing this manuscript. I am grateful to Kaushalendra Kumar, director of the Kanger Valley National Park, Jagdalpur, Bastar, India, who allowed me and my team to enter inside the Kotumsar Cave a number of occasions. I thank M.L. Naik, School of Life Sciences, Pt. Ravishankar Shukla University, Raipur, India, for providing me with the adequate research facilities. This research was supported by grants from the Department of Science and Technology, Ministry of Science and Technology, Government of India, New Delhi (No. SP/SO/C-17/93).

References cited

Adbel Latif, H., E.S. Hassan & C. von Campenhausen. 1990. Sensory performance of blind Mexican cave fish after destruction of the canal neuromasts. Naturwiss. 77: 237–239.

Behrens, M., T.G. Langecker, H. Wilkens & H. Schmale. 1997. Comparative analysis Pax-6 sequence & expression in the eye development of the blind cave fish *Astyanax fasciatus* and its epigean conspecific. Mol. Biol. Evol. 14: 299–308.

Biswas, J. 1990. Biospeleology: the behavioural and physiologic adaptations in a cavernicole. Ph.D. Thesis, Pandit Ravishankar Shukla University, Raipur. 136 pp.

Biswas, J. 1993. Kotumsar Cave ecosystem: an interaction between geophysical, chemical, and biological characteristics. Nat. Speleol. Soc. Bull. 54: 7–10.

Biswas, J. 1994. Effects of different light intensities on phototactic responses of a hill-stream loach, *Lepidocephalus guntea* (Ham.): a chronobiological approach. Proc. Indian Nat. Sci. Acad. 60B: 107–114.

Biswas, J., A.K. Pati & R.K. Pradhan. 1990a. Circadian and circannual rhythms in air-gulping behavior of cave fish. J. Interdisc. Cycle Res. 21: 257–268.

Biswas, J., R.K. Pradhan & A.K. Pati. 1990b. Studies on burying behavior in epigean and hypogean fish, *Oreonectus evezardi*: an example of behavioral divergence. Mem. Biospeleol. 17: 33–41.

Brown, F.A. 1961. Diurnal rhythm in cave crayfish. Nature 191: 929–930.

Cordiner, S. & E. Morgan. 1991. Endogenous locomotor activity patterns in the blind Mexican cave fish *Astyanax mexicanus*. J. Interdisc. Cycle Res. 22: 103.

De Prins, J., G. Cornélissen & W. Malbecq. 1986. Statistical procedures in chronobiology and chronopharmacology. Annu. Rev. Chronopharmacol. 2: 27–141.

Erckens, W. & W. Martin. 1982. Exogenous and endogenous control of swimming activity in *Astyanax mexicanus* (Characidae, Pisces) by direct light response and by a circadian oscillator. II. Features of time controlled behavior of a cave population and their comparison to a epigean ancestral form. Z. Naturforsch. Sect. C, Biosci. 37: 1266–1273.

Eriksson, L.O. 1975. Diel and annual locomotor activity rhythms in some fresh water fish species with special reference to the seasonal inversion in salmonids. Ph.D. Thesis, University of Umeå, Umeå. 178 pp.

Eriksson, L.O. & T. van Veen. 1980. Circadian rhythms in the brown bullhead, *Ictalurus nebulosus* (Teleostei). Evidence for an endogenous rhythm in feeding, locomotor, and reaction time behavior. Can. J. Zool. 58: 1899–1907.

Gibson, R.N. 1971. Factors affecting the rhythmic activity of *Blennius pholis* L. (Teleostei). Anim. Behav. 19: 336–343.

Helfman, G.S. 1986. Diel distribution and activity of American eels (*Anguilla rostrata)* in a cave-spring. Can. J. Fish Aquat. Sci. 43: 1595–1605.

Hüppop, K. 1986. Oxygen consumption of *Astyanax fasciatus* (Characidae, Pisces): a comparison of epigean and hypogean populations. Env. Biol. Fish. 17: 299–308.

Hüppop, K. 1989. Genetic analysis of oxygen consumption rate in cave and surface fish of *Astyanax fasciatus* (Characidae, Pisces): further support for the netural mutation theory. Mem. Biospeleol. 16: 163–168.

Mitchell, R.W., W.H. Russell & W.E. Elliott. 1977. Mexican eyeless characin fishes, genus *Astyanax*: environment, distribution and evolution. Sp. Publ. Mus. Texas Tech. Univ. 12: 3–89.

Motohashi, Y., A. Reinberg,, F. Lévi, J. Nouguier, O. Benoit, J. Foret & P. Bourdeleau. 1987. Axillary temperature: a circadian marker rhythm for shift workers. Ergonomics 30: 1235–1247.

Nelson, W., Y.L. Tong, J.K. Lee & F. Halberg. 1979. Methods for cosinor rhythmometry. Chronobiologia 6: 305–323.

Pati, A.K., S. Gupta & R. Maheshwari. 1998. Optoelectronic device, a useful tool to record surfacing behavior in air-breathing fishes. Curr. Sci. 74: 146–149.

Peters, N., V. Schacht, W. Schmidt & H. Wilkens. 1993. Cerebral proportions and degree of development of sense organs in *Astyanax mexicanus* (Pisces, Characinidae) comparison between river fish and its cave-living forms anoptichthys. Zeit. Zool. Syst. Evol. 31: 144–159.

Pradhan, R.K. 1984. Biochemical studies of some tissues of *Nemacheilus evezardi* Day from Kotumsar Cave. Ph.D. Thesis, Pt. Ravishankar Shukla University, Raipur. 195 pp.

Pradhan, R.K. & J. Biswas. 1994. Towards regressive evolution: the periodic color change behavior of a troglophilic fish, *Nemacheilus evezardi* (Day). Int. J. Speleol. 23: 191–201.

Pradhan, R.K., A.K. Pati & S.M. Agarwal. 1989. Meal scheduling modulation of circadian rhythm of phototactic behavior in cave dwelling fish. Chronobiol. Int. 6: 245–249.

Reichle, D.E., J.D. Palmer & O. Park. 1965. Persistent rhythmic locomotor activity in the cave cricket, *Hadenoecus subterraneus*, and its ecological significance. Amer. Midl. Nat. 74: 57–66.

Riedel, G. & L. Krug. 1997. The forebrain of the blind cave fish *Astyanax hubbsi* (Characidae). II. Projections of the olfactory bulb. Brain Behav. Evol. 49: 39–52.

Schwassmann, H.O. 1971. Biological rhythms. pp. 371–428. *In*: W.S. Hoar & D.J. Randall (ed.) Fish Physiology, Vol. 6, Academic Press, New York.

Teyke, T. & S. Schaerer. 1994. Blind Mexican cave fish *Astyanax hubbsi* respond to moving visual stimuli. J. Expt. Biol. 188: 89–101.

Wilkens, H., T.G. Langecker & J. Olcese. 1993. Circadian rhythms of melatonin synthesis in the pineal organ of cave-dwelling *Astyanax fasciatus* (Teleostei: Characidae). Mem. Biospeleol. 20: 279–282.

Zafar, N.P. & E. Morgan. 1992. Feeding entrains an endogenous rhythm of swimming activity in the blind mexican cave fish. J. Interdisc. Cycle Res. 23: 165–166.

Part 2. Ecology and conservation

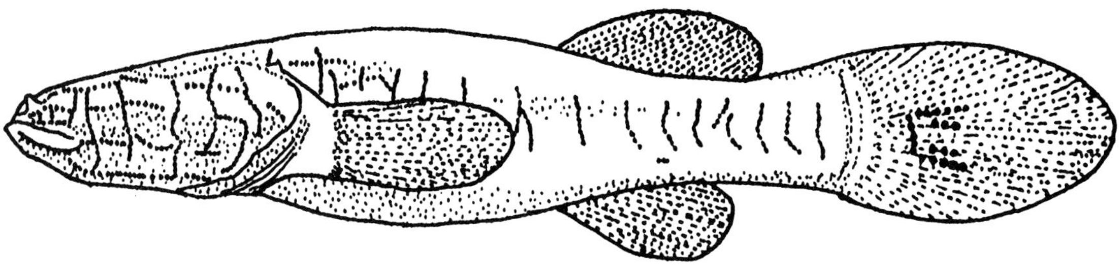

Environmental Biology of Fishes **62:** 133–160, 2001.
© 2001 *Kluwer Academic Publishers. Printed in the Netherlands.*

Ecology of subterranean fishes: an overview

Eleonora Trajano
Departamento de Zoologia, Instituto de Biociências da USP, C. P. 11461, 05422-970, São Paulo SP, Brazil (e-mail: etrajano@usp.br)

Received 26 April 2000 Accepted 14 January 2001

Key words: troglobitic fishes, population ecology, fish movements, life cycle, *Ancistrus*

Synopsis

A synthesis of ecological data available for subterranean fishes throughout the world is presented, and comparatively analyzed in an evolutionary context. Methods of ecological research are described, and their potential and limitations for the study of hypogean fishes are discussed. Ecology of troglobitic (exclusively subterranean) fishes is discussed with focus on distribution areas, population densities and sizes, use of habitat and movements, life cycle and feeding. When data are available, these species are compared with their epigean relatives. Putative ecological autapomorphies of troglobitic fishes, including habitat change, adaptations to cope with food scarcity, and precocial lifestyles, are interpreted in ecological and evolutionary contexts. Species interactions among subterranean species, including cases of syntopy and predation are briefly analyzed. Non-troglobitic hypogean fishes, with their ecological importance and evolutionary role, are also addressed. Problems of classification of subterranean fishes according to the Schiner-Racovitza system (troglobites, troglophiles and trogloxenes) are discussed, and a scenario of evolution of subterranean populations is presented.

Introduction

Troglobites, or species restricted to the subterranean habitat, usually show a series of autapomorphies (troglomorphisms), including morphological, physiological, behavioral and ecological features, which can be directly or indirectly related to the hypogean life, being either adaptive or not (for a discussion of theories about character evolution in the cave habitat, see Culver et al. 1995). The most common morphological traits characterizing troglobites are reduction of both eyes and pigmentation. Ecological traits frequently cited for troglobites include low population sizes, restricted geographic distribution, low tolerance to environmental stress, precocial life cycle features (e.g., Culver 1982, Trajano 1997a) related to the life under permanent darkness, few environmental cues, usually restricted space, and frequent food scarcity.

More than 86 species of subterranean troglobitic fishes belonging to 18 families in nine orders have been recorded throughout the world (Weber et al. 1998 among others), the majority being found in caves. Cave fishes are fairly easy to investigate in the field because they are relatively large, generally conspicuous, and not difficult to observe and collect, as many species have lost phobic reactions and show increased activity. Nevertheless, there have been few studies in this area, partly because ecological investigations require long-term permanence in the field and many troglobitic fishes occur in tropical regions, sometimes in caves of difficult access, where there are few local researchers.

The relatively abundant cavefish literature (about 1600 references, G. Proudlove personal communication) is dominated by studies on systematics, genetics, morphology, physiology and behavior in the laboratory and focuses on few taxa, especially the blind Mexican characins, genus *Astyanax*, followed by North American amblyopsids. A low proportion of these publications (fewer than 4%) dealt strictly with ecology of subterranean fishes.

One can trace ecological data scattered in the literature on cave fishes, including habitat descriptions and comments on habitat use, abundance, reproduction and behavior, back to Eigenmann's (1909) 'Cave Vertebrates of America', or earlier. However, detailed ecological studies with population analyses started with the investigation of the African cyprinid, *Caecobarbus geertsii*, by Heuts (1952). Some years earlier, Pavan (1946) published a population study on the Brazilian blind catfish, *Pimelodella kronei*, but essentially within an evolutionary context. Among the ecological studies published since then, we can mention the comprehensive study on the North American amblyopsids by Poulson (1963), also the object of recent studies by Means & Johnson (1995) and Brown (1996); the rare ecological studies on the otherwise intensively investigated Mexican blind tetras, *Astyanax* spp. (Mitchell et al. 1977, Parzefall 1983); and a series of ecological studies on Brazilian troglobitic catfishes published in the last decade (Trajano 1991, 1997a,b, Mendes 1995a,b, Trajano & Bockmann 2000).

These works included information on population sizes and densities, fish movements and life history aspects besides feeding (reproduction, growth rates, etc.). However, few authors studied feeding beyond general diet (as Trajano 1997b, focusing on temporal variation and seasonality). Likewise, few publications included a detailed description of cavefish habitats (as those by Heuts 1952, for *C. geertsi*, and Mitchell et al. 1977, for *Astyanax* spp.) and habitat preferences. Mitchell et al. (1977) made the most thoughtful and comprehensive description of habitats occupied by cave fishes, listing and describing all localities known for Mexican troglobitic *Astyanax*, and detailing the geomorphology and hydrology in the so-called Huastecan Province; such a publication provides a solid basis for discussion of the origin and evolution of these populations. The modern tendency is to greatly abbreviate habitat descriptions, which rarely include good maps, and information on fish microhabitats based on careful field and laboratory observation is scarce, making it difficult to get a proper notion of habitat use. Therefore, ecological aspects such as habitat change and specialization are usually overlooked.

Herein I present an overview of results from ecological studies on hypogean fishes throughout the world, comparatively analyzing data on population sizes and densities, use of space (habitat, movements, species syntopy), feeding and life history of different taxa, as a contribution to the knowledge of fish species restricted to this special habitat. When data are available, subterranean populations were compared to those of epigean close relatives living in the same area, in an attempt to distinguish between plesiomorphic character states and ecological autapomorphies of troglobites, resulting from specialization to the particular conditions in the subterranean habitat. Because few troglobitic species, and even fewer epigean relatives, have been studied with focus on ecological aspects, general patterns may be difficult to visualize. The present overview shows how much there still is to be learned about these fascinating animals.

Methods of ecological studies in subterranean fishes: a comparative analysis

Basically, three kinds of methods have been used in studies of the ecology of subterranean fishes: extensive collections of preserved samples, direct observation in the habitat, and capture-mark-recapture programs. Different methods are the most useful for investigating different aspects of ecology, and each one has its advantages and limitations. Therefore, they must be regarded as complementary.

Two methods have been applied to determine population sizes and densities in cave fishes: visual censuses (VC) and mark-recapture (MR) techniques. In visual censuses, 'instantaneous' counts of the number of fish seen in the habitat are made from the surface or during snorkeling or SCUBA diving. These censuses may encompass the total accessible habitat, as done for amblyopsids (Poulson[1], Willis & Brown 1985, Brown & Todd 1987) and the Brazilian catfish, *Taunayia* sp. (Trajano & Bockmann 2000). In other cases, selected sections of the cave habitat are inspected, such as small areas, for *Astyanax* spp. (Parzefall 1983), transects, for an undescribed genus and species of Brazilian heptapterine from NE Brazil (Mendes 1995a – the species is here cited as *Imparfinis* sp.), or stream sectors, for Thai balitorids, *Schistura oedipus* and *Cryptotora thamicola* (Trajano, Borowsky, Mugue & Smart unpublished data).

Visual censuses produce estimates of minimum population densities, from which one can calculate the total population size if: (1) the total habitat area occupied by the fish population is known, and (2) population density remains constant throughout this area. These two conditions are rarely met and such estimates generally provide only a rough comparative notion of minimum

[1] Poulson, T.L. 1969. Population size, density and regulation in cave fishes. pp. 4–5. *In*: Proc. 4th International Congress of Speleology, Ljubljana.

population sizes. We must keep in mind that VC produce underestimates, especially when done from the surface; as pointed out by Brown & Todd (1987), snorkeling greatly increases accuracy in VC.

Mark-recapture methods produce more accurate population data, but are time-consuming, demanding more visits to the study site and adequate fish marking techniques. In addition to assessment of population sizes, individual marking allows determination of fish movements and home ranges, and estimation of growth rates without the need for large preserved samples, frequently difficult to obtain due to the small size and ecological fragility of many troglobitic populations.

Few MR studies on cavefish have been carried out so far. Medium (one season) to long term (one to several years) studies comprising multiple marking and recapture occasions were done for Brazilian catfishes, *Pimelodella kronei*, in Areias Cave (Trajano 1991) and *T. itacarambiensis* in Olhos d'Água cave (Trajano 1997a), and the North American Ozark cavefish, *Amblyopsis rosae*, in Logan Cave (Means & Johnson 1995, Brown 1996); a study on Mexican ictalurids, *Prietella phreatophila* and *P. lundbergi*, is in progress (Krejca et al. 2000). Two-census MR studies were carried out for the Brazilian armored catfish, *A. cryptophthalmus* in Angélica and Passa Três caves (Trajano 2001), and troglobitic *Astyanax* in Pachon and Yerbaniz caves (Mitchell et al. 1977). Depending on the extension and difficulty of progression along the cave, and on population densities associated with difficulty in capturing fish (determining the number of fish to be marked), the whole accessible cave habitat may be studied, as was the case with *A. rosae* and *T. itacarambiensis*. In other cases, only part of the habitat was investigated (*P. kronei*, *A. cryptophthalmus*, *P. mexicana*). For the latter, estimates from studied sectors in the caves have been extended to the whole cave population, assuming constant population densities; such estimates are, therefore, more subject to errors.

Individual marking has been achieved using subcutaneous marks by, for example, injecting biocompatible pigments (tattooing), as in *P. kronei*, *T. itacarambiensis* and *Ancistrus cryptophthalmus*, or visually implanting tags, in *Amblyopsis rosae*. The fishes were captured in hand nets (*T. itacarambiensis*, *A. cryptophthalmus*, *A. rosae*) or minnow traps (*P. kronei*), anesthetized, measured, weighed, examined for individual natural marks and gonad development (seen by transparency), marked and released in the collection subsections. Usually only fish larger than a certain size (adults and large juveniles) have been included in this kind of study, either because young fishes are not located and captured, or because they are too small for marking. No methodological details were given for the *Astyanax* study.

This method produces a large amount of ecological data, including population size and density, spatial and temporal distribution of class sizes, weight and condition factor, and also of reduction of eyes and pigmentation (for variable populations), fish movements and home ranges, life cycle parameters (reproduction, growth rates), among other data. However, in some cases MR is not practical, either because population densities are too small to obtain the minimum number of marks needed for mathematical treatment (case of the Brazilian heptapterine, *Taunayia* sp.), or localities are too remote and difficult to access, hampering the necessary periodical visits, especially when there are no local researchers, as in the case with many African and Asian fishes.

In general, for VC and complementary observation on fish distribution inside caves, habitat preferences and behavior, as well as for hand-netting, the specimens were visually located from stream-side, as is the case with *Astyanax* spp. and amblyopsids in shallow habitats (Poulson[1], Parzefall 1983, Brown & Todd 1987), the balitorids *S. oedipus*, *C. thamicola* and *Nemacheilus troglocataractus* (Trajano, Borowsky, Mugue & Smart unpublished data), the catfishes *T. itacarambiensis* (Trajano 1997a,b) and *A. cryptophthalmus* (Trajano 2001); while snorkeling, for *Taunayia* sp. (Trajano & Bockmann 2000), *Astyanax* spp. and *A. rosae* in deeper habitats (Parzefall 1983, Brown & Todd 1987); or SCUBA diving, for the new heptapterine genus from NE Brazil (Mendes 1995a,b).

Most ecological studies were carried out only during dry seasons, because caves located in regions with seasonal climates may become inaccessible during the rainy periods, and there is always the danger of being caught by flash floods during those periods, a general feature of karst drainages. This restriction to part of the annual cycle imposes a limitation to the generalizations that can be taken from ecological studies of subterranean species. Such is the case with *T. itacarambiensis*, *A. cryptophthalmus*, *Astyanax* spp. and cave balitorids. On the other hand, *P. kronei*, which lives in an area with less accentuated seasonality, could be studied throughout the year (Trajano 1991). The same was true for *A. rosae* in Logan Cave (Brown 1996).

Population density is defined as number of individuals per unit area, per unit volume, or per unit of habitat

(Krebs 1994). Ideally, units of volume should be used for fishes, except for those strictly benthic. Many cave fishes, including those belonging to typically bottom-dwelling epigean taxa, explore not only the bottom but also the midwater (Trajano & Bockmann 1999). Thus, the number of individuals per unit volume is more representative of the real population densities. However, in the absence of precise data on habitat volumes due to the high spatial heterogeneity, with a great deal of variation in depth along the studied caves, population densities estimated for subterranean fishes have been expressed as number of fish per unit area (m^{-2}). This clearly produces a bias that should be taken into account when comparing populations from different caves.

Likewise, if the goal is to evaluate the impact of cavefish populations over the subterranean ecosystems, biomass is another parameter to be considered. Most troglobitic fishes are small, less than 10 mm SL (e.g., Greenwood 1967), and do not greatly differ in terms of biomass. The main exception among the species studied with focus on population ecology is *P. kronei*, a relatively large catfish (maximum size around 180 mm SL) showing a mode of weight = 18.0–22.0 g, with maximum weight = 52.0 g (Trajano & Britski 1992). For comparison, the mode of weight distribution in *T. itacarambiensis* (max. size = 83 mm SL) during the dry season was 1.0 2.0 g, max. weight = 6.0 g (Trajano 1997a, unpublished data) and the mode of weight in a small sample of *S. oedipus* from Mae Lana Cave (max. size = 74 mm SL) was 0.4–1.0, max. weight = 5.0 g (Trajano, Borowsky, Mugue, Smart & Krejca unpublished data).

Methodological discrepancies include different methods to estimate population parameters (e.g., VC versus MR) and different collection (hand-netting versus trapping) and marking (more or less permanent marks) techniques, associated with differences in habitat characteristics (more or less accessible and seasonal habitats) and fish behavior (shy, elusive versus non-shy species), which should always be made explicit. Such differences merit special concern when comparing data available for cave fishes.

Securing preserved samples is necessary not only in the faunistic survey step, for recognition of species and determination of distributional ranges, but also for the study of diet and reproduction. Observation of feeding behavior may provide evidence on the kind of food consumed by the fish, but unquestionable identification of items actually ingested depends on analysis of gut contents. Detection of developed gonads seen by transparency in living fish may produce useful data on reproduction, but limited to the later stages of gonad development and subject to errors due to difficulties in visualization.

Ecology of subterranean fishes

The problem of species and population delimitation

In the context of ecological studies, distinction of species and delimitation of populations is an important issue. When differences in habitat use, population densities and life cycle parameters are observed between localities, one must determine whether they reflect geographic variation in ecological conditions, or are species-specific characteristics. More or less isolated populations may differ in population parameters. For instance, Heuts (1952) detected distinct growth patterns and age-class distributions in samples of *C. geertsii* from different, separate caves located in a relatively small karst area in Congo. Likewise, a first condition to estimate population sizes is to circumscribe such populations based on biologically meaningful criteria.

Caves, defined as natural cavities large enough to be entered by humans (Monroe[2]), constitute biologically an arbitrary subdivision of the subterranean domain. In many cases, they encompass only a fraction of the total species habitat, and studies in caves only assess a peripheral part of this population, which may be strongly affected by the proximity to large contacts with the surface. This is especially true for species occupying the phreatic habitat, such as typical phreatobites (e.g., *Satan eurystomus*, *Trogloglanis pattersoni*, *Phreatobius cisternarum*, *Uegitglanis zammaranoi*, *Phreatichthys andruzzi*). It is also the case for hypogean fishes occurring in widely distributed caves, such as the troglobitic amblyopsids (except *Speoplatyrhinus poulsoni*, known from a single cave), characins, genus *Astyanax*, and the new heptapterine genus from NE Brazil, which are expected to occur in the aquifer(s) between the caves.

In many cases, the delineation of subterranean species is not clear and a detailed taxonomic revision may reveal a set of separate, cryptic species under the same designation. Molecular evidence shows that

[2] Monroe, W.H. 1970. A glossary of karst terminology. Geological Survey Water-Supply Paper 1899-K. 26 pp.

T. subterraneus is paraphyletic (Bergstrom & Noltie[3]), with two of its four lineages more closely related to *A. spelaea* than to the two other lineages, and that *A. rosae* correspond to four different lineages (possibly distinct species) because the divergence between populations from four watersheds is equivalent to, or greater than, the divergence among different species and genera in the family. Willis & Brown (1985) state that continuous cave passages linking all the populations of troglobitic amblyopsids is very unlikely, thus each species, and possibly each current population of cavefish, represents a separate evolutionary event. Molecular studies also indicate that *S. oedipus* comprises three separate groups (Borowsky & Mertz 2001).

Morphological, distributional and molecular data suggest that the cave populations of blind tetras, genus *Astyanax*, found in the Huastecan Province, resulted from at least two, probably three or more isolation events (Alvarez 1946, Mitchell et al. 1977, Borowsky & Espinasa[4], but see Espinasa & Borowsky 2001). Thus, they correspond to different diagnosable, phylogenetic species (sensu Pinna 1999), whose exact number and geographic limits are not yet clear. The cave population of *Poecilia mexicana* from Cueva del Azufre, in Mexico, presents several morphological and behavioral troglomorphisms, at least part of them genetically based (Parzefall 1993). Therefore, much as the troglomorphic subterranean populations of *Astyanax* (see Trajano 1993) deserves the status of separate species, but it is still treated as *P. mexicana*, the name of an epigean species. These are examples of nomenclature problems reflecting the lack of taxonomic resolution, which affects all areas of research interest.

Pimelodella kronei was recorded in two sets of limestone caves separated by a non-soluble, diabase dyke that prevents subterranean dispersion between those caves. Therefore, this taxon may comprise two distinct, cryptic species (Trajano 1997c). So far, we have been able to study only populations in the set of caves comprising the type-locality, Areias Cave. There is morphological and behavioral evidence that the populations of *A. cryptophthalmus* from Angélica-Bezerra caves and Passa Três Cave are at least partially isolated (Trajano 2001), and these populations will be treated separately herein.

In conclusion, it is necessary to define the boundaries of populations that operate in a homogeneous, integrated way, which should constitute the basic units for ecological studies. A starting point is to focus on cave systems defined on a hydrogeological basis (conduits extending continuously between the input and output points of a karst rock, including passages too small to be entered by humans, Ford & Williams 1989) as geographic units, until more detailed comparative morphological studies and assessment of genetic differentiation provide a better understanding of population limits, and a more precise view of species distributions.

Distribution and habitat of troglobitic fishes

Troglobitic fishes occur in all continents except Europe, and also in many islands (e.g., Cuba, Madagascar). The highest species richness is observed in China, Mexico, Brazil and southeast Asia, not necessarily corresponding to the areas with greatest total richness ('hot spots' of biodiversity) of aquatic troglobites. Most troglobitic fishes are siluriforms (found in the Americas, especially the tropics, and in Africa, with a few in Asia) or cypriniforms, mostly cyprinids or balitorids (located throughout tropical Asia, some cyprinids in Africa). Hence, these groups of freshwater fishes present a high potential for easy adaptation to subterranean life.

On the other hand, characiforms, other important group of freshwater fishes, are in general poorly represented among the troglobitic fauna: besides the poorly known Brazilian tetra characin, *Stygichthys typhlops*, whose epigean closest relatives have not been determined (Romero & McLeran 2000), only the Mexican species inhabiting the Huastecan Province, *Astyanax mexicanus* (= *A. fasciatus*), has successfully originated several troglobitic populations, due to preadaptations to the cave life; for example, a chemically stimulated spawning behavior, ability to feed in darkness, and a well-developed lateral line system (Wilkens 1988). Likewise, cichlids, another important group of freshwater fishes, have no troglobitic derivatives, although epigean specimens may be occasionally found in caves. Among the usually abundant cyprinodontiforms, only one Mexican species, *Poecilia mexicana*, has specialized to cave life, under very unusual conditions: an exceptionally high availability of food (Langecker et al.

[3] Bergstrom, D.E. & D.B. Noltie. 1998. The phylogeny, historical biogeography, and evolution of troglobitism in *Amblyopsis rosae* (Ozark cavefish) and *Typhlichthys subterraneus* (southern cavefish). Abstract 185. *In*: 78th Annual meeting Amer. Soc. Ichthyol. Herpetol., Guelph.

[4] Borowsky, R. & L. Espinasa. 1997. Antiquity and origins of troglobitic Mexican tetras, *Astyanax fasciatus*. pp. 359–361. *In*: Proc. 12th International Congress of Speleology, La-Chaux-des-Fonds.

1996). Those are good examples of low potential for subterranean life.

Preadaptation to hypogean life is clear for siluriforms, which are generally nocturnal, chemo-oriented fishes, most omnivores or generalist carnivores, but not so for cypriniforms which, like characiforms, include many diurnal, visually-oriented species. For the latter, preadaptations for subterranean life must be sought among the closest epigean relatives, which could have retained the preadaptive character states shown by the troglobite's ancestors, as in the epigean *A. mexicanus*.

There is a great variation in the extent of geographic distribution of each troglobitic species, without a phylogenetic/taxonomic correlation. Several species have been so far found in a single cave (e.g., *Sinocyclocheilus anophthalmus*, *Triplophysa ayunnanensis*, *Nemacheilus troglocataractus*, *Clarias cavernicola*, *Rhamdia reddelli*, *Taunayia* sp., *Trichomycterus chaberti*, *T. itacarambiensis*, *Ancistrus galani*, *Speoplatyrhinus poulsoni*, *Poecilia mexicana*), while others present a slightly wider distribution area, occurring in a few caves (e.g., *Caecobarbus geertsii*, *Schistura oedipus*, *Cryptotora thamicola*, *Pimelodella kronei*, the new heptapterine genus) to those who inhabit many, more or less widely distributed caves (e.g., *Astyanax* spp., *Lucifuga subterraneus*, *L. dentatus*, troglobitic amblyopsids except *S. poulsoni*). Troglobitic fishes inhabiting non karst regions include the Brazilian heptapterine, *Phreatobius cisternarum*, found in alluvial fan from a large area around the Amazon delta (Carvalho[5]).

Subterranean karstic habitats occupied by different species of troglobitic fishes are shown in Table 1. This table does not include all known hypogean fishes, because for many species such data are unavailable. My main purpose was to illustrate the variety of habitats where troglobitic fishes may be found. It is also an attempt to detect patterns related to the fish characteristics, such as systematic position, geographic and biological attributes, which could help to understand why and how fishes colonize and speciate in the subterranean biotope. For species occurring in a large number of caves, such as amblyopsids, *Astyanax* spp. and *Lucifuga* spp., it was difficult to define a typical habitat for the 'species'. In such cases, the habitat classification represents the main tendencies detected for the species, and exceptions will certainly be found. I did not include in Table 1 non-karst habitats such as granular, porous aquifers inhabited by *P. cisternarum*.

Three main karst habitat categories may be distinguished: (1) upper phreatic zone, represented by lakes or static water bodies, connecting to the surface through voids varying from impenetrable fissures to large caves, and generally characterized by lentic conditions; (2) zone of seasonal oscillation of water table, that during flooding may form pools and lakes that will remain isolated during the dry season or may contribute water to dramatically increase the discharge of underground streams; and (3) open channel stream passages, usually showing lotic conditions, that include base-level streams (which may or may not represent the phreatic zone) and vadose upper tributaries (A. Auler personal communication). In addition, some species may live in perched aquifers, as those forming water bodies in rimstone dams fed by percolation water and, occasionally, by water from the stream channel during floods.

The upper phreatic zone habitat may include flooded caves – caves formed above the water table which have been posteriorly flooded due to rise of the water table, caused, for instance, by subsidence of the whole region or elevation of the sea level (for caves situated not far from the coast). Elevation of the sea level affected several caves inhabited by *Lucifuga* spp. in Cuba (Juberthie[6]). Regional subsidence is probably the cause of flooding of caves from Mato Grosso do Sul, SW Brazil, inhabited by *Ancistrus formoso* and a new species of *Trichomycterus* (treated as *Trichomycterus* sp. 2 in Trajano 1997d). It is noteworthy that the original hypogean habitat, present when the fishes first colonized the caves, may be quite different from the recently observed. This is possibly the case of the catfishes from Mato Grosso do Sul, which belong to stream-adapted genera.

There is a great variation in the ecological conditions of habitats occupied by troglobitic fishes, from streams with well-oxygenated waters in riffles and rapids alternating with pools, to lentic environments such as water bodies formed by the water table inside caves or inaccessible phreatic habitats. Extreme adaptation to fast-moving waters is observed in the balitorid *Cryptotora thamicola*, generally found in waterfalls over flowstone (see below) in caves from Thailand.

[5] Carvalho, A.L. 1967. Novos dados para o conhecimento de 'Phreatobius cisternarum' Goeldi (Pisces, Pygidiidae, Phreatobiinae). Atas do Simpósio sobre a Biota Amazônica 3 (Limnologia): 83–88.

[6] Juberthie, C. 1973. Notes sur l'écologie des poissons souterrains de Cuba. pp. 285–293. *In*: Comptes Rendus du 96e Congrès National des Sociétés Savantes, tome III, Toulouse.

Table 1. Types of subterranean karst habitats occupied by some species of troglobitic fishes. (/) = syntopic species.

Habitat		Species	Country
Upper phreatic zone	Connected to surface through inaccessible fissures*	*Stygichthys typhlops*	Brazil
		Trogloglanis pattersoni/ *Satan eurystomus*	USA
		Horaglanis krishnai	India
	Connected to surface through resurgences, sinkholes, natural wells, cenotes, flooded caves, etc.	*Phreatichthys andruzzii*	Somalia
		Barbopsis devecchii	Somalia
		Typhlogarra widowsoni/ *Caecocypris basimi*	Iraq
		Iranocypris typhlops	Iran
		Prietella spp.	Mexico
		Uegitglanis zammaranoi	Somalia
		Ancistrus formoso	Brazil
		Lucifuga spp.	Cuba
		Oxyeleotris caeca	Papua/ New Guinea
	Connected to surface through caves	*Protocobitis typhlops*/ *Oreonectes* sp.	China
		New heptapterine genus	Brazil
Zone of seasonal oscillation of water table		*Astyanax* spp.	Mexico
		Indoreonectes evezardi	India
		Speoplatyrhinus poulsoni	USA
Open channel stream Passages	Base-level streams	*Sinocyclocheilus anophthalmus*/ *Triplophys ayunnanensis*/ *Paralepidocephalus* aff. *yui*	China
		Nemacheilus troglocataractus	Thailand
		Pimelodella kronei	Brazil
		Rhamdia reddelli	Mexico
		Trichomycterus chaberti	Bolivia
		T. itacarambiensis	Brazil
		Ancistrus crytophthalmus	Brazil
		Ancistrus galani	Venezuela
		Poecilia mexicana	Mexico
	Vadose tributaries	*Schistura oedipus*	Thailand
		Cryptotora thamicola	Thailand

*Collected in artificial wells.

Among cave stream-dwellers, habitat preferences have been observed, for instance, in *P. kronei* (Trajano personal observations) and *Rhamdia reddelli* (Miller 1984), that prefer pools to riffles; an opposite tendency is shown by *A. cryptophthalmus*, which prefers rapids to pools (Trajano 2001). In other cases, no habitat preferences were detected: *T. itacarambiensis*, for instance, occurs both in riffles and in pools (Trajano 1997a).

The widely distributed troglobitic amblyopsids may occupy habitats that fall into different categories depending on the locality, but always showing preference to stagnant or slow-moving water, including places in which the upper phreatic zone is connected to the surface through caves, the zone of seasonal oscillation of the water table, or pools and slowly flowing sectors in open channel streams (Poulson 1963, Romero 1998a,b). Thus, these species were not included in Table 1.

Other species live in typically phreatic waters, such as *Phreatobius cisternarum*, *Satan eurystomus*, *Trogloglanis pattersoni*, *Uegitglanis zammaranoi*, *Phreatichthys andruzzii*, *Barbopsis devecchii* and *Taunayia* sp. Some of these fishes show miniaturization, probably achieved by paedomorphosis (as proposed by Weitzmann & Vari 1988 for miniature epigean species), as an adaptation to life in small phreatic spaces: *S. eurystomus*, *T. pattersoni*

(Langecker & Longley 1993), *Prietella lundbergi* (Walsh & Gilbert 1995), *Taunayia* sp., and the new heptapterine genus (Trajano & Bockmann 1999). Small size may also be energetically adaptive for hypogean environments (Walsh & Gilbert 1995).

Many fishes living in phreatic waters, or in shallow habitats subject to periodical drought or to special conditions (e.g., a high concentration of bacteria), must be tolerant to hypoxia in view of the low oxygen concentration frequently recorded in these habitats. For instance, very low values of dissolved oxygen (DO), ranging from 0.7 to 2.1 $mg\,l^{-1}$, were recorded in the phreatic water body inhabited by *Taunayia* sp. (Trajano & Bockmann 2000), much below the normal values for streams, which is >6.4, frequently 8–10 $mg\,l^{-1}$ (S. Buck personal communication, Trajano & Bockmann 2000). Low oxygen content characterizes the sulfur spring cave inhabited by troglobitic *Poecilia mexicana* (Langecker et al. 1996). Low values of DO were recorded in wells where phreatobic fishes have been collected: *Barbopsis devecchii* (1.0–5.0 $mg\,l^{-1}$), *Phreatichthys andruzzii* (2.7–3.8 $mg\,l^{-1}$) and *Uegitglanis zammaranoi* (5.7 $mg\,l^{-1}$) from Somalia (Ercolini et al. 1982); *Trogloglanis pattersoni* and *Satan eurystomus* from Edwards Aquifer in Texas (4.3–5.1 $mg\,l^{-1}$ – Longley & Karney 1979).

Putative adaptations to cope with hypoxia include an intense skin irrigation suggestive of cutaneous respiration in *P. cisternarum* (Carvalho[5]) and probably also in *Taunayia* sp. (Trajano & Bockmann 2000). Reduction of respiratory intensity resulting in low oxygen consumption was demonstrated for some troglobitic fishes such as *C. geertsii*, *P. andruzzii*, *B. devecchii* and amblyopsids (Ercolini et al. 1987), but not for troglobitic *Astyanax* (Wilkens 1988). Lowered metabolic rates which have been interpreted as an adaptation to low and discontinuous food supply (Poulson 1963), may also aid life in non-cave aquifers. Subterranean populations of *Indoreonectes evezardi*, which lives in small pools with stagnant water during the dry seasons, exhibit air-gulping behavior; the epigean relatives, that live in permanent hill streams, do not show such behavior (Biswas et al. 1990). *P. mexicana* is reported as exhibiting a remarkable tolerance to low oxygen concentrations, and also to high sulfide contents (Langecker et al. 1996).

Tolerance to depth is highly variable among troglobitic fishes, and some species may be found at great depths in lakes, flooded caves or non-cave aquifers. For instance, the amblyopsid *T. subterraneus* is apparently adapted to shallow water and seems to be very sensitive to supersaturated conditions (Schubert et al. 1993); individuals observed in a deep lake in Sell's Cave were never seen at depths greater than 2 m (Cooper & Iles 1971). The Brazilian phreatobites, *P. cisternarum* and *S. typhlops*, have been obtained from wells drilled at 10–20 and 30 m, respectively (Carvalho[5], Romero & McLeran 2000). *Lucifuga subterraneus* and *L. dentatus* are found up to 30 m deep in Cuban flooded caves (Juberthie[6]). The new heptapterine genus from NE Brazil inhabits a big lake in Poço Encantado Cave, with maximum depth around 60 m, but the blind catfishes were observed until 35 m, concentrating between the surface and 10 m (Mendes 1995b). Catfishes living in flooded caves from Mato Grosso do Sul, SW Brazil, were found in sectors as deep as 32 m (*Ancistrus formoso*, Sabino & Trajano 1997) and 46 m (*Trichomycterus* sp. 2 – E.P. Costa Jr. personal communication). However, by far the species known to live at the greatest depths are the ictalurids from Edwards aquifer: *S. eurystomus* was recorded from 610 m deep artesian wells (Longley & Karney 1979). These species have a reduced swimbladder which is replaced by large fat deposits as an adaptation to food scarcity and to help maintain neutral buoyancy, reduced muscles and weak ossification, apparently a paedomorphic trait in *S. eurystomus* and *T. pattersoni*; all probably adaptations to cope with high hydrostatic pressure (Langecker & Longley 1993).

Population sizes and densities

Population densities reported for several cavefish species in different caves are presented in Table 2. For populations studied by MR methods, these numbers were calculated by dividing the estimated population size by total area (from topographic maps) of the studied sections; therefore, these numbers correspond to mean population densities throughout the caves. In the case with VC, some values correspond to total counts in the accessible habitat divided by its area (*Taunayia* sp., *N. troglocataractus*, amblyopsids), and therefore also represent mean densities, whereas others result from counts in selected small areas (*Astyanax* spp., *P. mexicana*, *S. oedipus*, the new heptapterine genus from NE Brazil) and are more subject to errors due to spatial variation in fish densities. Extreme values recorded by Willis & Brown (1985) were not included in the table, because they may represent very unusual situations (maximum of one individual in 1200 m^2 surveyed

Table 2. Approximate population densities (individual m^{-2}) reported for subterranean fish species, based on mark-recapture studies (MR) or visual censuses (VC) in different caves; in parentheses, maximum densities are recorded.

Species	Country-Cave	Population density	Method	References
Heptapterinae	Brazil			
Pimelodella kronei	Areias	0.04	MR	Trajano (1991)
New genus	Poço Encantado	0.03	VC	Mendes (1995a)
Taunayia sp.	Toca do Gonçalo	0.01	VC	Trajano & Bockman (2000)
Trichomycteridae	Brazil			
T. itacarambiensis	Olhos d'Água	0.15	MR	Trajano (1997a)
		0.05	VC	Trajano (1997a)
Loricariidae	Brazil			
A. cryptophthalmus	Angélica	0.90	MR	Trajano (2001)
	Passa Três	0.50	MR	Trajano (2001)
		0.20	VC	Trajano & Souza (1994)
Characidae	Mexico			
Astyanax spp.	La Chica	(5)	VC	Parzefall (1983)
	Pachon	(5)	VC	Parzefall (1983)
	Micos	(15)	VC	Parzefall (1983)
Balitoridae	Thailand			
Schistura oedipus	Mae Lana	0.02	VC	J. Spies personal communication
		(0.25)	VC	Trajano, Borowsky, Mugue & Smart unpublished data
	Nong Pha Cham	(60)	VC	Trajano, Borowsky, Mugue & Smart unpublished data
N. troglocataractus	Wang Ba Dan	0.01	VC	Trajano, Borowsky, Mugue & Smart unpublished data
Amblyopsidae	U.S.A.			
T. subterraneus	Several caves	0.03	VC	Poulson[1]
Amblyopsis spelaea	Several caves	0.05	VC	Poulson[1]
Amblyopsis rosae	Several caves	0.15	VC	Poulson[1]
A. rosae	11 caves	0.005–0.15	VC	Willis & Brown (1985)
Poeciliidae	Mexico			
Poecilia mexicana	Azufre	(100–200)	VC	Parzefall (1993)

during five visits to Turnback Creek Cave; three individuals in a pool of 1 m^2 in Jail Cave).

Some of these species are also found in localities other than those included in Table 2. Whereas *Taunayia* sp., *T. itacarambiensis*, *N. troglocataractus* and *P. mexicana* occur only in the studied caves, *P. kronei* was found in six localities (possibly two groups; the Areias group comprises three other localities, Trajano 1997c, unpublished data), the new heptapterine genus in four (Mendes 1995a unpublished data), *S. oedipus* in five (three groups; the Mae Lana group comprises populations from two other caves, Borowsky & Mertz 2001), *Astyanax* spp. in 29 (at least the Micos population corresponds to a separate group comprising three of those 29 caves), and, among the amblyopsids, the three nominal species were recorded in dozens of caves (Willis & Brown 1985), but these probably correspond to a higher number of phylogenetic species whose distribution was not clearly established yet. In the case of *Astyanax*, population densities greatly vary from cave to cave, but large populations seem to be more common (Mitchell et al. 1977).

Thus, distribution of many cavefishes encompass multiple caves that may be several kilometers distant from each other, presumably including intervening non-cave subterranean habitat. On the other hand, higher population densities are expected to occur in caves with greater food availability due to the larger contacts with the surface. In fact, the highest densities of *Lucifuga* spp. are observed in the twilight zone under cave entrances. Such higher density would be the result of a balance struck between the positive reaction to food and the negative reaction to light in these bythitids (Eigenmann 1909). Therefore, the values shown in Table 2 are probably close to the highest densities of the studied populations throughout their distribution area.

Non-quantitative data on population densities, based on observations in the cave habitat and on the size of samples obtained, are available for other troglobitic fishes. Among cave species showing apparently low population densities, we can mention the bythitid *O. pearsei*, from cenotes in Yucatan, Mexico (Chumba-Segura 1983); the armored catfish, *Ancistrus galani*, from Venezuela (Perez & Viloria 1994); the amblyopsid *S. poulsoni*, for which no more than 10 individuals have ever been observed on a single visit to Key Cave, with 3300 m of mapped passages (Romero 1998a).

On the other hand, the Bolivian trichomycterid, *T. chaberti*, is described as numerous (Durand 1968). Apparently, *Lucifuga subterraneus* and *L. dentatus* are also quite abundant. These species have been reported from several caves and may reach high population densities: as many as 10 individuals were observed by Eigenmann (1909) in a well about '6 feet in diameter and 10 feet deep'. According to Yang & Chen (1993a), the cyprinid *Sinocyclocheilus anophthalmus*, from a cave in Yunnan Province, China, has a large population, larger than that of the syntopic *Triplophysa ayunnanensis*.

A great deal of variation in population densities is observed among troglobitic fishes, without a taxonomic correlation. At first sight, such variation could be at least partly attributed to differences in study methods, because the lowest densities were generally recorded in VC. A comparison between the numbers of fish estimated by MR and those seen during visual inspection of the same habitat provide a measure of underestimation produced by VC. Even under the most favorable conditions for fish detection, like those observed at Passa Três Cave (small, shallow stream; non-cryptobiotic, exposed fishes), the population estimated by MR is larger than that seen by the above-water observers. Because, as already mentioned, in many cases it is not possible to carry out MR studies, it may be useful to establish an average proportion expressing the differences between estimated and observed populations which could be used as a correction factor for values obtained in VC. In view of the very limited number of studies using MR, this must be regarded as a first, very preliminary attempt to produce such indices.

The proportions between numbers of estimated and observed fish in different studies varied, but most were around three estimated for each captured/observed individual. In *T. itacarambiensis* and *A. cryptophthalmus* from Passa Três Cave, such proportions were 3 and 2.5, respectively (Table 2). From Brown's (1996) data on *A. rosae* at Logan Cave, I calculated 2.2 estimated individuals for each captured fish; the mean population size estimated by this author was three times larger than the maximum number of fish seen by Willis & Brown (1985) in this cave. All these fishes are easy to capture in hand-nets (Means &Johnson 1995, Trajano personal observation), therefore the number of captures would closely approach that of observed fishes. About 25% of the *T. itacarambiensis* catfishes observed in the cave escaped our capture attempts, and this has been added to the total number of captures (Trajano 1997a) to produce the value from VC shown in Table 2. In *P. kronei*, the proportion of estimated to trapped fishes, varying from 2.5 to 7.6 among capture occasions (Trajano 1991), also gives an idea of the percentage of the population unseen by observers. Therefore, as a first approach, it seems reasonable to multiply by three the values obtained in VC in order to get numbers more comparable to those produced by MR studies.

Even applying a 'correction factor' of three, population densities still greatly differed among the studied cave fishes, which may be arbitrarily divided into three groups (some species not included in Table 2 were also classified on the basis of literature data on numbers of fishes collected/observed): (1) species with low population densities (<0.1 individuals m^{-2}), for example *N. troglocataractus*, *T. subterraneus*, *Speoplatyrhinus poulsoni*, *Trichomycterus* sp. from El Guácharo Cave, *Taunayia* sp., possibly *P. kronei*, the new heptapterine genus; (2) species with median population densities (0.1–1.0 individuals m^{-2}), *S. oedipus*, *A. spelaea*, *A. rosae*, *T. itacarambiensis*, *A. cryptophthalmus*, and (3) species with high population densities (> 1.0individuals m^{-2}), *Astyanax* spp., *P. mexicana*, *C. geertsi* in some caves. If biomass was used as another correction factor, then *P. kronei* should be ranked with species showing higher population densities.

In the few cases in which comparative data are available, population densities estimated for epigean close relatives are, as expected in view of the higher food availability in this habitat, higher than that observed for troglobitic species, but not by a huge margin. Population densities of eyed *Ancistrus* sp. in epigean reaches of streams crossing caves where *A. cryptophthalmus* is found are 50% greater than

Table 3. Population sizes (number of individuals) estimated for some cave fishes using mark-recapture methods. The extension of stream occupied by these population is given in parenthesis.

Species	Cave (extension)	Population size	References
Pimelodella kronei	Areias (5000 m)	900–1200	Trajano (1991)
Trichomycterus itacarambiensis	Olhos d'Água (5000 m)	1500–2000	Trajano (1997a)
Ancistrus cryptophthalmus	Angélica (8000 m)	>20 000	Trajano (2001)
	Passa Três (2000 m)	1000	Trajano (2001)
Astyanax spp.	El Pachón	8500*	Mitchell et al. (1977)
	Yerbaniz	8700**	Mitchell et al. (1977)

*95 percent confidence limits of 1280 to 18280.
**95 percent confidence limits of 1810 to 15530.

those of this troglobite (Trajano 2001). Densities of 0.043–0.097 individuals m^{-2} were recorded in epigean streams for *P. transitoria* (Gerhard 1999), putative sister-species to *P. kronei* (population density in Areias Cave = 0.04 individuals m^{-2}, Table 2). Also as expected, *P. transitoria* present lower population densities inside caves, about ten times lower than that of *P. kronei* in Areias Cave, where these species are syntopic (Trajano 1991). Another troglophile, the amblyopsid *Chologaster agassizi*, also presents cave population densities which are lower than the observed densities for this species in the epigean habitat, and lower than the densities shown by its troglobitic relatives (Poulson[1]). Maximum densities of epigean *Astyanax mexicanus* recorded by Parzefall (1983, table 1) greatly varied among streams: some were comparable to the highest values reported for cave populations (15 individuals m^{-2}, as in Micos cave), others were much higher (up to 200 individuals m^{-2}); however, the numbers recorded in test areas both in epigean and cave habitats did not greatly diverge (Parzefall 1983, figures 1–2). Epigean *A. mexicanus* co-occur with troglobitic populations in some caves, where they are usually less abundant than eyeless tetras and frequently show signs of starvation (La Cueva Chica, characterized by a high input of energy represented by bat guano, is an exception); even so, extensive hybridization may occur (Mitchell et al. 1977).

Estimates of population sizes based on capture-mark-recapture are in Table 3. These estimates do not include individuals that were too young, too rarely found, and/or too small for marking. Thus, the values correspond to the average adult stock found in each studied cave (i.e. the accessible habitat). In the case of the stream-dweller *T. itacarambiensis*, the study encompassed almost all the cave habitat known for these fishes, and the values obtained probably closely approach the total population sizes. For *P. kronei* and *A. cryptophthalmus*, estimates from MR in sectors of the studied caves have been extended to the whole cave populations, assuming constant population densities. Because these populations are probably continuous with those inhabiting other caves nearby (Bombas, Córrego Seco and Águas Quentes, for *P. kronei*, Trajano 1997c, unpublished data; Bezerra for *A. cryptophthalmus,* Trajano 2001), the total populations in the respective cave systems may be a few times larger than that.

The rare studies based on MR of *Astyanax* (Table 3) well illustrate how underestimated population size values may be when based on VC, and the consequences of such errors. Avise & Selander (1972), based on inspection of part of the available fish habitat, estimated the populations from La Cueva Chica, El Pachón and Los Sabinos caves as 200–500 individuals each, founding their evolutionary discussion on these numbers. However, in El Pachón and probably in the other caves as well, the total numbers of fish are many times higher, and may invalidate at least part of Avise & Selander's (op. cit.) conclusions. Even considering the value at the lower limit of confidence (1280 individuals, Mitchell et al. 1977), the population estimated by MR in El Pachon cave is 3–6 times larger than that visually estimated. Likewise, the hybrid population from La Cueva Chica is probably thousands of individuals, in view of the high number of fish captured since the 30s (samples with up to 100 specimens, Romero 1983), apparently without affecting significantly the population.

Cryptobiotic fishes, such as several *Trichomycterus* species, are especially prone to underestimation in VC. From a VC by Romero (1987), a low population density of 0.01 individuals m^{-2} was calculated for the troglomorphic population of *Trichomycterus* (then referred as *T. conradi*) in Cueva del Guácharo (Venezuela), corroborated by the small samples usually obtained in the cave (Nalbant & Linares 1987, Andreani-Armas 1990). However, as noted by Romero (1987), some individuals tend to bury into the substrate, thus the number present in the cave was certainly higher.

The new species of *Trichomycterus* (treated as *Trichomycterus* sp. 1 in Trajano 1997d) syntopic with *A. cryptophthalmus* in Passa Três Cave illustrates the inadequacy of VC for cryptobiotic cave fishes. In a one-day collection based on visual inspection of the habitat, 15 specimens of *Trichomycterus* and 70 of *A. cryptophthalmus* were captured (Trajano & Souza 1994), suggesting a population density of the former was 4.5 times lower than the latter species. After careful collections, searching also under hiding places for a MR study in progress, M.E. Bichuette raised the proportion of captured specimens of *Trichomycterus* to 30 compared to 100 captured *A. cryptophthalmus*. However, none of 27 marked specimens of *Trichomycterus* sp. 1 were recaptured, whereas recapture rates of *A. cryptophthalmus*, marked using the same method, reached 44% in this cave. Mortality due to marking procedures is highly unlikely because *Trichomycterus* catfishes marked in the laboratory lived normally (Trajano personal observation) and tattooing with biocompatible pigments has been proven to be quite safe for fishes in general (Hill & Grossman 1987, Trajano 1997a). Therefore, absence of recaptures in *Trichomycterus* sp. from Passa Três indicates a population size higher than previously thought (perhaps even higher than that of *A. cryptophthalmus*) and/or frequent movements between the cave and inaccessible spaces contiguous to it.

Likewise, according to Brown (1996), specimens of *A. rosae* may go undetected for long periods, hidden under rocks and in crevices, that may contribute to negative bias in population estimates. *T. subterraneus* has the ability to burrow in interstitial spaces and disappear into the substrate, entering and exiting gravel interstices periodically (Schubert et al. 1993).

Thus, it is advisable to observe the fish behavior before carrying out population studies based on VC. Fortunately, most studied subterranean fishes are not cryptobiotic, even those belonging to typically cryptobiotic epigean taxa, just like other Brazilian *Trichomycterus* species (*T. itacarambiensis*, *Trichomycterus* sp. 2, from SW Brazil) and the heptapterines *P. kronei*, *Taunayia* sp. and new genus from NE Brazil (Trajano & Bockmann 1999).

Poulson (1963) estimated population sizes for amblyopsids in several caves, based on VC: *T. subterraneus*, around 40 individuals per cave (maximum 150 individuals); *A. spelaea* – 105 per cave (max. 175); *A. rosae*, about 75 per cave (max. 100). According to Willis & Brown (1985), population sizes of about 150 individuals are probably normal for the Ozark cavefish, *A. rosae*. However, these authors observed much lower numbers in 14 caves visited, finding between one and 12 individuals in 13 of these caves, and 100 in Cave Springs (where Poulson found a similar number). They attribute this to population losses caused by habitat destruction and collection. Mean population size of *A. rosae*, estimated by MR in Logan Cave between 1992 and 1995, was 34 individuals (considered an underestimate, Brown 1996), about three times the maximum number recorded in VC by Willis & Brown (1985).

Monthly fluctuations in the estimated size of the Logan Cave population were interpreted by Brown (1996) as reflecting frequent movements between the accessible habitat and the non-cave aquifer (see below). According to Brown, individuals in the cave are probably members of a large population living in the aquifer. The constant inflow of fish into the cave and the small, fluctuating number of observed fish indicate that the Logan 'population' is not self-sustaining. This may also be the case with other caves inhabited by *A. rosae*, explaining, along with their cryptobiotic habits, the low numbers of fish usually seen. As also pointed out by Poulson (1963), the Ozark cavefish exhibits a clear preference for still water: in Logan Cave, *A. rosae* is not found in sections with fast flowing water over bedrock substrate, and, in Cave Spring Cave, the majority of cavefishes are found in a single still, shallow pool. This, along with its occurrence in widely distributed caves, corroborates the notion that the main habitat for the species is the non-cave aquifer. Therefore, the actual total population size is far from being assessed, but may be much larger than currently thought. Such may also be the case of *T. subterraneus* and *A. spelaea*. However, it must be recalled that *T. subterraneus* and *A. rosae* are probably artificial taxa, thus a more realistic picture on population sizes depends on studies delimiting the natural taxa inside these nominal species.

It is noteworthy that epigean populations of *Chologaster agassizi* estimated by MR in southern Illinois were generally small (18 to 302 individuals per stream). In fact, this species is referred as rare to uncommon throughout most of its range, except for a large spring in Kentucky where it is exceptionally abundant (Smith & Welch 1978).

Parzefall (1993) refers to densities of 100–200 individuals m^{-2} for the troglobitic population of

P. mexicana in its only locality, Cueva del Azufre (also known as de las Sardinas or de Villa Luz), a remarkable cave due to the exceptional food availability represented by chemoautotrophic sulfur-bacteria. Such values probably correspond to localized, maximum densities. Short-term MR experiments in some sectors of this cave resulted in values around 1700 individuals in chamber X (~10–15% of the cave stream area) and 100–140 individuals in chamber XIII (see cave maps in Langecker et al. 1996, Hose & Pisarowicz 1999); a population of approximately 9000 individuals was estimated for the epigean reach of the spring downstream of the cave (J. Parzefall unpublished data). Assuming constant population densities throughout the cave, total population size of troglobitic *P. mexicana* would be around 15 000 individuals. However, such estimates were made before the annual traditional religious ceremony, at the end of the dry season, when native people collect and eat large quantities of fish (see Hose & Pisarowicz 1999), greatly depleting the population. This population constitutes an example of extraordinary recovering ability, probably unparalleled among troglobitic fishes.

Heuts (1952) provided estimates of population sizes based on VC of *C. geertsi* studied in seven caves from a small karst area in the Congo. He refers to a population roughly estimated at one thousand individuals in one locality, four thousands in a second one, and several thousands in a third locality. If all these populations are connected, representing a single species, then the total population size for *C. geertsii* is quite large, around tens of thousands of individuals.

Stream-adapted troglobitic species, which tend to be more restricted to fast-flowing waters in cave systems, offer more reliable population data. On the other hand, populations widely occupying the phreatic habitat are particularly difficult to estimate, even when there are estimates of population densities in accessible parts of this habitat (such as for *Taunayia* sp. and the new heptapterine genus, Table 2). The main problem is determining the total distribution area, especially when natural accesses (such as wells, small caves or springs) are rare or nonexistent. It is necessary to assume that population densities throughout wide subterranean areas are similar to those observed at the limited contacts with the surface. Therefore, only rough, highly tentative estimates may be made.

A low population size is referred to for the ictalurid *Prietella phreatophila* by Krejca et al. (2000), who estimated that less than 100 fish live in the accessible parts of Sótano de Amezcua, Mexico, where a MR study is in progress. These authors point to the vulnerability of such a small population to environmental disturbances, which could quickly eliminate this localized population. However, considering that this phreatobic species is distributed throughout a large area, from the type-locality to 160 km to the north (Hendrickson & Krejca personal communication), the total population may be considerably large.

Indirect estimates of population sizes are available for the phreatobic North American ictalurids, *Satan eurystomus* and *Trogloglanis pattersoni*, collected in waters pumped from several wells in the San Antonio pool area (103 600 hectares) of the Edwards aquifer. Based on the flow rates in a few artesian wells from a single pump station, Longley & Karney (1979) estimated a loss of about 460 specimens of *T. pattersoni* and 230 *S. eurystomus* from the subterranean population during one year (1977–1978); from pumped flow records for the period 1950 to 1977, a total loss of 3250 *T. pattersoni* and 1650 *S. eurystomus* was estimated. These numbers provide a crude estimate of total population sizes of tens of thousands for both species. Population density in the predatory *S. eurystomus* is, as expected, lower than (approximately half) that of the detritivorous *T. pattersoni*. These data reveal how questionable it is to make inferences on the relative abundance of phreatobites based on small collected samples: before that extensive program of collection in artesian wells and pumping stations with nets attached to the outlets, only six individuals of *T. pattersoni* and three *S. eurystomus* were captured (Langecker &Longley 1993), leading to the faulty conclusion that these species are very rare.

The phreatobic cyprinid *Barbopsis devecchii* also seems to present large populations. As a result of an extensive collecting program in Somalia during the 60s and 70s, more than 500 specimens were obtained. Lower, but still impressive numbers of cyprinids, *Phreatichthys andruzzii* (140 specimens) and clariids, *Uegitglanis zammaranoi* (82 specimens), were captured on those occasions (Ercolini et al. 1982). The latter is probably much more abundant than indicated by this number: Thinès (1958) opened an artificial canal fed by a subterranean stream and found about 150 individuals along 30 m of canal (section 1.5 × 1.0 m; 0.4 m water column). Such exceptionally high population density is explained by the occurrence of a large amount of plant debris as a food source inside the canal.

From the above, it may be concluded that population sizes estimated for troglobitic fishes vary greatly, from hundreds to tens of thousands of individuals, without a phylogenetic correlation. Based on such estimates, and also on population densities or number of fish collected/seen allied to fish distributions, I tentatively classified several troglobitic species into categories of total population sizes according to three orders of magnitude: (1) Very small populations (10^2 individuals): low population densities and restricted distributions: *C. thamicola*, *Taunayia* sp., *S. poulsoni*, and possibly *N. troglocataractus*. (2) Small populations (10^3 individuals): low densities and median distributions, or median to high densities and limited distributions: *P. kronei*, *T. itacarambiensis*, *A. cryptophthalmus* in Passa Três Cave, and, possibly, *A. rosae*. (3) Medium-sized to large populations (10^4 individuals): low densities and wide distributions, or median to high densities and limited to large distributions: *Astyanax* sensu lato, *C. geertsii*, *S. oedipus* sensu lato, *A. cryptophthalmus* in Angélica Cave, *S. eurystomus*, *T. pattersoni*, *T. subterraneus*, *P. mexicana*. The latter are within the lower range observed for many epigean freshwater populations (Trajano 1997b), except for very restricted species living in specialized habitats, such as temporary epigean pools.

It is interesting to note that species showing very small populations are among the most specialized troglobitic fishes, homogeneously anophthalmic and depigmented (except for pigmentation in *A. cryptophthalmus*), indicating speciation in peripheral isolates. The comparison between the populations of *A. cryptophthalmus* from Passa Três and from Angélica caves is particularly illustrative of a relationship between population size and degree of troglomorphism: the small population apparently isolated in Passa Três shows much more reduced eyes and cryptobiotic habits than that from the Angélica-Bezerra Cave system (Trajano 2001).

Fish movements and home ranges

In view of the contrasting ecological conditions under which epigean and subterranean populations live, differences are expected in the use of space, including the frequency and extensiveness of movements, home ranges, etc. For instance, food scarcity could determine an increase in movement rates or intensification of territorial behavior. Comparisons between epigean and subterranean related species, or between subterranean populations living under different conditions as a result of distinct regional climates (e.g., highly seasonal versus non-seasonal climates) may contribute to answering this question. Unfortunately, very few of such studies have been so far carried out.

The first evidence of movement in subterranean organisms is a temporal variation in the number of individuals observed at given sectors of the habitat (this is particularly evident for soil organisms). Eigenmann (1909) reported a decreasing number of *Lucifuga subterraneus* and *L. dentatus* individuals in the accessible parts of some Cuban caves following successive collections in a period of few months. Therefore, population loss was not compensated for by immigration from inaccessible underground spaces, indicating a low rate of movement within a relatively short period of three months.

Even in the absence of variations in population density, occurrence of movement cannot be ruled out because immigration and emigration rates may be similar, thus fish numbers will not change with time. The most reliable method to study fish movements is MR, which allows detecting and quantifying the movements of individual fish.

Detailed data on cavefish movements is available for Brazilian catfishes living in cave streams: *P. kronei* in Areias Cave (Trajano 1991), *T. itacarambiensis* in Olhos d'Água Cave (Trajano 1997b), *A. cryptophthalmus* in Angélica and Passa Três caves (Trajano 2001); and *A. rosae* from stream-like habitat in Logan Cave (Means & Johnson 1995, Brown 1996). In such studies, the investigated sectors were divided into subsections, inside which the fishes were grouped. The extension of such subsections depended on the total extent of habitat sampled, the number of fish to be marked in each occasion, and the collecting method. These factors determined the maximum number and extension of subsections viable for fieldwork: 5 m for *Amblyopsis rosae* (cave 1300 m long, low population density), 20 m for *Ancistrus cryptophthalmus* at Passa Três (cave 400 m long, high population density), 100 m for *T. itacarambiensis* (cave 5000 m long, medium population density) and *A. cryptophthalmus* at Angélica (cave 8000 m, but only 500 m studied due to the high population density), 200 m for *P. kronei* (cave system 5000 m, but only 800 m studied due to the use of large traps which were difficult to carry into the cave). Therefore, the spatial scales of discrimination of fish movements varied among these studies.

High recapture rates, indicating small population sizes and/or low movement rates, have been obtained in cave stream habitats: total recapture rate of *P. kronei* was 47.7%, varying between 15 and 100% among recapture occasions; total recapture rate of

T. itacarambiensis was 25.7%, with an increase from 13% in the first recapture occasion to 29% in the last one (Trajano 1997a); recapture rates of *A. cryptophthalmus* were 44% at Passa Três Cave and 15% at Angélica Cave (Trajano 2001); in *A. rosae*, recapture rates varied between 25 and 88% among recapture occasions (Means & Johnson 1995). Lower rates were generally obtained for stream-dwelling epigean species: total rates of 17–20% were recorded for *P. transitoria* in the epigean habitat (similar rates were also observed for cave populations of this species, Gerhard 1999); rates varying between 3 and 33% in samples with minimum 20 marked fishes were recorded by Hill & Grossman (1987) for three North American species.

Knapp & Fong (1999) obtained similar rates (mean 36%, ranging from 19 to 50%) for troglobitic amphipods *Stygobrommus emarginatus* at stream sites. On the other hand, recapture rates were extremely low (3.0%) at mud-bottomed pool sites, situated in the upper reaches of the stream and fed by seeps and by water from the mainstream during high water. These authors interpreted this low value as an evidence of movements between the epikarst and the studied pools, which would represent small windows into the epikarst inhabited by a much larger population of amphipods. M.E. Bichuette (personal communication) also obtained very low recapture rates (4.4%) for *Trichomycterus* species living at similar habitats in caves from São Domingos karst area in central Brazil, probably representing a similar case among cave fishes. Likewise, none of 17 specimens of *S. oedipus* marked in an upper vadose soft-bottomed, slow-moving water tributary in Mae Lana Cave was recaptured a few days later (Trajano, Borowsky, Mugue & Smart unpublished data).

Results obtained for the three studied Brazilian cave stream-dwelling siluriforms indicate a high degree of sedentariness, with small variations between species. The armored catfish, *A. cryptophthalmus*, as expected for a loricariid, is usually observed attached to rocky substrates scraping off the film of detritus covering the substrate. It seems to be very sedentary: within a period of one month, no fish was recaptured at Angélica more than 100 m from its initial capture, and, in Passa Três Cave, maximum movements 20–40 m along the stream were recorded (Trajano 2001).

Trichomycterus itacarambiensis, studied during the dry season, was less sedentary: 34% of recaptures were made in a section other than that of previous capture, corresponding to movements at distances greater than 100 m in periods of one to five months. Active, non-seasonal movements (both upstream and downstream) were recorded at maximum distances of 600 m; a few longer movements, up to 1500 m along the stream, were considered to be passive since they were always downstream movements. Passive movements are probably more frequent during the rainy season. Contrary to what was expected, *P. kronei*, a larger, potentially more mobile species studied during a one-year period, showed a higher degree of sedentariness, moving less frequently and for smaller distances: 20% of recaptures were made at a site different from the previous one, and only 2% of total recaptures were at distances greater than 300 m (Trajano 1997a).

Such differences may reflect variation in feeding styles associated with differences in the degree of food scarcity and species behavior, illustrating ecological and phylogenetic constraints determining patterns of cavefish movements. Armored catfishes, typically grazers and most territorial, are in general relatively sedentary, at least in short time-scales. *P. transitoria*, putative sister-species to *P. kronei*, is also highly sedentary: 93% of recaptures by Gerhard (1999) were made within the same 50 m long stream sections. *P. kronei* is a solitary, aggressive, possibly territorial fish, tending to stay in the same site for a long time. Because it has reduced cryptobiotic habits (Trajano & Bockmann 2000), defense of territory is probably centered on food resources (as also observed for *A. cryptophthalmus*, Trajano & Souza 1994), and not on hiding places as in *P. transitoria* and many other cryptobiotic fishes. Due to the absence of a well-defined dry season in the area, food availability for this cavefish is constant and apparently not greatly limiting, judging from the high percentage of specimens with stomach contents (Trajano 1989). This would justify the defense of relatively small feeding territories.

On the other hand, the degree of aggressiveness is lower and more variable in the few studied *Trichomycterus* catfishes, including the Brazilian cave species. Territoriality in these fishes seems to be restricted to particular situations, when individuals kept isolated are confronted with intruders. Besides, food scarcity in *T. itacarambiensis* habitat is, at least during part of the year, more accentuated than for *P. kronei*. Probably, the best style for *T. itacarambiensis*, representing the most economically favorable energetic balance between food demand and availability, is to have a relatively large home range, moving more frequently and for longer distances, at least during the dry season, in order to enhance the chances of finding food (Trajano 1997a).

Fluctuations in the estimated population size of *A. rosae*, associated with high recruitment rates at some occasions, with a peak during summer, were interpreted as an evidence of high rate of immigration from the aquifer into Logan Cave (Brown 1996). Inside the cave, the great majority of fish were captured in a 220 m long pool, and few movements of marked individuals upstream or downstream of this section were recorded. On the other hand, most fish moved throughout the pool, and a few individuals tended to remain within the lower 100 m of pool; as expected, larger fish tended to move longer distances and more frequently than small ones. In view of its relatively small size (max. TL = 63 mm), *A. rosae* has relatively large home ranges at Logan Cave when compared to the studied Brazilian cave catfishes. Periodic movements between the accessible habitat and the surrounding aquifer (including gravel interstices) was reported by Schubert et al. (1993) for *T. subterraneus* found in a boxed spring from Missouri.

Feeding

In view of the food scarcity generally prevalent in the hypogean environment, it is expected troglobitic fishes to be generalist feeders. As a matter of fact, hypogean fish species studied with a focus on this aspect are omnivores (e.g., *Astyanax* spp.) or, more frequently, generalist carnivores feeding on an opportunistic basis (amblyopsids, bythitids, siluriforms, etc.). Usual food items for cave fishes include crustaceans from copepods to crayfish, aquatic insects (juveniles and adults), mollusks (mainly gastropods), worms (annelids, nematodes), besides several allochthonous items (arachnids, terrestrial insects, etc.). For instance, *T. subterraneus*, *A. rosae* and small *A. spelaea* feed mainly on copepods, complemented by other crustaceans (cladocerans, isopods, amphipods, crayfish) and insect larvae; cannibalism on young fishes was recorded for *Amblyopsis* spp. (Poulson 1963). *Lucifuga subterraneus* and *L. dentatus* feed largely on crustaceans, including troglobitic shrimps, *Troglocubanus* sp. and *Typhlatya* (Eigenmann 1909, Nalbant 1981). Miller (1984) attributes the rarity of shrimps and mysids in pools inhabited by *R. reddelli* to predation by this catfish. Stream-dwelling trichomycterines such as *Trichomycterus conradi* from Cueva del Guácharo and *T. itacarambiensis*, and the heptapterine *P. kronei*, feed largely on insects, complemented by crustaceans and oligochaetes (Andreani-Armas 1990, Trajano 1989, Trajano 1997b). *P. mexicana*, which lives in an unusual chemoautotrophically based ecosystem, have an unusual diet, mainly composed of *Beggiatoa* bacteria and subsidiarily of insects (e.g., chironomid larvae) and other invertebrates (Langecker et al. 1996). Some cave species seem to prefer living prey (amblyopsids, *Taunayia* sp.), while others may also scavenge, feeding on both living and dead organisms (most catfishes such as *P. kronei*, *T. itacarambiensis*, etc.). The former have well-developed lateral line systems and probably use mechanical stimuli in the search for food, particularly moving prey.

There are records of cavefish using submerged bat guano as a food source (e.g., *Astyanax*, Parzefall 1993), but in many cases it is not clear whether these fishes are feeding directly on the guano or on the invertebrates living on it. For instance, based on field observations and examination of gut contents by transparency, Mendes (1995b) first concluded that the new heptapterine from NE Brazil fed on bat guano; however, the analysis of stomach contents revealed that this species is strictly carnivorous, preying on invertebrates found near or on the guano. Few troglobitic fishes are truly detritivorous, such as the ictalurid *T. pattersoni* (Langecker & Longley 1993) and the loricariid *A. cryptophthalmus*; the latter probably feeds on the thin film of detritus covering rocks inside caves (Trajano 2001).

The effect of food scarcity is made clear in the high proportion of specimens captured with no or few stomach contents observed in subterranean species like *Caecocypris basimi* (Banister & Bunni 1980), and *T. itacarambiensis* collected during dry season (Trajano 1997b). This, allied to the generally small size of samples, may hamper the investigation of individual, spatial and temporal variations in feeding habits. Small samples also render difficult a detailed analysis of diet allowing to detect more subtle, but nevertheless important, differences between species, especially relevant for syntopic troglobitic fishes (see below). Therefore, feeding is one of the least known aspects of cavefish biology, with a few exceptions.

In spite of the low frequency of *T. itacarambiensis* individuals with stomach contents, the large population size made it possible to get sufficient data for analysis. These data showed a decrease in the frequency of insects and an increase of oligochaetes as food items throughout the pronounced dry period, probably reflecting a decrease in insect availability as the season advanced (Trajano 1997b). Comparison with *P. kronei* illustrates the consequences of living under different climatic regimens: the contribution of allochthonous, epigean items to the diet was significantly higher for the latter, which is found in a less seasonal area without a typical dry season, in a larger cave stream with higher and relatively constant carrying capacity (Trajano 1997b).

In the few cases where direct comparisons were made (e.g., Poulson 1963, Trajano 1989), it was shown that troglobitic fishes belong to the same trophic guilds as their epigean relatives. This indicates that, concerning general diet, epigean ancestors were already preadapted to feed in the subterranean habitat. Differences in the proportion of items consumed by hypogean and epigean populations were tentatively attributed to differences in prey availability rather than food preferences (Andreani-Armas 1990, Trajano 1997b), although no quantitative analysis of such availability in the habitat was actually made. On the other hand, different foraging styles were observed for the troglobites, as an adaptation to the special conditions in the hypogean environment. For instance, troglobitic heptapterines also feed at midwater, adding surface picking to the behavior usually employed by the typically bottom-dwelling epigean relatives (Trajano 1989, Trajano & Bockmann 1999). Conversely, a stereotyped polarization to the bottom was described for troglobitic *Astyanax* spp. and *Caecobarbus geertsii* (Thinès & Wissocq 1972, Parzefall 1993). Amblyopsids illustrate a higher feeding efficiency in troglobitic species associated with behavioral specializations (Poulson 1963). The number of specimens with stomach contents in *P. kronei* were greater than in its eyed sister-species, *Pimelodella transitoria*, providing another example of high feeding efficiency in cave fishes (Trajano 1989).

Besides sensorial and behavioral specializations increasing feeding efficiency in troglobitic fishes, adaptation to the food scarcity generally prevalent in the hypogean habitat include the low metabolic rates recorded for some species and fat storage helping to survive long starvation periods. Large deposits of adipose tissue in different body parts were observed in many wild caught troglobitic fishes. Ridges of fat tissue along the base of adipose and anal fins are present in the troglobitic heptapterines, *Rhamdia reddelli*, *R. zongolicensis* and *R. macuspanensis* (Weber 1996, Weber & Wilkens 1998), and *P. kronei* (Trajano personal observation), but not in their epigean relatives, respectively *R. laticauda* and *P. transitoria*. The ictalurids *Prietella lundbergi*, *T. pattersoni* and *S. eurystomus* have large lipid deposits in subcutaneous areas and viscera, associated with mesenteries, and in the place of the reduced gasbladder (Langecker & Longley 1993, Walsh & Gilbert 1995); however, some individuals of *T. pattersoni* and *S. eurystomus* showed signs of starvation. Experiments with troglobitic *Astyanax* from Pachon Cave have shown that these fishes exhibit an improved ability to store fat and build up enormous reserves (Wilkens 1988).

Less specialized species, such as *T. itacarambiensis*, show these adaptations in lower degree or not at all, and may be heavily stressed during long periods of pronounced food scarcity: during the dry season, these fishes exhibit significant loss of weight, stop growing, and many individuals present negative growth – they actually shrink (Trajano 1997a).

Life cycle parameters: reproduction, growth rate and longevity

It is usually stated that many troglobites show a precocial lifestyle (sensu Balon 1981, 1999), including life history traits such as large and yolky eggs, low fecundity, delayed and infrequent reproduction, slow growth rates, and increased longevity, as adaptations to the food-poor subterranean habitat (Culver 1982). However, in the case with subterranean fishes, little has been done after the classical study on amblyopsids by Poulson (1963), and no other author encompassed all features, from reproduction to longevity, in such a comprehensive way. This author compared one epigean species, one troglophilic species, and the three troglobitic species known at that time, based on large preserved samples, and demonstrated that the latter present a more precocial lifestyle (similarly analogous to K-selected life cycle) when compared to their eyed relatives. This study became a paradigm for life cycle adaptation in subterranean organisms.

Other authors obtained data on reproduction of additional troglobitic fishes, including frequency and periodicity of reproduction and number and size of eggs. A few species have been studied with a focus on growth rates: *C. geersti*, *P. kronei*, *T. itacarambiensis*, *A. cryptophthalmus*, and *A. rosae*. The preliminary data available also point to precocial lifestyles in these populations, with low proportions of reproductive females producing relatively large eggs, slow individual growth rates and high longevity. These data indicate that precocial development may be indeed a common feature of troglobitic fishes. However, they are not necessarily autapomorphic character states representing specializations to the subterranean habitat. A precocial lifestyle may be a characteristic already present in epigean species and favoring the colonization of the hypogean environment, as seems to be the case with the Brazilian catfishes, *Pimelodella transitoria* and *P. kronei* (see below).

The study of reproduction in troglobitic fishes poses several problems. As already mentioned, many subterranean habitats are not accessible during the rainy periods, thus it is not possible to follow the whole annual cycle. Moreover, low population densities and infrequent reproduction result in small samples with very low numbers of mature individuals. Therefore, in contrast to epigean fishes, for which reproduction is one of the most intensively studied aspects of their biology, reproductive data for troglobitic fishes are limited.

Seasonal reproduction, with well-defined annual cycles, was reported for several species and seems to be more frequent than non-seasonal breeding in cavefishes. *A. spelaea*, and possibly *A. rosae* as well, breed during high water from February through April (Poulson 1963). *Ogilbia pearsei*, from Yucatan, which is viviparous like all other bythitids, display a reproductive period from December to February (Iliffe 1993). The Cuban bythitids, *L. dentatus* and *L. subterraneus*, seem to breed throughout the year, but a reproductive peak may occur in March; reproduction is also infrequent in these species (Eigenmann 1909). Epigean and troglobitic *Astyanax* show spawning peaks which differ among populations in the laboratory and this data are reinforced by sparse observation in the field: March for the epigean species, April for the La Cueva Chica population, May for the Sabinos population and August for the Pachón population (Sadoglu 1979). *U. zammaranoi* presumably reproduces during the dry, hot season from January to March (Ercolini et al.1982).

A small proportion (18%) of specimens of the new heptapterine taxon from NE Brazil had developed gonads (infrequent reproduction), and all mature females were found at the end of the rainy season in March (Mendes 1995b). In *T. itacarambiensis*, studied during the dry season, the highest proportion of mature females (18%) was also observed at the end of the rainy season (March), dropping to less than 3% in the subsequent months; this points to seasonal reproduction with a peak during or just at the end of the rainy period (Trajano 1997b). Both *T. itacarambiensis* and the new heptapterine genus live under seasonal climates, subject to pronounced dry periods. The low proportion of ripe fish, which is expected in view of the food scarcity prevailing in the subterranean habitat, indicates that not all individuals reproduce every year, as observed by Poulson (1963) for troglobitic amblyopsids (see below).

Non-seasonal reproduction was suggested for the syntopic *Typhlogarra widdowsoni* and *Caecocypris basimi* (Bannister & Bunni 1980). A very low number of *P. kronei* catfishes with mature gonads was found (Trajano 1991), hampering any conclusion about reproductive patterns.

Fewer and larger eggs were observed for the troglobitic amblyopsids, *Typhlichthys subterraneus* (~50 mature ova, size 2.0–2.3 mm, per female), *Amblyopsis spelaea* (~70 ova, 2.0–2.3 mm) and *A. rosae* (23 ova, 1.9–2.2 mm), than in the troglophilic *Chologaster agassizi* (~150 ova, 1.5–2.0 mm) and in the epigean *C. cornuta* (93 ova, 0.9–1.2 mm). On average, these species first reproduce at ages of 24, 40, 37, 12 and 12 months, respectively, and the estimated proportion of females breeding every year is 50, 10, 20, 100 and 100% respectively, resulting in infrequent reproduction by a small proportion of troglobitic females. Therefore, the effective stock, which is the part of the population genetically contributing to the next generation, determining the rates of genetic drift (Futuyma 1986), is much smaller than the total populations estimated by MR or VC (see above).

Eggs with an increased amount of yolk, producing larger embryos, were reported for several troglobitic populations of *Astyanax* (Wilkens 1988). Large eggs have been observed for *A. cryptophthalmus* (L. Finley personal communication). Production of a small number of eggs (23–28 oocytes ready for spawning), which may be retained for long periods (up to six months), was described for *Horaglanis krishnai* (Mercy et al. 1982). Such retention ability indicates that favorable conditions for spawning occur only during limited periods.

Low individual growth rates and high longevity are other related life cycle traits of troglobitic amblyopsids. Based on examination of marks on scales regarded as annual growth annuli, Poulson (1963) estimated maximum ages of 4.2 years for *T. subterraneus* (maximum size = 62 mm SL), 7.0 years for *A. spelaea* (max. size = 85 mm) and 4.8 years for *A. rosae* (max. size = 52 mm), in contrast to 2.3 years for *C. agassizi* (max. size = 61 mm) and 1.3 years for *C. cornuta* (max. size = 57 mm). Growth rates calculated based on the greatest lifespans and sizes recorded (Poulson 1963; table 5) and on growth curves (Poulson 1963; figures 1 and 2) were around 1.0–1.25 mm month^{-1} for *T. subterraneus*, 1.0 mm month^{-1} for *A. spelaea*, and

0.9 mm month^{-1} for *A. rosae*; such values are lower than those calculated for *C. agassizi* (1.7 mm month^{-1} calculated from growth curves, 2.2 mm month^{-1} as calculated from max. lifespan) and much lower than the calculated growth rate for *C. cornuta* (2.4 and 3.8 mm month^{-1}, respectively). Differences in life history result in distributions of age frequencies skewed towards older age classes in *A. rosae* and *A. spelaea*. Differences in growth rates and allometry would also account for differences in morphology (e.g., size of head, cerebellum and inner ear, fin length) between these species.

Mean growth rates estimated for *A. rosae* studied by MR in Logan Cave (Brown 1996) corresponded to values lower than those recorded by Poulson (1963) for this species: for size class 30–39 mm TL, mean growth rate = 0.7 mm month^{-1}; size class 40–49 mm TL, 0.3 mm month^{-1}; size class 50+ mm TL, 0.06 mm month^{-1}. Thus, it would take 3.9 years for a 30 mm fish to attain 50 mm. However, some individuals exhibited much higher growth rates, of 1 mm month^{-1} or more (that may be the case of the largest individuals examined by Poulson 1963). Growth spurts were not necessarily correlated with the increase of food availability represented by bat guano during the summer. Therefore, growth could be controlled by other factors, such as age, sex and reproductive condition, in addition to exogenous factors (Brown 1996).

Smith & Welch (1978), who carried out an MR study of epigean *C. agassizi* in southern Illinois, obtained growth rates varying during the spring studied, but they were generally low (overall growth = 0.8–1.7 mm month^{-1}). Some individuals were found to attain relatively large sizes (more than 70 mm SL), which were probably older than three years.

Annual growth in pulses was indirectly recorded for *T. itacarambiensis*. Very low rates were recorded in a MR study carried out during the dry season: mean growth rate was 0.06 mm month^{-1} for the whole studied population (30–83 mm SL) and 0.17 mm month^{-1} for size class 30–45 mm SL (immatures); several cases of negative growth were recorded (Trajano 1997a). This was followed by a significant decrease in condition factor during the studied dry period (Trajano personal observation), demonstrating the role of food limitation in the life history of this species. Growth rates under laboratory conditions that simulated the rainy season as regards food availability (food ad libitum) were much higher: 1.2 mm month^{-1} for adult catfishes, 2.1 mm month^{-1} for young ones. Based on these rates, longevity of seven years or more was estimated for the species. Comparison with a few studied Chilean trichomycterids suggests a precocial lifestyle for *T. itacarambiensis* (Trajano 1997a).

Average growth rate of adult individuals of *P. kronei* (more than 70 mm SL) from Areias Cave, Upper Ribeira karst area, was 1.0 mm month^{-1}. Based on this, longevity of 10–15 years was estimated for the species, considered high when compared to the few other catfishes studied at that time (Trajano 1991). However, average growth rates similar to those reported for *P. kronei* were calculated by Gerhard (1999) for its putative sister-species, *Pimelodella transitoria*, in epigean streams of that karst area during the rainy season (1.0 mm month^{-1}), and lower rates were recorded during the dry season (0.5 mm month^{-1}). Growth rates of 0.7 to 0.9 mm month^{-1} were observed for individuals of *P. transitoria* living in caves. Cases of negative growth were recorded for this species, suggesting that feeding stress may also occur in epigean habitats. Longevity estimated for *P. transitoria* was 4.1–12.4 years. *Rhamdioglanis frenatus*, another heptapterine syntopic with *P. transitoria*, shows higher growth rates (3.2 mm month^{-1} in the rainy season; 0.8 mm month^{-1} in the dry season) and lower estimated longevity (4.0–8.4 years) (Gerhard 1999).

Therefore, a precocial lifestyle is already present in *P. transitoria*, possibly as a preadaptation to the cave life. This may be a species-specific characteristic. *Pimelodella pappenheimi*, that also lives in a stream from the Atlantic rainforest to the south of the Upper Ribeira Valley, presents a less precocial lifestyle: with similar maximum lengths, average growth rates were higher and maximum longevity recorded (four years) was lower in *P. pappenheimi* than in *P. transitoria* (Amaral et al. 1999). In this case, the low growth rates and high longevity reported for *P. kronei* may be not a consequence of isolation and further specialization to the subterranean habitat, but just a plesiomorphic trait inherited from the epigean ancestors.

Some records in laboratory confirm the high longevity for cavefishes: three years for males and five years for females of the small-sized and rather unspecialized *P. mexicana*, 10 years for *Astyanax* spp. (J. Parzefall personal communication), at least ten years is estimated for *T. itacarambiensis* (specimens collected 7.5 years ago with 50–60 mm SL are still alive), 18 years for *P. kronei* (a specimen collected when 115 mm long lived more than 10 years in laboratory) (Trajano personal observation), more than 15 years for *U. zammaranoi* and more than 23 years for *P. andruzzi* (specimens of these species collected respectively in 1985 and 1977 are still alive, R. Berti personal communication).

Heuts (1952) performed a detailed study of growth in *C. geertsii* based on examination of scale rings of preserved samples, presuming that such rings were real year-rings due to seasonality of the hydrological system. Longevities of 9 to 13 years were estimated for fishes attaining 80–90 mm, with mean growth rates varying among populations from 0.25 to 0.60 mm month^{-1}. Compared to the epigean *Barbus holotaenia*, found in the neighborhood of the studied caves (four years to attain 80 mm, growth rates 250–300% higher than those recorded for *C. geertsi*), *C. geertsi* has a precocial lifestyle.

Heuts (1952) detected a variety of growth patterns among populations of *C. geersti*, associated with pronounced differences in the distribution of age classes. One of the studied populations exhibited a constant, but very small geometric growth rate, while the others showed deviations from this simple type, by accelerations and retardations in given periods of life. Such differences would reflect local ecological conditions. Growth patterns in *C. geersti* differ from the usual patterns observed for epigean fishes. Heuts interprets these unusual patterns as a result of heterochronic development, hypothesizing that regressive character states are a consequence of retardation of body growth.

Heterochrony seems to be a frequent phenomenon in the evolution of subterranean fishes and may explain at least part of the special morphological and behavioral traits of troglobites: longer fins, larger heads and greater exposure of sense organs in amblyopsids (Poulson 1963); small size and weak ossification in *T. pattersoni* and *S. eurystomus* (Langecker & Longley 1993); small size, reduction of cryptobiotic habits and increased midwater activity in Brazilian heptapterines (Trajano & Bockmann 1999). However, each of these character states have different adaptive meanings, and were probably achieved by different processes: progenesis in the case of ictalurids and heptapterines (Trajano & Bockmann 1999) and neoteny in amblyopsids (Poulson 1963), or possibly hypermorphosis (sensu Chaline 1987), at least in *A. spelaea*. The latter exhibits a particularly extended lifespan, attaining much larger sizes. For example, there are records of individuals 200 mm long (Greenwood 1967), while other amblyopsids do not surpass 70 mm. Therefore, those traits are not the result of a common causal phenomenon.

Species interactions: syntopy of troglobitic fishes and predators

Among more than 86 species of subterranean troglobitic fishes known throughout the world, there are 15 cases of coexistence of troglobitic fishes in the same cave system (Proudlove[7], personal communication): 13 cases involve two species and two involve three syntopic species. Differences between these figures and those reported by Proudlove are due to the fact that the existence of a second, undetermined species, troglomorphic *Rhamdia* catfish syntopic with *R. reddelli* in Mexico could not be confirmed afterwards (A. Weber personal communication), and that the Brazilian *A. cryptophthalmus*, *Trichomycterus* sp. 1 and *Eigenmannia vicentespelaea* were reported as syntopic by mistake in the São Vicente system (Trajano 1997d); only the former two are really syntopic in Passa Três Cave, part of that system (Trajano 2001). Half the cases of syntopy correspond to species belonging to the same families, ecologically close. The case of the bythitid *Lucifuga* is particularly interesting: three species of this genus are reported as syntopic in one cave and two in a second cave from Cuba.

If all these cases actually correspond to syntopy and all the species recorded are valid and truly troglobitic, almost 40% of the troglobitic fishes share at least part of their habitat with other troglobitic fish species, and a higher proportion share habitat with any fish species if we also consider non-troglobitic fishes that may coexist with the troglobites (see below). Therefore, a considerable proportion of troglobitic fishes may interact with other fish species.

Habitat partitioning or differences in diet, allowing coexistence with minimum competition, may explain some of the well-documented cases of syntopy of subterranean troglobitic fishes. Different habitat preferences have been observed for the cave balitorids, *S. oedipus* and *C. thamicola*, which are syntopic in Mae Lana Cave in NW Thailand. In this cave, *S. oedipus* occurs in slow moving waters over silty substratum in two upper tributaries of the base-level stream and in rimstone dams atop flowstone; *C. thamicola*, on the other hand, is found in waterfalls formed by runoff waters from those tributaries, which cascade over flowstones, and in one section with rapids on limestone bedrock of the mainstream (Trajano, Borowsky,

[7] Proudlove, G.S. 1997. A synopsis of the hypogean fishes of the World. pp. 351–353. *In*: Proc. 12th International Congress of Speleology, La-Chaux-des-Fonds.

Mugue & Smart unpublished data). The occurrence of *C. thamicola* in specialized, highly discontinuous habitats in two different caves (Tham Mae Lana and the type-locality, Tham Susa; Kottelat 1988) poses an interesting question about the fish dispersion and connectedness of populations.

On the other hand, *A. cryptophthalmus* and *Trichomycterus* sp. 1, syntopic at Passa Três Cave, central Brazil, present different feeding habits: while *A. cryptophthalmus* is detritivorous, *Trichomycterus* sp. 1 is a generalist carnivore (Trajano & Bichuette unpublished data), much as their respective epigean relatives. Unfortunately, for the majority of syntopic fish species, there is no detailed data on diet and microhabitat occupation for an analysis of mechanisms allowing coexistence. In some cases of closely related hypogean species that are syntopic in some caves, such as *Lucifuga* spp., morphology and biology are so similar that they are treated together (e.g., Eigenmann 1909, Juberthie[6]) suggesting direct competition when these species meet each other.

Some syntopic species present similar population densities, but the majority seem to differ. Based on the size of collections, population density of the carnivorous *S. eurystomus* is half that of the detritivorous *T. pattersoni* (Longley & Karney 1979). The same is true for *C. basimi* in relation to *T. widdowsoni* (Banister & Bunni 1980). The Chinese *S. anophthalmus* has a larger population size (the proportion is not given) than *T. ayunnanensis* and *Paralepidocephalus* aff. *yui* (Yang & Chen 1993a). On the other hand, similar population densities are reported for *Protocobitis typhlops* and *Oreonectes* sp., also from China (Yang & Chen 1993b).

There is a general notion that, as a consequence of food scarcity, subterranean food pyramids would not support more than one level of predators (e.g., Mohr & Poulson 1966). Therefore, hypogean fishes, usually predaceous, would not generally be subject to important predation pressures, and some specializations of troglobitic fishes have been interpreted in this context, such as the absence of generalized phobic reactions and reduction of cryptobiotic habits allied to increased midwater activity in Brazilian siluriforms (Trajano & Bockmann 2000), and reduction of dorsal aculeum in *P. kronei* (Trajano & Britski 1992) and some *Rhamdia* species (Weber 1996). Refractoriness to disturbing stimuli, probably related to lowered metabolism (energy economy), and increased activity in troglobitic amblyopsids and *T. widdowsoni* (Poulson 1963) are specializations that can develop in the absence of important predation pressures.

As a matter of fact, there are few cases of potential cavefish predators identified in the natural habitat, and rare confirmed reports of predation on cave fishes. Epigean fishes were pointed out as potential predators of amblyopsids such as *A. rosae* (Brown 1996, Poly & Boucher 1996) and *Taunayia* sp. (Trajano & Bockmann 2000). The erythrinid *Hoplerythrinus unitaeniatus* was observed preying on the troglobitic armored catfish, *Ancistrus cryptophthalmus*, in a cave in central Brazil (M.E. Bichuette personal communication). Langecker & Longley (1993) suggested that *Satan eurystomus* may be a predator of *Trogloglanis pattersoni*, which, if confirmed, would constitute the first example of such kind of interaction between troglobitic fish species. Crabs, *Epilobocera* sp., were observed attacking *Lucifuga* spp. in Cuban caves (Díaz-Pérez 1988), and, according to Romero (1998a), crayfishes are known predators of cave fishes. Decapod crustaceans, frequently co-inhabiting caves with fishes, could be important predators, especially of juvenile fishes. Conversely, young decapods are food for *L. dentatus* (Eigenmann 1909), *P. kronei* (Trajano 1989) and other cavefishes, configuring a complex food web.

Extensive fieldwork showed that fish predators are actually absent for the great majority of studied species or, if they do exist, these predators are so rare that their impact on the resident fish population is negligible. This is the case for *Astyanax* spp, *P. mexicana*, *P. kronei*, *T. itacarambiensis*, the new heptapterine genus from NE Brazil, and amblyopsids, among others.

On the other hand, cannibalism, observed for amblyopsids (Poulson 1963) and *Taunayia* sp. (Trajano & Bockmann 2000) but possibly more common among subterranean fishes (the scarcity of feeding data for most species may account for the absence of additional records), may be an important ecological factor regulating population densities.

Non-troglobitic cave fishes

Non-troglobitic fishes regularly found in the subterranean biotope, although rarely forming large populations in this habitat, also deserve a place in this synthesis. Most importantly, these are also hypogean fishes and are subject to the same ecological constraints as troglobites. In addition, they may coexist and interact with troglobitic fishes, as predators, prey, competitors, or simply sharing space. Finally, they illustrate the kind of fish community that can be isolated in the subterranean habitat, originating troglobitic species. These fishes may be troglophiles, organisms able to complete their life cycle both in hypogean and in epigean habitats (Holsinger & Culver 1988), trogloxenes, organisms found regularly in the subterranean habitat, but which must return periodically to the surface to complete their life cycle (op. cit.), or just transient occurrences in the subterranean environment.

Such an emphasis has been understandably given to the troglobitic fauna that little attention is paid to non-troglobitic fishes, and little data are available for such communities beyond species records. As a consequence, there is little evidence allowing classification of most of these fishes as troglophiles, trogloxenes or accidentals. Rare presumed troglophiles have being studied in detail and compared to troglobitic relatives. For instance, the troglophilic amblyopsid *C. agassizi* is so frequent in caves from south-central United States that it received the same emphasis as troglobitic amblyopsids (e.g., Poulson 1963, Smith & Welch 1978). *C. agassizi* is clearly preadapted to the hypogean life, not only in relation to behavior and sense organs (much as the exclusively epigean *C. cornuta*), but it is also adjusted endocrinologically to live under permanent darkness (in contrast to *C. cornuta*). The heptapterine *P. transitoria*, putative sister-species of *P. kronei* and syntopic with the latter, was object of stud ies by Pavan (1945, 1946) and Trajano and colleagues (Trajano 1989, 1991, 1994, Hoenen & Trajano 1995, Trajano & Gerhard 1997); this species also exhibits several preadaptations to the cave life, as discussed above. *Synocyclocheilus* cyprinids, with five troglobitic species in China (Proudlove[7]), are described as having 'half-cave-dwelling' habits, usually occurring and breeding in underground rivers or caves connected to epigean rivers or lakes (Chen & Yang[8]), thus are typical troglophiles.

On the other hand, the status of eyed tetra characins *Astyanax mexicanus* found in caves is unclear, especially when they co-occur with troglobitic populations, hybridizing to a greater or lesser extent (Mitchell et al. 1977). In the literature, it is implied that epigean tetras are frequently introduced in caves, possibly during high water. However, if such fishes are able to breed with blind forms, they could also breed among themselves. The point is: how self-sustained must these cave populations of eyed tetras be in order to fit into the troglophile category? (Self-sustainability is not mentioned in any definition of troglophilic populations, but it seems to be a logical inference.) A similar situation is observed in Brazilian caves: groups of normally eyed and pigmented *Astyanax* fishes, sometimes very large (dozens to hundreds of individuals), have been found in caves from Mato Grosso do Sul, SW Brazil (E.P. Costa Jr. personal communication) and central Brazil (see below). The status of these groups remain undetermined, but no troglobitic *Astyanax* was found in these areas (although troglobitic siluriforms do exist in both areas, indicating that there were opportunities for isolation), suggesting that such groups do not form self-sustained populations.

I will illustrate non-troglobitic ichthyofaunas with three examples showing how diverse this fauna may be under certain circumstances, and the kind of community that can be isolated in the subterranean habitat and originate troglobitic populations: fish communities in caves from western Thailand and central Brazil (unpublished data) and non-troglobitic fishes in caves from West Virginia, U.S.A, where troglobitic species are absent (Poly & Boucher 1996). Although generally not numerous in caves, those large animals may exert a considerable impact on subterranean communities, preying on invertebrates including troglobitic species, such as the endangered crayfish *Cambarus aculabrum* from Arkansas, possibly also on troglobitic fishes, and conversely as an abundant food source for cavernicoles.

A considerable number of fish species belonging to different families and orders were found in streams crossing caves from West Virginia (and in other North American caves as well), at distances of hundreds of meters from the known entrances, but only a few were found on a regular basis and in reasonable numbers (constituted by several dozen): *Cottus* spp. (Cottidae), *Semotilus atromaculatus* and *Pimephales notatus* (Cyprinidae), *Lepomis cyanellus* (Centrarchidae), *Catostomus commersoni* (Catostomidae). *Cottus* spp. are one of the non-troglobitic fishes most commonly found in caves throughout the United States. In other regions of North America, catfishes such as *Ameiurus*, which are rare in the caves studied in West Virginia, may be quite common; 100 individuals were found in a cave in Florida.

Poly & Boucher (1996) mention the occurrence of depigmentation, including albinism (albinos may comprise 2–3% of the subterranean populations), and other morphological abnormalities in non-troglobitic fishes occurring in North American caves. They suggest

[8] Chen, Y. & J. Yang. 1993. Species and origin of cave-dwelling *Sinocyclocheilus* fishes. pp. 123–124. *In*: Proc. 11th International Congress of Speleology, Beijing.

that such abnormalities, and also the appearance of out-of-season reproductive features, may be a consequence of hormonal imbalance. This hypothesis was based on authors like Rasquin & Rosenbloom (1954), who failed to rear epigean *Astyanax mexicanus* in permanent darkness for months, although this could not be confirmed by later studies (Wilkens 1988).

Cavers in western Thailand are familiar with non-troglobitic fishes frequently seen in caves crossed by large streams, characterized by high water volumes and, in many cases, strong currents. These fishes, mainly cypriniforms, are found all the way through most caves, often many kilometers from the nearest entrance, in the typical aphotic subterranean habitat. They are observed swimming in isolation or in small groups, up to ten individuals. Such fishes include the cyprinids *Garra* sp., *Danio* sp., *Tor* sp. and *Poropuntius* sp., the balitorids *Schistura* sp., *Homaloptera* sp., *Balitora* sp., and *Hemizyon* sp., plus a sisorid, *Glyptothorax* sp., and an unidentified silurid (Borowsky[9]). The cyprinids, including large bodied species, were usually seen in slow-moving water of variable depths; both adults and juveniles were recorded. Collected specimens seemed to be well fed and in good healthy condition. The eyed balitorids found inside caves were in habitats typical for these fishes – swimming over or maintaining position on rocks in fast-flowing, turbulent water. We never observed epigean fishes and troglobitic balitorids in syntopy. Representatives of other fish taxa, such as siluriforms and synbranchids were rarely seen during our visits (Trajano, Borowsky, Mugue & Smart unpublished data).

A similarly rich fauna of non-troglobitic fishes was recorded in São Domingos karst area, Goiás State, central Brazil, in caves crossed by large streams similar to those studied in Thailand. This area harbors the richest subterranean ichthyofauna in Brazil, including at least four (possibly six) troglomorphic species and several non-troglobitic ones. This may be partly due to the fact that the area is part of the Tocantins River basin, with a diversified Amazonian epigean fauna as source of potential colonizers of the hypogean environment, and partly because many cave entrances are sinkholes favoring energy input. As a matter of fact, large amounts of detritus are observed inside these caves, so food does not seem to be a limiting factor.

[9] Borowsky, R. 1998. Survey of the cave fishes of south-western and peninsular Thailand. Report to the Thai Royal Forest Department, Web version <http://pages.nyu.edu/~rb4/thaicave.htm>.

Several non-troglobitic species were regularly found in São Domingos caves since the 80s (Dessen et al. 1980, Trajano 1987, Trajano & Bichuette unpublished data), including characiforms such as *Hoplerythrinus unitaeniatus* (Erythrinidae), *Brycon* spp. and *Astyanax* sp. (Characidae), siluriforms as *Imparfinis hollandi*, *I. minutus* and *Rhamdia quelen* (Pimelodidae, Heptapterinae), *Pseudocetopsis plumbeus* (Cetopsidae), *Hypostomus* sp., *Loricaria* sp., *Rineloricaria* sp. (Loricariidae), gymnotiformes as *Apteronotus* sp., *Sternachorhinchus* sp. and *Sternopygus* sp., and the cichlid *Cichlasoma araguaiensis*; semi-aquatic tetrapods such as alligators and turtles were also occasionally found in these caves. A small population of *Pimelodella* catfishes apparently showing some reduction of eyes, pigmentation and phobic reactions (a character state also observed for *P. kronei* – Trajano 1989), was found in a small tributary in one of those caves. The commonest non-troglobitic fish species in São Domingos caves are *H. unitaeniatus*, *Imparfinis* spp., *Astyanax* sp. and gymnotiforms. There is a record of a small group formed by the same individuals of *B. brevicauda* living for at least three years in the same site in São Mateus Cave (R.H. dos Santos personal communication), demonstrating that these fishes are able to survive in healthy conditions for long periods inside a cave.

Non-troglobitic fishes co-occur with troglobitic species in Angélica-Bezerra Cave system (*A. cryptophthalmus*), in Passa Três Cave (*A. cryptophthalmus* and *Trichomycterus* sp. 1), where a considerable number of *H. unitaeniatus* is found in pools in the small stream crossing the cave (see Trajano 2001), along with *Astyanax* sp. (a depigmented specimen was found in 1988, Trajano & Souza 1994), and in São Vicente I Cave (*Eigenmannia vicentespelaea*). A study based on the analysis of stomach contents of those non-troglobitic fishes is in progress to determine their possible role as predators of troglobitic organisms. Nevertheless, in view of their relative abundance and diversity, there is little doubt that non-troglobitic fishes are important components of the subterranean ichthyofauna at the São Domingos karst area.

General discussion

According to Vandel (1964), Schiner originally defined troglobites as organisms exclusively found in caves. However, because caves constitute an artificial subdivision of the hypogean domain based on human parameters, this category must be redefined to encompass all subterranean habitats, for example, interconnected spaces of subsoil characterized by permanent darkness and tendency to climatic stability (Juberthie 1983), which determine the most important ecological traits of strictly hypogean species. More recently, the term 'stygobite' came into use to designate aquatic troglobites. Considering that, in accordance with the expanded definition of troglobites comprising species restricted to subterranean habitats in general, 'aquatic troglobites' and 'stygobites' are synonyms, and to maintain coherence with the term 'troglomorphisms' for the autapomorphies shown by such species, I herein use the term troglobites for exclusively hypogean fish species.

Because applying the geographic criterion of exclusion from epigean habitats is frequently unfeasible, particularly in poorly studied tropical regions, the presence of troglomorphisms, especially reduction of eyes and/or pigmentation in relation to epigean relatives, is commonly used to deduce the status of troglobite for subterranean species. Although 'troglomorphic traits' may occur in species living in epigean habitats which share important ecological characteristics with the hypogean ones (e.g., deep, turbid river waters) (see Thinès & Proudlove 1986 and Trajano 1997d for lists of troglomorphic fishes including non-hypogean species), the presence of troglomorphisms is generally a useful criterion to recognize troglobites and has been used in the case of species discussed herein.

Representatives of epigean fish species found in the subterranean environment are more difficult to accommodate into the Schiner-Racovitza classification (cf. Holsinger & Culver 1988), as pointed out by Poly & Boucher (1996). Few eyed and pigmented species, such as *C. agassizi* and probably *P. transitoria*, have been recognized as troglophiles. For the great majority of non-troglomorphic fishes found in caves, there is no indisputable evidence allowing their classification as troglophiles, trogloxenes or accidentals. Many appeared to be quite healthy, although it was usually impossible to determine the length of time these fishes had been in caves, indicating that they are able to orient themselves and to get some food under permanent darkness and food scarcity. In the case of Mexican tetras, *A. mexicanus*, epigean individuals entering some caves may reproduce with the resident troglobitic tetras, producing hybrid populations (Mitchell et al. 1977, Romero 1983). However, evidence of establishment of reproductive, self-sustained populations, defining the troglophilic condition, is usually lacking. It is also unclear how sensitive each one of these non-troglobitic species are to the absence of photoperiods and how they cope with hormonal imbalance and other physiological consequences of life under constant darkness. Inability to adjust endocrinologically to such conditions may explain the absence in caves of fish apparently preadapted to the hypogean life, as seems to be the case with *C. cornuta* (Poulson 1963), and *T. bifasciata* (epigean sister-species of the troglobitic *Taunayia* sp.), that survived shortly under constant darkness in laboratory (Trajano personal observation).

Regardless of their status as troglophiles, trogloxenes or accidentals, non-troglomorphic fishes in general are not numerous in caves, frequently showing population densities lower than those of troglobitic fishes. This is probably because few individuals among the epigean populations are able to cope with the harsh ecological conditions in the subterranean biotope, and accidental introduction of fishes into this habitat is a relatively rare event.

Population data on non-troglobitic fishes, although scarce and rarely quantitative, indicate that most cavefish populations, when first isolated in the subterranean habitat, are small and prone to genetic drift due to bottleneck effects. Such founding populations are expected to be small because they had not yet acquired the special adaptations to cope with food scarcity, as those shown by most troglobites, and many of them were probably even more food-limited and habitat-restricted than now. Several authors hypothesize that isolation may occur during dry periods, as a consequence of paleoclimatic changes, when previously continuous epigean/hypogean water bodies fragmented due to breakdown of hydrological continuity (Willis & Brown 1985, Trajano 1995, Cobolli-Sbordoni et al. 1996). During these periods, energy input to caves would decrease due to the interruption of epigean drainage, and aquatic communities would be isolated in discontinuous hypogean water bodies. Some of these isolated populations may survive periods of climatic stress, accumulating significant changes that may be eventually recognized as distinct phylogenetic troglobitic species. The others may survive without significant change, or simply not survive such periods at all.

In isolation, genetic drift and other evolutionary processes would lead to transformation of characters such as eyes and pigmentation, resulting in troglobitic speciation. After the appearance of derived character states that improve efficiency of orientation and feeding in the hypogean environment, the new troglobitic populations may increase, attaining the densities herein recorded for several fish species. However, effective population sizes may remain low due to a decrease in reproductive rates, a common specialization of troglobites, which may be associated with a decrease in the proportion of individuals reproducing at least once during their lifespans.

Therefore, effective population sizes of epigean or troglophilic fishes phylogenetically related to the troglobitic species would constitute the best approximations for the initial 'N' modeling the evolution of troglomorphic characters (see, for instance, Culver 1982). Unfortunately, there are no reliable estimates of effective population sizes either for non-troglobitic or for troglobitic fish species.

As aforementioned, most of the 86+ species of troglobitic fishes so far recorded are siluriforms or cypriniforms, found in all continents but Europe. The highest species richness is observed in karst areas from China, Mexico, Brazil and southeast Asia. Such areas do not necessarily correspond to regional or local hotspots of biodiversity of troglobites. For instance, no other confirmed troglobites, aquatic or terrestrial, were found in São Domingos karst area, in the Brazilian region with highest richness of troglobitic fishes.

Interestingly, only three out of 20 caves and wells identified by Culver & Sket (2000) as local hotspots of biodiversity (those with 20 or more troglobitic species) harbor troglobitic fishes: Shelta (*T. subterraneus*) and Mammoth (*T. subterraneus* and *A. spelaea*) caves, in the U.S.A., and Gua Salukkang Kallang, in Indonesia. However, 14 of these caves are located in Europe, where no troglobitic fish has ever been found. The scarcity of high diversity tropical caves can be explained by the short history of biospeleology in the tropics, and related to the paucity of specialists dedicated to the taxonomy of tropical subterranean groups. With the progress of biospeleological surveying and taxonomic studies in tropical localities, many more hotspots of biodiversity are expected to be identified in these regions. The real puzzle is the absence of troglobitic fishes from Europe, especially from the Dinaric karst, which has a high diversity of other hypogean fauna. Considering the extension of this karst area and its rich geological history explaining the occurrence of hotspots of biodiversity (Culver & Sket 2000), and the presence of troglobitic aquatic salamanders (ecologically close to fishes), at least in theory, conditions are good for colonization of caves by fishes and for isolation ultimately originating troglobitic species in this and other European karst areas.

No strong phylogenetic/taxonomic or geographic correlation was found for population parameters in troglobitic fishes: related species may differ considerably in relation to the extent of distribution, habitat, population density and size, etc. Apparently, such traits correspond mostly to responses to local conditions, as exemplified by the comparison between the Brazilian catfishes, *P. kronei* and *T. itacarambiensis*, which live in karst areas which are quite distinct climatically. Local ecological conditions would determine the kind and extent of specializations in troglobitic fishes, limited by biological constraints proper to each group; detritivores are not expected to become carnivores while adapting to the subterranean habitat, just as the not usually numerous heptapterines are not expected to show high population densities in caves. Nevertheless, some amazing cases of high degree of evolutionary modification affecting a large number of characters have been described, especially among species that managed to adapt to extreme conditions in deep phreatic habitats (e.g., *T. pattersoni*, *S. eurystomus*, *U. zammaranoi*, *P. cisternarum*).

In some cases, habitat and habitat preferences observed in troglobitic fishes correspond to plesiomorphic character states retained from epigean ancestors, in others there was an evolutionary change after speciation in the subterranean biotope. Epigean *Trichomycterus* catfishes are typical stream-dwellers, as are the troglobitic *T. chaberti*, *T. itacarambiensis*, *Trichomycterus* sp. 1, from central Brazil; on the other hand, *Trichomycterus* sp. 2, from SW Brazil, inhabits flooded caves, a specialized habitat. Habitat preferences of *P. kronei* and *A. cryptophthalmus* are similar to those of their respective epigean close relatives: *P. transitoria*, sister-species to *P. kronei*, is also found preferentially in pools (Gerhard 1999), whereas epigean *Ancistrus* prefer rapids, like *A. cryptophthalmus* (Trajano 2001).

Probable cases of habitat change were reported for *Taunayia* sp. and the new heptapterine genus (Trajano & Bockmann 1999), *S. oedipus* (Trajano, Borowsky, Mugue & Smart unpublished data), and possibly also *Ancistrus formoso* and *Trichomycterus* sp. 2, all of them found in the phreatic zone habitat. These fishes belong to the stream-adapted epigean taxa (the subclade *Nemuroglanis* for the two former, balitorids for the latter), but have secondarily adapted to lentic conditions.

Precocial phenotypes are products of stability and the resulting competition, or specialization or both (Balon 1981, 1993). Eggs with a larger amount and density of yolk develop longer but directly into juveniles that are comparatively advanced at the time of the first feeding. This represents a clear survival advantage, especially important in food-poor and stable habitats such as caves. There seems to be an evolutionary trend to shorten the vulnerable larva period in competitive environments (Balon 1990, 1999). Therefore, a precocial lifestyle is probably an adaptive specialization shown by troglobitic fishes.

Acknowledgements

I am greatly indebted to Jakob Parzefall and to Maria Elina Bichuette for sharing unpublished population data on cave fishes, and to Augusto Auler for his invaluable help with definition of fish habitats and additional geospeleological information. Thanks also to Richard Borowsky and Rosely Rodrigues for revision of the English version of the manuscript. I am partially supported by the Conselho Nacional de Desenvolvimento Científico e Tecnológico (CNPq n. 306066/88-2).

References cited

Alvarez, J. 1946. Revisión del género *Anoptichthys* com descripción de una especie nueva (Pisc., Characidae). Anal. Esc. Nac. Cienc. Biol. 4: 263–282.

Amaral, M.F., J.M.R. Aranha & M.S. Menezes. 1999. Age and growth of *Pimelodella pappenheimi* (Siluriformes, Pimelodidae) from an Atlantic Forest stream in southern Brazil. Braz. Arch. Biol. Technol. 42: 449–453.

Andreani-Armas, L.E. 1990. Estudio comparativo de dos poblaciones, una hipogea y outra epigea de *Trichomycterus* sp. (Siluriformes, Trichomycteridae). Bol. Soc. Venez. Speleol. 24: 7–14.

Avise, J.C. & R.K. Selander. 1972. Evolutionary genetics of cave-dwelling fishes of the genus *Astyanax*. Evolution 26: 1–19.

Balon, E.K. 1981. Saltatory processes and altricial to precocial forms in the ontogeny of fishes. Amer. Zool. 21: 573–596.

Balon, E.K. 1990. Epigenesis of an epigeneticist: the development of some alternative concepts on the early ontogeny and evolution of fishes. Guelph Ichthyol. Rev. 1: 1–42.

Balon, E.K. 1993. Dynamics of biodiversity and mechanisms of change: a plea for balanced attention to form creation and extinction. Biol. Conservation 66: 5–16.

Balon, E.K. 1999. Alternative ways to become a juvenile or a definitive phenotype (and on some persisting linguistic offenses). Env. Biol. Fish. 56: 17–38.

Banister, K.E. & M.K. Bunni. 1980. A new blind cyprinid fish from Iraq. Bull. Br. Mus. Nat. Hist. (Zool.) 38: 151–158.

Biswas, J., A.K. Pati & R.K. Pradhan. 1990. Circadian and circannual rhythms in air gulping behavior of cave fish. J. Interdiscipl. Cycle Res. 21: 257–268.

Borowsky, R.B. & L. Mertz. 2001. Genetic differentiation among populations of the cave fish *Schistura oedipus* (Cypriniformes: Balitoridae). Env. Biol. Fish 62: 225–231 (this volume).

Brown, J.Z. 1996. Population dynamics and growth of Ozark cavefish in Logan Cave National Wildlife Refuge, Benton County, Arkansas. M.Sc. Thesis, University of Arkansas, Fayetteville. 105 pp.

Brown, A.V. & C.S.Todd. 1987. Status review of the threatened Ozark cavefish (*Amblyopsis rosae*). Proc. Ark. Acad. Scien. 41: 99–100.

Chaline, J. 1987. Paléontologie des Vertébrés. Dunod, Paris. 177 pp.

Chumba-Segura, L. 1983. Brotulidae: *Typhliasina pearsei*. Fauna Cenotes Yucatan (4): 1–9.

Cobolli-Sbordoni, M., E. De Matthaeis, M. Mattoccia, R. Berti & V. Sbordoni. 1996. Genetic variability and differentiation of hypogean cyprinid fishes from Somalia. J. Zoo. Syst. Evol. Res. 34: 75–84.

Cooper, J.E. & A. Iles. 1971. The southern cavefish, *Typhlichthys subterraneus*, at the southeastern periphery of its range. Nat. Speleol. Soc. Bull. 33: 45–49.

Culver, D.C. 1982. Cave life: evolution and ecology. Harvard University Press, Cambridge. 189 pp.

Culver, D.C. & B. Sket. 2000. Hotspots of subterranean biodiversity in caves and wells. J. Cave Karst Stud. 62: 11–17.

Culver, D.C., T.C. Kane & D.W. Fong. 1995. Adaptation and natural selection in caves. Harvard University Press, Cambridge. 223 pp.

Dessen, E.M.B., V.R. Eston, M.S. Silva, M.T. Temperini-Beck & E. Trajano. 1980. Levantamento preliminar da fauna de cavernas de algumas regiões do Brasil. Ciênc. Cult., São Paulo 32: 714–725.

Díaz-Pérez, P.A. 1988. *Lucifuga teresinarum* sp. n., nueva especie cubana de peces troglobios (Ophidiiformes, Bythitidae). Rev. Biol. (La Habana) 2: 37–43.

Durand, J.P. 1968. Étude des poissons récoltés dans la grotte de Umayalanta (Bolivie), *Trichomycterus chaberti* sp. n. Ann. Spéléol. 23: 343–353.

Eigenmann, C.H. 1909. The Cuban blind fishes. pp. 185–207. *In*: Cave Vertebrates of America, Carnegie Inst. Wash. Publ. 104: 1–241.

Ercolini, A., R. Berti, L. Chelazzi & G. Messana. 1982. Researches on the phreatobic fishes of Somalia: achievements and prospects. Monit. Zool. Ital. Suppl. 17: 219–241.

Ercolini, A., R. Berti, L. Chelazzi & G. Messana. 1987. Oxygen consumption in hypogean cyprinids (Pisces). Monit. Zool. Ital. Suppl. 22: 23–30.

Espinasa, L. & R.B. Borowsky. 2001. Origins and relationship of cave populations of the blind Mexican tetra, *Astyanax fasciatus*, in the Sierra de El Abra. Env. Biol. Fish. 62: 233–237 (this volume).

Ford, D.C. & P.W. Williams. 1989. Karst geomorphology and hydrology. Unwin Hyman, London. 601 pp.

Futuyma, D.J. 1986. Evolutionary biology. Sinauer Associates, Sunderland. 600 pp.

Gerhard, P. 1999. Ecologia de populações e comportamento de quatro espécies de bagres Heptapterinae (Teleostei: Siluriformes) em riachos do Alto Vale do rio Ribeira (Iporanga, São Paulo). M.Sc. Thesis, Universidade de São Paulo, São Paulo. 129 pp.

Greenwood, P.H. 1967. Blind cave fishes. Stud. Speleol. 1: 262–275.

Heuts, M.J. 1952. Ecology. Variation and adaptation of the blind African cave fish *Caecobarbus geertsi* Blgr. Ann. Soc. R. Zool. Belg. 82: 155–230.

Hill, J. & G.D. Grossman. 1987. Effects of subcutaneous marking on stream fishes. Copeia 1987: 492–495.

Hoenen, S.M.M. & E. Trajano. 1995. Locomotor responses of *Pimelodella* spp. from southeastern Brazil to chemical signals of other individuals (Teleostei: Siluriformes: Pimelodidae). Mém. Biospéol. 22: 57–63.

Holsinger, J.R. & D.C. Culver. 1988. The invertebrate cave fauna of Virginia and a part of eastern Tennessee: zoogeography and ecology. Brimleyana 14: 1–162.

Hose, L.D. & J.A. Pisarowicz. 1999. Cueva de Villa Cruz, Tabasco, Mexico: reconnaissance study of an active sulfur spring cave and ecosystem. J. Cave Karst Stud. 61: 13–21.

Iliffe, T.M. 1993. Fauna troglobia acuática de la Península de Yucatán. pp. 673–686. *In*: S.I. Salazar-Vallejo & N.E. González (ed.) Biodiversidad Marina y Costera de México, Com. Nal. Biodiversidad Y CIQRO, México.

Juberthie, C. 1983. Le milieu souterrain: étendue et composition. Mém. Biospéol. 10: 17–65.

Knapp, S.M. & D.W. Fong. 1999. Estimates of population size of *Stygobromus emarginatus* (Amphipoda: Crangonyctidae) in a headwater stream in Organ Cave, west Virginia. J. Cave Karst Stud. 61: 3–6.

Kottelat, M. 1988. Two species of cavefishes from northern Thailand in the genera *Nemacheilus* and *Homaloptera* (Osteichthyes: Homalopteridae). Rec. Austr. Mus. 40: 225–231.

Krebs, C.J. 1994. Ecology. HarperCollins College Publ., New York. 801 pp.

Krejca, J., S. Taylor & D. Hendrickson. 2000. The Mexican blindcat research team. Nat. Speleol. Soc. News 58: 165–171.

Langecker, T.G. & G. Longley. 1993. Morphological adaptations of the Texas blind catfishes *Trogloglanis pattersoni* and *Satan eurystomus* (Siluriformes: Ictaluridae) to their underground environment. Copeia 1993: 976–986.

Langecker, T.G., H. Wilkens & J. Parzefall. 1996. Studies on the trophic structure of an energy-rich Mexican cave (Cueva de las Sardinas) containing sulfurous water. Mém. Biospéol. 23: 121–125.

Longley, G. & H. Karney. 1979. Status of *Trogloglanis pattersoni* Eigenmann, the toothless blindcat and status of *Satan eurystomus* Hubbs and Bailey, the widemouth blindcat. Endangered Species Report 5 (Part 1), Special Publication, U.S. Fish and Wildlife Service, Albuquerque. 54 pp.

Means, M.L. & J.E. Johnson. 1995. Movement of threatened Ozark cavefish in Logan Cave National Wildlife Refuge, Benton County, Arkansas. SWtern Natur. 40: 308–313.

Mendes, L.F. 1995a. Observations on the ecology and behavior of a new species of troglobitic catfish from northeastern Brazil (Siluriformes, Pimelodidae). Mém. Biospéol. 22: 99–101.

Mendes, L.F. 1995b. Ecologia populacional e comportamento de uma nova espécie de bagres cavernícolas da Chapada Diamantina, BA (Siluriformes, Pimelodidae). M.Sc. Thesis, Universidade de São Paulo, São Paulo. 86 pp.

Mercy, T.V.A., K.G. Padmanabhan & N.K. Pillai. 1982. Morphological studies of the oocytes of the blind catfish *Horaglanis krishnai* Menon. Zool. Anz. 209: 211–223.

Miller, R.R. 1984. *Rhamdia reddelli*, new species, the first blind pimelodid catfish from Middle America, with a key to the American species. Trans. San Diego Soc. Nat. Hist. 20: 135–144.

Mitchell, R.W., W.H. Russell & W. R. Elliott. 1977. Mexican eyeless characin fishes, genus *Astyanax*: environment, distribution, and evolution. Spec. Publ. Mus. Texas Tech Univ. 12: 1–89.

Mohr, C.E. & T.L. Poulson. 1966. The life of the cave. McGraw-Hill Book Co., New York. 232 pp.

Nalbant, T.T. 1981. A study of the subterranean species of *Lucifuga* from Cuba, with the description of *Lucifuga simile* sp. n. (Pisces, Ophidiiformes, Bythitidae). pp. 185–190. *In*: T. Orghidan, A. Núñez Jiménez, V. Decou, S.T. Negrea & N. Viña Bayés (ed.) Résultats des Expéditions Biospéologiques Cubano-Roumaines à Cuba, 3, Ed. Acad. Republ. Social. România, Bucureşti.

Nalbant, T.T. & O. Linares. 1987. A study of a subterranean population of *Trichomycterus guianense* (Eigenmann, 1909) from Venezuela (Pisces, Siluriformes, Trichomycteridae). pp. 211–217. *In*: V. Decu, T. Orghidan, D. Dancau, C. Bordon, O. Linares, F. Urbani, J. Tronchoni & C. Bosque (ed.) Fauna Hipogea y Hemiedáfica de Venezuela y de Otros Paises de América del Sur, Ed. Acad. Republ. Social. România, Bucureşti.

Parzefall, J. 1983. Field observation in epigean and cave populations of the Mexican characid *Astyanax mexicanus* (Pisces, Characidae). Mém. Biospéol. 10: 171–176.

Parzefall, J. 1993. Behavioral ecology of cave-dwelling fishes. pp. 573–606. *In*: T.J. Pitcher (ed.) Behavior of Teleost Fishes, Chapman & Hall, London. 715 pp.

Pavan, C. 1945. Os peixes cegos das cavernas de Iporanga e a evolução. Bol. Fac. Filos. Ciênc. Letr., Biol. Geral 6, 79: 1–104.

Pavan, C. 1946. Observations and experiments on the cave fish, *Pimelodella kronei*, and its relatives. Amer. Natur. 80: 343–361.

Perez, A. & A. Viloria. 1994. *Ancistrus galani* n. sp. (Siluriformes: Loricariidae), with comments on biospeleological explorations in western Venezuela. Mém. Biospéol. 21: 103–107.

Pinna, M.C.C. de. 1999. Species concepts and phylogenetics. Review Fish Biol. Fisheries 9: 353–373.

Poly, W.J. & C.E. Boucher. 1996. Nontroglobitic fishes in caves: their abnormalities, ecological classification and importance. Amer. Midl. Natur. 136: 187–198.

Poulson, T.L. 1963. Cave adaptation in amblyopsid fishes. Amer. Midl. Natur. 70: 257–290.

Rasquin, P. & L. Rosenbloom. 1954. Endocrine imbalance and tissue hyperplasia in teleosts maintained in darkness. Bull. Amer. Mus. Nat. Hist. 104: 359–426.

Romero, A. 1983. Introgressive hybridization in the *Astyanax fasciatus* (Pisces: Characidae) population at La Cueva Chica. Nat. Speleol. Soc. Bull. 45: 81–85.

Romero, A. 1987. Observaciones preliminares sobre la conducta en individuos cavernicolas de *Trichomycterus conradi* (Pisces: Trichomycteridae). Bol. Soc. Venez. Cienc. Nat. 41(144): 205–216.

Romero, A. 1998a. Threatened fishes of the World: *Speoplatyrhinus poulsoni* Cooper & Kuehne, 1974 (Amblyopsidae). Env. Biol. Fish. 53: 293–294.

Romero, A. 1998b. Threatened fishes of the World: *Typhlichthys subterraneus* (Girhard, 1860) (Amblyopsidae). Env. Biol. Fish. 53: 74.

Romero, A. & A. McLeran. 2000. Threatened fishes of the World: *Stygichthys typhlops* Brittan & Bölke, 1965 (Characidae). Env. Biol. Fish. 57: 270.

Sabino, J. & E. Trajano. 1997. A new species of blind armored catfish, genus *Ancistrus*, from caves of Bodoquena region, Mato Grosso do Sul, southeastern Brazil (Siluriformes, Loricariidae, Ancistrinae). Revue Fr. Aquariol. 24: 73–78.

Sadoglu, P. 1979. A breeding method for blind *Astyanax mexicanus* based on annual spawning patterns. Copeia 1979: 369–371.

Schubert, A.L.S., C.D. Nielsen & D.B. Noltie. 1993. Habitat use and gas bubble disease in southern cavefish (*Typhlichthys subterraneus*). Int. J. Speleol. 22: 131–143.

Smith, P.W. & N.M. Welch. 1978. A summary of the life history and distribution of the spring cavefish, *Chologaster agassizi* Putnam, with population estimates for the species in southern Illinois. Illinois Nat. Hist. Survey, Biol. Notes (104): 1–8.

Thinès, G. 1958. Observations sur les habitats de l'*Uegitglanis zammaranoi* Gianferrari 1923, Clariidae aveugle de la Somalie Italienne. Rev. Zool. Bot. Afr. 57: 117–124.

Thinès, G. & G. Proudlove. 1986. Pisces. pp.709–733. *In*: L. Botosaneanu (ed.) Stygofauna Mundi, E. J. Brill, Leiden.

Thinès, G. & N. Wissocq. 1972. Étude comparée du comportement alimentaire de deux poissons cavernicoles (*Anoptichthys jordani* Hubbs et Innes et *Caecobarbus geertsi* Blgr.). Int. J. Speleol. 4: 139–169.

Trajano, E. 1987. Fauna cavernícola brasileira: composição e caracterização preliminar. Rev. Bras. Zool. 3: 533–561.

Trajano, E. 1989. Estudo do comportamento espontâneo e alimentar e da dieta do bagre cavernícola, *Pimelodella kronei*, e seu provável ancestral epígeo, *Pimelodella transitoria* (Siluriformes, Pimelodidae). Rev. Bras. Zool. 49: 757–769.

Trajano, E. 1991. Populational ecology of *Pimelodella kronei*, troglobitic catfish from southeastern Brazil (Siluriformes, Pimelodidae). Env. Biol. Fish. 30: 407–421.

Trajano, E. 1993. Considerations about systematics and the evolution of behavior in siluriform cave fishes. Mém. Biospéol. 20: 257–260.

Trajano, E. 1994. Comparative study on the brain and olfactory organ of the troglobitic catfish, *Pimelodella kronei* (Ribeiro, 1907), and its putative ancestor, *P. transitoria* (Ribeiro, 1912) (Siluriformes, Pimelodidae). Trop. Zool. 7: 145–160.

Trajano, E. 1995. Evolution of tropical troglobites: applicability of the model of Quaternary climatic fluctuations. Mém. Biospéol. 22: 203–209.

Trajano, E. 1997a. Population ecology of *Trichomycterus itacarambiensis*, a cave catfish from eastern Brazil (Siluriformes, Trichomycteridae). Env. Biol. Fish. 50: 357–369.

Trajano, E. 1997b. Food and reproduction of *Trichomycterus itacarambiensis*, cave catfish from south-eastern Brazil. J. Fish Biol. 51: 53–63.

Trajano, E. 1997c. Threatened fishes of the World: *Pimelodella kronei* (Ribeiro, 1907) (Pimelodidae). Env. Biol. Fish. 49: 332.

Trajano, E., 1997d. Synopsis of Brazilian troglomorphic fishes. Mém. Biospéol. 24: 119–126.

Trajano, E. 2001. Habitat and population data of troglobitic armored cave catfish, *Ancistrus cryptophthalmus* Reis, 1987, from central Brazil (Siluriformes: Loricariidae). Env. Biol. Fish. 62: 195–200 (this volume).

Trajano, E. & F.A. Bockmann. 1999. Evolution of ecology and behavior in Brazilian cave Heptapterinae catfishes, based on cladistic analysis (Teleostei: Siluriformes). Mém. Biospéol. 26: 123–129.

Trajano, E. & F.A. Bockmann. 2000. Ecology and behavior of a new cave catfish of the genus *Taunayia* from northeastern Brazil (Siluriformes, Heptapterinae). Ichthyol. Explor. Freshwaters 11: 207–216.

Trajano, E. & H.A. Britski. 1992. *Pimelodella kronei* (Ribeiro,1907) e seu sinônimo *Caecorhamdella brasiliensis* Borodin, 1927: morfologia externa, taxonomia e evolução (Teleostomi, Siluriformes). Bol. Zool., São Paulo 12: 53–89.

Trajano, E. & P. Gerhard. 1997. Light reaction in Brazilian cave fishes (Siluriformes: Pimelodidae, Trichomycteridae, Loricariidae). Mém. Biospéol. 24: 127–138.

Trajano, E. & A.M. Souza. 1994. The behavior of *Ancistrus cryptophthalmus*, an armored blind catfish from caves of central Brazil, with notes on syntopic *Trichomycterus* sp. (Siluriformes, Loricariidae, Trichomycteridae). Mém. Biospéol. 21: 151–159.

Vandel, A. 1964. Biospéologie. Gauthier-Villars, Paris. 619 pp.

Walsh S.L. & C.R. Gilbert. 1995. New species of troglobitic catfish of the genus *Prietella* (Siluriformes: Ictaluridae) from northeastern México. Copeia 1995: 850–861.

Weber, A. 1996. Cave dwelling catfish populations of the genus *Rhamdia* (Pimelodidae, Siluroidei, teleostei) in Mexico. Mém. Biospéol. 23: 73–85.

Weber, A., G.S. Proudlove, J. Parzefall, H. Wilkens & T.T. Nalbant. 1998. Pisces (Teleostei). pp. 1179–1190. *In*: C. Juberthie & V. Decu (ed.) Encyclopaedia Biospeologica, Tome II, Société de Biospéologie, Moulis.

Weber, A. & H. Wilkens. 1998. *Rhamdia macuspanensis*: a new species of troglobitic pimelodid catfish (Siluriformes: Pimelodidae) from a cave in Tabasco, Mexico. Copeia 1998: 998–1004.

Weitzman, S.H. & R.P. Vari. 1988. Miniaturization in South American freshwater fishes: an overview and discussion. Proc. Biol. Soc. Wash. 101: 444–465.

Wilkens, H. 1988. Evolution and genetics of epigean and cave *Astyanax fasciatus* (Characidae, Pisces). Evol. Biol. 23: 271–367.

Willis, L.D. & A.V. Brown. 1985. Distribution and habitat requirements of the Ozark cavefish, *Amblyopsis rosae*. Amer. Midl. Natur. 114: 311–317.

Yang, J. & Y. Chen. 1993a. The cavefishes from Jiuxiang limestone cave of Yunnan, China with reference to the character evolutions. pp. 126–128. *In:* Proc. 11th International Congress of Speleology, Beijing.

Yang, J. & Y. Chen. 1993b. The cavefishes from Duan, Guangxi, China with comments on their adaptations to cave habitats. pp. 124–126. *In*: Proc. 11th International Congress of Speleology, Beijing.

Environmental Biology of Fishes **62:** 161–169, 2001.
© 2001 *Kluwer Academic Publishers. Printed in the Netherlands.*

Population biology and growth of Ozark cavefish in Logan Cave National Wildlife Refuge, Arkansas

J. Zack Brown[a] & James E. Johnson
U.S. Geological Survey, Biological Resources Division, Arkansas Cooperative Fish & Wildlife Research Unit, Department of Biological Sciences, University of Arkansas, Fayetteville, AR 72701, U.S.A.
[a]*Present address: West Virginia Department of Natural Resources, 1310 Virginia Avenue, Fairmount, WV 26554, U.S.A. (e-mail: ozarkjohnson@aol.com)*

Received 7 December 1999 Accepted 11 December 2000

Key words: Amblyopsis rosae, threatened species, aquifer migration

Synopsis

Ozark cavefish, *Amblyopsis rosae*, is a threatened species endemic to the Springfield Plateau of the Ozark Highlands in Arkansas, Missouri, and Oklahoma. One of the largest known Ozark cavefish populations, located in Logan Cave, Arkansas, was surveyed 25 times over a two-year period between 1993 and 1995. During the study, 147 Ozark cavefish >30 mm (TL) were marked with visual implant tags and 140 Ozark cavefish were available for recapture; 68 were recaptured 189 times and the rest (72) were never recaptured. Individual Ozark cavefish persisted in Logan Cave for a relatively short time. Only 14% of 80 fish tagged during a previous study in 1992 were recaptured during this study, and half of all recaptured fish disappeared within three months. However, if a fish persisted for at least seven months in the cave, its probability of being recaptured over an additional year was high. Maximum persistence of a tagged fish was 28 months, suggesting these fish have a maximum life-span of 4–5 years. Growth averaged 0.6 mm per month, with maximum recorded growth of 6 mm per month and a maximum size of 65 mm TL. Smaller fish grew faster than larger fish but growth rates were sporadic, with several mid-sized fish (45–49 mm) showing little growth (0–3 mm year^{-1}) while some fish >50 mm grew up to 12 mm year^{-1}. Most fish gained in length during April–October, the same period a maternity colony of gray bats occupied the cave. Gross Ozark cavefish movement over the study period ranged up to 1002 m, with a mean movement of 1.2 m day^{-1}; movement was positively correlated with Ozark cavefish total length. Death seemed the most likely explanation for loss of tagged Ozark cavefish, including fish that emigrated out of the cave. Little up-stream movement was recorded between reaches and did not account for loss of tagged fish. Reproduction within the cave and immigration from the aquifer accounted for persistence of Ozark cavefish in Logan Cave.

Introduction

Cavefishes of the family Amblyopsidae consist of four genera and six species found only in the eastern United States (Woods & Inger 1957, Robins et al. 1991); all but the swampfish, *Chologaster cornuta*, are associated with caves or cave springs. The Ozark cavefish, *Amblyopsis rosae*, is a small (maximum total length [TL] 65 mm), eyeless, unpigmented fish endemic to aquifers in the karst topography of the Springfield Plateau of Arkansas, Missouri, and Oklahoma (Willis & Brown 1985). Over 90% of known individuals are reported from two caves in northwest Arkansas: Cave Springs Cave and Logan Cave (Means & Johnson 1995).

The Ozark cavefish has been known to biologists for a century (Eigenmann 1898), during which time systematic studies have outpaced ecological knowledge.

Poulson (1963) studied amblyopsid biology and estimated Ozark cavefish lived for 4–5 years by counting circuli on scales. Robison and Buchanan (1988) suggested recent findings might increase the estimated life span to 60 years or more. Means (1993) studied growth of Ozark cavefish from Logan Cave and found one individual that increased in total length from 32 to 52 mm in nine months (2.2 mm $month^{-1}$). Means and Johnson (1995) marked Ozark cavefish with visual implant tags to study movement and found Ozark cavefish continually immigrate into Logan Cave from upstream, presumably from the aquifer.

The U.S. Fish & Wildlife Service listed Ozark cavefish as a threatened species in 1984 (Johnson 1987) based on decline in both numbers and populations. A recovery plan, written in 1989, concentrated on limiting human access to caves. Although this is an obvious start towards recovery of what was considered a cave-dwelling species, limited access may prove more beneficial to other cave-dwelling species than to Ozark cavefish if their population centers are located in aquifers and the caves are population sinks (Means & Johnson 1995, Brown 1996).

This study concentrated on Ozark cavefish biology in Logan Cave. Specific objectives were to determine survival, growth rate, and movement of Ozark cavefish. If these animals do live for decades, then marked individuals should persist in Logan Cave and remain available for recapture. In addition, fish that survive for decades are often relatively slow growers (Bond 1996), so growth rate of Ozark cavefish should also provide perspective on longevity.

Another widely held hypothesis is that Ozark cavefish populations are independent and isolated, restricted to the caves they inhabit (Willis & Brown 1985). However, Means and Johnson (1995) noted immigration of unmarked fish into Logan Cave, and Woods and Inger (1957) suggested widespread dispersal through aquifers during wetter epochs. If Ozark cavefish maintain populations in aquifers, then fish observed in caves might be of limited value in determining overall population trends and health.

Methods

Study area

Logan Cave and its subterranean stream are the principal components of 49 ha Logan Cave National Wildlife Refuge in northwestern Arkansas. The surface area around the cave mouth was purchased by the U.S. Fish & Wildlife Service in 1989 to protect threatened Ozark cavefish, endangered cave crayfish, *Cambarus aculabrum*, and a maternity colony of endangered gray bats, *Myotis grisescens*. Recharge area for the cave stream is 3100 ha of privately owned pasture and woodland located north and east of the cave (unpublished data, T. & C. Aley, Ozark Underground Laboratory, Protem, MO).

There are two entrances to Logan Cave: the mouth from which the subterranean stream issues and a sinkhole 365 m upstream from the cave mouth. The cave is a tube-like passage through the Boone-St. Joe Formation that allows access to 1.3 km of cave stream. For this study, the stream in Logan Cave was divided into three reaches separated by physical features. The Lower Reach, 365 m long and extending from cave mouth to the sinkhole, was typified by shallow (10–30 cm deep) riffles and rocky runs. The Middle Reach (220 m) was a large pool formed behind the collapsed ceiling of the sinkhole that averaged ca. 1.5 m deep with a maximum depth of 2.5 m. The Upper Reach extended 685 m upstream from the pool to a large ceiling collapse that limited further upstream access. This reach was composed of long, shallow bedrock runs with a few short, deep plunge pools, backwaters, and eddies. The accessible portion of Logan Cave was marked at 5 m intervals to identify Ozark cavefish capture localities.

Flow in Logan Cave varied with precipitation events on the watershed. At base flow, volume at the cave mouth was about 0.1 $m^3\ s^{-1}$ and the water was clear. Annual mean water temperature was 14.5 ± 0.6°C (2 SE) (range 12.7–16.0°C), conductivity 245.6 ± 29.1 uS cm^{-1}, dissolved oxygen 7.0 ± 1.2 mg l^{-1}, and fecal coliform counts were high in summer months (up to 70 000 colonies 100 ml^{-1}) due both to the presence of bats in the cave and agriculture on the watershed (Means 1993). Precipitation events were delayed about 24 h at the cave mouth and resulted in increased flow, reduced water clarity, and fluctuating water temperature for a few days. A maximum storm event on 8 May 1995 resulted in a flow of 1.4 $m^3\ s^{-1}$ and high turbidity (V. Brahana personal communication), but flows subsided sufficiently for us to collect Ozark cavefish two weeks later.

Collections

Starting in September 1993, Logan Cave was entered monthly. Ozark cavefish collecting began at the mouth and progressed upstream to the furthest accessible point. A survey team consisted of three people, two collectors and a data recorder, moving upstream with submersible lights and aquarium dip nets. Shallow areas were waded and pools were searched using mask and snorkel. When an Ozark cavefish was captured, its location was recorded and TL measured to the nearest 1 mm with a ruler. Ozark cavefish >30 mm TL were tagged with an alpha-numeric subcutaneous visual implant tag (Haw et al. 1990) injected forward of the caudal peduncle; all Ozark cavefish were released at their point of capture. Minimum tagging size and tag placement followed Means & Johnson (1995). Gross movement was recorded as the summed distances each time an individual fish was recaptured and net movement as distance between first and last capture event. Movement records of less than 2 m were considered no movement. Fish tagged by Means (1993) in 1992 were incorporated into the present study as newly tagged fish when first captured. Ozark cavefish could not be sexed externally except for a few fecund females.

Statistical analyses

Non-parametric data were evaluated using a Mann–Whitney test and parametric data using unpaired t-tests and ANOVA (Zar 1984). Data from Ozark cavefish tagged during the 1992 study (Means 1993) and fish that lost tags (identified by a scar on the caudal peduncle) were used for growth and movement analyses but not for frequency of recapture estimates.

Results

Between 23 September 1993 and 9 September 1995, 25 sampling trips were conducted through Logan Cave that yielded 388 Ozark cavefish captures. Those trips were combined into 23 monthly surveys (Table 1). Numbers of Ozark cavefish captured per month ranged from 10 to 29 with a mean of 16.8 ± 1.96. A total of 147 Ozark cavefish >30 mm TL was marked with visual implant tags, but seven marked fish had to be excluded from recapture results because they were tagged on the final collecting trip. Mean total length of tagged Ozark cavefish was 43.5 ± 1.2 mm. Twenty three Ozark cavefish lost their tags, identified by an injection scar on the caudal peduncle, and had to be retagged. Mean total length of retagged fish (42.6 ± 2.8 mm) did not differ from all tagged Ozark cavefish ($t = 0.21$, $df = 150$, $p = 0.83$). The 47 Ozark cavefish marked during the first three collections (1993) were larger (mean length $= 49.3 \pm 2.0$) than Ozark cavefish tagged during the remainder of the study in 1994 and 1995 ($N = 100$, mean total length $= 40.8 \pm 1.4$ mm) ($t = 6.954$, $df = 145$, $p < 0.01$). Ozark cavefish <30 mm were captured 52 times during the present study but could not be identified as individuals because they were not tagged.

Between June 1992 and September 1995, 220 Ozark cavefish were tagged in Logan Cave and theoretically available for recapture. The previous study in Logan Cave (Means 1993) tagged 80 Ozark cavefish between June and December 1992 (Means 1993, Means & Johnson 1995) and we tagged an additional 140 fish. Our study started in September 1993 and only 5 of 23 Ozark cavefish captured during that first trip bore tags from the previous study. The recaptured fish were all large (mean 52.0 ± 2.0 mm, range 45–56 mm), and all but one had been recaptured repeatedly during the earlier study. Mean total length of fish from the earlier study (41.9 ± 1.6 mm) did not differ from fish collected during the present study ($t = 1.515$, $df = 225$, $p = 0.13$). In total, we recaptured 11 Ozark cavefish (14%) from the earlier study.

Persistence

Approximately half ($n = 72$) of the 140 tagged Ozark cavefish during this study were never recaptured (Table 1) whereas the rest ($n = 68$) were recaptured at least once. Mean size of lost fish (41.4 ± 1.7 mm) was significantly smaller than recaptured fish (mean $= 46.0 \pm 2.0$ mm; $t = 3.56$, $df = 138$, $p < 0.01$; Figure 1).

The 68 recaptured fish were recaptured a total of 189 times (mean $= 2.8 \pm 0.29$ fish^{-1}). Thirty-seven percent of recaptured fish were recaptured only once, 58% 2–8 times, and 4% 9–11 times. Most recaptured fish (62%) were collected during the next trip through the cave, and 82% were recaptured on the first or second trips after tagging. Six fish (9%) avoided recapture for three months after being tagged and one individual eluded first recapture for six months.

After the first recapture event, 95% of recaptured fish were caught within three months. This provided a level of confidence for persistence; if a tagged Ozark cavefish went uncaptured for three consecutive trips, its chance of being lost from Logan Cave was 95%. Using this criterion, a 95% persistence curve for Ozark

Table 1. Numbers and total lengths of Ozark cavefish collected in Logan Cave National Wildlife Refuge, Arkansas (±2SE). Numbers marked includes nine fish that were originally marked in an earlier study (Means & Johnson 1995). Fish <30 mm were not marked.

Date	Total no. cavefish captured	Numbers of recaptured cavefish	Unmarked cavefish >30 mm			Numbers of cavefish <30 mm	Numbers of marked fish never recaptured
			Numbers marked[c]	Range of lengths (mm)	Mean lengths (mm)		
Sep 93	23		23	33–58	49.4±2.8	[a]	18
Nov 93	13	2	11	38–62	48.4±4.3	[a]	2
Dec 93	19	5	13	34–59	49.7±3.8	1	1
Jan 94	26	17	5	34–52	42.8±7.4	4	0
Feb 94	19	15	4	36–53	47.8±8.0	0	1
Mar 94[a]	29	23	4	33-52	42.0±9.4	2	1
Apr 94	15	10	2	32–45	38.5±13.0	3	1
May 94	16	11	2	35–39	37.0±4.0	3	1
Jun 94	11	5	6	31–52	43.2±6.2	0	6
Jul 94	18	5	11	32–55	38.8±4.6	2	6
Aug 94	15	7	6	33–59	41.7±7.8	2	4
Sep 94	16	11	4	33–54	39.5±9.7	1	3
Oct 94[b]	16	8	4	32–39	35.8±3.3	4	2
Nov 94	11	7	2	30–38	34.0±8.0	2	2
Dec 94	17	6	6	34–49	39.3±5.0	5	5
Jan 95	10	2	4	32–55	41.8±10.1	4	3
Feb 95	13	6	5	34–51	41.2±5.7	2	3
Mar 95	18	9	3	38–49	42.7±6.5	6	1
May 95	13	5	5	33–53	44.0±6.7	3	2
Jun 95	13	7	6	35–45	38.7±3.5	0	4
Jul 95	22	7	11	37–55	41.4±3.1	4	4
Aug 95	16	11	3	34–42	38.7±4.8	2	2
Sep 95	19	10	7	31–54	40.0±5.9	2	
Sums	388	189	147			52	72
Means	16.8	8.6	6.3			2.5	3.3
2 SE	(±11.96)	(±2.08)	(±2.00)			(±0.73)	(±0.79)

[a]Cavefish <30 mm were not counted during the first two trips.
[b]Two collections combined for these months.
[c]Includes 23 cavefish that had to be retagged.

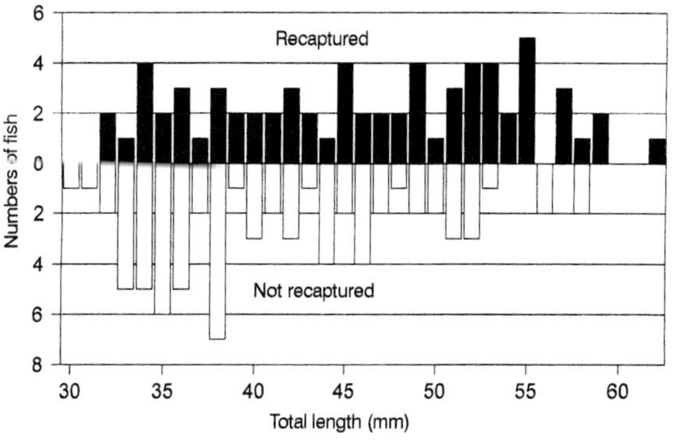

Figure 1. Length frequencies of 68 Ozark cavefish tagged and recaptured in Logan Cave NWR Arkansas, between September 1993 and August 1995, and 72 cavefish that were tagged but not recaptured.

cavefish recaptured at least once indicated half of recaptured Ozark cavefish disappeared from collections after three months, but if a fish endured for at least seven months, its chance of continued persistence was excellent (Figure 2).

The longest time an Ozark cavefish was known to persist in Logan Cave was 28 months. That fish was tagged by M.L. Means in June 1992 at 33 mm TL and recaptured eight times during his study and four times by us, with the last recapture in October 1994 at 56 mm TL. In captivity, two Ozark cavefish captured as taggable-sized fish in February 1992 survived for 4.1 and 4.5 years; the longest-lived fish was tagged with a visual implant tag.

During the first year of this study, 87 Ozark cavefish were tagged and 46 of them recaptured at least once.

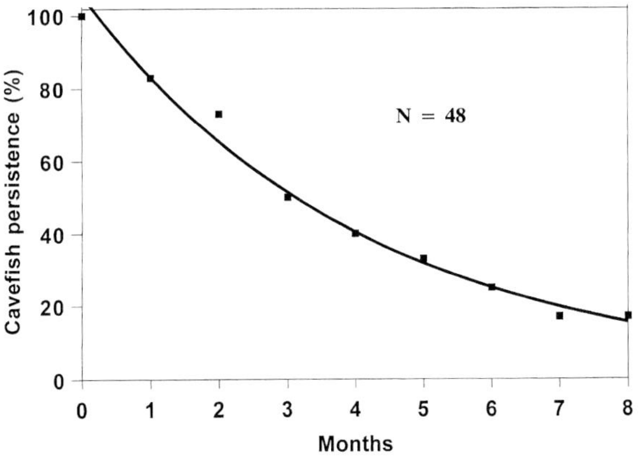

Figure 2. Exponential curve of the probability a tagged Ozark cavefish would be recaptured in Logan Cave over the next four consecutive trips. Data indicate at least 95% of all cavefish still in Logan Cave were likely to be recaptured during this interval.

Only 16 (35%) of those fish were recaptured during the second year and persisted a mean of 19.3 ± 1.0 months (range 13–28 months) between their first and last capture events. The remaining 30 fish (65%) were not recaptured during the second year of study and were presumed lost from the cave population.

Capture of untagged Ozark cavefish remained constant after the first three collections that tagged resident fish (mean = 5.0 ± 1.1 new fish $month^{-1}$). The only exception was a significant increase of smaller fish in July of each year (Mann–Whitney test, Z = 2.29, p = 0.03; Figure 3).

Growth

Twenty recaptured fish decreased in total length (range −1 to −5 mm, mean = −2.0±0.5 mm). Several larger fish showed eroded caudal fins, a condition also noted in Ozark cavefish retained in captivity for several years, but this did not account for all of the length loss. Fish exhibiting negative growth were excluded from growth rate calculations.

The absolute growth rate for the remaining 48 Ozark cavefish was 0.6 ± 0.2 mm $month^{-1}$ (Figure 4), with smaller fish growing more rapidly than larger ones. When separated into size classes, 30–39 mm fish had mean growths of 1.1±0.3 mm $month^{-1}$, 40–49 mm fish 0.5±0.2 mm $month^{-1}$, and 50+ mm fish 0.4±0.3 mm $month^{-1}$ (ANOVA, df = 47, F = 7.25, p = 0.001).

We determined a maximum growth rate of 2.7 ± 0.7 mm $month^{-1}$ from a subsample of 11 Ozark

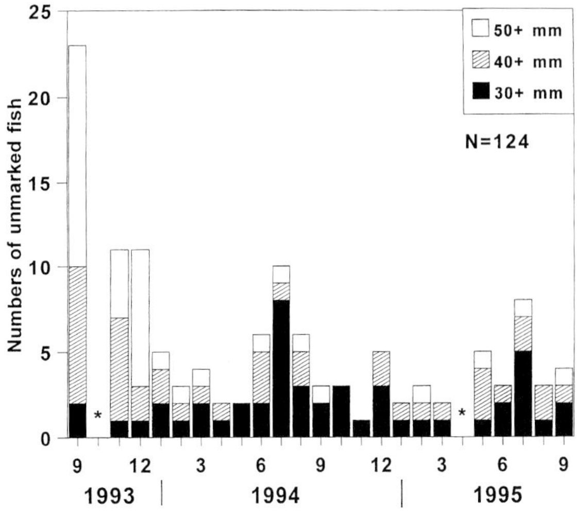

Figure 3. Number of cavefish captured in Logan Cave per trip. Asterisks denote months when no collections were made. Twenty three fish that were retagged were not included.

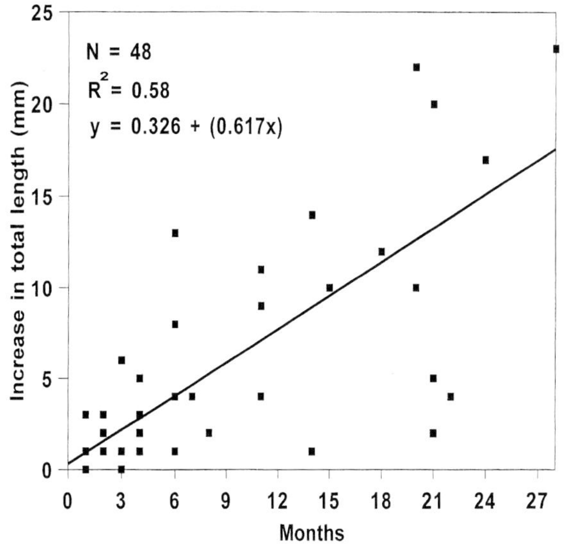

Figure 4. Absolute growth rate of 48 Ozark cavefish that demonstrated increased lengths between 1993 and 1995.

cavefish (TL range 32–60 mm) that were recaptured after only a single month in the wild. One of these fish grew from 35 to 41 mm between May and June 1994. The annual growth period was between April and October, with all but one of the 48 fish that recorded at least a 2 mm increase in size increasing in length during that period.

Movement

Most Ozark cavefish (82%) were captured in the pooled Middle Reach of Logan Cave, 11% were found in the Upper Reach and 7% in the Lower Reach. Most recaptured Ozark cavefish (65%) moved <100 m from their point of first capture, but a few moved long distances. Gross movement ranged up to 1002 m over the 24 month study, and net movement up to 673 m. Gross mean daily movement was 1.2 ± 0.5 m day^{-1}.

Most recaptures (97%) took place in the reach in which the Ozark cavefish was first tagged. In the Upper Reach, 16 Ozark cavefish were marked and eight recapture events occurred; 10 fish were never recaptured. Most captures/recaptures occurred in two areas that contained small backwater pools. In the Middle Reach, 120 fish were marked, 57 fish were recaptured 174 times, and 63 fish were never recaptured. Mean movement was 0.9 ± 0.2 m day^{-1}, with 47% of movements upstream, 39% downstream, and 14% less than 2 m in either direction ($t = 0.677$, $df = 93$, $p = 0.27$). In the Lower Reach, 11 fish were tagged, 5 fish were recaptured 7 times, and 6 fish were never recaptured. Four recaptured fish remained in a backwater area at the cave mouth for four months before disappearing. The remaining individual moved downstream 55 m over three months to within 13 m of the cave mouth before disappearing.

Only four recaptured Ozark cavefish moved from one reach of Logan Cave to another. These fish ranged from 40 to 59 mm TL (mean $= 48.3 \pm 8.0$ mm), not significantly different from the norm ($t = 1.67$, $df = 145$, $p = 0.25$). No marked fish moved downstream out of the Middle Reach into the Lower Reach, one recaptured fish moved from the Middle Reach 28 m upstream into the Upper Reach, and three fish moved from the Upper Reach downstream to the Middle Reach. The four fish that changed reaches moved more than the fish that remained in a single reach (mean $= 7.7 \pm 7.9$ m day^{-1} ($t = 6.84$, $df = 63$, $p < 0.01$).

Fish size influenced movement. Ozark cavefish between 30 and 39 mm TL displayed a mean daily movement of 0.5 ± 0.5 m day^{-1}, while Ozark cavefish 40 mm or more moved 1.3 ± 0.7 m day^{-1}; these differences were significant (Bonferroni t-test, $t = 2.08$, $df = 41$, $p = 0.04$).

Untagged fish continued to be captured in Logan Cave throughout the 2 year study (Figure 3). These fish could have grown into the taggable cohort, remained hidden within crevices of the cave stream during earlier collections, or immigrated into the cave from the aquifer. To determine if immigration was occurring in Logan Cave, we considered only those untagged Ozark cavefish 40 mm or greater. This size was selected to insure that fish <30 mm from the previous month could not have grown into the taggable fish cohort in a single month. Ninety-one larger Ozark cavefish were captured in Logan Cave throughout the study, but 15 were retagged fish leaving 76 large Ozark cavefish available for study. These larger Ozark cavefish were captured during every trip but three, and 50+ mm fish were captured in 61% of these trips, including the last trip in September 1995 (Figure 3). If we also exclude the first three collections that tagged 41 larger 'resident' fish, it leaves 35 larger Ozark cavefish tagged between January 1994 and September 1995 that were the likely result of immigration. Capture rate of these fish averaged 1.7 ± 0.6 fish month^{-1} and suggests an annual immigration rate of 20 ± 7.2 Ozark cavefish into Logan Cave.

Discussion

Our data suggest that once an Ozark cavefish enters Logan Cave, either via reproduction or immigration, it remains there for a relatively short time before disappearing. About half of tagged fish were never recaptured and the rest of recaptured fish disappeared within three months. After seven months, more than 80% of fish recaptured at least once were gone, but most tagged fish that endured beyond that period persisted for another year or more. Loss of tags, along with unequal capture rate of different individuals discouraged use of models like Jolly–Seber to estimate population size (Pollock et al. 1990, Guy et al. 1996).

Tag loss, hiding, death, and emigration could account for the disappearance of tagged Ozark cavefish. Means (1993) reported retagging four Ozark cavefish and we retagged 23 Ozark cavefish; the difference was due to lengths of the two studies rather than technique. Ozark cavefish were capable of avoiding capture by concealing themselves, but this avoidance appeared passive rather than active. In captivity, Ozark cavefish sometimes remained wedged between rocks for days. In Logan Cave, we also observed Ozark cavefish in crevices or resting vertically between rocks. Four tagged Ozark cavefish were missed for six or more consecutive trips and one individual was recaptured on two consecutive trips and then missed for 15 trips before being recaptured. This indicated Ozark cavefish could be missed by collectors and not every tagged fish in the cave would be collected each trip. However, continued misses of an individual fish were unusual events, and more than 80% of tagged Ozark cavefish were recaptured within two months of their original tagging if they were going to be recaptured at all. By the end of the study, 220 Ozark cavefish had been tagged in Logan Cave and it seems unlikely that all but 10 of them were hidden when we went through the cave on the last collecting trip.

Death, either before emigration, or after leaving the cave via its mouth, seems the most logical answer for the loss of most tagged fish. Only one dead Ozark cavefish (untagged) was found in Logan Cave (Brown 1996). However, dying fish may seek cover in crevices and fissures and be missed by collectors. Emigration out of the cave also seemed to occur. Several fish were observed close to the mouth of Logan Cave before disappearing. No Ozark cavefish has ever been reported outside a cave, but efficient visual epigean predators like banded sculpin, *Cottus carolinae*, likely account for this lack. Banded sculpin also enter Logan Cave and prey upon troglobitic cave crayfish, *Cambarus aculabrum*, and possibly Ozark cavefish (Means & Johnson 1995). All of this suggests a relatively short life-span of 4–5 years for Ozark cavefish in Logan Cave rather than the prolonged period (60 years) suggested by Robison & Buchanan (1988).

We disagree with Means & Johnson (1995) who stated that Ozark cavefish emigrate out of Logan Cave upstream, back into the aquifer. The majority of Ozark cavefish in both studies was first tagged in the Middle Reach of Logan Cave. Brown (1996) believed Logan Cave to be a population sink, with Ozark cavefish unable or unwilling to emigrate upstream. In more than four years and 37 collection trips through Logan Cave, only one tagged Ozark cavefish was found to have moved from the Middle to the Upper reach, and that fish moved only 28 m. Although Means and Johnson (1995) reported no marked fish moved from Middle to Upper reach, they believed that it was the only explanation for loss of tagged fish over their 6 month study. Migration remains a possibility to account for loss of fish downstream, but we find no evidence to support emigration upstream out of the cave and back into the aquifer.

We agree with Means and Johnson (1995) that Ozark cavefish immigrate into Logan Cave from the aquifer. Continued capture of untagged fish throughout the study could have resulted from failure to recognize fish that had lost tags, fish hiding, growth of small fish into the taggable size cohort or immigration. Tag loss did not account for the continued presence of untagged fish as we were able to identify fish that had lost tags. However, duration of visible scar tissue in Ozark cavefish is unknown.

Several clues suggest Ozark cavefish reproduce in Logan Cave and remain there to grow into the taggable cohort. We found no Ozark cavefish brooding eggs, embryos or young in the gill chambers (Poulson 1963), but this could have been overlooked during tagging and measuring efforts. There were indications of an extended spawning season with a peak in late summer. We continued to capture Ozark cavefish <30 mm throughout the year. Smaller fish were less mobile and thus less likely to immigrate than larger fish, suggesting these fish come from eggs deposited in Logan Cave. Poulson (1963) suggested it took more than two years for Ozark cavefish to reach 30 mm. We believe Ozark cavefish in Logan Cave took about one year to reach 30 mm.

Boyd (1997) observed what she believed to be fecund Ozark cavefish in Logan Cave in August 1996. Poulson (1963) and Jegla and Poulson (1970) suggested increased flows during spring might initiate metabolic increases leading to reproduction, and Boyd (1997) confirmed increased metabolism of Ozark cavefish in Logan Cave in summer and autumn, even though water temperatures remained relatively constant. Boyd (op. cit.) also observed one large and five small (ca. 10 mm) Ozark cavefish in July 1996. This group remained together for nearly two months until dispersed by flooding. This differs from Poulson (1963), Robison and Buchanan (1988), and Pflieger (1997) who suggest Ozark cavefish likely spawn in spring, during or immediately after high water periods, and may cannibalize their young.

Most Ozark cavefish growth occurred between April and October. Water temperatures generally differed less than 1°C throughout the year, leaving an increase in nutrients the most likely correlate. An estimated 20 000 gray bats occupied Logan Cave as a maternity colony between April and October (Harvey 1992, and personal communication) and guano was abundant in the cave and stream. Organic nutrients from the watershed also increased during this period and may have affected Ozark cavefish growth as well. Increased nutrient input presumably results in greater production of copepods, amphipods, and salamanders which in turn are consumed by Ozark cavefish.

It seems reasonable to suggest that the 35 larger fish tagged throughout the study were the result of immigration from the aquifer. Over 21 months, this yielded an immigration rate of 20 fish year^{-1}. The 1992 study also concluded untagged Ozark cavefish continually entered Logan Cave, but estimated a higher rate of 72 fish per year. Means and Johnson (1995) did not account for the possibility of growth into the taggable size, that would reduce their immigration estimate.

We suggest Ozark cavefish inhabiting Logan Cave are just a small portion of the true population that inhabits the associated aquifer, and Logan Cave serves as a window that provides partial access to that realm. Aquifer occupancy would explain why Ozark cavefish are sometimes collected from wells and springs not associated with caves (Robison & Buchanan 1988). It would also explain why most caves appear to support non-viable numbers of Ozark cavefish. Pflieger (1997) reported 12 caves in Missouri contained Ozark cavefish but observed no more than eight fish at the same time and place; half of occupied sites were based on observations of three fish or less. Willis and Brown (1985) noted 13 caves that contained Ozark cavefish, but only in Logan and Cave Springs caves did they ever observe more than five fish, even after multiple visits.

The idea that Ozark cavefish might be able to survive totally within the aquifer and migrate into Logan Cave was furthered when video cameras lowered down newly-dug wells in the Boone Formation photographed extensive mats of organic, bacteria-like material (V. Brahana personal communication), a potential energy source for Ozark cavefish, especially when combined with organic debris carried in from the watershed. The Ozark cavefish population in Logan Cave is sustained in part through immigration from this source, and along with reproduction in the cave itself, appears to have maintained a relatively stable population for nearly two decades.

Acknowledgements

We thank Arkansas Game & Fish Commission for funding this project under Section 6 of the Endangered Species Act, and for providing collecting permits. The U.S. Fish & Wildlife Service provided endangered species and special use permits. R. King, M. Revels, K. Pierson, C. Shepherd, and D. Catanzaro assisted in the field, and M. Means, G. Boyd, and V. Brahana shared data and ideas on Ozark cavefish and Logan Cave. We also thank T. Poulson, who read and commented on an earlier draft of this paper. This paper is a contribution from Arkansas Cooperative Fish & Wildlife Research Unit – U.S. Geological Survey, Biological Resources Division; Arkansas Game & Fish Commission; University of Arkansas; and Wildlife Management Institute Cooperative Unit.

References cited

Bond, C.E. 1996. Biology of fishes. Harcort Brace, New York. 750 pp.

Boyd, G.L. 1997. Metabolic rates and life history of aquatic organisms inhabiting Logan Cave stream in Northwest Arkansas. MS Thesis, University of Arkansas, Fayetteville. 98 pp.

Brown, J.Z. 1996. Population dynamics and growth of Ozark cavefish in Logan Cave National Wildlife Refuge, Benton County, Arkansas. MS Thesis, University of Arkansas, Fayetteville. 105 pp.

Eigenmann, C.H. 1898. A new blind fish. Proc. Indiana Acad. Sci. 1897: 231.

Guy, C.S., H.L. Blakenship & L.A. Nielsen. 1996. Tagging and marking pp. 353–383. *In:* B.R. Murphy & D.W. Willis (ed.) Fisheries Techniques, Amer. Fish. Soc., Bethesda.

Harvey, M.J. 1992. Bats of the eastern United States. Tennessee Tech. Univ., Cookeville. 22 pp.

Haw, F., P.K. Bergman, R.D. Fralick, R.M. Buckley & H.L. Blankenship. 1990. Visible implant fish tag. pp. 311–316. *In:* N.C. Parker et al. (ed.) Fish-Marking Techniques, Amer. Fish. Soc. Symp. 7, Bethesda.

Jegla, T.C. & T.L. Poulson. 1970. Circadian rhythms, I: reproduction in cave crayfish *Orconectes pellucidus intermis*. Comp. Biochem. Physiol. 33: 347–355.

Johnson, J.E. 1987. Protected fishes of the United States and Canada. Amer. Fish. Soc., Bethesda. 42 pp.

Means, M.L. 1993. Population dynamics and movement of Ozark cavefish in Logan Cave NWR, Benton County, Arkansas, with additional baseline water quality information. MS Thesis, University of Arkansas, Fayetteville. 126 pp.

Means, M.L. & J.E. Johnson. 1995. Movement of threatened Ozark cavefish in Logan Cave National Wildlife Refuge, Arkansas. Southwest. Nat. 40: 308–313.

Pflieger, W.L. 1997. The fishes of Missouri. Missouri Cons. Comm., Jefferson City. 371 pp.

Pollock, K.H., J.D. Nichols, C. Brownie & J.E. Hines. 1990. Statistical inference for catch-recapture experiments. Wildlife Monog. (107): 1–97.

Poulson, T.L. 1963. Cave adaptation in amblyopsid fishes. Amer. Midl. Nat. 70: 257–290.

Robison, H.W. & T.M. Buchanan. 1988. Fishes of Arkansas. Univ. of Arkansas Press, Fayetteville. 536 pp.

Robins, C.R., R.M. Bailey, C.E. Bond, J.R. Booker, E.A. Lachner, R.N. Lea & W. B. Scott. 1991. Common and scientific names of fishes from the United States and Canada. Amer. Fish. Soc. Special Publ. (20): 1–182.

Willis, L.D. & A.V. Brown. 1985. Distribution and habitat requirements of the Ozark cavefish, *(Amblyopsis rosae)*. Amer. Midl. Nat. 114: 311–317.

Woods, L.P. & R.F. Inger. 1957. The cave, spring, and swamp fishes of the family Amblyopsidae of the central and eastern United States. Amer. Midl. Nat. 58: 232–256.

Zar, J.H. 1984. Biostatistical analysis. Prentice Hall, Englewood Cliffs. 718 pp.

Environmental Biology of Fishes **62:** 170, 2001.
© 2001 *Kluwer Academic Publishers. Printed in the Netherlands.*

Threatened fishes of the world: *Ophisternon infernale* (Hubbs, 1938) (Synbranchidae)

Roger Medina-González[a], Graham Proudlove[b], Lizbeth Chumba-Segura[a] & Thomas Iliffe[c]
[a]*Departamento de Ecología and Departamento de Zoología, Facultad de Medicina Veterinaria y Zootecnia, Universidad Autónoma de Yucatán, Mérida, Yucatán 97100, México*
[b]*Department of Zoology, Manchester Museum, Manchester, M13 9PL, U.K. (e-mail: g.proudlove@umist.ac.uk)*
[c]*Department of Marine Biology, Texas A & M University, Galveston, TX 77553, U.S.A.*

Common name: Anguila ciega, anguila ciega yucateca (S). **Conservation status:** Endangered (IUCN). **Identification:** Total lack of external eyes and melanin pigment. Muzzle spatulate. Up to 325 mm SL. Lack of scales. Head long and bulb shaped with sensitive pores. Caudal region 37–54% longer than in other synbranchids (drawing by Rhian Hicks, 325 mm TL). **Distribution:** Cenotes and groundwater on the Yucatán Peninsula, Mexico (20–21°N, 88–90°W), including anchialine caves on the Caribbean coast. This species is sympatric in some areas with *Ogilbia pearsei* (Bythitidae) and *Rhamdia guatemalensis* (Pimelodidae) (Chumba-Segura 1984). **Abundance:** Very few specimens have been observed (maximum two at the same time) in very few localities. Underwater exploration should increase the number of recorded sites. It is of note that several locals in rural areas have seen *Ogilbia pearsei* but not *Ophisternon infernale*. **Habitat and ecology:** Builds mucus-lined burrows in organic rich sediments within the freshwater part of the aquifer. Has been observed under stones in muddy bottoms in the shallow parts of cave cenotes, and at 24 m in depth. The principle source of energy in this environment is bat and swallow guano. *Ophisternon infernale* feeds on the troglobitic shrimp *Creaseria morleyi*, tolerates low oxygen levels and is able to breathe atmospheric oxygen. **Reproduction:** Reported to be oviparous. **Threats:** The main threat is from water pollution, principally faecal coliforms and excess nitrate (Cruickshank et al. 1980). Populations in shallow caves are exposed to human disturbance, by modification and use of the caves, and by waste and wastewater disposal. Human populations obtain their water from the aquifer and, until recently, disposed of wastewater directly back into the same aquifer, a few metres from where it was obtained. This situation is particularly acute under the largest city, Mérida, although some piped water is now available. In rural areas there is little or no sewage treatment and industrial and domestic waste is discarded into caves. The population of the peninsula is growing: 2.3×10^6 in 1990, 2.9×10^6 in 1995. Additionally tourist numbers are also rising. This will place growing strain on the water resources of the area. For example, freshwater removal will lead to the incursion of saline waters into the freshwater zone. **Conservation action:** Wastewater in Mérida is now collected and pumped to treatment plants. The resulting treated product is injected into the saline groundwater 200 m below the surface. *O. infernale* has been reported from caves in the Reserva de la Biósfera de Sian Ka'an in Quintana Roo (Navarro-Mendoza 1988). It may also be present at Reserva Especial de la Biósfera de Ría Lagartos on the north coast of Yucatán, although it has not yet been reported. Its potential presence in such areas could be advantageous in terms of conservation actions (Navarro-Mendoza & Valdés-Casillas 1990). **Conservation recommendations:** Human water supply should be augmented by the collection (and subsequent treatment) of rainfall to supplement that obtained from the aquifer. All wastewater should be collected and properly treated. It should not be injected into the aquifer which contains is a very delicate ecosystem. The best economic solution is probably to pipe it out to sea although this will of course have its own (detrimental) effects. The continental shelf is wide and it will be expensive to dispose of wastewater far enough to avoid the immediate effect on the coastal systems. Detailed studies of the aquifer need to be made to monitor its health. It is necessary to survey the occurrence of cave and groundwater species such as *O. infernale* in order to have up to date distribution records. **Remarks:** This species is closely related to *Ophisternon aenigmaticum* and probably invaded freshwater caves from wetlands during the Pliocene.

Chumba-Segura, L. 1984. Synbranchidae: *Ophisternon infernale*. Fauna Cenotes Yucatán (6): 1–9.

Cruickshank, G., J. Aguirre, D. Kramer & E. Craviota. 1980. Bacterial contamination of the limestone aquifer beneath Mérida, México. pp. 341–345. *In*: Aquifer Contamination and Protection, UNESCO Publication No. 30.

Navarro-Mendoza, M. 1988. Inventario íctico y estudios ecológicos preliminares en los cuerpos de agua continentales en la reserva de la Biosfera de Sian Ka'an y áreas circumvecinas en Quintana Roo, México.

Navarro-Mendoza, M. & C. Valdés-Casillas. 1990. Peces cavernícolas de la península de Yucatán en peligro de extinción, con nuevos registros para Quintana Roo. pp. 218–241. *In*: J.L. Camarillo & F. Rivera A. (ed.) Areas Naturales Protegidas en México y Especies en Extinción, Nacional Autónoma de México, Ciudad de México.

Environmental Biology of Fishes **62:** 171–194, 2001.
© 2001 *Kluwer Academic Publishers. Printed in the Netherlands.*

How hydrogeology has shaped the ecology of Missouri's Ozark cavefish, *Amblyopsis rosae*, and southern cavefish, *Typhlichthys subterraneus*: insights on the sightless from understanding the underground

Douglas B. Noltie[a] & Carol M. Wicks[b]
[a]*Department of Fisheries and Wildlife Sciences, The School of Natural Resources, 302 Anheuser-Busch Natural Resources Building, University of Missouri-Columbia, Columbia, MO 65211-7240, U.S.A. (e-mail: noltied@missouri.edu)*
[b]*Department of Geological Sciences, 101 Geology Building, University of Missouri-Columbia, Columbia, MO 65211-1380, U.S.A. (e-mail: wicksc@missouri.edu)*

Received 2 May 2000 Accepted 10 January 2001

Key words: Amblyopsidae, ecology, hydrology, geology, habitat, conservation

Synopsis

Two troglobitic fishes of conservation concern that inhabit Missouri are the Ozark and southern cavefishes, *Amblyopsis rosae* and *Typhlichthys subterraneus*, respectively. These species inhabit the groundwater of karstified bedrock in the Springfield and Salem plateau regions of the state, respectively. These two areas differ substantially – geographically, geologically, and hydrologically. This paper explores how these differences interact in shaping (1) the habitat in which these two species dwell, (2) the resulting ecological constraints that the fishes face, (3) how their ecologies and populations reflect these constraints, and (4) how conservation and management efforts may need to be tailored to best compliment the unique challenges that each species presents.

Introduction

Zoogeographically, North America's freshwater fish fauna numbers approximately 1100 species in 50 families (Mayden 1992, Moyle & Cech 2000). Of these, only the family Amblyopsidae (the North American freshwater cavefishes) includes species in which the eggs are incubated in the gill chamber by a parent (Eigenmann 1899 a,b, Cooper & Keuhne 1974). Besides the uniqueness of their parental care, the North American amblyopsids are notable in the varying degrees to which they have become adapted to subterranean life. Six species comprise the family (Table 1), forming a purported continuum from least to most cave adapted (Poulson 1960, 1963, 1964), with the southern cavefish, *Typhlichthys subterraneus*, being a 'younger' troglobite than is the Ozark cavefish, *Amblyopsis rosae* (Poulson & White 1969). The Amblyopsidae are one of only three families of freshwater fishes in the U.S. to exhibit a subterranean existence: two species of Texas Ictaluridae are troglobitic, the widemouth blindcat, *Satan eurystomus*, and the toothless blindcat, *Trogloglanis pattersoni*[1,2] (Hubbs & Bailey 1947, Suttkus 1961, Langecker & Longley 1993 a,b), as are some cave-dwelling Missouri sculpins, *Cottus* spp. (G. Adams & B. Burr personal communication). For the amblyopsids, the traits which are interpreted as adapting them to subterranean life (small sizes, long

[1] Longley, G. & H. Karnei, Jr. 1978a. Status of *Satan eurystomus* Hubbs & Bailey, the widemouth blindcat. Report of Contract #14-16-0002-77-035 to the U.S. Fish and Wildlife Service, Albuquerque. 48 pp.
[2] Longley, G. & H. Karnei, Jr. 1978b. Status of *Trogloglanis pattersoni* Eigenmann, the toothless blindcat. Report of Contract #14-16-0002-77-035 to the U.S. Fish and Wildlife Service, Albuquerque. 54 pp.

Table 1. Members of the family Amblyopsidae, arranged in increasing order of designated degree of cave adaption (Poulson 1985).

Common name	Species name	Ecological designation
swampfish	*Chologaster cornuta*	trogloxene
springfish	*Chologaster* (*Forbesichthys*) *agassizi*	troglophile
southern cavefish	*Typhlichthys subterraneus*	troglobite
northern cavefish	*Amblyopsis spelaea*	troglobite
Ozark cavefish	*Amblyopsis rosae*	troglobite
Alabama cavefish	*Speoplatyrhinus poulsoni*	troglobite

Table 2. Summary of the status of southern *Typhlichthys subterraneus*, and Ozark *Amblyopsis rosae*, cavefish in various U.S. jurisdictions. Distribution data from Lee et al. (1980). Status data taken from Schmidt (1995).

Jurisdiction	Status of species by jurisdiction	
	Southern cavefish *Typhlichthys subterraneus*	Ozark cavefish *Amblyopsis rosae*
Federal (U.S.)	not listed	Threatened
Arkansas	Inventory element	Inventory element
Missouri	Watch	Endangered
Oklahoma	Extirpated	Threatened
Alabama	Threatened	never recorded as present here
Georgia	Rare	never recorded as present here
Indiana	Endangered	never recorded as present here
Kentucky	Special concern	never recorded as present here
Tennessee	Deemed in need of management	never recorded as present here

lifespans, infrequent reproduction, low fecundities, low metabolic rates, small population sizes, high levels of population endemism; Poulson 1958, 1960, 1963, 1964) render the species susceptible to extirpation. Consequently, all species in the family have been vested with protective status by some or all the states comprising their respective ranges (Table 2, Romero 1998a,b).

Among U.S. states, Missouri is notable for the amount of karst topography that occurs within its borders (Ford & Williams 1989). Owing to this, Missouri is also home to three of the six amblyopsid species: the springfish, *Forbesichthys* (=*Chologaster*) *agassizi*, occurs in southeastern Missouri's 'Bootheel'[3] (McDonald & Pflieger 1979), the southern cavefish in south-central Missouri, and the Ozark cavefish in southwestern Missouri (Woods & Inger 1957, Tafenelli & Russell 1972, Lee et al. 1980, Mayden & Cross 1983, Jones & Taber 1985). Historically, Missouri has figured prominently in the development of our knowledge of amblyopsid biology: *A. rosae* was first described from a Missouri site (Eigenmann 1898), some of the early amblyopsid research was conducted on Missouri specimens (Garman 1889, Eigenmann 1899a–d, Charlton 1933), a substantial fraction of Poulson's (1960, 1963, 1964) classic investigation addressed Missouri populations, and Missouri's springfish population remains the only one west of the Mississippi River (McDonald & Pflieger 1979).

For the amblyopsids in general, relatively little attention has been paid to the obvious linkage between specifics of the species' ecologies and specifics of the geologic/hydrologic environment in which they dwell. In this paper, we attempt a bridging of this divide. We focus our attention on Missouri, given our familiarity with the state's piscine and hydrogeological situation. We also limit our considerations to only two of Missouri's amblyopsid species (southern and Ozark cavefish, respectively), given that the single known springfish population in Missouri precludes adequate comparison.

To provide a foundation for this discussion, we need first to lay the appropriate groundwork. For the purposes of this paper, that groundwork will comprise a brief preliminary review of the Ozark cavefish and the sites in which it is found, and then a discussion of the geology and hydrogeology of the Springfield Plateau where it occurs (to help in developing perspective, and because details of the geology may be unfamiliar to the global biological community, our initial explanations here will be somewhat pedestrian). We follow with a parallel presentation for the southern cavefish and the Salem Plateau where it dwells; considering the plateaus separately is necessary because they are geologically and geographically separate karst formations; the parallels are intended to facilitate comparison. We then attempt a bridging of the biological–geological divide using several key issues as abutments. Finally, we discuss some of the implications that these perspectives raise.

[3] Aley, T., C. Aley & P. Moss. 2000. Delineation of the recharge area for the spring cavefish (*Forbesichthys agassizi*) springs, Scott County, Missouri. Ozark Underground Laboratory contract report to the Missouri Department of Conservation. 48 pp.

Background: species-specific ecology and karst

Ozark cavefish

Description

The Ozark cavefish, *Amblyopsis rosae*, is a small fish (65 mm total length, Means & Johnson 1995) that lacks pelvic fins (Eigenmann 1899c, Nelson 1989). Young are born with the anal–genital pore positioned just anterior to the anal fin; with growth, the pore's location shifts forwards until in adults it occupies a jugular position (Woods & Inger 1957). The species is depigmented; it lacks melanophores in the skin (Woods & Inger 1957), and any apparent coloration is imparted by the underlying circulatory system or body cavity organs (Robison & Buchanan 1988). The eyes are vestigial (Eigenmann 1899d, 1909). The head shape is elongate and flattened (Robison & Buchanan 1988). The species' lateral line is elaborate, and branches extensively over the head (Eigenmann 1899c). In addition, the species is characterized as having four to six rows of neuromasts on the caudal fin (Robison & Buchanan 1988), and it lacks a post-cleithrum bone (Miller & Robison 1973). The neuromasts project from the skin's surface, and their number and structure is thought to enhance the detection of displacements of the water (e.g., prey, conspecific, or predator movements, current, proximity to surroundings; see Baker 1972). Their brain structuring indicates that the receipt and integration/interpretation of visual stimuli have been de-emphasized, whereas those from the olfactory and acoustico-lateralis system have been emphasized (Poulson 1960).

Ecology

For a small fish, Ozark cavefish are long-lived and slow-growing; breeding is not annual, maturation likely requires many years to attain, and fecundities are low[4] (Poulson 1960, 1964, Robison & Buchanan 1988). The species is thought to be a gill-chamber brooder (Breder & Rosen 1966, Balon 1975): non-brooded eggs would be susceptible to consumption by other energy-limited cave-dwellers. Population densities and sizes are perceived as being low[5] (Means 1993, Means & Johnson 1995), and together this suite of attributes renders populations susceptible to extirpation because of their inability to rapidly repopulate after die-offs (usually anthropogenic in origin). The species is energetically frugal, exhibiting a low metabolic rate (Poulson 1963, 1964, Boyd 1997), long glides between swimming strokes (Poulson 1963), and minimal overt aggressive behavior (Bechler 1980, 1981, 1983). The species' rheotaxic responses demonstrate an attraction to low flows; thigmotaxis is also seen (Poulson 1963). The species feeds mainly on copepods, invertebrates, and the aquatic young of salamanders and conspecifics from on the substrate and under rocks (Poulson 1963).

Genetically, the species' range can be quartered, with genetically differentiable forms (mtDNA) being affiliated with four river basins, respectively (Bergstrom 1997): (1) northwest Arkansas' Illinois River (drains west into the Arkansas), (2) southwest Missouri's White River (drains northwards from Arkansas into Missouri and then back south through Arkansas into the Mississippi), (3) southwest Missouri's arm of the Neosho River (drains west into Oklahoma's Neosho), and (4) northeast Oklahoma's Neosho River mainstem (drains south into the Arkansas). Together with the additional site-to-site differences that were detected within river basins, it appears that many of the sites are genetically isolated (experience minimal gene flow), and thus constitute discrete endemic populations.

Distribution

The Ozark cavefish is known (either presently or historically) from 41 sites occurring on the Springfield Plateau in the tri-state area of southwest Missouri (20 sites), northwest Arkansas (10 sites), and northeast Oklahoma[5] (11 sites) (Brown & Todd 1987, K.B. Lister unpublished data). At all of these locations, the water table is intersected (groundwater is present). Access to the sites is through either a cave stream, a karst window, or a well. These sites occur in the main portion of the plateau (no sites are known from the portion of the plateau that extends southeast into Arkansas), although no sites are known from the central area in and around McDonald County, Missouri.

Site characterization

Based on our observations of the locations within which Ozark cavefish are seen, there are two types of sites (Figure 1). 'Lateral' sites are cave streams which

[4] Brown, A.V., K.B. Brown, L.D. Willis, D.C. Jackson & P.P. Brussock. 1982. Distribution and abundance of the Ozark cavefish *Amblyopsis rosae* (Eigenmann) in Missouri. Final Report to the Missouri Department of Conservation, Division of Natural Resources. 20 pp.

[5] Brown, A.V. & L.D. Willis. 1984. Cavefish *(Amblyopsis rosae)* in Arkansas: populations, incidence, habitat requirements and mortality factors. Final report to the Arkansas Game and Fish Commission. 61 pp.

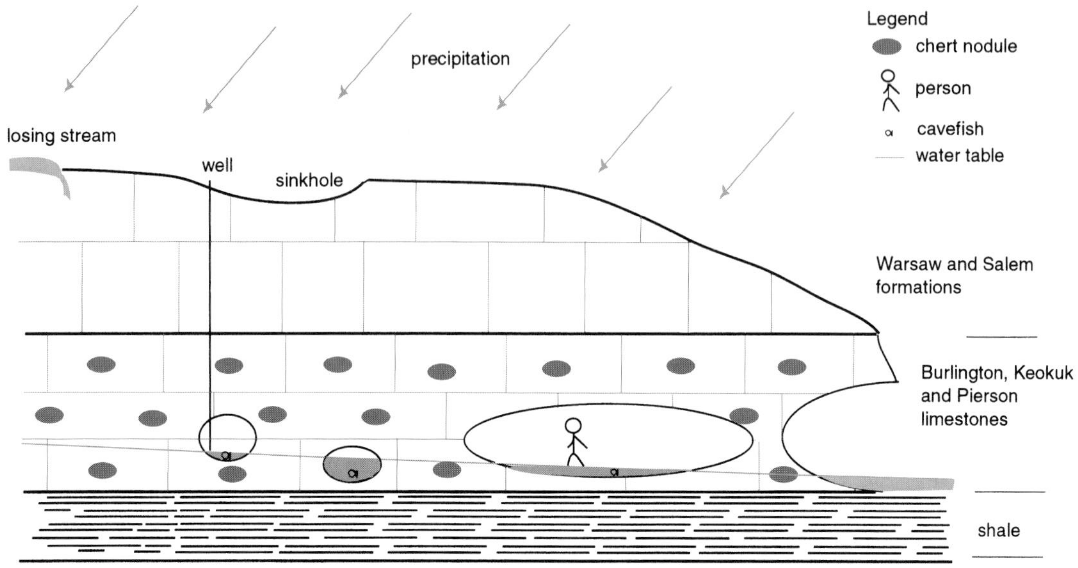

Figure 1. An illustration of the Springfield Plateau showing the geologic setting of the plateau, the development of karst within the plateau, and the two types of cavefish sites – lateral and vertical. The illustration is not drawn to scale.

occupy relatively large (i.e., penetrable by humans) passageways (conduits) that intersect the land surface near the base of either the plateau bluff line or of the river valley walls that have incised the plateau perimeter (escarpment). A 'vertical' site (i.e., a karst window or well) is one where the downwards opening penetrates the land surface and the underlying rock units, and then intersects a conduit (usually one smaller than is penetrable by humans). These sites are 'interior' sites, in that they are not associated with the plateau margin.

Within both kinds of sites, the in-stream substrates vary: clean-scoured bedrock occurs in shallow areas with fast flows (riffles), areas of 'breakdown' (fallen ceiling or wall rock or transported sediment) present coarse substrates that are eroded and transported with time, and areas of clay (fine-grained material) are common where the flow slows enough to allow deposition (pools).

Geology of the Springfield plateau

Where Ozark cavefish sites have or have not been found can best be understood by developing an appreciation of the geologic setting of the Springfield Plateau.

Location and stratigraphic column. The Springfield Plateau in southwestern Missouri, northwestern Arkansas, and northeastern Oklahoma is a physiographic province wherein rocks of the Mississippian Period (320 to 360 million years old) are exposed at the earth's surface. At the land surface, the rock exposed over the western portion of the plateau are of the Salem and Warsaw formations[6,7] (Table 3, Reams 1968, Thompson 1995, Unklesbay & Vineyard 1992, Fanning 1994, Adamski & Pugh 1996). The rocks exposed over the eastern portion of the plateau are the Keokuk, Burlington, and Pierson limestones. Vertically, the Salem and Warsaw formations overlie the rocks of the Keokuk, Burlington, and Pierson limestones; beneath these formations are the rocks of the Chouteau Group and the Chattanooga Shale. As an analogy, this sequence of rocks can be envisioned as being a large layer cake (Figure 1). The bottom layer is the Chattanooga Shale and Chouteau Group; the middle layer is rocks of the Keokuk, Burlington, and Pierson limestones; the top layer is rocks of the Salem and Warsaw formations. Across the Springfield Plateau, the

[6] Adamski, J.C., J.C. Petersen, D.A. Friewald & J.V. Davis. 1995. Environmental and hydrologic setting of the Ozark Plateaus study unit, Arkansas, Kansas, Missouri, and Oklahoma. U.S. Geological Survey, Water Resources Investigation Report 924022.

[7] Thompson, T.L. 1986. Paleozoic succession in Missouri. Part 4. Mississippian system. Missouri Department of Natural Resources, Division of Geology and Land Survey, Report of Investigation 70, 182 pp.

Table 3. Stratigraphic sequence of the Springfield Plateau region including a description of the hydrologic properties (modified from Robertson 1963, Feder 1973, Unklesbay & Vineyard 1992, Imes & Emmett 1994, Thompson 1995). The sequence is based on descriptions from southwestern and central Missouri and uses the terminology of Unklesbay & Vineyard (1992).

Age	Formation (thickness in m)	Lithologic description	Hydrologic characteristics	Regional hydrologic unit
Mississippian	Salem Formation (10)	An oolitic limestone with a small amount of chert and with crossbedding	Yields variable, most less than 3.5×10^{-3} m^3 s^{-1}	Springfield Plateau aquifer
	Warsaw Formation (15)	Medium-grained crystalline limestone	Yields highly variable, most less than 3.5×10^{-3} m^3 s^{-1}; many large springs issue from this formation	Springfield Plateau aquifer
	Burlington, Keokuk, and Pierson limestones (130)	A limestone with abundant chert and crinoid fragments	Yields highly variable, most less than 3.5×10^{-3} m^3 s^{-1}; many small springs issue from this formation, dry valleys in this formation	Springfield Plateau aquifer
	Chouteau Group (6)	Dense limestone	Very low well yields, 0.7×10^{-3} m^3 s^{-1}	Ozark confining unit
Devonian	Chattanooga Shale (9)	Black shale with a basal sandstone	Very low well yields, 0.7×10^{-3} m^3 s^{-1}	Ozark confining unit

majority of the top layer has been eroded away, exposing the underlying rocks of the Keokuk, Burlington, and Pierson limestones. Using our cake analogy, the plateau appears as if one had horizontally sliced off a large portion of the cake's upper layer along an irregular line. Along the eastern edge of the Springfield Plateau, the rocks of the Salem and Warsaw formations have been further eroded (again cut through irregularly and removed), exposing the underlying rock units. The regional dip (incline of the rock beds) is about 2° to 3° W, and can be considered hydrologically negligible over the length and width of the Plateau.

Strata characterization. To more fully appreciate where Ozark cavefish dwell, each of the Springfield Plateau's rock units needs to be described because the differences among the units are greater than the differences within any one unit. All of these rocks were deposited or formed below sea level, first the rocks of the Chattanooga Shale and Chouteau Group, then the Keokuk, Burlington, and Pierson limestones, and finally the Salem and Warsaw formations. Starting at the bottom, the Chattanooga Shale is predominantly a shale composed of very fine-grained detrital fragments of platy clay minerals that were deposited in deep, still water. Over geologic time, compression caused by the weight of the overlying rock forced the platy clay minerals to orient horizontally. The thickness of the unit ranges from 3 to 15 m (Thompson 1995), and the pores between the platy clay minerals are microscopic. The rocks of the Chouteau Group are composed of crystalline limestone which is dense and fine-grained. The Keokuk, Burlington, and Pierson formations are predominantly composed of limestones containing abundant bedded and nodular chert. The limestone formed from the fossilized parts of crinoids (sea lilies, subphylum Crinozoa) that lived attached to the bottom of the shallow sea. These limestones are composed nearly entirely of broken crinoids (lengths of broken stem, individual columnals, a few calyxes) stuck together with a calcitic cement. The thickness of the limestones can be as great as 130 m. The pores between the crinoid pieces are small, particularly after the calcite cement filled the available openings. The rocks of the Salem and Warsaw formations are composed of a crystalline limestone which is dense and exceedingly fine-grained. Its primary pore spaces are very small.

Karstification of the Springfield Plateau. Karstification is the process by which the 'primary porosity' (the amount and distribution of the pores at the time the rock was first formed) and the 'secondary porosity' (joints and bedding plane separations) of limestones and dolomites is solutionally (and sometimes mechanically) enlarged to become the features constituting 'tertiary porosity' (that porosity that is formed subsequent to formation of the rock unit) such as caves and conduits (Ford et al. 2000). The karstification process is an ongoing one that occurs over millions of years and that varies from location to location.

Two well-known attributes of fluid dynamics are fundamental to understanding the development of karst on the Springfield Plateau. The first attribute is that water flows down a gradient from locations of high fluid potential to areas of low potential, with fluid potential being a measure that integrates the energy available from three effects: (1) elevation above sea level, (2) pressure caused by the weight of the overlying water that is confined by overlying low-permeability layers, and (3) energy due to the velocity of water. The second attribute is that water follows the path of least resistance; it is thus along this path that limestone is most prone to dissolution. For the Springfield Plateau, the paths of least resistance were through the Keokuk, Burlington, and Pierson limestones wherein the dissolution-resistant chert tended to channel water movement (Hargrove 1968), resulting in dissolution and pore enlargement along these flow paths over geologic time. This allowed more carbon dioxide-rich water to flow along these paths, resulting in yet further dissolution and enlargement. In this way, karst formation can be seen as a process involving positive feedback: more dissolution begets yet more.

Understanding how the cavernous porosity developed in the Springfield Plateau, and the sizes to which the conduits developed is important in gaining an appreciation of where Ozark cavefish live. For the Springfield Plateau, the end result of the process above, as expressed in the modern day, is a single rock unit (the Keokuk, Burlington, and Pierson limestones) within which major cavernous porosity exists. Beneath this remains the impermeable Chouteau Group and Chattanooga Shale which impede water drainage from the overlying limestones. Overlying the limestone remains are the relatively thin limestones of the Salem and Warsaw formations, which are densely crystalline and also difficult for water to permeate.

Of the five 'types' of caves (defined by their geomorphic characteristics; see Palmer 1991), two are common: branchwork caves (tree-like structure) and network caves (maze-like structure). Given the geologic setting and Palmer's (1991) classification, it is reasonable to expect that a majority of the karst development in the Springfield Plateau to be of the branchwork type. In branchwork systems, the 'caves' in the upgradient reaches are very small in diameter and large in number; as the water moves down the hydrologic gradient, the 'caves' decrease in number but increase in diameter to the point where they may become large enough for humans to enter. In branchwork systems, there are usually only single pathways from any given input location to an individual discharge point (usually a spring). The downgradient reach of these 'cave' passages may be effluent (a spring or cave stream) if the water table is intersected, or a dry passage if the water table has fallen and remains below the level of the cave development. It is in the former that we are able to observe Ozark cavefish.

To aid in visualizing the pattern of cave development that occurs within the Keokuk, Burlington, and Pierson limestones series of the Springfield Plateau, and to help the reader 'see' into the rock, a reasonable (but not perfect) analogy is that of the human circulatory system. Distally, the capillaries are very small in diameter and large in number. As the heart is approached, the venules join to form fewer, yet larger, vessels. These join to form yet larger and fewer veins, which funnel blood into the heart. This pattern of branching and joining is an appropriate mental model of a 'cave' system in the limestones. Whereas humans can enter only the largest of the passages, Ozark cavefish are capable of negotiating the smaller, waterfilled, outlying passageways.

Caves in the Springfield Plateau are commonly stream caves (contain flowing streams); their structure and function are intimately related to the topography and substrate characteristics of the drainage basins in which they occur. Within these caves are abandoned stream passages (30 m above present cave stream levels) formed earlier in cave development. The karstification and resulting downcutting here was a prolonged process; speleothems within these caves date from tens to hundreds of thousands of years old, evidence of when their deposition began following sufficient initial karstification (T. Aley personal communication).

Modern hydrologic setting of the Springfield Plateau. In the modern day, the flow of water through the Springfield Plateau surface rocks is largely controlled by the karstification of the rocks below. The dimpled topography (sinkholes) funnels the water underground into the subterranean dissolutional voids. These are the 'vertical' sites within which Ozark cavefish are encountered. We also find losing streams where water is present in the upgradient reaches but absent in the downgradient reaches because it now flows below ground. Losing stream reaches can be thought of as storm water drains in city streets: when it rains, water flows down the street as a surface 'stream'; once down the drain, the water flows underground. If there is more water than the drain can accommodate, the excess flows over the land surface in existing channels which, under most conditions, are 'dry valleys'. Both of these surface features quickly funnel accumulating surface water into the subsurface, and provide point sources (rather than diffuse sources) of groundwater recharge.

Water infiltration

Because Ozark cavefish dwell in groundwater, understanding the present-day hydrology of the Springfield Plateau is critical to understanding their ecology and patterns of occurrence. Hydrologically, the way in which water moves into and through the strata of the Springfield Plateau is relatively straightforward (Figure 1). In a very general sense, the rocks are exposed at the land surface beneath a thin soil cap. Precipitation falling onto the land surface can enter into the rock units below and the conduits therein in two ways, the first being through losing streams and sinkholes (this input is considered discrete, see Ford & Williams 1989), the second being through movement of water down through the overlying soil horizons (this input is considered diffuse, see Ford & Williams 1989).

Water table

Critical to considerations regarding the Ozark cavefish is the location of the water table, the fish being restricted in their occurrence to the continuously saturated phreatic zone. The water table in the Springfield Plateau is relatively close to the land surface (Vandike 1993). Thus, cave passages in the Springfield Plateau often contain actively flowing streams. This contrasts with the contention of Brown et al.[4] (1982: 16) that Ozark cavefish sites 'receive water from deep groundwater sources'.

Recharge basins

Because water flows from high regional fluid potential to low regional fluid potential both atop and beneath the land surface, the concept of a groundwater drainage basin or watershed (= a recharge basin, the area from which all water upgradient is derived) can be applied to water within the rock. In delineating the Springfield Plateau's groundwater drainage basins, we must consider not just the drainages on the surface (a relatively two-dimensional perspective), but the entire volume of the rock matrix and the drainage paths therein (a three-dimensional perspective) down to the depth of the Chattanooga Shale and Chouteau Group through which neither fish nor appreciable amounts of water can pass (being impervious to water, it is also impervious to fish movement). Surface streams on the Springfield Plateau are generally incised to sufficient depths that the movement of both water and fish between basins is severely limited. The typical size of Springfield Plateau recharge basins is on the order of 10 to 100 km^2 of land surface[8,9] and generally dozens of meters in thickness/depth (Table 3). Base levels in the region are defined by the four major downcut rivers that radiate from the Plateau and into which most of the Plateau's water ultimately flows (broadly speaking, southwest Missouri's White River and Neosho River arm, northwest Arkansas' Illinois River, and northeast Oklahoma's Neosho River mainstem).

Subterranean flow paths

Water below the water table resides in the pores of the bedrock; the pores range in size from microscopic to conduits large enough for humans to enter, and exhibit a high level of connectivity (especially the conduits). A helpful adage is that '90% of the water flows through 10% of the pore space', those being the largest pores. In other words, most of the water flows through the conduits (Ford & Williams 1989). Thus, water in the rock strata can be visualized as existing in streams much like those on the surface, except that they are rock-roofed: the tributaries get smaller as one moves upgradient, the stream reaches contain pool and riffle sequences, and the stream bottoms are covered with gravels, sands, silts, and exposed bedrock. The composition of the underlying shales (confining units) is such that the pores therein are so small and poorly connected that water movement through them is almost completely impeded.

Rates of subterranean water movement

The conduit system that has developed in the Springfield Plateau allows influxes of water to move rapidly through the underground. Rate measurements have been determined for the Devils Icebox-Connors Cave system in central Missouri's Boone County

[8] Aley, T. & C. Aley. 1991. Hydrogeology of Ozark cavefish caves in Oklahoma. Phase 2 investigations. Final Report to the Oklahoma Chapter of The Nature Conservancy. 49 pp.

[9] St. Ivany, G. 1988. Geologic and hydrologic characteristics of the Gans, Bonne Femme, and Clear Creek watersheds and the Pierpont karst plain near Rock Bridge Memorial State Park. Missouri Department of Natural Resources Report. 23 pp. and 1 map.

(Wicks 1997); whereas this system does not contain cavefish, it can be considered representative of the conduit systems developed within the Springfield Plateau, given its occurrence in a disjunct 'island' of the same Mississippian-aged limestones.

For the Devils Icebox-Connors Cave system, variation through time in the hardness of water rising from karst springs was used to determine transit or flow-through times for water in the conduit system (Pitty 1968, Ternan 1972). Here, the time required for water entering the underground from a losing stream to re-emerge as spring efflux was three to four days; this was insufficient for thermal or chemical equilibrium to be achieved, as was indicated by the coefficients of variation of the water hardness (typically 18%) and of the water temperature (typically 15%) (Figure 2). For diffuse sources, the groundwater residence time was nearly eight days; because this water experienced a longer period of contact with the surrounding soil and bedrock, it more nearly approached a thermal and chemical equilibrium. This resulted in a lowering of the coefficients of variation of the water hardness (typically 12%) and of the water temperature (5%). Even though inflowing water from diffuse sources remained underground longer, its residence time was still insufficient for it to become chemically saturated or thermally equilibrated with the bedrock; the near absence of temporal variation in thermal or chemical attributes would be evidence of equilibration (Wicks 1997).

Geologic perspectives on where Ozark cavefish have been found

Within the geologic context, the occurrence of Ozark cavefish sites can be better appreciated. Ozark cavefish reside within the dissolutionally-enlarged voids of the Keokuk, Burlington, and Pierson limestones at depths below the water table but above the confining layers of the shales in the vertical rock sequence. In some areas of the Springfield Plateau, we speculate that the absence of a karstic layer (or its thinness and lack of phreatic water) precludes the occurrence of cavefish. For example, karstic rocks providing suitable habitat have been eroded away in and around McDonald County, Missouri (near the center of the Springfield Plateau) and at the eastern edge of the plateau, exposing the underlying shales; neither area seems geologically

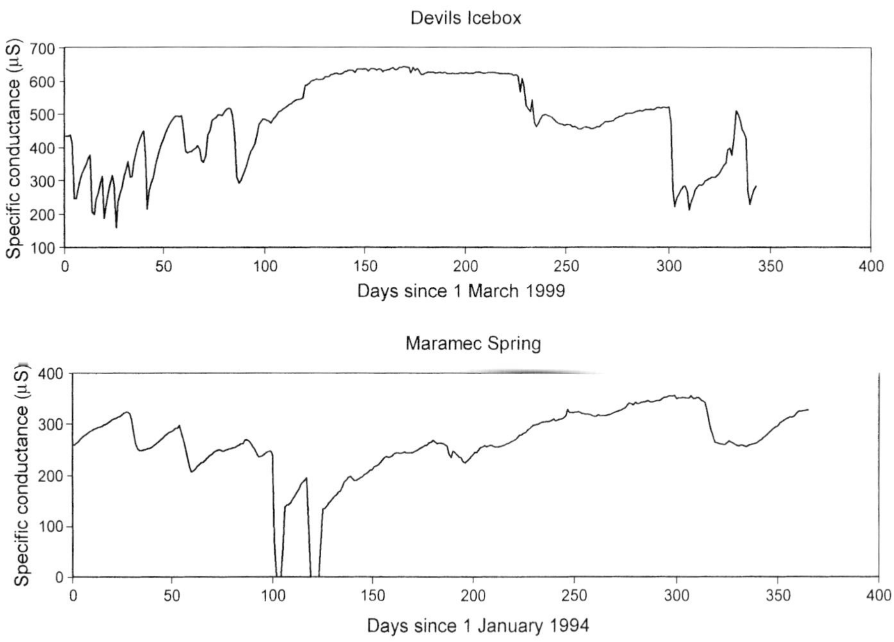

Figure 2. A diagram showing the variation in specific conductance of spring water emitted from the Devils Icebox–Connors Cave system (Springfield Plateau) and from Maramec Spring (Salem Plateau).

conducive to the occurrence of Ozark cavefish (i.e., their inground habitat is lacking), and indeed, the species has not been found in either location.

In addition, finding sites where Ozark cavefish occur is limited by the size and interests of the people looking for them. Many sites have been located by cavers who purposely look for caves accessible to humans. However, because cavefish are small, they can occupy passageways much smaller than can accommodate humans. The requirement that these passageways be phreatic is also a deterrence to human access. In some cases, cavefish have been found by landowners in hand-dug wells that are connected to small karst springs (T. Aley personal communication). As such, human factors (size, reliance on air, lack of interest in or access to wells) may be artificially limiting the number and types of sites where Ozark cavefish have been seen.

On the Springfield Plateau, Ozark cavefish are seen in either 'lateral' or 'vertical' sites (see above). This observation is easily understood once one considers that the rocks of the Springfield Plateau are riddled with completely or partially water-filled conduits, most (but not all) of which are too small for human access (Figure 1). The 'lateral' sites constitute cave streams flowing through human-sized conduit systems that become evident and also accessible because their intersection with the plateau margin results in an obvious opening. The 'vertical' sites (natural karst windows and man-made wells) constitute shafts that penetrate downwards and intersect an in situ conduit large enough to accommodate a cavefish but too small to accommodate human entry.

Much like surface streams, cave streams in areas of known Ozark cavefish occurrence do contain gravel, sand, and silt deposits[10] (White 1988). This bed material derives both from material collapsing in from the cave ceiling and from the transportation of material from upstream. The latter requires that solid materials be carried into the conduit either by a losing stream (allogenic input) or via diffuse flows (autogenic input). In either case, some (but not all) of the larger sediment particles will be filtered out as the sediment-laden water moves through the losing stream's gravel pack or the porous soil cover. In the case of the losing stream, the gravel pack represents a relatively coarse filter, such that the filtrate may contain micro-organisms and suspended particles (organic detritus and fine inorganic particles) that are consequently delivered into the cave stream and which are available for subsequent redeposition/redistribution.

[10] Dogwiler, T.J. & C.M. Wicks. 1999. Characterizing the sediments in the hyporheic zone in a karst system. Abstract, 1999 Geological Society of America Meeting 31: 155.

Southern cavefish

Description

As a fellow member of the family Amblyopsidae, many of the attributes of the southern cavefish, *Typhlichthys subterraneus*, parallel those of the Ozark cavefish. Thus, only where substantial differences occur will additional/alternative information be provided here. Like the Ozark cavefish, the southern cavefish is small (75 mm total length; Robison & Buchanan 1988) and lacks pelvic fins (Packard & Putnam 1872, Nelson 1989). The species appears depigmented; however, exposure to light has prompted the development of epidermal melanophores in some fish (Woods & Inger 1957). Again, the eyes are vestigial, although the dissimilarities to the Ozark cavefish are appreciable (Eigenmann 1899d, 1909, Poulson 1964, Mohr & Poulson 1966) and suggestive of a briefer history of in-ground inhabitation. The head shape is more elongate and flattened than is that of the Ozark cavefish (Robison & Buchanan 1988), the lateral profile of the head is somewhat concave, and the species' fins are less elongate than are those of the Ozark cavefish. The lateral line of the southern cavefish branches extensively over the head (Cooper & Kuehne 1974), but is somewhat less elaborate than is that of the Ozark cavefish (Woods & Inger 1957, Baker 1972), although the neuromast count is higher (Poulson 1963). In contrast to the Ozark cavefish, the species is characterized as having only two rows of neuromasts on the caudal fin (Robison & Buchanan 1988) and the presence of a post-cleithrum bone (Miller & Robison 1973). The neuromasts of southern cavefish are less elaborate than those of the Ozark cavefish (Poulson 1963). Structurally, their brains feature enlarged olfactory and acoustico-lateralis and reduced optic centers (Mohr & Poulson 1966).

Ecology

Southern cavefish appear to be long-lived and slow-growing. One author (DBN) has maintained individuals in the laboratory for a decade. At collection, these were considered adult-sized individuals. The smaller individuals have not grown to sizes approaching that of the larger ones (which might be expected if the larger ones had attained asymptotic size), and specimen deaths have not been restricted to the largest individuals. Consequently, the suggestion that the species' lifespan be counted in decades rather than years seems defensible. Reproductively, the species differs from the Ozark cavefish in that the eggs and clutches are larger, and that the females invest more energy in egg production and parental care (Poulson 1985). Some southern cavefish sites harbor many individuals: one site in Missouri repeatedly yields censuses approaching or exceeding 100 individuals; a contaminant spill caused nearly a thousand individuals to exit another Missouri site[11] (Crunkilton 1985). Energetically, the species differs from the Ozark cavefish in that it has a higher standard and routine metabolic rate (Poulson 1985), and that it exhibits a more diverse agonistic behavioral repertoire within which act durations were longer (Bechler 1980, 1981, 1983). The attraction of individuals to point sources of water efflux (simulated spring seeps) has been demonstrated experimentally (Schubert & Noltie 1995), although differences between sites in the intensity of this response may parallel differences in local flow regimes. Southern cavefish exhibit less pronounced rheotaxis than do Ozark cavefish. The species' diet is dominated by midwater and substrate invertebrates (Poulson 1963). This species has experimentally demonstrated a preference for coarse substrates over fine, and for occupying the interstitial spaces within a cobble layer (Schubert & Noltie 1995). Based on allozyme data, the species exhibits greater between-population levels of genetic variability than does the Ozark cavefish, although a lesser number of populations of the latter were considered (Swofford et al. 1980).

Distribution

West of the Mississippi River, the southern cavefish is known (either presently or historically) from 31 sites (historic or active) occurring on the Salem Plateau in southcentral Missouri (28 sites, MDC Natural Heritage Inventory Database, unpublished data) and northcentral Arkansas (three sites, Woods & Inger 1957, Paige et al. 1981, Robison & Buchanan 1988). The Missouri and Arkansas sites are arrayed roughly in a ring, in that they occur clockwise in Missouri's Phelps, Dent, Carter, Wayne, Ripley, Oregon, Howell, Ozark, Dallas, Laclede, and Camden counties, whereas no occurrences are known from counties within the ring (Pulaski, Texas, Wright, Webster, and Douglas). The Arkansas sites occur on the southern margin of this ring in Randolph, Fulton, and Stone counties (Robison & Buchanan 1988). Importantly, this distribution pattern matches that of the distribution of Missouri's 'big' (= first magnitude) springs, these being limited to the periphery of the Salem Plateau (Vinyard & Feder 1982).

East of the Mississippi, populations occur from southern Indiana through Kentucky and Tennessee, to northern Alabama and northwestern Georgia (Lee et al. 1980). Because of this broad distribution, the species has been claimed as possibly being 'the most widely distributed troglobite in North America' (Barr & Holsinger (1985: 330). For reasons detailed below, we would not be surprised to find genetic and/or morphological differences between fish from different sides of the Mississippi. Indeed, Barr & Holsinger (1985: 330) suggest that *Typhlichthys subterraneus* may constitute 'a morphospecies composed of several local biological species'.

Site characterization

Site inspections indicate that most occupied locations have, as a characteristic feature, the presence of a permanently effluent spring, either one that discharges at the surface into a spring pool, or that contributes to a cave stream or the filling of a sink hole. Alternatively, pumped wells occasionally bring specimens to the surface (Smith 1980). There is no clear site type dichotomy, as is the case for Ozark cavefish sites (lateral vs. vertical; see above).

Among southern cavefish sites, the immersed substrates vary from exposed bedrock to clays, depending on the local erosive or depositional environment. Coarser substrates are generally found where the spring upwelling occurs, especially where the effluent flows are large and the spring throat expansive. The strata exposed above the baseflow water line often show evidence of extensive fracturing. For a number of these sites (e.g., Maramec Spring), fish are seen only when they have been 'flushed' or 'washed out' from below ground; they appear not to be long-term inhabitants of the spaces that are humanly accessible.

Geology of the Salem Plateau

The distribution of southern cavefish sites can best be understood by developing an appreciation of the geologic setting of the Salem Plateau.

[11] Vandike, J.E. 1982. The effects of the November 1981 liquid-fertilizer pipeline break on groundwater in Phelps County, Missouri. Water Resources Data and Research Report, Division of Geology and Land Survey, Missouri Department of Natural Resources. 27 pp.

Location and stratigraphic column. The Salem Plateau is a physiographic province in south-central Missouri and north-central Arkansas wherein the rocks of late Cambrian-Ordovician periods (440 to 530 millions years old) are exposed at the Earth's surface. The rocks predominantly exposed at the land surface are of the Jefferson City Dolomite, Roubidoux Formation, and Gasconade Dolomite[6,7] (Table 4) (Reams 1968, Thompson 1995, Unklesbay & Vineyard 1992, Fanning 1994, Adamski & Pugh 1996). Near the St. Francois Mountains, the rocks exposed at the land surface are of the Eminence Dolomite and Potosi Dolomite.

Vertically, the Jefferson City Dolomite, Roubidoux Formation, and Gasconade Dolomite overlie the Eminence and Potosi dolomites; beneath the Potosi Dolomite are the rocks of the Derby-Doe Run Dolomite and Davis Formation. As an analogy, this sequence of rocks can be thought of as another large layer cake (Figure 3). In this case, the bottom layer is the DerbyDoe Run Dolomite and Davis Formation; the middle layer is the rocks of the Eminence and Potosi dolomites, and the top layer is the Jefferson City Dolomite, Roubidoux Formation, and Gasconade Dolomite.

Across the Salem Plateau, in a few locations, the top layers have been eroded away, exposing the underlying Eminence and Potosi dolomites. Using our cake analogy, the plateau appears as if one had gouged out

Table 4. Stratigraphic sequence of the Salem Plateau region including a description of the hydrologic properties (modified from Feder 1973, Thompson 1995).

Age	Formation (thickness in m)	Lithologic description	Hydrologic characteristics	Regional hydrologic unit
Ordovician	Jefferson City Dolomite (60–90)	Dolomite, fine- to medium-grained, argillaceous, cherty	Yields highly variable, most less than 1.3×10^{-3} $m^3 s^{-1}$; a few small springs issue from this formation	Ozark Plateau aquifer
	Roubidoux Formation (40–60)	Dolomite, fine-grained cherty, quartz sandstone beds	Reliable aquifer, average yields about 1.3×10^{-3} $m^3 s^{-1}$ in central Ozarks; many small springs issue from this formation	Ozark Plateau aquifer
	Gasconade Dolomite (75–90)	Dolomite, fine- to coarse grained, cherty	Well yields are highly variable, between 0.7×10^{-3} and 6.3×10^{-3} $m^3 s^{-1}$; many large springs issue from this formation	Ozark Plateau aquifer
Cambrian	Eminence Dolomite (45–90)	Dolomite, medium- to coarse-grained, medium- to massively-bedded cherty	Well yields are highly variable, between 0.7×10^{-3} and 1.3×10^{-3} $m^3 s^{-1}$; many large springs issue from this formation, dry valleys in this formation	Ozark Plateau aquifer
	Potosi Dolomite (75–90)	Dolomite, fine- to medium-grained, massively bedded, contains abundant quartz druse, vuggy	Best aquifer in Ozarks; most well yields are between 1.3×10^{-2} and 3.2×10^{-3} $m^3 s^{-1}$,with yields over 3.2×10^{-2} $m^3 s^{-1}$ not uncommon	Ozark Plateau aquifer
	Derby-DoeRun Dolomite (30–60) and	Dolomite, fine-grained, argillaceous, silty	Acts as a confining unit throughout most of the Ozarks	St. Francois confining unit
	Davis Formation (40–70)	Shale, dolomite, thin bedded, contains edgewise conglomerate	Acts as a confining unit throughout most of the Ozarks	St. Francois confining unit
	Bonneterre Formation (60–140)	Dolomite, fine- to medium-grained, glauconitic in places, contains dark green to black, thin shale beds	May yield up to 6.3×10^{-3} $m^3 s^{-1}$ in eastern Ozarks; no large springs issue from this formation	Saint Francois aquifer
	Lamotte Sandstone (0–150)	Sandstone and conglomerate, quartzose, arkosic	May yield up to 1.6×10^{-3} $m^3 s^{-1}$ in eastern Ozarks; no large springs issue from this formation	Saint Francois aquifer
Precambrian	Crystalline rocks	Igneous and metamorphic rocks	Very low well yields	Basement crystalline rocks

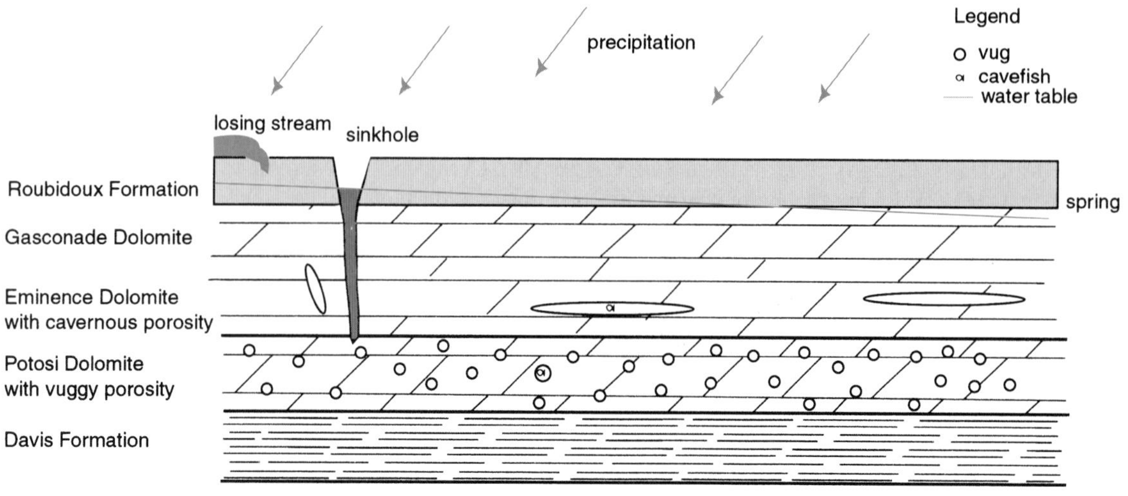

Figure 3. An illustration of the Salem plateau showing the geologic setting of the plateau, the result of hydrothermal alteration of the Potosi, the development of karst within the plateau, and the two types of cavefish sites – springs and erosion features. The illustration is not drawn to scale.

portions of the upper layer. Along the eastern edge of the plateau (near the St. François Mountains), the Jefferson City Dolomite, Roubidoux Formation, and Gasconade Dolomite have been completely eroded, exposing the underlying rock units. The regional dip (inclination of the beds) is modest (between 2° and 5°) and can be considered hydrologically negligible (effectively flat) over the length and breadth of the plateau.

Strata characterization. To more fully appreciate where southern cavefish dwell, each of the Salem Plateau's rock units needs to be described because the differences among the units are greater than the differences within any unit. All of these rock units were deposited or formed below or near sea level.

Starting at the bottom, the oldest rocks are the Davis Formation, a shale composed of very fine-grained detrital fragments of platy clay minerals that were deposited in deep, still water. The thickness of the Davis Formation ranges from 50 to 70 m (Thompson 1995). The pores between the platy clay minerals are microscopic due to the aligning of the clay minerals that occurred under the weight of the overlying geologic materials.

The sandwiched Potosi and Eminence Dolomites were deposited as limestones during the Cambrian Period. The Potosi Dolomite is a massively-bedded unit that contains chert, quartz, and substantial barite (economic deposits in some locations) (Leach 1979). The thickness of the Potosi Dolomite ranges from 75 to 90 m, and it thins laterally from its outcrop area (i.e., the layer is thicker towards the east and thinner towards the west). The Eminence Dolomite is a massively-bedded unit with quartz and chert (less chert than observed in the Potosi) and little to no barite. The average thickness of the Eminence Dolomite is 65 m (Thompson 1995).

The Jefferson City Dolomite, Roubidoux Formation, and Gasconade Dolomite were deposited conformably (deposition occurred continuously, with no evidence of erosion or sub-aerial exposure) on the top of the Eminence Dolomite. The 75 to 90 m-thick Gasconade Dolomite is a cherty dolomite (up to 50% chert by volume). The contact between the Gasconade Dolomite and the Eminence Dolomite is marked by the presence of the 10 m-thick Gunter Sandstone member which is included in the Gasconade Dolomite. Overlying the Gasconade Dolomite is the Roubidoux Formation, which ranges from 30 to 85 m thick. The Roubidoux Formation is a mix of dolomite, cherty dolomite, sandy dolomite, and sandstone. The uppermost unit is the Jefferson City Dolomite, which ranges from 60 to 100 m thick.

During the Pennsylvanian-Permian (250 to 320 million years ago) (Symons & Sangster 1991), a salty (14–22 percent by weight equivalent sodium chloride, NaCl) and hydrothermal (approximately 150°C) solution flowed through the rock units of interest in the Salem and Springfield plateaus (see Schmidt 1962, Roedder 1977, Leach 1979, Leach et al. 1975).

Compared to the Salem Plateau, the Springfield Plateau was farther away laterally and stratigraphically higher than the source of this hydrothermal fluid. Consequently, the hydrothermal alteration of the Salem Plateau rocks was much more dramatic (Gregg & Shelton 1989, Appold & Garven 1999).

Exposure to these warm brines at depth changed the calcite (specific gravity = 2.71 g cm^{-3}) in the limestones into the dolomite/dolostone (specific gravity= 2.85 g cm^{-3}) observed in the rocks today, and also deposited the barite that is commonly observed in the Potosi Dolomite (Leach 1979). Exposure to the warm brines also had the effect of dramatically increasing the porosity of the rocks, especially that of the deepest Potosi Dolomite. Following the period of hydrothermal alteration and dolomitization, the porosity of the Potosi more closely resembled that of Swiss cheese (i.e., larger pores, less uniformly distributed, but still common). (The barite is deposited within some, but not all of the pores or holes in the Swiss cheese.) However, this alteration of the porosity of the Salem Plateau's Potosi occurred via a fundamentally different process than that which rendered porous the rocks of the Springfield Plateau.

Karstification of the Salem Plateau. A timeline specifically for the karstification of the Salem Plateau has not been established. However, these rocks have been exposed since the end of the Paleozoic, and had been hydrothermally altered during the Paleozoic.

The fundamental processes involved in the development of karst on the Salem Plateau are the same as for the Springfield Plateau. Water enriched in carbon dioxide flowed from areas of high fluid potential to areas of low fluid potential following the path of least resistance. Within the Salem Plateau, this path of least resistance was likely the Potosi and Eminence dolomites that had been previously altered (opened up) by hydrothermal solutions. As the pathways enlarged, more carbon dioxide rich water flowed along them, enhancing the porosity even more.

Understanding how the cavernous porosity developed in the Salem Plateau and the sizes to which the conduits developed is important to gaining an appreciation of where the southern cavefish live. In the Salem Plateau, the end result of the processes of hydrothermal alteration and karstification is the sequence of rock units, the Eminence and Potosi dolomites, within which cavernous and vuggy porosity exists, respectively. Beneath these units is a thick impermeable Davis Formation which impedes movement of water out of the Potosi dolomite. Overlying these rocks, the sandy dolomites and sandstone of the Jefferson City Dolomite, Roubidoux Formation, and Gasconade Dolomite allow downward percolation of water, both as slow diffuse movement (autogenic recharge) and as discrete inputs (losing streams and sinkholes; allogenic recharge).

Given the five 'types' of caves (defined by their geomorphic characteristics, see Palmer 1991), and the geologic setting of the Salem Plateau (sandstone and overlying dolomite), the conduit patterns that developed in the Eminence dolomite are likely to be network (mazelike structures). In network systems, the overall size and number of the conduits is similar to that developed in branchwork systems. The difference between branchwork and network systems is that in network systems more than one flowpath may exist between any given input location and discharge point (usually a spring). The downgradient reach of these conduits may be effluent (a spring or a cave stream) if the water table is intersected, or a completely water-filled conduit when the water table is above the top of the conduit. It is in the latter that southern cavefish are observed.

To aid in visualizing the patterns of porosity development that occurs within the Eminence and Potosi dolomites of the Salem Plateau and to help the reader 'see' into the rock, two analogies will be drawn, one to the Eminence and one to the Potosi dolomites. The porosity of the Potosi dolomite is best thought of as being similar in size and shape to that observed in a very large block of Swiss cheese. Some pores are connected to other pores, and some pores are dead-ends. The pores range in diameter from millimeters to tens of centimeters. In the Eminence dolomite, the conduit development probably resembles a network pattern (similar to the gridwork of many city streets) or branchwork pattern. In the Salem Plateau, this pattern lies above (stratigraphically higher than) the Swiss cheese pattern. Humans can enter only the largest of the passages, whereas southern cavefish can negotiate all water-filled pores within the both the Eminence and Potosi dolomites.

Modern hydrologic setting of the Salem Plateau. Because southern cavefish dwell in groundwater, understanding the hydrology of the Salem Plateau is critical to understanding their ecology and patterns of occurrence. Hydrologically, the way in which water moves into and through the strata of the Salem Plateau is relatively straightforward (Figure 3).

The uppermost rocks, and to a lesser extent the Eminece and Potosi dolomites, are exposed at the land surface (beneath a thin soil cap). Precipitation falling onto the land surface can enter the rock units below and conduits and vugs therein in two ways, the first being via discrete recharge zones such as losing steams and sinkholes (allogenic recharge) and the second being via the diffuse movement of water through overlying sandstone and sandy dolomites (autogenic recharge). Both means deliver water, sediment, and microorganisms into and through the ground and thus to the springs that drain the Salem Plateau (Vineyard & Feder 1982). The Eleven Point River and the Current River in southern Missouri are illustrative of this relationship: the Eleven Point River is a losing river and the Current River is a gaining spring-fed river. The water in the Eleven Point River flows across the surface, and then down through losing reaches to enter the groundwater environment. The water moves continuously down the gradient of fluid potential and discharges back at the surface in the Current River.

Water table

In the Salem Plateau, the water table is relatively close to the land surface and is often within the horizon of the Roubidoux Formation or Jefferson City Dolomite (Vandike 1993). Consequently, the pore spaces in all the underlying formations are completely filled with water. The cavernous porosity of the Eminence Dolomite and vuggy porosity of the Potosi Dolomite are completely filled with water.

Recharge basins

Within the karstified and hydrothermally altered Salem Plateau, the groundwater basins are three-dimensional entities that include both the modern-day surface streams and rivers, and all the through-rock drainage that converges to a point of interest. The big springs that ring the Salem Plateau (Vineyard & Feder 1982) are notable features here. The typical size of the recharge basins on the Salem Plateau is hundreds of square kilometers (a thousand is not uncommon) and hundreds of meters in thickness/depth.[12] Base level in the region is defined by the Missouri River to the north and by the Arkansas River to the south. Most surface drainages radiate outwards from the Plateau center.[13]

When considering the movement of water and southern cavefish within the Salem Plateau, we have to separate the respective abilities of the cavefish and of water to move from place to place. Generally speaking, groundwater is constrained to remain within a groundwater basin and flows from regions of high fluid potential (upland areas of the Plateau) to regions of low fluid potential (springs). The horizon in which we find the cavernous and vuggy porosity (large enough for the southern cavefish to be in) is significantly below the water table. As such, the cavefish can swim wherever there are connected water-filled pores, even if the cavefish swim across a groundwater divide (an imaginary boundary between the two groundwater basins). As far as a cavefish is concerned, the Eminence and Potosi dolomites constitute a continuous and laterally extensive matrix of water-filled, connected pores; groundwater divides are a human construct.

Subterranean flow paths

Water below the water table within the Salem Plateau occurs in the pores of rocks. These pores range in size from microscopic spaces between individual sand grains within the Roubidoux Formation, to centimeter-sized spaces in the Potosi Dolomite, to large cavernous pores within the Eminence Dolomite. The vugs and cavernous pores are well-connected and most of the water that flows through the Salem Plateau flows through these two formations, as well yields indicate.[14] The pores within the other formations are small enough, for the most part, to exclude cavefish; the inhabitable units are thus effectively limited to the Eminence and Potosi dolomites.

Rates of subterranean water movement

Relying on an empirical relation between the coefficient of variation of hardness of water and of the temperature of the water as developed by Pitty (1968) and Ternan (1972), and used by Wicks (1997), we calculated the flow-through times of water in the Salem Plateau based on data collected by Vandike (1996). The flow-through time was 30 days (Figure 2). In addition,

[12] Imes, J.L. & L.F. Emmett. 1994. Geohydrology of the Ozark Plateau aquifer system in parts of Missouri, Arkansas, Oklahoma, and Kansas. USGS Professional Paper 1414-D. 127 pp.

[13] Femmer, S.R. 1997. Water-quality assessment of the Ozark Plateaus study unit, Arkansas, Kansas, Missouri, and Oklahoma habitat data and characteristics at selected sites, 1993–1995. U.S. Geological Survey Open-File Report 97–236. 44 pp.

[14] Robertson, C.E. 1963. Water well yield map of Missouri. Groundwater Maps of Missouri. State of Missouri. 1 p.

Dreiss (1989) calculated the mean residence time of the water in several spring basins during storm events based on the method of moments, and found the mean residence times to be 16 to 30 days, which agrees with the findings of Wicks & Bohn (2000). These flow-through times for water in the Salem Plateau recharge basins are longer than those of the Springfield Plateau. However, they are still considered hydrologically fast.

Geologic perspectives on where southern cavefish have been found

The distribution pattern of southern cavefish occurrence locations matches that of the distribution of Missouri's big springs, i.e. the periphery of the Salem Plateau (Vineyard & Feder 1982). Within this geologic context, the presence and absence of cavefish sites can be understood.

Southern cavefish are found at sites of two main types: first, in the discharge pools of the Salem Plateau's big springs, and second, where substantial erosion has cut far down into the Eminence Formation and Potosi Dolomite, these strata also being the sources of water for the big springs (Unklesbay & Vineyard 1982, Thompson 1995, Vandike 1996) (Figure 3). Similarly to the Ozark cavefish, our ability to locate southern cavefish is similarly limited by our ability to 'see' through the rock and into the unit(s) where the cavefish reside. The only opportunity humans are afforded to view southern cavefish is in the springs that drain these formations, and at erosional features sufficiently deep that they actually penetrate the formations. The absence of large springs or deep erosional features in the central portion of the Salem Plateau precludes seeing southern cavefish there; however, this in no way implies that southern cavefish are not present there, below the land's surface. While the Potosi Dolomite and Eminence Dolomite contain pores of sufficient size and inter-connectedness to provide habitat for southern cavefish, the lack of pores of sufficient sizes to allow humans access to the species' underground world artificially limits our appreciation of where they reside: we know them only from where they are 'delivered' to us for viewing.

Synthesis: biology meets geology

The following paragraphs represent our attempt to address some common questions and/or perceptions concerning the ecology of Missouri's Ozark and southern cavefishes. In so doing, we use evidence garnered by the geological sciences to propose explanations for the between-species differences that biologists have observed.

The depths at which they dwell

Ozark cavefish

The stratigraphy of the Springfield Plateau in general and the widespread shales in particular, confine Ozark cavefish to areas near the surface. The karstic layer lies at or immediately adjacent to the land surface (or, in portions of McDonald county, has been completely eroded away), and is underlain by a shale layer above which the fish are confined. Thus, the fish's occurrence is by necessity restricted to this stratigraphic unit (Figure 1). The occurrence of visible tree roots penetrating the interior of a number of Ozark cavefish sites reinforces perceptions of proximity to the surface. There are several consequences of this near-surface occurrence.

First, infiltration of the underground by surface waters will tend to be rapid within the recharge basins in which Ozark cavefish dwell (three to four days, Wicks 1997). Major rains falling on saturated soils can result in sufficiently rapid infiltration that the contributed water fails to thermally and chemically equilibrate prior to reaching the habitat occupied by the fish. Consequently, the physicochemical properties of the medium are more variable for the Ozark cavefish than for the southern cavefish. It should be noted that the popularly held notion of the environmental constancy of groundwater is inaccurate – although dampened compared to surface waters, fluctuations in groundwater temperature and other attributes are appreciable. Coupled with this is a greater potential for drought-related lowering of the water table, resulting in greater vertical (and/or lateral) reductions in the proportion of the stratum in which fish can reside.

Second, proximity to the surface, the presence of losing streams, and the permeability of the strata together facilitate the influx of surface-derived allochtonous materials into the subsurface. The subterranean food webs within which Ozark cavefish function are thus apt to be fueled by comparatively greater amounts and larger particles of detrital material, because it has not been filtered out in the process of its downward passage through the overlying rock layers and soil horizons. The potential thus exists for these subterranean food webs to be comparatively populous and/or diverse. While such inputs might normally be perceived as advantageous, these conditions also render Ozark cavefish habitats susceptible to potentially deleterious inputs (e.g., chemical residues or spills, urban or agricultural runoff, silt). However, with rapid influx comes the potential for rapid outflow.

Third, the closer the cavefish habitat is to the surface, the greater the risk for 'physical' impacts to Ozark cavefish habitats. Quarrying, well drilling, and road/highway construction all have the potential to directly impinge on the water-filled rock matrix within which the Ozark cavefish dwells.

Fourth, it is geologically interesting to contemplate the western margin of the Springfield Plateau. Here, the strata comprising the plateau are below layers of Pennsylvanian rocks, isolating the Ozark cavefish-containing layers from the surface and from dynamic freshwater recharge. While karstic strata may well occur at depths west of the Springfield Plateau, it seems less likely that Ozark cavefish will occur therein because such locations would require lateral inputs of allochtonous material and fluids. In addition, the deep groundwater west of the Springfield Plateau is brackish,[12] indicating that significant freshwater recharge is not occurring.

Southern cavefish

The geology of the karstic strata of the Salem Plateau suggests that southern cavefish must typically reside at appreciable depths, on the order of hundreds of meters below the land surface (Vandike 1996, Crunkilton 1985). Coring and stratigraphic studies indicate that the Potosi Dolomite and Eminence Dolomite can lie some 175 to 240 m below the land surface across the lateral extent of the Salem Plateau (Table 4), and these are the only strata that have been rendered sufficiently porous (vuggy and/or cavernous) that fish would be capable of negotiating their way through the rock matrix. In addition, below this layer lies a shale layer (Davis Formation) above which the fish are confined.

Additional evidence of the deep occurrence of the southern cavefish is derived from the storied Maramec fertilizer pipeline break. In November 1981, a pipeline carrying liquid ammonium fertilizer (a mixture of nitrate and ammonium compounds) developed a significant leak, releasing an estimated 71 000 l of nitrogen rich fertilizer into a losing stream 21 km from Maramec Spring. This resulted in a severe, yet short-term, contamination of the spring water (Vandike 1985, Crunkilton 1985, Weaver 1994, Wicks & Hoke 2000). Less than two weeks after the break, the dissolved oxygen concentration of the water discharging from the spring dropped to near zero and stayed near zero for several days (the contaminant traveled 21 km in 12 days; 1.75 km day^{-1} or 2 cm s^{-1}). During this period, many hundreds of dead or dying southern cavefish were observed boiling up from the spring into the spring pool. We infer that the hundreds of cavefish that emerged or floated out of the spring throat had an origin that was relatively close to the spring and not near the location of the pipeline break, as these fish do not swim as quickly and/or continuously as the contaminant plume migrated through the basin (2 cm s^{-1}). Thus the plume could have overtaken many more hundreds of cavefish along the path, leaving them dying and decaying within the aquifer. Only those cavefish near the spring were likely to have been flushed alive from the aquifer in advance of the contaminant plume, owing to the tight, convoluted flow paths therein. Therefore, we believe that the estimate of hundreds of dead cavefish probably greatly underestimated the number of fish actually killed by the plume.

Subsequently, Vandike (1996) reported that the Maramec Spring throat extends downward at approximately a 6° angle for at least 58 m vertically and approximately 520 m horizontally (the spring throat at Bennett Spring is estimated to have a similar geometry and extent, Vandike 1992); the true extent of the spring throat passages of both systems have not been detailed, because cave divers have not completed their explorations). The lessons learned here are that the conduit passages extend for at least 21 km, that movement of water through the karstic basin is very rapid, and that the spring throat from which the cavefish emerged has its origin deep within the rock units.

Together, these circumstances allow us to infer that southern cavefish are likely living at depth beneath the Salem Plateau within either the Eminence or Potosi dolomites. In consideration of this inference, the specific name of the southern cavefish, *subterraneus*, seems especially fitting. Also consistent is the observation by Schubert et al. (1993) of 'gas bubble disease' in southern cavefish collected at depth from a Missouri spring site (the fish equivalent of the 'decompression sickness' or the 'bends' that caisson workers and SCUBA divers experience, respectively). Such symptoms would be unlikely to occur in specimens restricted to near-surface waters. Considerations by other authors lend further credence to this inference. Woods & Inger (1957), Poulson (1958), and Armstrong & Williams (1971) contemplated sub-water-table dispersal (see Moneymaker & Rhoades 1945). Bretz

(1953) recognized that the water draining from Welch Spring Cave (a southern cavefish site) derives from strata 150 m below the surface. Poulson & White (1969) suggest that *Typhlichthys* occupies areas remote from cave entrances where less food is available. Mohr & Poulson (1966) describe the discovery of southern cavefish from deep subterranean channels beneath a riverbed during a drilling operation.

The consequences of this deep subterranean occurrence are several. First, the depth to which rainfall must penetrate the subsurface to recharge the strata of the Salem Plateau suggests that it will have achieved a greater degree of thermal and chemical equilibrium before it encounters the fish than is the case for the Springfield Plateau. As an example, flow-through times of 16 to 30 days are expected for the Salem Plateau where southern cavefish occur (Dreiss 1989), as opposed to the three to seven days measured for Springfield Plateau systems where Ozark cavefish occur (Wicks 1997). Consequently, the physico-chemical environment of the southern cavefish is likely less variable than that of the Ozark cavefish. However, because large losing streams and sinkholes on the Salem Plateau quickly deliver great quantities of water, suspended sediment, nutrients, and bacteria down to the Potosi Dolomite, and because even 30 days is a hydrologically short period of time, water in the Salem Plateau aquifer rarely reaches complete chemical or thermal equilibrium with the bedrock, as is evidenced by specific conductance data (Vandike 1996). Should chemical or biological pollutants be introduced into losing streams on the Salem Plateau, they could be rapidly carried deep into the underground. Larger spills into smaller losing streams would likely yield greater below-ground concentrations than would the dilution of smaller spills into larger streams. Greater below-ground dissipation might also be expected within a larger groundwater basin for which the flow-through times are greater.

Second, distance from the surface may tend to impede the influx of large, surface-derived allochtonous particulate materials into the subsurface (except those delivered via losing streams), although the percolation of dissolved nutrients and microbial loads would be relatively unaffected. The food webs within which southern cavefish occur thus may be sustained by lesser quantities and smaller particles of detrital material because of the greater potential for filtration along the flowpath. The potential thus exists for these subterranean food webs to be somewhat less enriched, both in regard to species richness and population densities.

Third, southern cavefish will be relatively immune to modifications to the near-surface geology. Road building usually only 'scratches the surface', and deep-pit limestone quarrying is uncommon in Missouri because surface quarries abound. However, timbering at the land surface might cause siltation of the losing rivers and degradation of the fish's habitat. In addition, concern for the southern cavefish arises when one considers the potential impact of the lead mining that is common in this part of the state. These mines extract ore from the underlying Bonneterre Dolomite (Table 4), a potential consequence of which is degradation of surface and sub-surface environments.

Fourth, the substantial vertical and lateral extent of the Potosi Dolomite and Eminence Dolomite occupied by southern cavefish suggests that these fish have a relatively large and comparatively continuous volume of habitat in which to dwell and through which to disperse.

Characterization of cavefish habitat

Ozark cavefish

From the work of Means (1993), Means & Johnson (1995), and Brown (1996), we know that marked Ozark cavefish will frequently disappear from the cave stream from which they were captured, and may re-appear later in another part of the stream, or disappear from further view. Outwashing from the cave remains one possible explanation; however, one would expect to observe such displaced individuals, given the number of specimens involved, and to see some association with spate/freshet occurrences. Instead, a more plausible scenario is that the fish are 'retreating' either downwards into the interstitial spaces within the substrate matrix lining the cave floor, or laterally into water-filled fractures too small for humans to enter.

Based on the geology of the karst strata in which Ozark cavefish occur, we can conceptualize their habitat as follows. The basins in which the fish reside are relatively small, covering only tens of square kilometers of land surface.[8] The fish's penetration of the rock would be primarily through fractures that have been enlarged via dissolutional processes. The fracture network would exhibit a degree of 'blockiness' (not necessarily right angle bends, although many orthogonal examples exist) in predominantly two dimensions; movement through the fractures would involve a series of right and left movements. Conduit-type passageways might be encountered periodically; fish could choose to occupy such locations for periods of time (especially if food resources are plentiful therein), or could negotiate their way across or along (= traverse) them to the other side. Consequently, fish seen in such conduits that are accessible by humans will be viewed as being in habitat in which they truly reside and not, as has been suggested by some, as locations to which they have been accidentally displaced.

Southern cavefish

The geology of the Salem Plateau dictates that the southern cavefish's habitat be appreciably different than that of the Ozark cavefish. The stratum occupied by the former has a 'Swiss cheese' texture to it; it is riddled with voids of varying sizes, the connections between which are sometimes sufficiently large that fish can pass through them. However, functional 'dead ends' or 'squeezes' are not uncommon, and the resulting three-dimensional network of passable 'tubes' is convoluted and tortuous.

Although southern cavefish clearly seem to be living at depth, our human perception of their habitat has been shaded (and misconstrued) by the nature of the sites in which we encounter the species near the surface. These sites are generally characterized by the presence of a continuously effluent spring that supplies either a surface outlet, a cave stream, or a sinkhole/karst window (Figure 3). Consequently, in situ observations have been restricted to these near-surface localities. However, the preponderance of evidence from the mapping of spring throats by cave divers, from inadvertent fertilizer spills, and from an understanding of the location of fish-sized pores suggests that the fish reside at depth within the groundwater basins. The data from the spill within the Maramec Spring basin suggest instead that each of these springs (sites) represents the 'terminus' of a hydrogeologic system comprising a large surface-recharge area (approximately 1000 km^2; Vandike 1992, 1996, Wicks & Hoke 2000), a deep and laterally extensive karst aquifer wherein the fish reside, and effluent springs conduct water (and fish) from depth to surface or near-surface outlet points.

Consequently, the sites at which we see southern cavefish are not representative of the habitat in which most individuals occur; instead, they represent atypical habitats into which fish have been flushed/washed/carried or transported. Fish washed out of the aquifer may thus be inclined to return to depth. This is supported by the observation of a cave diver who, when returning southern cavefish to Maramec Spring after the fertilizer spill, stated that 'I let the fish go; it swam around for a bit, orienting itself, then it dove into the spring throat' (J. diPardo personal communication). Typically, fish flushed from the aquifer cannot return due to the complexity of the flow path, or to the high velocity of the water discharging from the spring throat against which they must swim. Thus, the fish seen in such near-surface locations are likely (a) a small portion of those populating the hydrogeologic system in toto, and (b) individuals which might be expected to make minimal impacts on the population as a whole (either through reproduction or via consumption of scarce food resources). As a result, monitored numerical fluctuations at such sites (periodic censuses) may be but tenuously related to numerical fluctuations in the population proper. Additionally, while protecting these sites is of obvious merit, it must be conceded that these localities are but a fraction of the total volume of habitat comprising these hydrogeologic systems. Consequently, southern cavefish systems in Missouri are analogous to those in Texas' Edwards Aquifer wherein the widemouth blindcat, *Satan eurystomus*, and the toothless blindcat, *Trogloglanis pattersoni*, occur (Hubbs & Bailey 1947), although the Texas systems are apparently chemosynthetic[1] (Grossman et al. 1995).

Laboratory experiments concerning southern cavefish habitat use has produced results consistent with the perceptions presented above. Southern cavefish demonstrate a clear preference for occupying the interstitial spaces between the elements of a matrix of large, cobble-sized particles (Schubert & Noltie 1995). This same study provides evidence of rheotaxis and preference for the deeper portions of the water column. The species exhibits a clear propensity to explore previously unfamiliar habitat, and will readily enter and traverse water-filled tubes mimicking those that might occur underground (Hapner & Noltie unpublished data).

Genetics and evolution

Ozark cavefish

Work regarding the genetic relatedness of Ozark cavefish populations (Bergstrom 1997) provides evidence of four differentiable genetic 'groupings' of fish associated respectively with four major Springfield Plateau surface drainage basins: (1) northwest Arkansas' Illinois River (drains west into the Arkansas), (2) southwest Missouri's White River (drains northwards from Arkansas into Missouri and then back south through Arkansas into the Mississippi), (3) southwest Missouri's arm of the Neosho River (drains west into Oklahoma's Neosho), and (4) northeast Oklahoma's Neosho River mainstem (drains south into the Arkansas). Additional genetic differentiability was evident on a site-to-site basis. Together, these data are indicative of there being restricted gene flow between even adjacent sites, and that the effects of such restrictions become more apparent as one moves from the within-river drainage to the between-river drainage scale. Consequently, we can conceive of these populations as being true populations or demes, and not just as 'windows' on a single, large, panmictic population that inhabits the complete breadth of the aquifers of the Springfield Plateau. For these discrete populations to attain the significant levels of endemicity that they display requires that fish movement between sites be minimal, and that impediments to such movement exist between the below-ground drainage basins in which the sites occur.

Southern cavefish

Less work has been completed on the genetics of the southern cavefish, due in part to perceptions that the species' status is less precarious than that of the Ozark cavefish. However, Bergstrom's (1997) work examined fish from five Missouri and one Alabama sites, and found relative to the Ozark cavefish (1) that there was more similarity among fish across sites, and (2) that some sites contained fish that were genetically very different. One way to reconcile these differences is to conceive of the karst wherein southern cavefish dwell as being laterally extensive and continuous, with few dry or faulted barriers to dispersal between areas. Recharge basins exist, but they are of large surface area and are contiguous. Consequently, fish can remain within the confines of individual recharge basins and attain a level of endemicity (i.e., develop initially unique genotypes). However, opportunities exist for fish to move between recharge basins, allowing new and different genotypes to become established therein. This would explain, for example, why Bergstrom (1997) found fish with substantially different genotypes within individual sites.

Amblyopsid phylogeny

Bergstrom's (1997) phylogeny of the family Amblyopsidae hypothesizes that the lineages containing the Ozark and southern cavefishes, respectively, diverged about eight million years before the present (mybp); Bergstrom (1997) also provides an historical biogeographic scenario consistent with this and subsequent 'splits' within the respective lineages. Geologically, a time has not been fixed for when Missouri karst had developed sufficiently to become inhabitable by cavefish. However, the timing of the first lineage splitting within the *Amblyopsis rosae* and *Typhlichthys subterraneus* clades (Bergstrom op. cit.) could be used to suggest possible minima: karstification of the Springfield and Salem plateaus, respectively, would have had to precede invasion of the underground by fish and the subsequent reproductive isolation reflected by the lineage splitting. The prediction for the Salem Plateau would likely be conservative, since invasion of the underground by the *T. subterraneus* clade may have been delayed by the depths at which the karstified strata occur (100 to 200 m below the land surface). Erosion rates for the Springfield and Salem plateaus (stable platforms with well-integrated drainage) would typically range from one to ten cm per 1000 years (Bloom 1998). Downcutting to depths of 100 to 200 m would thus require between 100 000 and 20 million years, with five million years being an intermediate value.

Behavior

Based on the notions above, one might surmise that Ozark cavefish populations might be denser and energetically richer, but smaller spatially and in total, whereas southern cavefish populations might be of lesser densities and more energetically impoverished, but larger spatially and in total. In addition, the Ozark cavefish may tend to move through the underground mainly using a network of relatively large interconnected fissures of the shallow aquifers underlying the Springfield Plateau. In contrast, southern cavefish instead utilize the haphazard array of interconnected pore spaces that permeate the deeply-occurring formations beneath the Salem Plateau. Ozark cavefish thus may have more frequent opportunities to meet and interact with conspecifics, whereas southern cavefish may rarely meet and interact with conspecifics because they occur at lower population densities and in a more honeycombed habitat. This infrequency of encounter may explain why southern cavefish display more behavior types, more behavioral diversity, more energetic territorial charges, and longer bouts during conspecific interactions than do Ozark cavefish (Poulson 1985).

Ecology

If we accept the notion of southern cavefish being more 'subterranean', then one must wonder whether their biology is similarly attuned. Southern cavefish have more neuromasts but smaller olfactory rosettes than do Ozark cavefish (Poulson 1985); in their 'swiss cheese' world, the ability to negotiate and sense prey and current movements may be more important than the ability to smell the presence of prey. Southern cavefish are more adroit at capturing benthic prey, which is more apt to dwell at depth; Ozark cavefish are more adept at capturing midwater prey, which is more likely to occur in relatively superficial waters (Poulson 1985). The heads of southern cavefish are narrower than are those of Ozark cavefish (Poulson 1985); this would facilitate swimming through the complex network of pores deep beneath the Salem Plateau. When escaping a potential threat, southern cavefish exhibit more side-to-side turning than do Ozark cavefish (Poulson 1985), which would be required in their habitat. Reproductively, southern cavefish females make more and bigger eggs that contain more energy and consequently cost the female more to produce and care for than is the case for Ozark cavefish (Poulson 1985): putting more effort into reproduction may be necessary for *Typhlichthys* to guarantee the survivorship of sufficient progeny.

Implications

We have attempted to articulate our shared perception of the nature of the subterranean habitats wherein Ozark and southern cavefish dwell in Missouri. In so doing, we have also examined how aspects of the species' ecologies 'fit' with these two contrasting hydrogeologic scenarios and the differing constraints that they respectively impose on the species. Apart from honing our perceptions of where these fishes dwell, our interpretations raise some interesting issues, challenges, and implications for amblyopsid conservation and management.

First, although the biological community has come to an understanding of the necessity of managing these species on a recharge-area basis and scale, we now see that the scale of such efforts must differ between the species: southern cavefish recharge areas exceed those of Ozark cavefish typically by one or more orders of magnitude. It may be feasible to apply 'best management practices' to a significant proportion of the land surface within an Ozark cavefish recharge area; in contrast, doing so on what might approach a county scale for southern cavefish is a daunting, but no less critical, challenge. The need to delineate these recharge areas, and the high hazard and vulnerability areas within them, is crucial; even though the fish may dwell below the land's surface, it is human activity on the surface and its consequent impact on the subsurface that will determine the fate of these species.

Second, given the biological and geologic evidence that Ozark cavefish populations are smaller in area and in total numbers, and are more isolated than are those of the southern cavefish, we must now contend with the ramifications of this population structuring. The genetic evidence (Bergstrom 1997) suggests a degree of cave-to-cave and inter-watershed endemism, such that each population may constitute an Evolutionary Significant Unit (ESU; sensu Waples 1991, 1995) meriting status under the Endangered Species Act. The isolation responsible for these genetic differences suggests that the extirpation of an Ozark cavefish population is likely irreversible; in contrast, the lack of dissection of the southern cavefish's habitat leads one to suspect that a localized extirpation may be more easily repopulated, given time and correction of the circumstances causing the extirpation. In this respect, Ozark cavefish populations are much more susceptible to environmental mishaps such as chemical spills, although such effects would tend to be more localized. For southern cavefish, the effects would be more widespread, but might be less catastrophic due to the potential for post-remediation repopulation.

Third, there have been substantial investments of time and effort made in attempting to enumerate and monitor through time the numbers of Ozark cavefish at particular locations[5] (Means 1993, Means & Johnson 1995, Brown 1996). Our understanding of how the hydrogeology of the Springfield Plateau recharge basins tends to localize these populations suggests that such censing efforts, although challenging, may ultimately prove fruitful. In contrast, the expansive and contiguous nature of the recharge basins of the Salem Plateau suggests that efforts to census, monitor, and infer population trends for southern cavefish will be much more challenging and less likely to yield accurate counts using current methodologies.

Fourth, some research has suggested that southern cavefish might serve as an ecological surrogate for the Ozark cavefish (e.g., Schubert & Noltie 1995). Future research will need to be predicated upon an appreciation of the substantial differences in the subterranean habitats of the two species (that is, southern cavefish occur at depth in honeycombed environments; Ozark cavefish are more surficial and occur in conduit, bedding plane, and fracture openings), and in the hydrogeologic characteristics of the contrasting groundwater basins (Salem Plateau = large and deep; Springfield Plateau = small and shallow).

Fifth, southern cavefish have been designated as being less 'cave-adapted' than are Ozark cavefish (Poulson 1960, 1963, 1964, 1985, Culver 1982), based on comparisons of the species' morphological, physiological, ecological, and behavioral traits. However, because southern cavefish appear to be far more 'subterranean' in their occurrence, one is prompted to question whether past interpretive assessments of the relative merits of their traits adequately portray the degree of troglobitism.

Sixth, we have confined our considerations to Missouri amblyopsids. However, troglobitic confamilials occur over much of the east-central United States (Lee et al. 1980), and it would be interesting to examine in greater detail the geological and hydrological context in which populations of these other species occur. In particular, a presumably disjunct group of southern cavefish populations occurs east of the Mississippi River in Alabama, Georgia, Indiana, Kentucky, and Tennessee (Lee et al. 1980). Bergstrom (1997) considered a single Alabama specimen in his study of amblyopsid phylogeny, and it proved to be markedly different than the multiple Missouri *Typhlichthys* specimens he considered. This raises questions about when this more easterly karst was inhabitable by cavefish, how deeply buried the inhabitable stratum is, and how discrete and of what size the recharge basins are, among other questions.

Seventh, biologists express concern regarding the population impacts of the introduction of epigean species into cave streams, the closure of cave mouths, human disturbance of cave streams due to spelunker traffic, and human removals of individual cavefish[15] While these 'peripheral' impacts are by no means negligible, it is our sense that their importance pales in comparison with impacts that occur at the recharge-area scale. When faced with time and resource allocation decisions, managers would be well advised to focus their attention on threats to groundwater quality, the through-ground infiltration of allochtonous nutrients and silt, and the physical integrity of the karst itself. Although cavefish physically dwell below ground, the land surface is an integral and inseparable component of their habitat. It is to these ecosystem-level changes that cavefishes are potentially most susceptible.

[15] Missouri Department of Conservation. 1999. An action plan for the Ozark cavefish (*Amblyopsis rosae*) in Missouri. Missouri Department of Conservation, Jefferson City. 26 pp.

Conclusion

We wish to make it clear that while this paper has focused on the Ozark and southern cavefish, attempts to understand, protect, and manage these species will also benefit the other members of the subterranean communities within which they occur. These communities contain other stygobites that are themselves rare (e.g., Koppelman & Figg 1995); in all likelihood, taxa also occur that have yet to be examined or described scientifically[16] . Although the cavefish have benefitted from protected status, this in and of itself does little to perpetuate the species. By fostering an enhanced understanding of these enigmatic fishes and their habitat, we hope to facilitate future conservation efforts.

Acknowledgements

The authors thank Tom Aley, Toby and Katy Dogwiler, Jim Johnson, Jim Vandike, Jon Van Brahana for criticizing the manuscript. Funding for this work was provided to D.B.N. by the Missouri Agricultural Experiment Station, the Missouri Cooperative Fish and Wildlife Research Unit, and The School of Natural Resources at the University of Missouri, and to C.M.W. by the National Science Foundation and the University of Missouri Research Board. This publication is a contribution from the Missouri Agricultural Experiment Station (Journal Series 13, 131) and from the Missouri Cooperative Fish and Wildlife Research Unit (the U.S. Geological Survey, the Missouri Department of Conservation, The School of Natural Resources of the University of Missouri-Columbia, and the Wildlife Management Institute cooperating).

[16] Missouri Department of Conservation. 2000. A new species of cave crayfish. Missouri Conservationist 61: 16.

References cited

Adamski, J.C. & A.L. Pugh. 1996. Occurrence of pesticides in ground water of the Ozark Plateaus province. Water Res. Bull. 32: 97–105.

Aley, T. 1978. A predictive hydrologic model for evaluating the effects of land use and management on the quantity and quality of water from Ozark springs. Missouri Speleol. 18: 1–185.

Appold, M.S. & G. Garven. 1999. The hydrology of ore formation in southeast Missouri district; numerical models of topography-driven fluid flow during the Ouachita Orogeny. Econ. Geol. 94: 913–935.

Armstrong, J.G. & J.D. Williams. 1971. Cave and spring fishes of the southern bend of the Tennessee River. J. Tenn. Acad. Sci. 46: 107–115.

Baker, C.D. 1972. The cephalic lateral line systems of amblyopsid blind cavefishes and other percopsiform fishes (Pisces: Percopsiformes). Ph.D. Dissertation, University of Louisville, Louisville. 113 pp.

Balon, E.K. 1975. Reproductive guilds of fishes: a proposal and definition. J. Fish. Res. Board. Can. 32: 821–864.

Barr, T.C., Jr. & J.R. Holsinger. 1985. Speciation in cave faunas. Ann. Rev. Ecol. Syst. 16: 313–337.

Bechler, D.L. 1980. The evolution of agonistic behavior in amblyopsid fishes. Ph.D. Dissertation, St. Louis University, St. Louis. 160 pp.

Bechler, D.L. 1981. Agonistic behavior in the Amblyopsidae, the spring, cave, and swamp fishes. Proc. Int. Cong. Speleol. 8: 68–69.

Bechler, D.L. 1983. The evolution of agonistic behavior in amblyopsid fishes. Behav. Ecol. Sociobiol. 12: 35–42.

Bergstrom, D.E. 1997. The phylogeny and historical biogeography of Missouri's *Amblyopsis rosae* (Ozark cavefish) and *Typhlichthys subterraneus* (southern cavefish). M.S. Thesis, University of Missouri, Columbia. 63 pp.

Bloom, A.L. 1998. Geomorphology: a systematic analysis of late cenozoic landforms. 3rd edition. Prentice Hall, Upper Saddle River. 482 pp.

Boyd, G.L. 1997. Metabolic rates and life history of aquatic organisms inhabiting Logan Cave stream in northwest Arkansas. M.S. Thesis, University of Arkansas, Fayetteville. 107 pp.

Breder, C.M. Jr. & D.E. Rosen. 1996. Modes of reproduction in fishes. Natural History Press, Garden City. 941 pp.

Bretz, J.H. 1953. Genetic relations of caves to peneplains and big springs in the Ozarks. Amer. J. Sci. 251: 1–24.

Brown, A.V. & C.S. Todd. 1987. Status review of the threatened Ozark cavefish (*Amblyopsis rosae*). Proc. Ark. Acad. Sci. 41: 99–100.

Brown, Z.J. 1996. Population dynamics and growth of Ozark cavefish in Logan Cave National Wildlife Refuge, Benton County, Arkansas. M.S. Thesis, University of Arkansas, Fayetteville. 105 pp.

Charlton, H.H. 1933. The optic tectum and its related fiber tracts in blind fishes. A. *Troglichthys rosae* and *Typhlichthys eigenmanni*. J. Comp. Neurol. 57: 285–325.

Cooper, J.E. & R.E. Kuehne. 1974. *Speoplatyrhinus poulsoni*, a new genus and species of subterranean fish from Alabama. Copeia 1974: 486–493.

Crunkilton, R. 1985. Subterranean contamination of Maramec Spring by ammonium nitrate and urea fertilizer and its implication on rare cave biota. Missouri Speleol. 25: 151–158.

Culver, D.C. 1982. Cave life: evolution and ecology. Harvard University Press, Cambridge. 189 pp.

Dreiss, S.J. 1989. Regional scale transport in a karst aquifer: 1. Linear systems and time moment analysis. Water Res. Res. 25: 126–141.

Eigenmann, C.H. 1898. A new blind fish. Proc. Indiana Acad. Sci. 1897: 231.

Eigenmann, C.H. 1899a. The blind fishes of North America. Pop. Sci. Month. 56: 473–486.

Eigenmann, C.H. 1899b. The blind fishes. Biol. Bull. 8: 113–126.

Eigenmann, C.H. 1899c. A case of convergence. Science 9: 280–282.

Eigenmann, C.H. 1899d. The eyes of the blind vertebrates of North America. I. The eyes of the Amblyopsidae. Arch. Entwick. Org. 8: 545–617.

Eigenmann, C.H. 1909. Cave vertebrates of America: a study in degenerative evolution. Carnegie Institution of Washington Publication (104): 1–241.

Fanning, B.J. 1994. Geospeleologic analysis of cave and karst development within the Boone and St. Joe formations of Benton and Madison counties, northwest Arkansas. M.S. Thesis, University of Arkansas, Fayetteville. 144 pp.

Feder, G.L. 1973. A conceptual model of the hydrologic system supplying the large springs in the Ozarks. Ph.D. Dissertation, University of Missouri, Columbia. 149 pp.

Ford, D.C. & P.W. Williams. 1989. Karst geomorphology and hydrology. Unwin Hyman Ltd., London. 601 pp.

Ford, D.C., S.-E. Lauritzen & R. Ewers. 2000. Hardware and software modeling of initial conduit development in karst rocks. pp. 175–183. *In:* A.B. Klimchouk, D.C. Ford, A.N. Palmer & W. Dreybrodt (ed.) Speleogenesis: Evolution of Karst Aquifers, National Speleological Society, Huntsville.

Garman, S. 1889. Cave animals from southwestern Missouri. Bull. Mus. Comp. Zool. 17: 232.

Gregg, J.M. & K.L. Shelton. 1989. Minor- and trace-element distributions in the Bonneterre Dolomite (Cambrian), southeast Missouri: evidence for possible multiple-basin fluid sources and pathways during lead–zinc mineralization. Bull. Geol. Soc. Amer. 101: 221–230.

Grossman, E.L., J.L. Schlichenmeyr, K.T. Clemence, G.A. Ulrich, J.M. Suflita, D. Martino & J.W. Ammermon. 1995. Microbial sulfur cycling in a shallow aquifer system. EOS, Trans. Amer. Geophys. Union 76: F221.

Hargrove, G. 1968. Relation of a chert zone to development of Devils Icebox, Boone County, Missouri. Missouri Speleol. 10: 15–20.

Hubbs, C.L. & R.M. Bailey. 1947. Blind catfishes from artesian waters of Texas. Occas. Papers Mus. Zool. Univ. Mich. 499: 1–15.

Koppelman, J.H. & D.E. Figg. 1995. Genetic estimates of variability and relatedness for conservation of an Ozark cave crayfish species complex. Conserv. Biol. 9: 1288–1294.

Langecker, T.G. & G. Longley. 1993a. Morphological adaptations of the Texas blind catfishes *Trogloglanis pattersoni* and *Satan eurystomus* (Siluriformes: Ictaluridae) to their underground environment. Copeia 1993: 976–986.

Langecker, T.G. & G. Longley. 1993b. Blind catfish (*Trogloglanis pattersoni* and *Satan eurystomus*) from deep artesian waters: a study on convergent adaptation to cave and deep sea biota. pp. 239–255. *In:* J.H. Schroder, J. Bauer & M. Schartl (ed.) Trends in Ichthyology: An International Perspective, GSF Forschungszentrum für Umwelt und Gesundheit GmbH, in association with Blackwell Scientific Publications, Oxford.

Leach, D.L. 1979. Temperature and salinity of the fluids responsible for minor occurrences of sphalerite in the Ozark region of Missouri. Econ. Geol. 74: 931–937.

Leach, D.L., R.C. Nelson & D. Williams. 1975. Fluid inclusion studies in the northern Arkansas zinc district. Econ. Geol. 70: 1084–1091.

Lee, D.S., C.R. Gilbert, C.H. Hocutt, R.E. Jenkins, D.E. McAllister & J.R. Stauffer, Jr. 1980. Atlas of North American freshwater fishes. North Carolina State Museum of Natural History, Raleigh. 854 pp.

Mayden, R.L. (ed.). 1992. Systematics, historical ecology, and North American freshwater fishes. Stanford University Press, Stanford. 969 pp.

Mayden, R.L. & F.B. Cross. 1983. Re-evaluation of Oklahoma records of the southern cavefish, *Typhlichthys subterraneus* Girard (Amblyopsidae). Southwest. Nat. 28: 471–473.

McDonald, E.F. Jr. & W.L. Pflieger. 1979. The spring cavefish, *Chologaster agassizi* (Pisces: Amblyopsidae) in southeastern Missouri. Amer. Midl. Nat. 102: 194–196.

Means, M.L. 1993. Population dynamics and movement of Ozark cavefish in Logan Cave NWR, Benton County, Arkansas with additional baseline water quality information. M.S. Thesis, University of Arkansas, Fayetteville. 126 pp.

Means, M.L. & J.E. Johnson. 1995. Movement of threatened Ozark cavefish in Logan Cave National Wildlife Refuge, Arkansas. Southwest. Nat. 40: 308–313.

Miller, R.J. & H.W. Robison. 1973. The fishes of Oklahoma. Oklahoma State University Press, Stillwater. 246 pp.

Mohr, C.E. & T.L. Poulson. 1966. The life of the cave. McGraw Hill Book Company, New York. 232 pp.

Moneymaker, B.C. & R.F. Rhoades. 1945. Deep solution channel in western Kentucky. Bull. Geol. Soc. Amer. 56: 39–44.

Moyle, P.B. & J.J. Cech, Jr. 2000. Fishes – an introduction to ichthyology, 4th ed. Prentice Hall, Upper Saddle River. 612 pp.

Nelson, J.S. 1989/90. Analysis of the multiple occurrence of pelvic fin absence in extant fishes. Matsya (India) 15/16: 21–38.

Packard, A.S. Jr. & P.W. Putnam. 1872. The Mammoth Cave and its inhabitants, or descriptions of the fishes, insects and crustaceans found in the cave: with figures of the various species, and an account of allied forms, comprising notes upon their structure, development and habits, with remarks upon subterranean life in general. Naturalists' Agency, Salem Press, Salem. 62 pp.

Paige, K.N., C.R. Tumlinson & V.R. McDaniel. 1981. A second record of *Typhlichthys subterraneus* (Pisces: Amblyopsidae) from Arkansas. Southwest. Nat. 26: 67–92.

Palmer, A.N. 1991. Origin and morphology of limestone caves. Bull. Geol. Soc. America 103: 1–21.

Pitty, A.F. 1968. Some features of calcium hardness fluctuations in two karst streams and their possible value in geohydrological studies. J. Hydrol. 6: 202–208.

Poulson, T.L. 1958. Cave fishes (Amblyopsidae). The Bloomington Indiana Grotto Newsletter 1: 29–37.

Poulson, T.L. 1960. Cave adaptation in amblyopsid fishes. Ph.D. Dissertation, University of Michigan, Ann Arbor. 185 pp.

Poulson, T.L. 1963. Cave adaptation in amblyopsid fishes. Amer. Midl. Nat. 70: 257–290.

Poulson, T.L. 1964. Animals in aquatic environments: animals in caves. pp. 749–771. *In:* D.B. Dill, E.F. Adolph & C.G. Wilber (ed.) Handbook of Physiology: A Critical, Comprehensive Presentation of Physiological Knowledge and Concepts, Section 4: Adaptation to the Environment, American Physiological Society, Washington, D.C.

Poulson, T.L. 1985. Evolutionary reduction by neutral mutations: plausibility arguments and data from amblyopsid fishes and linyphiid spiders. Nat. Speleol. Soc. Bull. 47: 109–117.

Poulson, T.L. & W.B. White. 1969. The cave environment. Science 165: 971–981.

Reams, M.W. 1968. Cave sediments and the geomorphic history of the Ozarks. Ph.D. Dissertation, Washington University, St. Louis. 113 pp.

Robison, H.W. & T.M. Buchanan 1988. Fishes of Arkansas. The University of Arkansas Press, Fayetteville. 536 pp.

Roedder, E. 1977. Fluid inclusion studies of the ore deposits of the Viburnum Trend, southeast Missouri. Econ. Geol. 72: 474–479.

Romero, A. 1998a. Threatened fishes of the world: *Typhlichthys subterraneus* (Girard, 1860) (Amblyopsidae). Env. Biol. Fish. 53: 74.

Romero, A. 1998b. Threatened fishes of the world: *Amblyopsis rosae* (Eigenmann, 1898) (Amblyopsidae). Env. Biol. Fish. 52: 434.

Schmidt, K. 1995. Endangered, threatened, and special status fishes of North America, 4th ed. American Currents, special issue. North American Native Fishes Association. 77 pp.

Schmidt, R.A. 1962. Temperatures of mineral formation in the Miami-Pitcher district as indicated by liquid inclusions. Econ. Geol. 57: 1–20.

Schubert, A.L.S., C.D. Nielsen & D.B. Noltie. 1993. Habitat use and gas bubble disease in southern cavefish (*Typhlichthys subterraneus*). Int. J. Speleol. 22: 131–143.

Schubert, A.L.S. & D.B. Noltie. 1995. Laboratory studies of substrate and microhabitat selection in the southern cavefish (*Typhlichthys subterraneus* Girard). Ecol. Freshw. Fish 4: 141–151.

Smith, V.J. 1980. Some aspects of the life history of the southern cavefish (*Typhlichthys subterraneus* Girard) in Missouri. M.S. Thesis, University of Missouri, Columbia. 123 pp.

Suttkus, R.D. 1961. Additional information about blind catfishes from Texas. Southwest. Nat. 6: 55–64.

Swofford, D.L., B.A. Branson & G.A. Sievert. 1980. Genetic differentiation of cavefish populations. Isozyme Bull. 13: 109–110.

Symons, D.T.A. & D.F. Sangster. 1991. Paleomagnetic age of the central Missouri barite deposits and its genetic significance. Econ. Geol. 86: 1–12.

Tafanelli, R. & J.E. Russell. 1972. A extension of the range of the blind cavefish *Amblyopsis rosae* (Eigenmann). Southwest. Nat. 17: 310.

Ternan, J.L. 1972. Comments on the use of a calcium hardness variability index in the study of carbonate aquifers, with reference to the central Pennines, England. J. Hydrol. 16: 317–321.

Thines, G. 1969. L'Evolution regressive des poissons cavernicoles et abyssaux. Masson et. Cie., Paris. 394 pp.

Thompson, T.L. 1995. The stratigraphic succession in Missouri. Missouri Department of Natural Resources Division of Geology and Land Survey, volume 40 (2nd series) revised, Columbia. 188 pp.

Unklesbay, A.G. & J.D. Vineyard. 1992. Missouri geology: three billions years of volcanoes, seas, sediments, and erosion. University of Missouri Press, Columbia. 189 pp.

Vandike, J.E. 1985. Hydrogeologic aspects of the November 1981 liquid fertilizer pipeline break on groundwater in the Maramec Spring recharge area, Phelps County, Missouri. Missouri Speleol. 25: 93–101.

Vandike, J.E. 1992. The hydrogeology of the Bennett Spring area, Laclede, Dallas, Webster, and Wright counties, Missouri. Missouri Department of Natural Resources, Water Resources Report 38: 1–111.

Vandike, J.E. 1993. Groundwater level data for Missouri water year 1991–1992. Missouri Department of Natural Resources, Division of Geology and Land Survey, Water Resources Report 42: 1–95.

Vandike, J.E. 1996. The hydrology of Maramec Spring. Missouri Department of Natural Resources, Water Resources Report 55: 1–104.

Vineyard, J.D. & G.L. Feder. 1982. Springs of Missouri. Missouri Geological Survey and Water Resources, Water Resources Report 29: 1–267.

Waples, R.S. 1991. Pacific salmon, *Oncorhynchus* spp., and the definition of 'species' under the Endangered Species Act. Mar. Fish. Rev. 53: 11–22.

Waples, R.S. 1995. Evolutionary significant units and the conservation of biological diversity Species Act. pp. 8–27. *In:* J.L. Nielsen & D.A. Powers (ed.) Evolution and the Aquatic Ecosystem, American Fisheries Society, Bethesda.

Weaver, H.D. 1994. Highway and pipeline corridors: risk zones for Missouri caves. Missouri Cave and Karst Conserv. Dig. July: 7–12.

Wicks, C.M. 1997. Origins of groundwater in a fluviokarst basin: Bonne Femme basin in central Missouri, USA. Hydrogeol. J. 5: 89–96.

Wicks, C.M. & B. Bohn. 2000. Application of unit hydrograph technique to the discharge record at Big Spring, Carter County, Missouri. pp. 31–41. *In:* I.D. Sasowsky & C.M. Wicks (ed.) Groundwater Flow and Contaminant Transport in Carbonate Aquifers, Balkeman, Rotterdam.

Wicks, C.M. & J.A. Hoke. 2000. Prediction of the quality and quantity of Maramec Spring water. Ground Water 38: 218–225.

Willis, L.D. & A.V. Brown. 1985. Distribution and habitat requirements of the Ozark cavefish, *Amblyopsis rosae*. Amer. Midl. Nat. 114: 311–317.

Woods, L.P. & R.F. Inger. 1957. The cave, spring, and swamp fishes of the family Amblyopsidae of central and eastern United States. Amer. Midl. Nat. 58: 232–256.

Environmental Biology of Fishes **62:** 195–200, 2001.
© 2001 *Kluwer Academic Publishers. Printed in the Netherlands.*

Habitat and population data of troglobitic armored cave catfish, *Ancistrus cryptophthalmus* Reis, 1987, from central Brazil (Siluriformes: Loricariidae)

Eleonora Trajano
Departamento de Zoologia, Instituto de Biociências da USP, C. P. 11461, 05422-970, São Paulo SP, Brazil (e-mail: etrajano@usp.br)

Received 19 April 2000 Accepted 14 January 2001

Key words: Angélica Cave, capture-mark-recapture census, karst, Passa Três Cave, population ecology

Synopsis

Population ecology of the troglobitic armored catfish, *A. cryptophthalmus*, from caves in São Domingos karst area of central Brazil, was studied in a two-census capture-mark-recapture study carried out during the dry season of 1999. Like many loricariids, including epigean *Ancistrus* catfishes from that area, *A. cryptophthalmus* is a typical bottom-dweller with a strong preference for rocky substrates in fast-flowing stream sectors. Relatively high population densities, when compared to other cavefishes, were recorded in the studied cave sections: 0.9 individuals m^{-2} in Angélica Cave and 0.5 individuals m^{-2} in Passa Três Cave. Passa Três Stream is a small tributary of the São Vicente Cave system, and the estimated population for Passa Três Cave (ca. 2000 m long) is around 1000 individuals. On the other hand, the large Angélica Cave Stream (~8 km long) harbors a very large population, conservatively estimated at 20 000 individuals. No fish movements of distances longer than 100 m were recorded in the 40-day period. In Passa Três Cave, a few movements of distances up to 60 m were detected. Proportion of mature females in Angélica Cave was higher than in Passa Três Cave, possibly reflecting a greater food availability in the first cave. Mean individual growth rates in Angélica Cave (1.35 mm $month^{-1}$ for size class <41 mm SL, 0.17 mm $month^{-1}$ for size class >41 mm SL) point to relatively high longevities, possibly attaining 15 years.

Introduction

Brazil has a remarkably rich subterranean ichthyofauna, composed of several troglobitic species (restricted to the subterranean habitat, usually showing reduction of eyes, pigmentation, and other troglomorphisms) occurring in karst areas throughout the country. Currently, more than twelve Brazilian subterranean troglomorphic fish species have been found, including several undescribed taxa (Trajano 1997a, unpublished data). Most are siluriforms, belonging to the taxa 'Pimelodidae' Heptapterinae (*Phreatobius cisternarum*, *Pimelodella kronei*, *Taunayia* sp., and an undescribed taxon, probably a new genus and species, from NE Brazil, besides possibly troglomorphic *Rhamdiopsis* sp. and *Pimelodella* sp.), *Trichomycterus* (Trichomycteridae) (*T. itacarambiensis* and three, possibly four undescribed species), and *Ancistrus* (Loricariidae) (*A. cryptophthalmus* and *A. formoso*).

A great deal of variation, without a phylogenetic/taxonomic correlation, is observed in the degree of reduction of eyes, pigmentation, and other troglomorphisms characterizing these fishes. Some species are totally anophthalmic and depigmented (e.g., *A. formoso*, *Taunayia* sp., the new heptapterine genus), others present variably reduced eyes and pigmentation (e.g., *A. cryptophthalmus*, *T. itacarambiensis*, *P. kronei*). The fish distribution and habitat also vary: some species are restricted to one or a few caves within relatively small areas, others are widely distributed, and they may be found in typically phreatic habitats (directly accessible to humans or not) or in open channel streams. Likewise, population parameters, such as population density and size, greatly differ among species, from rare to moderately abundant or very abundant populations (Trajano 2001). This rich and diverse ichthyofauna offers good opportunities for studies focusing on different biological aspects. However, various methods, from relatively simple to very specialized ones (as those involving SCUBA diving, or water pumping in non-cave habitats), are required to deal with each of these species.

Stream-dwelling cave fishes, such as *P. kronei* and *T. itacarambiensis*, and those found in pools and lakes inside accessible caves, as is the case with *Taunayia* sp. and the new heptapterine genus, are relatively easy to sample and study. These fishes have already been studied with a focus on ecology, behavior and other aspects (e.g., Trajano 1991a,b, 1997b,c, Mendes 1995, Trajano & Bockmann 2000). With the recent development of speleodiving in Brazil, species living in flooded caves (*A. formoso*, *Trichomycterus* sp. from SW Brazil) are currently under study.

At present, we are investigating three troglomorphic fish species which are sympatric in cave-stream habitats from São Domingos karst area, Tocantins River basin, central Brazil: the armored catfish, *A. cryptophthalmus*, an undescribed species of *Trichomycterus*, and *E. vicentespelaea* (M. E. Bichuette personal communication).

A. cryptophthalmus was described based on a sample collected in the 1970s in Angélica, Bezerra and Passa Três caves (Reis 1987). Cavers working in the area described the armored catfishes as very abundant in these caves. In 1988, we visited Passa Três Cave and obtained a large sample, part of which was used for behavioral study; on this occasion, we gathered some ecological data (Trajano & Souza 1994). In view of its abundance, distribution and easy collection, *A. cryptophthalmus* was an excellent candidate for population studies based on capture-mark-recapture techniques. However, like other species living in highly seasonal habitats, subject to flash floods during the rainy period (e.g., *T. itacarambiensis*), its study had to be restricted to the dry season.

Presented herein are the results of a two-census mark-recapture study on *A. cryptophthalmus*, carried out during the dry season of 1999, focusing on population sizes and densities, habitat use and movements, and growth rates.

Study site

São Domingos karst area, part of the Bambuí Speleological Province (Trajano & Sanchez 1994) is situated in the Cerrado domain (savannah-like vegetation), characterized by a tropical semi-humid climate, with 4–5 dry months (Nimer 1989). The studied caves are located within the limits of the Terra Ronca State Park, São Domingos County, Goiás State.

In spite of our efforts and those of other speleologists surveying the main stream caves in the area, *A. cryptophthalmus* catfish have been found in only three caves from two different cave systems (conduits extending continuously between the input and output points of a karst rock, Ford & Williams 1989), located in the same limestone outcrop: Angélica and Bezerra caves, which form the Angélica–Bezerra system, and Passa Três Cave, a tributary conduit of São Vicente I Cave, that belongs to the São Vicente system. These streams, with epigean and hypogean reaches, run parallel westwards, contributing to form the Paranã River, one of the main tributaries of the Tocantins River.

The Passa Três Stream is a small tributary of the São Vicente system (13° 25′ S, 46° 22′ W). After an epigean course, it sinks into the Passa Três conduit, more than 2000 long (stream length about 2000 m; mean width around 1.0 m). Approximately 100 m distant from the cave entrance, there is a major waterfall (~5 m high during the dry season); upstream and downstream from this waterfall, the stream is nearly horizontal, with shallow riffles, showing moderate water current over a rocky and graveled bottom, alternating with some deep, soft-bottomed pools. A very low, but short underwater passage separates the first 400 m from the remaining 1500 m long reach, that joins the large São Vicente Stream inside São Vicente I Cave. The population study was carried out in the 400 m long reach upstream from the underwater connection.

The Angélica is a large stream, of the same stream order (sensu Mattheus 1998) as the São Vicente Stream. The large entrance of Angélica Cave (13° 21′ S, 46° 23′ W) is the sinkhole of the Angélica Stream. The cave has 13 800 m of mapped passageways; the stream conduit is about 8000 m long, average width is 5 m and depth varies from 0.5 to 2+ m. In most of its extension, it is a fast-flowing stream with strong current, running over rocky substrate (blocks, gravel and sand). Due to the cave size and large number of catfishes found throughout it, our study was limited to sections in the first 900 m downstream from the sinkhole (main entrance).

According to our own observations and local information, the subterranean water level, specially in large streams such as the Angélica–Bezerra and São Vicente streams, is generally little affected during dry seasons. Even in Passa Três Cave, the water level slightly decreased through the studied period, which was particularly dry, significantly affecting the epigean drainage in the area. Therefore, carrying capacity of these streams and, consequently, energy input as organic matter carried into caves, has little seasonal variation.

Materials and methods

There is some evidence that the populations from Passa Três (the type-locality) and Angélica–Bezerra caves are at least partially isolated from each other because they have a mosaic of morphological and behavioral differences, and may correspond to separate subterranean species. For instance, the degree of eye reduction is higher in the Passa Três fish, but the reduction of pigmentation is more accentuated in the Angélica fish. As well, cryptobiotic (hiding) habits are reduced in the former (cf. Trajano & Souza 1994), but not in the latter. Therefore, these populations have been treated separately.

A two-census mark-recapture study was carried out in the three caves within a time interval of approximately 40 days during the dry season of 1999 (July and September). Due to the occurrence of flash floods, it is not safe to extend the population study to the rainy period in this karst area. The 400 m long stream reach studied in Passa Três Cave was divided into 20 m long sections. All catfishes captured in the same section were handled together. The studied reach of Angélica Cave was divided into sections about 100 m long. In five of these sections, we hand-netted and marked all fish we could capture. In one of these sections, about 600 m far from the entrance, a 10 m^2 area with rapids showing high density of fish was selected for a localized survey; the individuals captured and marked at this particular site were treated separately. In Bezerra Cave, armored catfishes were captured in two 50 m long sections.

To individually mark the armored catfishes, we used 'photonics marking', from New West Co., combining different colors and position of marks for individual recognition. For this, the specimens were anesthetized in MS-222 or benzocaine solution, measured, weighed, examined for individual natural marks and gonad development (seen by transparency), marked and released in their collection sections. Due to technical difficulties, only specimens larger than 20 mm SL were marked; catfishes smaller than that were measured, weighed, and released. A few specimens were killed by anesthesia overdose, and preserved for study of morphology, molecular biology, and analysis of diet.

Population sizes were estimated by Petersen estimator (= Lincoln index) (Begon 1979, Blower et al. 1981), except for Bezerra Cave, where no catfish were recaptured. This method, which is applied for closed populations, is adequate in the present case in view of the short interval between captures (35 days on average) and the apparent stability of the populations, with evidences of negligible mortality and migration within this interval (high recapture rates, long survival and low growth rates in the laboratory).

Results

Like loricariids in general, *A. cryptophthalmus* is a typical bottom-dweller. Both cave and eyed armored catfishes, *Ancistrus* sp., from epigean streams in the area, showed clear preferences for rocky substrates (from pebbles with a minimum diameter of 10 cm to boulders and rocky walls) in fast-flowing waters. Rarely, *A. cryptophthalmus* specimens have been observed in low current, soft bottomed pools. In the cave habitat, many armored catfishes are seen exposed on the upper surface of rocks. However, in the laboratory, specimens from Angélica Cave exhibit cryptobiotic habits, hiding most of time under limestone blocks, much as epigean *Ancistrus* sp. from the same region. Armored catfishes in Bezerra Cave are particularly elusive, avoiding most collection attempts, which may explain the relatively low number of fish captured. On the other hand, the specimens from Passa Três Cave have reduced hiding habits, as previously reported (Trajano & Souza 1994). The population data for *A. cryptophthalmus*, relative to specimens more than 20 mm SL, studied in July and September 1999 at Angélica, Bezerra and Passa Três Caves, are shown in Table 1.

High recapture rates (44%) were obtained in Passa Três Cave, pointing to low population size and/or low rate of movements. A stock of about 200 individuals was estimated for the 400 m long studied reach. The average stream width is about 1 m, thus population

Table 1. Number of specimens of troglobitic armored catfishes, *Ancistrus cryptophthalmus*, captured and released (r) in July 1999, and total number of specimens captured (n) in September 1999, including those previously marked (m), in three caves from São Domingos karst area, Central Brazil. N = population size (in total number of individuals), SE = standard error, estimated by Petersen method (notation according Begon 1979).

	Passa Três Cave	Angélica Cave	Bezerra Cave
Captured	106	181	23
Released (r)	101	173	23
Captured (n)	84	181	34
Marked (m)	44	14	0
N	192.8	2236.6	—
SE	19.3	502.7	—

density in the studied section is 0.5 individuals m^{-2}. The total population in Passa Três Cave (about 2000 m) is estimated as 1000 individuals.

The results relative to the Angélica Cave population presented in Table 1 exclude data from the selected 10 m^2 area with rapids, where the recapture rate was exceptionally high (r = 50, n = 34, m = 19, N = 89.5, SE = 12.5). Excluding this area, a low recapture rate (8%) was obtained in the cave, probably reflecting a large population size: a population size of approximately 2200 individuals was estimated for 500 m of the Angélica Cave stream. The average stream width is 5 m, thus a population density of 0.9 individuals m^{-2} was recorded in this reach. Interestingly, the localized survey in that selected area also points to a density of 0.9 individuals m^{-2}. This roughly corresponds to that which was observed during visual searching. According to local cavers, the population density does not visibly decrease downstream. Even considering an error of 50% in this visual estimate, with population dropping to near half that value (0.5 individuals m^{-2}), the population size of *A. cryptophthalmus* in Angélica Cave is exceptionally high for a cave fish, conservatively estimated at 20 000 or more individuals. Considering that the species also occurs at Bezerra Cave, more than 3.5 km long, although in lower densities (it was not possible to get a more precise figure due to the absence of recaptures), the total population size in this cave system may reach tens of thousands armored catfishes.

Average population densities of eyed *Ancistrus* catfishes found in epigean reaches of streams passing through caves inhabited by *A. cryptophthalmus* are up to 50% higher than those shown by this troglobite. Densities of 3 individuals m^{-2} are common in the epigean streams, and maximum densities of 10 individuals m^{-2} have been recorded, which are close to densities observed for the small section with rapids selected for study in Angélica Cave.

The capture-mark-recapture study corroborates the notion that *Ancistrus* catfishes, like loricariids in general, are sedentary. In Angélica Cave, no catfish was recaptured in a section different from that of the previous capture, i.e., no movement of distances greater than 100 m were recorded in the 40-day interval. Site fidelity is illustrated by the high recapture rates (38%) at the 10 m^2 selected area in Angélica Cave. In Passa Três Cave, where sections were shorter (20 m), seven out of 44 recaptures were found in a section different from that of the previous capture. Five among these seven recaptures were made at sections contiguous to those of the first capture, corresponding to movements of maximum distances of 40 m, and two were made at non-contiguous sections, corresponding to maximum movements of 60 m, within 40 days.

Among 231 and 215 specimens captured in Angélica Cave in July and September, respectively, 17 and 20 were probably females with developed gonads (five specimens in July and eight in September were certainly females with clearly visible oocytes; in the other cases, some doubts persisted). Assuming a sex ratio of 1 : 1, this would correspond to relatively high percentages of probably reproductive females during the study period in Angélica Cave: 15% in July and 18% in September. On the other hand, only one out of 106 armored catfishes captured in Passa Três Cave in July and three out of 84 captured in September were in a reproductive state. None of the specimens examined in Bezerra Cave had developed gonads.

The smallest female probably having developed gonads was about 41 mm SL; the smallest one that was certainly mature was 44 mm SL. For the Angélica Cave population, mean growth rate for size class <41 mm SL (N = 6) was 1.35 mm $month^{-1}$, and for size class >41 mm SL (N = 27), it was 0.17 mm $month^{-1}$; mean growth rate for size class <44 mm (N = 17) was 0.54 mm $month^{-1}$. In general, the subterranean water level is not much affected during the dry season, and food does not seem to be limiting during this period. Therefore, much higher growth rates are not expected for the rainy season (that, in other situations, would represent a period of greater energy input and higher growth rates, Trajano 1997b). The maximum size recorded was 78 mm SL for a specimen from Passa Três Cave, but this seemed to be an exceptional case. Five other individuals attained 70 mm, which seems to be the average maximum size for the troglobitic armored catfishes. Based on the growth rates reported for Angélica Cave, maximum longevity is estimated to be 15 years or more. Laboratory data are consistent with this figure: one specimen (>40 mm SL when collected) from Passa Três lived for nine years in captivity (Trajano unpublished data).

Discussion

Habitat preferences in *A. cryptophthalmus* do not differ from those exhibited by epigean relatives, *Ancistrus* sp., living in the same area. The latter are found in the epigean reach just upstream the Angélica Cave, and there is no geographic barrier between the epigean and cave armored catfishes at the present day. On the other hand, no epigean *Ancistrus* was found in the epigean reach of Passa Três Stream upstream of the cave. This suggests that the latter is more isolated than the Angélica population.

Population densities reported for *A. cryptophthalmus* (which do not greatly differ from the observed for the epigean relatives) may be considered high when compared to other cave fishes (Trajano 2001). *A. cryptophthalmus* probably feeds on the thin film of detritus covering rocks inside caves. The origin of this fine-grained organic matter is the great amount of detritus carried into the caves by the sinking streams, which flow throughout the year—large underwater deposits of vegetal debris were observed at several points in the caves, especially in pools. This high and continuous food availability, especially in Angélica Cave, probably explains the high population densities found for the species. The large habitat extension in the Angélica–Bezerra Cave system, associated with the high population density, result in a particularly numerous cave population. The low proportion of reproductive females and lower population densities (although still relatively high) in Passa Três Cave indicate that this population is more food-limited than the one from Angélica Cave.

A high degree of site fidelity is expected for small-sized detritivorous bottom-dwellers, especially when food is not limiting, as seems to be the case with the studied populations. Relatively low growth rates pointing to high longevities, are suggestive of a precocial life style, probably related to environmental stability (e.g., Balon 1988, 1990).

Acknowledgements

I am greatly indebted to Maria Elina Bichuette, Edmundo P. D. da Costa Jr., Alberto Akama, Marcelo Britto, Danilo Allegrini, Flávio Lima, Dirk Möller, Yvonne Meyer-Lucht, Jens Poschadel and Jakob Parzefall, for their valuable field assistance, and also to Ramiro Hilário dos Santos, local guide and tireless protector of the caves. I thank the Fundação Estadual do Meio Ambiente de Goiás (FEMAGO), for permission to develop the study in Terra Ronca State Park, allowing the use of the park facilities, and especially to Regina B. Schulz, a local park official. This study was supported by the Fundação de Amparo à Pesquisa do Estado de São Paulo (FAPESP grant n. 99/0376-1) and by the Conselho Nacional de Desenvolvimento Científico e Tecnológico (CNPq fellowship n. 306066/88-2). Permission for collection was given by IBAMA.

References cited

Balon, E. 1988. Tao of life: universality of dichotomy in biology. 1. The mystic awareness. Rivista di Biologia/Biology Forum 81: 185–230.

Balon, E. 1990. Epigenesis of an epigeneticist: the development of some alternative concepts on the early ontogeny and evolution of fishes. Guelph Ichthyol. Rev. 1: 1–42.

Begon, M. 1979. Estimating animal abundance, Edward Arnold, London. 97 pp.

Blower, J.G., L.M. Cook & J.A. Bishop. 1981. Estimating the size of animal populations. George Allen & Unwin, London. 128 pp.

Ford, D.C. & P.W. Williams. 1989. Karst geomorphology and hydrology. Unwin Hyman, London. 601 pp.

Mattheus, W.J. 1998. Patterns in freshwater fish ecology. Chapman & Hall, New York. 756 pp.

Mendes, L.F. 1995. Observations on the ecology and behavior of a new species of troglobitic catfish from northeastern Brazil (Siluriformes, Pimelodidae). Mém. Biospéol. 22: 99–101.

Nimer, E. 1989. Climatologia do Brasil, IBGE, Rio de Janeiro. 421 pp.

Reis, R.E. 1987. *Ancistrus cryptophthalmus* sp. n., a blind mailed catfish from the Tocantins River basin, Brazil (Pisces, Siluriformes, Loricariidae). Rev. fr. Aquariol. 14: 81–84.

Trajano, E. 1991a. Populational ecology of *Pimelodella kronei*, troglobitic catfish from southeastern Brazil (Siluriformes, Pimelodidae). Env. Biol. Fish. 30: 407–421.

Trajano, E. 1991b. Agonistic behavior of *Pimelodella kronei,* a troglobitic catfish from southeastern Brazil (Siluriformes, Pimelodidae). Behav. Processes 23: 113–124.

Trajano, E. 1997a. Synopsis of Brazilian troglomorphic fishes. Mém. Biospéol. 24: 119–126.

Trajano, E. 1997b. Population ecology of *Trichomycterus itacarambiensis*, a cave catfish from eastern Brazil (Siluriformes, Trichomycteridae). Env. Biol. Fish. 50: 357–369.

Trajano, E. 1997c. Food and reproduction of *Trichomycterus itacarambiensis*, cave catfish from south-eastern Brazil. J. Fish Biol. 51: 53–63.

Trajano, E. 2001. Ecology of subterranean fishes: an overview. Env. Biol. Fish. 62: 00–00 (this volume).

Trajano, E. & F.A. Bockmann. 2000. Ecology and behavior of a new cave catfish of the genus *Taunayia* from northeastern Brazil (Siluriformes, Heptapterinae). Ichthyol. Explor. Freshwaters 11: 207–216.

Trajano, E. & L.E Sanchez. 1994. Brésil. pp. 527–540. *In*: C. Juberthie & V. Decu (ed.) Encyclopaedia Biospeologica, Société de Biospéologie, Moulis.

Trajano, E. & A.M. Souza. 1994. The behavior of *Ancistrus cryptophthalmus*, an armored blind catfish from caves of Central Brazil, with notes on syntopic *Trichomycterus* sp. (Siluriformes, Loricariidae, Trichomycteridae). Mém. Biospéol. 21: 151–159.

Environmental Biology of Fishes **62:** 201–213, 2001.
© 2001 *Kluwer Academic Publishers. Printed in the Netherlands.*

The conservation status of hypogean fishes

Graham S. Proudlove
Department of Zoology, The Manchester Museum, Manchester University, Manchester, M13 9PL, U.K.
(e-mail: g.proudlove@umist.ac.uk)

Received 24 March 2000   Accepted 9 January 2001

Key words: cave fishes, subterranean habitat, threats, Red List, captive breeding

Synopsis

Hypogean fishes are susceptible to five main threats: habitat degradation, hydrological manipulations, environmental pollution, overexploitation, and introduced alien species. Because they are endemic to small areas and have small populations, any threat could have serious consequences. For these reasons most populations have been considered 'threatened' or 'vulnerable'. International Union for the Conservation of Nature, and Natural Resources has produced five Red Lists of threatened animals since 1977 and have listed seven, eight, 31, 31 and 63 hypogean fishes, respectively, in some threat category. Before 1994 the categories were qualitative and workers could assess the same species in different ways. In 1994, a new set of quantitative categories were published. The 1996 list was the first to use these categories and three hypogean species were listed critically endangered, two endangered, 46 vulnerable and 12 data deficient. Some of these categories are incorrect or inappropriate and most of the data deficient ones can now be categorised. All 85 known species are categorised here and reasons given for their placements. The current list has three critically endangered, none endangered, 73 vulnerable, four near threatened, one least concern and four data deficient. Some species are relatively well known, others are poorly known. Some are receiving conservation attention, most receive none. A few captive populations have been established, some of which are breeding. Few populations could be restocked if extirpated. Knowing where to direct conservation efforts is a problem since morphological 'species' are proving to be from more than one epigean invasion and therefore represent more than one species. Hypogean fishes merit protection but they may not receive more than cursory protection.

Introduction

This contribution (1) examines the types of threat to which hypogean fishes are subjected, (2) reviews the history of attempts by the International Union for the Conservation of Nature and Natural Resources (IUCN) to assess their conservation status, and (3) examines the level of knowledge we have of them and of conservation measures being taken. Two different schemes have been used by IUCN to summarise conservation status. The old scheme (e.g. IUCN 1993), used from 1977 to 1994, was not particularly objective and different threat criteria were applied by different workers. The new scheme (IUCN 1994) relies on measurements of relevant parameters to assess risk levels. The last list (IUCN 1996) had 63 hypogean species. Proudlove[1] lists 85 described species of hypogean fishes, all of which are assessed here.

Threat types and some examples

Bruton (1995a) lists eight broad types of threat that fishes can face but three are not relevant here. Those that affect hypogean fishes are listed below.

[1] Proudlove, G.S. 1997. A synopsis of the hypogean fishes of the world. pp. 351–354. *In:* Proc. 12th Int. Congr. Spel., La Chaux de Fonds, Switzerland.

Habitat degradation

Examples of habitat degradation include rock removal (e.g. quarrying) and alteration (e.g. siltation due to dams or logging). Wilson (1996) reports that deforestation in the area upstream of Ankarana, Madagascar may threaten *Glossogobius ankaranensis* by increasing the level of catastrophic scouring floods in some areas and siltation in others.

Hydrological manipulations

These can include water removal for human consumption or irrigation and damming. *Satan eurystomus*, *Trogloglanis pattersoni*, *Prietella phreatophila*, *Uegitglanis zammaranoi*, *Barbopsis devecchii*, *Phreatichthys andruzzii* and *Clarias cavernicola* inhabit sites from which water is removed for human use. The first two live in deep artesian waters that may be retained for up to 10000 years (Winograd & Robertson 1982). Groundwater pumping may impact *P. phreatophila* (Hendrickson personal communication). *Clarias cavernicola* has been classed as endangered since 1977 and critically endangered since 1996. The water level has fallen 20 m in the subterranean habitat of this species followed by the near-exposure of underwater ledges that appear to be the main feeding area. If the level falls further, the feeding ability of this species may be impaired (Skelton 1990, Bruton 1995b). Two species from Australia, *Milyeringa veritas* and *Ophisternon candidum*, live in an area where 'the water table is declining and becoming more saline' (Humphreys & Adams 1991). The erection of dams on Green River in Kentucky raised the water level in Mammoth Cave and this may have affected *Amblyopsis spelaea* either because of silt deposition on previously sandy habitats and/or because of the reduction or loss of particulate organic matter, a major food supply in the aquatic environment (see Lisowski & Poulson 1981, Poulson 1992 for further details). Edwards et al. (1989) discuss the water resources of Texas and conclude that 12% of the endemic fauna in the Balconian biotic province is endangered, including *Satan eurystomus* and *Trogloglanis pattersoni*.

In a paper that primarily considers human population growth over the next 50 years, Fischer & Heilig (1997) discuss the water resources available to the increasing human population. Their table 9 provide water supply data for 1995 with projections for 2030 and 2050, and this shows that water available *for humans* will *decrease* over the next 50 years in many areas of the world, including those which are known to have cave fishes. Any area in which the human population is short of water is likely to be very unhealthy for aquatic organisms of all types (see also Vorosmarty et al. 2000). Cave animals in general may be especially vulnerable if they live in areas where the recharge time is long compared to the discharge time. For example, it may be that the aquifers of Somalia only get replenished rarely. The July and January average rainfall are both less than 25 mm and the area is considered arid. Without discharge, or active removal by humans, the water remains on the ground for a long period, long enough for the survival of a rich hypogean community (Ercolini et al. 1982). Active removal of water from these aquifers (which is taking place already) is likely to be to the detriment of the whole community. Of specific interest here are the three hypogean fishes *Barbopsis devecchii*, *Phreatichthys andruzzii* and *Uegitglanis zammaranoi*.

Environmental pollution

Pollution includes eutrophication and contamination from agricultural and industrial runoff (e.g. fertilisers, pesticides and heavy metals). *Prietella phreatophila*, *Speoplatyrhinus poulsoni*, *Satan eurystomus* and *Trogloglanis pattersoni* are all vulnerable to this threat (Culver 1986). The population of *Typhlichthys subterraneus* in Sloans Valley Cave, Kentucky, which is different from other populations and may be a new taxon (Cooper & Beiter 1972), is at risk from heavy metal runoff from a landfill site (Tercafs 1992) (but see Hopper & Hansen[2] for an optimistic outcome). Another population of *T. subterraneus* was extirpated from Hidden River Cave in Kentucky (along with all other cave life) after gross pollution lasting several decades (Lewis[3], see below for recent improvements

[2] Hopper, H.L. & W. Hansen. 1996. Sloans Valley cave system: managing an 'open system'. pp. 164–170. *In*: Proceedings of the 1995 National Cave Management Symposium, Indiana Karst Conservancy, Indianapolis.

[3] Lewis, J.H. 1996. The devastation and recovery of caves and karst affected by industrialisation. pp. 214–227. *In*: Proceedings of the 1995 National Cave Management Symposium, Indiana Karst Conservancy, Indianapolis.

at this site). Vandike[4] (1984) and Crunkilton[5] reported the death of 1000 *T. subterraneus* after ammonium salts caused catastrophic deoxygenation at Meramec Spring, Missouri. The human population of the Yucatan Peninsula in Mexico is increasing, and is augmented also by a rising tourist presence. Until recently the only water available to these people was from the groundwater, and the only place for waste water was into the same groundwater. Cruickshank et al. (1980) document a coliform contamination of the aquifer which is also suffering from elevated nitrate levels. An area of Matanzas, Cuba, is affected by sugar-waste disposal directly into the karst aquifer (Varela et al. 1980). This is in the vicinity of caves containing *Lucifuga* species.

This type of threat may be acute in nature, e.g. a toxic spill could kill many organisms very quickly; the Meramec Spring incident is of this type. Or the threat may be chronic, e.g. bacterial inwash from chicken faeces over many months or years. This may exist at a low level for any period but could lead to a number of deleterious conditions in the natural fauna, e.g. opportunistic troglophiles may increase in number at the expense of troglobites. Different approaches are required to protect against these two types of threat. In Indiana, the state Department of Transportation commissioned a report (Keith et al.[6]), which proposes mechanisms available to stop toxic chemicals leaking into the karst aquifer. Graening[7] has studied the bacteria of an Ozark cave stream and now has baseline data against which changes in the bacterial load can be gauged.

[4] Vandike, J.E. 1981. The effects of the November 1981 liquid fertiliser pipeline break on groundwater in Phelps County, Missouri. Report for Water Resources Data and Research, Missouri Department of Natural Resources, Division of Geology and Land Survey, Rolla. 24 pp.

[5] Crunkilton, R. 1985. Subterranean contamination of Meramec Spring by ammonium nitrate and urea fertiliser and its implications for rare cave biota. pp. 151–158. *In*: Proceedings of the 1984 National Cave Management Symposium, Missouri Speleology 25(1–4).

[6] Keith, J.H., J.L. Bassett & J.A. Duwelius. 1996. INDOT implementation of a memorandum of understanding to reduce the impacts of highway construction in Indiana karst. pp. 203–211. *In*: Proceedings of the 1995 National Cave Management Symposium, Indiana Karst Conservancy, Indianapolis.

[7] Graening, G.O. 1998. Biospeleology as the basis of groundwater management. pp. 61–61. *In:* Abstracts Nat. Speleol. Soc. Meeting, Sewanee.

Overexploitation

The cave form of *Astyanax fasciatus* and *Caecobarbus geertsii* have been subject to collection for the aquarium trade. The former is easily bred in captivity and probably all specimens sold today are captive bred. This is not true of *C. geertsi* which cannot be captive bred and all specimens must come from wild stocks. It is now listed in CITES Appendix II (see Discussion). In 1993, 1500 wild caught individuals were reported as imports by the U.S.A., country of origin Uganda. However, it seems possible that the species was misidentified as it does not occur in Uganda (John Caldwell, World Conservation Monitoring Centre personal communication). *Clarias cavernicola* has been recommended as a member of CITES Appendix III (C. Hay in Bruton 1995b). Culver (1986) records that *Amblyopsis rosae* (see also Canaday & Vitello[8]) and *Prietella phreatophila* are threatened by overcollection. A different sort of exploitation is suffered by *Lucifuga* species in Cuba. Tourist scuba diving in the anchialine caves is detrimental to these populations (A. Garcia Debros personal communication).

Impacts of introduced aquatic animals

Bruton (1995b) suggests that *Clarias cavernicola* may be in danger from the introduction of *Clarias gariepinus* to its only known location. *C. gariepinus* is already introduced into Harasib Cave lake in Namibia and it is not impossible that it could spread, either by further human-induced introductions or by active dispersal within the aquifer. The introduction of epigean specimens of *Astyanax fasciatus* into the Cueva de El Pachon (Mexico) lead to hybridisation with the troglomorphic and isolated hypogean population (Langecker et al. 1991). Introgressive hybridisation has also occurred in La Cueva Chica, the type locality of the cave form of *A. fasciatus* (Romero 1983). Some marine fish species have been introduced into the habitats of *Lucifuga* species in Cuba, where they not only compete for resources (mainly food) but also actively prey on the cave fishes (A. Garcia Debros personal communication).

[8] Canaday, B.D. & C.B. Vitello. 1996. Ozark cavefish public outreach and habitat management project. pp. 62–65. *In:* Proceedings of the 1995 National Cave Management Symposium, Indiana Karst Conservancy, Indianapolis.

IUCN Red Book listing

Pre-1994 threat categories

The categories relevant to the present discussion (IUCN 1993) are: endangered (E) = in danger of extinction, survival unlikely if causal factors continue operating; vulnerable (V) = likely to enter E category in near future if causal factors continue operating; rare (R) = small world population, at risk, but not in V or E categories; indeterminate (I) = known to be E, V or R but with insufficient information accurately to place; insufficiently known (K) = suspected but not known to be E, V, R or I.

1994 threat categories

The relevant categories (IUCN 1994) are: critically endangered (CR), endangered (EN), vulnerable (VU), lower risk (LR), and data deficient (DD). These are defined by various criteria which include: reduction in population size, small or fragmented extent of occurrence, known small population size and known high probability of extinction (see IUCN 1994 for full details). Some details of the scheme are to be amended and details are provided in IUCN/SSC Criteria Review Working Group (1999). In particular, the lower risk (LR) category may be removed and below vulnerable there would be near threatened (NT) and least concern (LC). I have used the modified scheme here though it does not take official status until October 2000.

Commentary on the placements made in this work (TW)

The Red List placements from the past five lists are given in Table 1. A dollar ($) sign appended to a species name signifies that this species is known only from its type locality. The D2 criterion under VU is applicable to all species with a very limited distribution which these species clearly (currently) have. All thus marked are classed as VU(D2) and are not considered further here. It should be noted that this is a minimum category and that some of these species may merit higher categories when more data are available. For species not so considered, for those not listed in the last Red List, and for those whose categories (and/or criteria) in the last list were inappropriate, I have provided an assessment. All 85 known species have now been assessed. Clearly, cave biologists and fish biologists need to critically evaluate these assessments. Table 2 summarises the various listings.

The cave form of *Astyanax fasciatus* (Characidae). This species is known from at least 30 sites over ca. 4000 km^2 (Mitchell et al. 1977). It cannot therefore be given a D2 criterion. I retain the rest of the 1996 assessment pending further data. Borowsky & Espinasa[20] present evidence that there may be three different populations of this animal and if this is so we will need to assess them individually.

Prietella phreatophila (Ictaluridae). Work over the past five years by Hendrickson and colleagues (e.g. Hendrickson[9], Garcia de Leon et al.[10], Hendrickson et al.[11]) has increased our knowledge of this species. In particular, its known distribution has increased considerably, and it can no longer be assessed as EN under any of the 1996 Red List criteria which applied to the type locality, its only known site at that time. It is now known from 12 sites over an area of perhaps 5–7000 km^2. The threats to it within this area are groundwater extraction and contamination. Taking these factors into account the best current assessment is VU(B1 + 2c).

Amblyopsis spelaea (Amblyopsidae). Historically known from six counties in two states (Indiana and Kentucky) (Culver & Hobbs III unpublished [www.karstwaters.org/troglist.htm]). Keith (1988) reported an additional two counties in Kentucky and made a thorough study of the populations in these two states. It is recorded from 45 sites in Indiana and 17 in Kentucky. Eleven of the Indiana sites are in the Blue River basin which was listed in the 1998 Top Ten most endangered karst ecosystems (Mylroie & Tronvig 1998; see Table 5). Clearly, this species requires very

---

[9] Hendrickson, D.A. 1998. Blindcats of the genus *Prietella* (Ictualuridae) of north-eastern Mexico: an overview of recent explorations and studies. Abstract 1860 of the 78th Ann. Meet. Amer. Soc. Ichthy. Herp., University of Guelph, Guelph.

[10] Garcia de León, F.J., D.A. Hendrickson & D.M. Hillis. 1998. Molecular phylogeny of Ictaluridae with emphasis on the Mexican blindcats, genus *Prietella*. Abstract 363 of the 78th Ann. Meet. Amer. Soc. Ichthy. Herp., University of Guelph, Guelph.

[11] Hendrickson, D.A., J. Krejca & J.M. Rodriguez Martinez. 1998. Blind catfish (*Prietella*: Ictaluridae) phylogenies and troglobitic invertebrates as indicators of international (Mexico–USA) subterranean aquatic interconnections, and as tools for environmental risk assessment. Abstract 566 of the 78th Ann. Meet. Amer. Soc. Ichthy. Herp., University of Guelph, Guelph.

Table 1. A summary of the threat levels applied to hypogean fishes under the pre-1994 threat categories and under the 1994 categories. I77 = Miller (1977), I88 = IUCN (1988), I90 = IUCN (1990), I94 = IUCN (1993), I96 = IUCN (1998). Entries in the TW (= this work) column are suggestions for species omitted from the 1996 list or not described then, those classed DD and for those which are certainly incorrect. Comments on the TW placements are given in the text (∗ = monotypic, hypogean-restricted genus, # = multitypic, hypogean-restricted genus, $ = known only from the type locality). The threat abbreviations are described in the text (nd = not described at the time of that Red List, — = described at the time but not included in that Red List, = = same category in this work as in I96). The species list is from Proudlove[1].

	I77	I88	I90	I94	I96	TW
1. *Astyanax fasciatus* (cave form)	—	—	R	—	VU (A1ac + 2c, B1 + 2c, D2)	Not D2
2. *Astyanax aeneus*[$]	nd	nd	nd	nd	nd	VU(D2)
3. *Stygichthys typhlops*[*,$]	—	—	—	—	DD	VU(D2)
4. *Caecobarbus geertsii*[*]	—	—	R	E	VU(A1e, B1 + 2e, D2)	=
5. *Phreatichthys andruzzii*[*]	—	—	—	—	VU(D2)	=
6. *Barbopsis devecchii*[*]	—	—	—	—	VU(D2)	=
7. *Iranocypris typhlops*[*,$]	—	—	R	R	VU(D2)	=
8. *Typhlogarra widdowsoni*[*]	—	—	R	R	VU(D2)	=
9. *Caecocypris basimi*[*,$]	nd	—	R	R	VU(D2)	=
10. *Garra barreimiae*[$]	nd	—	R	R	VU(D2)	=
11. *Garra dunsirei*[$]	nd	—	R	R	VU(D2)	=
12. *Typhlobarbus nudiventris*[*,$]	nd	—	R	R	VU(D2)	=
13. *Sinocyclocheilus anatirostris*[$]	nd	—	—	—	VU(D2)	=
14. *Sinocyclocheilus anophthalmus*[$]	nd	nd	R	—	VU(D2)	=
15. *Sinocyclocheilus microphthalmus*[$]	nd	nd	—	—	VU(D2)	=
16. *Sinocyclocheilus angularis*[$]	nd	nd	nd	—	VU(D2)	=
17. *Sinocyclocheilus hyalinus*[$]	nd	nd	nd	nd	VU(D2)	=
18. *Sinocyclocheilus cyphotergus*[$]	nd	nd	—	—	VU(D2)	=
19. *Poropuntius speleops*	nd	nd	nd	—	VU(D2)	=
20. *Troglocyclocheilus khammouanensis*[*,$]	nd	nd	nd	nd	nd	VU(D2)
21. *Nemacheilus smithi*[$]	—	—	R	R	VU(D2)	=
22. *Nemacheilus troglocataractus*	nd	nd	R	R	VU(D2)	=
23. *Troglocobitis starostini*[*,$]	nd	—	R	R	DD	VU(D2)
24. *Oreonectes anophthalmus*[$]	nd	—	R	R	VU(D2)	=
25. *Triplophysa gejiuensis*[$]	nd	—	R	R	VU(D2)	=
26. *Triplophysa xiangxiensis*[$]	nd	—	R	R	VU(D2)	=
27. *Triplophysa yunnanensis*[$]	nd	nd	nd	—	—	VU(D2)
28. *Triplophysa shilinensis*[$]	nd	nd	nd	—	—	VU(D2)
29. *Cryptotora thamicolav*[$]	nd	nd	—	—	VU(D2)	=
30. *Schistura sijuensis*	nd	—	R	R	VU(D2)	=
31. *Schistura oedipus*[$]	nd	nd	R	R	VU(D2)	=
32. *Schistura jarutanini*[$]	nd	nd	R	R	VU(D2)	=
33. *Sundoreonectes tiomanensis*[$]	nd	R	R	V	U(D2)	=
34. *Indoreonectes evezardi*[$]	—	—	—	—	DD	VU(D2)
35. *Paracobitis longibarbatus*[$]	nd	nd	nd	nd	nd	VU(D2)
36. *Paralepidocephalus yui*[$]	nd	nd	nd	nd	—	VU(D2)
37. *Protocobitis typhlops*[*,$]	nd	nd	nd	nd	VU(D2)	=
38. *Trogloglanis pattersoni*[*]	V	R	V	V	VU(D2)	=
39. *Satan eurystomus*[*]	V	R	V	V	VU(D2)	=
40. *Prietella phreatophila*[#]	E	I	R	E	EN(A1ace + 2ce, B1 + 2bc)	VU(B1 + 2c)
41. *Prietella lundbergi*[#]	nd	nd	nd	nd	VU(D2)	=
42. *Pterocryptis cucphuongensis*	nd	—	—	—	—	VU(D2)
43. *Pterocryptis buccata*[$]	nd	nd	nd	nd	nd	VU(D2)
44. *Clarias cavernicola*[$]	E	E	E	E	CR(B1 + 2c, E)	=
45. *Horaglanis krishnai*[*]	—	—	R	R	VU(D2)	=
46. *Uegitglanis zammaranoi*[*]	—	—	—	—	VU(B1 + 2c)	=
47. *Phreatobius cisternarum*	—	—	—	—	DD	DD
48. *Pimelodella kronei*	—	—	—	—	DD	VU(D2)
49. *Rhamdia quelen*[$] (cave form)	—	—	—	—	—	VU(D2)
50. *Rhamdia laticauda typhla*[$]	nd	—	—	—	—	VU(D2)

Table 1. (*continued*)

	I77	I88	I90	I94	I96	TW
51. *Rhamdia reddelli*	nd	nd	V	V	VU(C2b, D2)	=
52. *Rhamdia zongolicensis*[$]	nd	nd	nd	nd	VU(D2)	=
53. *Rhamdia macuspanensis*[$]	nd	nd	nd	nd	nd	VU(D2)
54. *Trichomycterus chaberti*[$]	—	—	—	—	DD	VU(D2)
55. *Trichomycterus itacarambiensis*[$]	nd	nd	nd	nd	nd	VU(D2)
56. *Ancistrus cryptophthalmus*	nd	—	—	—	—	VU(D2)
57. *Ancistrus galani*[$]	nd	nd	nd	nd	nd	VU(D2)
58. *Ancistrus formoso*	nd	nd	nd	nd	nd	VU(D2)
59. *Astroblepus pholeter*[$]	—	—	—	—	—	VU(D2)
60. *Astroblepus riberae*	nd	nd	nd	nd	nd	VU(D2)
61. *Eigenmannia vicentespelaea*[$]	nd	nd	nd	nd	nd	VU(D2)
62. *Amblyopsis spelaea*[#]	V	V	V	V	VU(D2)	NT
63. *Amblyopsis rosae*[#]	V	V	V	V	VU(D1 + 2)	VU(C2a)
64. *Typhlichthys subterraneus*[*]	—	—	—	—	VU(D2)	NT
65. *Speoplatyrhinus poulsoni*[*]	E	V	E	E	CR(C2b)	=
66. *Lucifuga subterraneus*[#]	—	—	—	—	VU(D2)	=
67. *Lucifuga dentatus*[#]	—	—	—	—	VU(D2)	=
68. *Lucifuga spelaeotes*[#]	—	—	—	I	VU(A1ce, B1 + 2bc, D2)	LC
69. *Lucifuga simile*[#]	nd	—	—	—	VU(D2)	=
70. *Lucifuga teresinarum*[#]	nd	nd	—	—	VU(D2)	=
71. *Ogilbia pearsei*	—	—	E	E	VU(D2)	VU(B1 + 2c)
72. *Ogilbia galapagosensis*	—	—	—	—	DD	DD
73. *Poecilia mexicana*[$]	—	—	—	—	—	VU(D2)
74. *Ophisternon infernale*	—	—	E	E	EN(A1ac + 2c, B1 + 2c)	VU(B1 + 2c)
75. *Ophisternon candidum*	—	I	I	K	DD	NT
76. *Monopterus eapeni*[$]	—	—	—	—	DD	VU(D2)
77. *Monopterus roseni*[$]	nd	nd	nd	nd	nd	VU(D2)
78. *Typhleotris madagascariensis*[#]	—	—	—	—	VU(D2)	=
79. *Typhleotris pauliani*	—	—	—	—	VU(D2)	=
80. *Luciogobius pallidus*	—	—	—	—	DD	DD
81. *Luciogobius albus*	—	—	—	—	DD	DD
82. *Milyeringa veritas*[*]	—	—	—	R	DD	NT
83. *Caecogobius cryptophthalmus*[*,$]	nd	nd	nd	—	—	VU(D2)
84. *Glossogobius ankaranensis*[$]	nd	nd	nd	nd	CR(B1 + 2c)	=
85. *Oxyeleotris caeca*[$]	nd	nd	nd	nd	nd	VU(D2)

Table 2. A summary of the number of hypogean fish species listed in the past five IUCN Red Lists and the threat categories they were placed in[a].

Year	DS	E	V	R	I	K	Σ(%)	NL (%)	
1977	38	3	4	0	0	0	7 (18)	31 (82)	
1988	52	1	3	2	2	0	8 (15)	44 (85)	
1990	63	4	5	21	1	0	31 (49)	32 (51)	
1994	68	6	5	18	1	1	31 (46)	37 (54)	
	DS	CR	EN	VU	LR	DD	Σ (%)	NL	
1996	73	3	2	46	0	12	63 (86)	10 (14)	
	DS	CR	EN	VU	NT	LC	DD	Σ (%)	NL (%)
TW	85	3	0	73	4	1	4	85 (100)	0(0)

[a]DS = number of described species, E = endangered, V = vulnerable, R = rare, I = indeterminate, K = insufficiently known (all pre-1994 threat categories). CR = critically endangered, EN = endangered, VU = vulnerable, NT = near threatened, LC = least concern, LR = lower risk, DD = data deficient (all 1994 threat categories). Σ = total species listed in all categories, NL = number of species described but not listed.

careful observation and management but at present it can only be placed in NT as it does not qualify for any of the VU criteria.

Amblyopsis rosae (Amblyopsidae). Historically known from 14 counties in four states (Arkansas, Kansas, Missouri and Oklahoma) (Culver & Hobbs III unpublished [www.karstwaters.org/troglist.htm]). Recent surveys (Willis & Brown 1985, Brown & Todd 1987) suggest substantial losses of former range, particularly in south-western Missouri. These surveys also suggest that there is only one large population, in Cave Springs Cave, Arkansas. The reduction in range, and the very small population sizes, together with their fragmented nature, suggest a current category of VU(C2a).

Typhlichthys subterraneus (Amblyopsidae). Historically known from 21 counties in seven States (Alabama, Arkansas, Georgia, Kentucky, Missouri, Oklahoma and Kentucky) (Culver & Hobbs III unpublished [www.karstwaters.org/troglist.htm]). This species is still widespread but in common with all hypogean species is susceptible to a number of threats. It cannot merit more than NT at the moment but its status should be monitored.

Lucifuga spelaeotes (Bythitidae). The category and criteria applied to this species in 1996 presumably were based on the status of the species at its type locality, Mermaids Pool, New Providence, Bahamas. This is certainly polluted and the species has not been seen there in recent years although it is now recorded from another site on New Providence (R. Attrill personal communication). Other islands also have populations of this species: Andros Island (Proudlove 1984), Grand Bahama (Yager 1981, Yager & Williams 1988, Cunliffe 1985), Berry Island (specimen UMMZ 213989), Abaco (specimen ANSP 148497, Anonymous 1979) and Long Island (specimen AMNH 57448). Given this widespread distribution and the relatively unspoiled nature of most of these islands, the current status of this species is probably LC. This is the only hypogean fish to merit this 'safe' category.

Ogilbia pearsei and *Ophisternon infernale* (Bythitidae and Synbranchidae). These species are sympatric on the Yucatan Peninsula, Mexico, which is under heavy pressure from human population growth and tourism. The whole peninsula is underlain with a freshwater aquifer lying atop a seawater baselayer. The freshwater is under threat from pumping and from pollution by coliform bacteria and nitrates. The 1996 list has *Ophisternon infernale* as EN(A1ac + 2c, B1 + 2c) while *Ogilbia pearsei* is 'only' VU(D2), although both species had been classed E in the previous two lists. The known decrease in water quality (Cruickshank et al. 1980), together with the small number of known sites for both species indicate a current assessment as VU(B1 + 2c). This may have to be increased if the threats become more severe.

Ophisternon candidum and *Milyeringa veritas* (Synbranchidae and Gobiidae). Humphreys (personal communication) has collected a considerable amount of information on these sympatric species in the past decade and they can no longer be described as data deficient. *Milyeringa veritas* is now known from 25 locations (Humphreys 1999), including Barrow Island, 170 km NE from mainland North West Cape. It may well be that this species can be safely placed in the LC category as it has a sufficiently large population, spread over a large enough area, to survive for the indefinite future. However, of the 25 sites, 19 are insecure and 6 already lost through infilling, drying or siltation. Ten are close to planned development or within periurban areas. Only six are within the Cape Range National Park and even here infrastructure developments are planned. Taking these factors into account a status of NT seems appropriate since urbanisation could rapidly lead to deterioration in water quality which could be detrimental to the whole stygobitic ecosystem of the Cape Range peninsula. As Humphreys (op. cit.) says: 'The stygal community of the Cape Range peninsula is unusual in having two species of sympatric cave fish within a rich community of stygal Crustacea, many belonging to relict groups. As the fish and crustaceans are part of the same food web, the broad conservation implications are clear – the stygofauna is a functional community, changes to part of which may affect the remainder'. Although there is still very little known about *Ophisternon candidum* it has been much less common than *M. veritas* in samples taken over the past 40 years (11 known sites in 1994 vs. 18 for *M. veritas*, Humphreys 1999). Since it is one of the components of the Cape Range fauna and would suffer the same fate as any of its other components, an assessment of NT seems a minimum at the moment. Further details of the current conservation measures planned for this area are given in Humphreys (1997).

Species classed as DD. I have retained DD for four species. *Phreatobius cisternarun* (Pimelodidae) is known from a large area in the mouth of the Amazon but is rarely collected and its true habitat is unknown. It does not seem likely that it is threatened at the moment but we have too few data to know. *Ogilbia galapagosensis* (Bythitidae) has not been studied since its description in 1965. Its habitat may be extensive and probably pristine but these require confirmation. The two Japanese species *Luciogobius albus* and *L. pallidus* (Gobiidae) are both known from numerous prefectures but we have no other information about them or their habitat.

Discussion

Level of knowledge

Three primary pieces of information are required to accurately assess the status of any population: its total population size (including details of population structure), the area (or areas) over which it is distributed, and the nature of any threat (or threats) to which it is subjected. These data need to be collected through time to see if they change, and are ideally monitored on a regular basis. We have insufficient information for almost all hypogean fishes. There are a number of reasons for this absence of data. Many species are found in remote locations which are not often visited. Others reside in third world countries where funding for monitoring is lacking, and hypogean fishes are unknown to most people. This problem is exacerbated in some countries by the fact that it is not easy for Western cavers and scientists to visit them because of political problems. None of these animals are easy to collect or to mark (for mark, release and recapture programs for instance). The following species are the only ones about which we have a reasonable amount of information. The cave form of *Astyanax fasciatus* in Mexico (Mitchell et al. 1977, Wilkens 1988); the amblyopsids in North America (e.g. Poulson[12], Keith 1988, Brown & Todd 1987, Means & Johnson 1995); *Poecilia mexicana* in Mexico (Walters & Walters 1965, Langecker et al. 1996); *Pimelodella kronei* in Brazil (e.g. Trajano 1991); *Barbopsis devecchii*, *Phreatichthys andruzzii* and *Uegitglanis zammaranoi* in Somalia (Ercolini et al. 1982); the Texan ictalurids *Satan eurystomus* and *Trogloglanis pattersoni* (Longley & Karnei[13, 14], Langecker & Longley 1993) and the two Australian species *Milyeringa veritas* and *Ophisternon candidum* (Humphreys & Adams 1991, Humphreys 1993). Only the first three of these have been studied in any depth. The cave form of *Astyanax fasciatus* is probably the only hypogean species that can be described as totally safe. It is easy to breed in captivity and is commercially available in many parts of the world. However, we currently know little about its genetic structure and this knowledge is necessary if we want to protect the, possibly numerous, subpopulations.

Conservation measures

The most active conservation measures are being taken in the United States. Many individual states and the Federal Government have passed laws recently that offer some protection to caves and karst, and these will in many cases be beneficial to cave fauna (Huppert 1995). *Speoplatyrhinus poulsoni*, described by the U.S. Fish & Wildlife Service[15] as 'the rarest American cavefish and one of the rarest of all freshwater fish', lives in Key Cave, Alabama, which has been proposed as a National Wildlife Refuge. This species is also the subject of a U.S. Fish & Wildlife Service Recovery Plan (U.S. Fish & Wildlife Service[16], see also U.S. Fish & Wildlife Service[17]). *Amblyopsis rosae* is being actively studied (Brown & Todd 1987, Means & Johnson 1995) and it has been shown that its habitat extends into the aquifer surrounding the caves. The populations are not, therefore, as small as was thought, but this means that we have to be protective of a larger area. A public education program (Canaday & Vitello[8]) and a U.S. Fish & Wildlife Service Recovery Plan

[12] Poulson, T.L. 1969. Population size, density and regulation in cave fishes. pp. 189–192. *In:* Proc. 4th Int. Congr. Speleology, Yugoslavia.

[13] Longley, G. & H. Karnei. 1979a. Status of *Trogloglanis pattersoni* Eigenmann, the toothless blindcat. US Fish and Wildlife Service, Albuquerque. 54 pp.

[14] Longley, G. & H. Karnei. 1979b. Status of *Satan eurystomus* Hubbs and Bailey, the widemouth blindcat. US Fish and Wildlife Service, Albuquerque. 48 pp.

[15] U.S. Fish & Wildlife Service. 1996. Final environmental assessment and land protection plan. Proposed establishment of Key Cave National Wildlife Refuge, Lauderdale County, Alabama, U.S. Fish and Wildlife Service, Southeast Region. Unnumbered pp.

[16] U.S. Fish & Wildlife Service. 1990. Alabama cavefish (*Speoplatyrhinus poulsoni*) Cooper and Kuehne, 1974 recovery plan, second revision. U.S. Fish & Wildlife Service. 65 pp.

[17] U.S. Fish & Wildlife Service. 1982. Gray bat recovery plan. U.S. Fish & Wildlife Service. 65 pp.

Table 3. Hypogean fish species covered in the Environmental Biology of Fishes 'Threatened Fishes of the World' series.

Species	Family	Country	Reference
Clarias cavernicola	Clariidae	Namibia	Bruton (1995b)
Iranocypris typhlops	Cyprinidae	Iran	Coad (1996)
Pimelodella kronei	Pimelodidae	Brazil	Trajano (1997)
Amblyopsis spelaea	Amblyopsidae	U.S.A.	Romero & Bennis (1998)
Amblyopsis rosae	Amblyopsidae	U.S.A.	Romero (1998a)
Speoplatyrhinus poulsoni	Amblyopsidae	U.S.A.	Romero (1998b)
Milyeringa veritas	Gobiidae	Australia	Romero & Vanselow (2000a)
Ophisternon candidum	Synbranchidae	Australia	Romero & Vanselow (2000b)
Stygichthys typhlops	Characidae	Brazil	Romero & McLeran (2000)
Caecobarbus geertsii	Cyprinidae	Congo	Proudlove & Romero (2001)
Ogilbia pearsei	Bythitidae	Mexico	Proudlove et al. (2001)
Ophisternon infernale	Synbranchidae	Mexico	Medina-Gonzalez et al. (2001)

(U.S. Fish & Wildlife Service[18]) are also positive moves. *Amblyopsis spelaea* has been studied by Keith (1988) who concludes that it may still be widespread through Kentucky and Indiana but that more field data are needed. The possibly unique population of *Typhlichthys subterraneus* in Sloans Valley Cave (see above) may be better protected now that the Sloans Valley Conservation Task Force is actively managing the cave (Hopper & Hansen[2]). Improvements to wastewater treatment, and reduction in toxic inputs from various sources, have led to a very significant amelioration in water quality in Hidden River Cave where cave life is returning. The Cave Research Foundation, which is currently exploring the cave, reports that schools of cave fishes have been seen (www.caveresearch.org/eodisc.html). *Trogloglanis pattersoni* and *Satan eurystomus* have both been the subject of endangered species reports (Longley & Karnei[13,14]). Elliott (2000) summarises the status of cave and karst conservation in North America (see also a review by Walsh 2000).

The level of research and protection in other countries is not as good, although there are some positive moves. *Caecobarbus geertsii* (Congo) is included in Appendix II of the CITES Convention which means that specimens cannot be traded without the granting of an export permit[19]. The owner of Aigamas Cave, Namibia is aware of the status of *Clarias cavernicola* and does not provide access to the cave. This species is proposed to be included in Appendix III of CITES (Bruton 1995b). The Oman cavefish, *Garra barreimiae*, has several captive breeding stocks in the United Kingdom (information from Chester Zoo, see also Banister et al. 1992). A captive stock of this species in the laboratory of Horst Wilkens, in Hamburg, Germany, has reproduced several times (Horst Wilkens personal communication; other captive stocks are discussed below). The North West Cape in Australia, home of *Milyeringa veritas* and *Ophisternon candidum*, has been the subject of much recent research and there is now available a body of work upon which any future developments in the area can be based (Humphreys 1993, 1997). Jane Wilson and her associates (Fowler et al. 1989, Wilson 1996) have proposed a management plan for the Ankarana area of Madagascar which would protect its cave fishes and all of its other unique life forms. A recent report (P. Chapman personal communication) suggests that the protection of Ankarana is high and its future good. The journal *Environmental Biology of Fishes* runs short synopses on the threatened fishes of the world and so far nine hypogean fishes have been covered (Table 3). The IUCN subgroup World Commission on Protected Areas (WCPA) has prepared a set of guidelines for the protection of caves and karst, which is aimed at management and conservation agencies (Watson et al. 1997). If the recommendations in this volume are followed some of the threats listed above will be ameliorated or removed. Although with a British slant, the National Caving Association Cave Conservation Handbook (National Caving Association 1997) provides much useful information on many aspects of cave conservation and is a most useful document applicable globally.

[18] U.S. Fish & Wildlife Service. 1989. Ozark cavefish recovery plan. U.S. Fish & Wildlife Service, Atlanta. 15 pp.

[19] CITES. 1981. Proceedings of the 3rd meeting of the Conference of the Parties, New Delhi, India, 25 Feb–8 Mar, 2: 1107–1112.

Table 4. Captive populations of hypogean fishes.

Species	Where	Status	Responsible	References
Astyanax fasciatus	Germany	Breeding	Wilkens	Wilkens (1988)
Astyanax fasciatus	World-wide	Breeding	'Aquarists'	—
Caecobarbus geertsii	U.S.A., etc.	Non-breeding	'Aquarists'[a]	CITES
Phreatichthys andruzzii	Germany	Breeding	Wilkens	Pers. comm.
Phreatichthys andruzzii	Italy	Non-breeding	Sbordoni	Sbordoni et al. (1996)
Phreatichthys andruzzii	Italy	Non-breeding	Ercolini et al.	Ercolini et al. (1982)
Garra barreimiae	U.K.×3	Bred once only	Bell/Crumpler	Banister et al. (1992)
Garra barreimiae	Germany	Breeding	Wilkens[b]	Pers. comm.
Barbopsis devecchii	Italy	Non-breeding	Sbordoni	Sbordoni et al. (1996)
Barbopsis devecchii	Italy	Non-breeding	Ercolini et al.	Ercolini et al. (1982)
Prietella phreatophila	U.S.A.	Non-breeding	Hendrickson	Pers. comm.
Clarias cavernicola	Namibia	Non-breeding	Marais	Pers. comm.
Uegitglanis zammaranoi	Italy	Non-breeding	Ercolini et al.	Ercolini et al. (1982)
Pimelodella kronei	Brazil	Non-breeding	Trajano	Pers. comm.
Rhamdia reddelli	Germany	Breeding	Wilkens	Pers. comm.
Rhamdia zongolicensis	Germany	Breeding	Wilkens	Pers. comm.
Typhlichthys subterraneus	U.S.A.	Non-breeding	Various	Tom Poulson pers. comm.
Amblyopsis spelaea	U.S.A.	Non-breeding	Various	Tom Poulson pers. comm.
Amblyopsis rosae	U.S.A.	Non-breeding	Various	Tom Poulson pers. comm.
Poecilia mexicana	Germany	Breeding	Wilkens	Parzefall pers. comm.
Milyeringa veritas	Australia	Non-breeding	Young	Young (1986)

[a]Despite aquarist interest, and presumably expertise, it does not seem that anyone has successfully bred this species.
[b]Horst Wilkens sometimes uses injection of pituitary extract, but often a change of water is sufficient to precipitate breeding.

Current captive stocks of hypogean fishes

A valuable conservation tool now used in zoos around the world is captive breeding; species are kept extant if they are endangered or extinct in the wild and individuals for reintroduction can be maintained. Table 4 summarises the current status of captive cave fish stocks. Of 16 species in captivity, only six are breeding regularly. Significant conservation progress could be made in this area. Fishes are small and easily maintained in large numbers. If ways to breed them could be found, we would be able to reintroduce them in the event of extinction in the wild. There are, however, a number of problems and pitfalls. Some species (e.g. *Speoplatyrhinus poulsoni*) are probably too rare to risk removing any individuals from the natural site. A careful examination of the genetic structure of the population(s) would be required to assure that sufficient, and accurate, genetic diversity was being maintained. Very careful attention to stock control would be needed for the same reason. It would be vital to maintain replicate stocks around the world for backup purposes in case of disaster.

Problems

All of our currently recognised hypogean species are based on phenotypic characters, principally external morphology, but with examples from behaviour and physiology. In a few instances genotypic characters have been examined and in these cases the evidence suggests that there may be separate populations with similar morphology. Borowsky & Espinasa[20] present evidence for three populations of the cave form of *Astyanax fasciatus*. The currently accepted systematics of the Amblyopsidae has been questioned by Bergstrom et al.[21,22] *Phreatichthys andruzzii* consists of two non-interbreeding populations (Sbordoni et al. 1996). *Schistura oedipus* has three non-interbreeding groups where the 'separation of groups is of substantial age' (Borowsky & Mertz 2001 this volume). We

[20] Borowsky, R. & L. Espinasa. 1997. Antiquity and origins of troglobitic Mexican tetras, *Astyanax fasciatus*. pp. 359–361. *In:* Proceedings of the 12th Int. Congress of Speleology, La Chaux de Fonds, Vol. 3.

[21] Bergstrom, D.E., D.B. Noltie & T.P. Holtsford. 1995. Ozark cavefish genetics: the phylogeny of Missouri's Ozark cavefish (*Amblyopsis rosae*) and southern cavefish (*Typhlichthys subterraneus*). Final report to the Missouri Department of Conservation: Endangered species project SE-01-27. 43 pp.

[22] Bergstrom, D.E., D.B. Noltie & T.P. Holtsford. 1997. Molecular phylogenetics and historical biogeography of the family Amblyopsidae. pp. 4–5. *In:* Sasowsky, I.D., D.W. Fong & E.L. White (ed.) Conservation and Protection of the Biota of Karst, Karst Waters Institute Special Publication 3, Charles Town, West Virginia.

Table 5. Cave ecosystems containing fishes identified by the Karst Waters Institute as very endangered (Mylroie & Tronvig 1998, Tronvig & Belson 1999).

Ecosystem	Country	Species
Blue River Basin	U.S.A.	*Amblyopsis spelaea*
Cape Range Peninsula	Australia	*Milyeringa veritas*, *Ophisternon candidum*
South Central Kentucky Karst	U.S.A.	*Amblyopsis spelaea*, *Typhlichthys subterraneus*
Edwards Aquifer	U.S.A.	*Satan eurystomus*, *Trogloglanis pattersoni*
Snail Shell Cave	U.S.A.	*Typhlichthys subterraneus*
Cambodian caves	Cambodia	'Fishes'[a]

[a]Tronvig & Belson (1999) report that the ecosystem is extremely rich and diverse. No cave fishes are yet known from this area but it is very likely that one or more exist since there are several in Thailand and one in Vietnam which lie to either side of Cambodia.

need to study the genetic diversity of hypogean fish populations if we are to obtain sufficient information to conserve different subgroups.

Prospects

The Karst Waters Institute (Mylroie & Tronvig 1998, Tronvig & Belson 1999) has assessed the globally most endangered karst ecosystems and produced two 'Top Ten' lists. Of 18 sites (two are in both lists), six are ecosystems that contain cave fishes (Table 5). Lawton & May (1995) and Quammen (1996) both provide graphic evidence for the unprecedented decline in biodiversity in historical times. We are facing an extinction crisis and cave faunas will suffer just like other more well known animals. Overall, the prospects are at best neutral and quite possibly very poor. Over the medium term (decades) the number of described hypogean species will rise to perhaps 200 and there will be few or no losses. Over the longer term (centuries) the inexorable increase in the human population will lead to more and more habitat loss (perhaps the primary threat) and degradation. The prospects for many, if not most, species is not good and hypogean fishes, if they survive at all, will do so as a by-product of other conservation efforts (e.g. the maintenance of aquifers for drinking water). It seems most unlikely that these animals will ever be any sort of priority.

Acknowledgements

Bill Humphreys and Richard Borowsky both provided unpublished manuscripts which were very helpful. Dean Hendrickson helped with the assessment of *Prietella phreatophila*. Tristram Besterman and Mike Hounsome allowed my affiliation with the Manchester Museum for which I am very grateful.

References cited

Anonymous. 1979. Cave fish. Baham. Nat. 4: 22.

Banister, K.E., J. Bell & M. Crumpler. 1992. Omani blind cave fish. Aquar. Pondk. (July): 38–40.

Borowsky, R.B. & L. Mertz. 2001. Genetic differentiation among populations of the cave fish *Schistura oedipus* (Cypriniformes: Balitoridae). Env. Biol. Fish. 62: 225–231 (this volume).

Brown, A.V. & C.S. Todd. 1987. Status review of the threatened Ozark cavefish (*Amblyopsis rosae*). Proc. Ark. Acad. Sci. 41: 99–100.

Bruton, M.N. 1995a. Have fishes had their chips? The dilemma of threatened fishes. Env. Biol. Fish. 43: 1–27.

Bruton, M.N. 1995b. Threatened fishes of the world: *Clarias cavernicola* Trewavas, 1936 (Clariidae). Env. Biol. Fish. 43: 162.

Coad, B.W. 1996. Threatened fishes of the world: *Iranocypris typhlops* Bruun & Kaiser, 1944 (Cyprinidae). Env. Biol. Fish. 46: 374.

Cooper, J.E. & D.P. Beiter. 1972. The southern cavefish, *Typhlichthys subterraneus* (Pisces, Amblyopsidae) in the eastern Mississippian plateau of Kentucky. Copeia 1972: 879–881.

Cruickshank, G., J. Aguirre, D. Kraemer & E. Craviota. 1980. Bacterial contamination of the limestone aquifer beneath Merida, Mexico. pp. 341–345. *In:* R.E. Jackson (ed.), Aquifer Contamination and Protection, Project 8.3 of the International Hydrological Programme, UNESCO.

Culver, D.C. 1986. Cave faunas. pp. 427–442. *In:* M. Soule (ed.) Conservation Biology, The Science of Scarcity and Diversity, Sinauer Associates, Sunderland.

Cunliffe, S. 1985. The flora and fauna of Sagittarius, an anchialine cave and lake in Grand Bahama. Cave Sci. 12: 103–109.

Edwards, R.J., G. Longley, R. Moss, J. Ward, R. Matthews & B. Stewart. 1989. A classification of Texas aquatic communities with special consideration toward the conservation of endangered and threatened taxa. Texas J. Sci. 41: 231–240.

Elliott, W.R. 2000. Conservation of the North American cave and karst biota. pp. 665–689. *In:* H. Wilkens, D.C. Culver & W.F. Humphreys (ed.) Ecosystems of the World, Vol. 30, Subterranean Ecosystems, Elsevier, Amsterdam.

Ercolini, A., R. Berti, L. Chelazzi & G. Messana. 1982. Researches on the phreatobitic fishes of Somalia: achievements and prospects. Moni. Zool. Ital. (Suppl.) 17: 219–241.

Fischer, G. & G.K. Heilig. 1997. Population momentum and the demand on land and water resources. Phil. Trans. Roy. Soc. Lond. B 352: 869–889.

Fowler, S.V., P. Chapman, D. Checkley, S. Hurd, M. McHale, G.S. Ramangasan, J.E. Randraimasy, P. Stewart, R. Walters & J.M. Wilson. 1989. Survey and management proposals for a tropical deciduous forest reserve at Ankarana in northern Madagascar. Biol. Conserv. 47: 297–313.

Humphreys, W.F. (ed.) 1993. The biogeography of Cape Range, Western Australia. Records of the Western Australian Museum (Suppl.) 45: 1–248.

Humphreys, W.F. 1997. Protection of arid karstlands on the Cape Range peninsula, Western Australia. pp. 28–32. *In:* I.D. Sasowsky, D.W. Fong & E.L. White (ed.) Conservation and Protection of the Biota of Karst, Karst Waters Institute Special Publication 3, Charles Town.
Humphreys, W.F. 1999. The distribution of Australian cave fishes. Rec. West. Austr. Mus. 19: 469–472.
Humphreys, W.F. & M. Adams. 1991. The subterranean aquatic fauna of the North West Cape peninsula, Western Australia. Rec. West. Austr. Mus. 15: 383–411.
Huppert, G.N. 1995. Legal protection for caves in the United States. Environ. Geol. 26: 121–123.
IUCN. 1988. 1998 IUCN Red list of threatened animals. International Union for the Conservation of Nature and Natural Resources, Gland. 154 pp.
IUCN. 1990. The IUCN Red list of threatened animals. IUCN – The World Conservation Union, Gland. 192 pp.
IUCN. 1993. 1994 IUCN Red list of threatened animals. IUCN, Gland. 286 pp.
IUCN. 1994. IUCN Red list categories. IUCN, Gland. 21 pp.
IUCN. 1996. 1996 IUCN Red list of threatened animals. IUCN, Gland. 368 pp.
IUCN/SSC Criteria Review Working Group. 1999. IUCN Red List criteria review provisional report: draft of the proposed changes and recommendations. Species (31–32): 43–57.
Keith, J.H. 1988. Distribution of northern cavefish, *Amblyopsis spelaea* DeKay, in Indiana and Kentucky and recommendations for its protection. Nat. Areas J. 8: 69–79.
Langecker, T., H. Wilkens & P. Junge. 1991. Introgressive hybridisation in the Pachon Cave population of *Astyanax fasciatus* (Teleostei, Characidae). Ichth. Explor. Freshw. 2: 209–212.
Langecker, T.G. & G. Longley. 1993. Morphological adaptations of the Texas blind catfishes *Trogloglanis pattersoni* and *Satan eurystomus* (Siluriformes, Ictaluridae) to their underground environment. Copeia 1993: 976–986.
Langecker, T.G., H. Wilkens & J. Parzefall. 1996. Studies on the trophic structure of an energy-rich Mexican cave (Cueva de las Sardinas) containing sulphurous water. Mem. Biospel. 23: 121–125.
Lawton, J.H. & R.M. May (ed.) 1995. Extinction rates. Oxford University Press, Oxford. 233 pp.
Lisowski, E.A. & T.L. Poulson. 1981. Impacts of lock and dam six on base level ecosystems in Mammoth Cave. Cave Res. Found. Ann. Rept.: 48–54.
Means, M. & J.E. Johnson. 1995. Movement of threatened Ozark cavefish in Logan Cave national wildlife refuge, Arkansas. Southw. Nat. 40: 308–313.
Medina-González, R., G. Proudlove, L. Chumba-Segura & T. Iliffe. 2001. Threatened fishes of the world: *Ophisternon infernale* (Hubbs, 1938) (Synbranchidae). Env. Biol. Fish. 62: 170 (this volume).
Miller R.R. 1977. IUCN Red Data Book 4, Freshwater fishes. IUCN, Gland. 77 pp.
Mitchell, R.W., W.H. Russell & W.R. Elliott. 1977. Mexican eyeless characin fishes, genus *Astyanax*: environment, distribution and evolution. Sp. Publ. Mus. Tex. Tech. Univ. (12): 1–89.
Mylroie, J. & H. Tronvig. 1998. Karst Waters Institute creates top ten list of endangered karst ecosystems. Nat. Spel. Soc. News 56: 41–43.
National Caving Association. 1997. Cave conservation handbook. National Caving Association, London (various paging).
Poulson, T.L. 1992. The Mammoth Cave ecosystem. pp. 569–611. *In:* A.I. Camacho (ed.) The Natural History of Biospeleology, Monografias del Museo Nacional de Ciencias Naturales, Madrid.
Proudlove, G.S. 1984. Preliminary observations on the biology of inland Blue Holes, Andros Island, Bahamas. Cave Sci. 11: 53–56.
Proudlove, G.S. & A. Romero. 2001. Threatened fishes of the world: *Caecobarbus geertsii* Boulenger, 1921 (Cyprinidae). Env. Biol. Fish. 62: 238 (this volume).
Proudlove, G., R. Medina-González, L. Chumba-Segura & T. Iliffe. 2001. Threatened fishes of the world: *Ogilibia pearsei* (Hubbs, 1938) (Bythitidae). Env. Biol. Fish. 62: 214 (this volume).
Quammen, D. 1996. The song of the dodo. Island biogeography in an age of extinctions. Pimlico, London. 702 pp.
Romero, A. 1983. Introgressive hybridisation in the *Astyanax fasciatus* (Pisces, Characidae) population at La Cueva Chica. Bull. Nat. Spel. Soc. 45: 81–85.
Romero, A. 1998a. Threatened fishes of the world: *Amblyopsis rosae* (Eigenmann, 1898) (Amblyopsidae). Env. Biol. Fish. 52: 434.
Romero, A. 1998b. Threatened fishes of the world: *Speoplatyrhinus poulsoni* Cooper & Kuehne, 1974 (Amblyopsidae). Env. Biol. Fish. 53: 293–294.
Romero, A. & L. Bennis. 1998. Threatened fishes of the world: *Amblyopsis spelaea* DeKay, 1842 (Amblyopsidae). Env. Biol. Fish. 51: 420.
Romero, A. & P.B.S. Vanselow. 2000a. Threatened fishes of the world: *Milyeringa veritas* Whitley, 1945 (Eleotridae). Env. Biol. Fish. 57: 36.
Romero, A. & P.B.S. Vanselow. 2000b. Threatened fishes of the world: *Ophisternon candidum* (Mees, 1962) (Synbranchidae). Env. Biol. Fish. 58: 214.
Romero, A. & A. McLeran. 2000. Threatened fishes of the world: *Stygichthys typhlops* Brittan & Bohlke, 1965 (Characidae). Env. Biol. Fish. 57: 270.
Sbordoni, M.C., E. Dematthaeis, M. Mattoccia, R. Berti & V. Sbordoni. 1996. Genetic variability and differentiation of hypogean cyprinid fishes from Somalia. J. Zool. Syst. Evol. Res. 34: 75–84.
Skelton, P.H. 1990. The status of fishes from sinkholes and caves in Namibia. Namibia Scient. Soc. J. 42: 75–83.
Tercafs, R. 1992. The protection of the subterranean environment. Conservation principles and management tools. pp. 481–524. *In:* A.I. Camacho (ed.), The Natural History of Biospeleology, Monografias del Museo Nacional de Ciencias Naturales, Madrid.
Trajano, E. 1991. Population ecology of *Pimelodella kronei*, troglobitic catfish from south-eastern Brazil (Siluriformes, Pimelodidae). Env. Biol. Fish. 30: 407–421.
Trajano, E. 1997. Threatened fishes of the world: *Pimelodella kronei* (Ribeiro, 1907) (Pimelodidae). Env. Biol. Fish. 49: 332.
Tronvig, K. & C. Belson. 1999. Karst Waters Institute's second annual top ten list of endangered karst ecosystems. Nat. Spel. Soc. News 57: 265–267.

Vandike, J.E. 1984. Hydrogeologic aspects of the November 1981 liquid fertiliser pipeline break on groundwater in the Meramec Spring recharge area, Phelps County, Missouri. *In:* Proceedings of the 1984 National Cave Management Symposium, Missouri Speleology 25: 93–101.

Vorosmarty, C.J., P. Green, J. Salisbury & R.B. Lammers. 2000. Global water resources: vulnerability from climate change and population growth. Science 289: 284–288.

Varela, E., O. Barros & S. Jimenez. 1980. The effects of sugar-waste disposal on a karstic limestone aquifer. pp. 329–339. *In:* R.E. Jackson (ed.) Aquifer Contamination and Protection, Project 8.3 of the International Hydrological Programme, UNESCO, Paris.

Walsh, S.J. 2000. Conservation of aquatic karst biotas: shedding light on troubled waters, essay 15. pp. 106–108. *In:* R.A. Abell, D.M. Olson, E. Dinerstein, P.T. Hurley, J.T. Diggs, W. Eichbaum, S. Walters, W. Wettengel, T. Allnutt, C.J. Loucks & P. Hedao (ed.) Freshwater Ecoregions of North America: A Conservation Assessment, Island Press, Washington, D.C.

Walters, L.H. & V. Walters. 1965. Laboratory observations on a cavernicolous poeciliid from Tabasco, Mexico. Copeia 1965: 214–223.

Watson, J., E. Hamilton-Smith, D. Gillieson & K. Kiernan (ed.) 1997. Guidelines for cave and karst protection. IUCN, Gland. 63 pp.

Wilkens, H. 1988. Evolution and genetics of epigean and cave *Astyanax fasciatus* (Characidae, Pisces): support for the neutral mutation theory. Evol. Biol. 23: 271–367.

Willis, L.D. & A.V. Brown. 1985. Distribution and habitat requirements of the Ozark cavefish, *Amblyopsis rosae*. Amer. Midl. Nat. 114: 311–317.

Wilson, J.M. 1996. Conservation and ecology of a new blind fish *Glossogobius ankaranensis* from the Ankarana caves, Madagascar. Oryx 30: 218–221.

Winograd, I.J. & F.N. Robertson. 1982. Deep oxygenated groundwater: anomaly or common occurrence? Science 216: 1227–1230.

Yager, J. 1981. Remipedia, a new class of crustacea from a marine cave in the Bahamas. J. Crust. Biol. 1: 328–333.

Yager, J. & D. Williams. 1988. Predation by grey snapper on cave bats in the Bahamas. Bull. Mar. Sci. 43: 102–103.

Young, M. 1986. Bringing the blind gudgeon into captivity. J. Austr. New Guin. Fishes Ass. 4: 145–148.

Environmental Biology of Fishes **62:** 214, 2001.
© 2001 *Kluwer Academic Publishers. Printed in the Netherlands.*

Threatened fishes of the world: *Ogilbia pearsei* (Hubbs, 1938) (Bythitidae)

Graham Proudlove[a], Roger Medina-González[b], Lizbeth Chumba-Segura[b] & Thomas Iliffe[c]
[a]*Department of Zoology, Manchester Museum, Manchester M13 9PL, U.K. (e-mail: g.proudlove@umist.ac.uk)*
[b]*Departamento de Ecología and Departamento de Zoología, Facultad de Medicina Veterinaria y Zootecnia, Universidad Autónoma de Yucatán, Mérida, Yucatán 97100, México*
[c]*Department of Marine Biology, Texas A & M University, Galveston, TX 77553, U.S.A.*

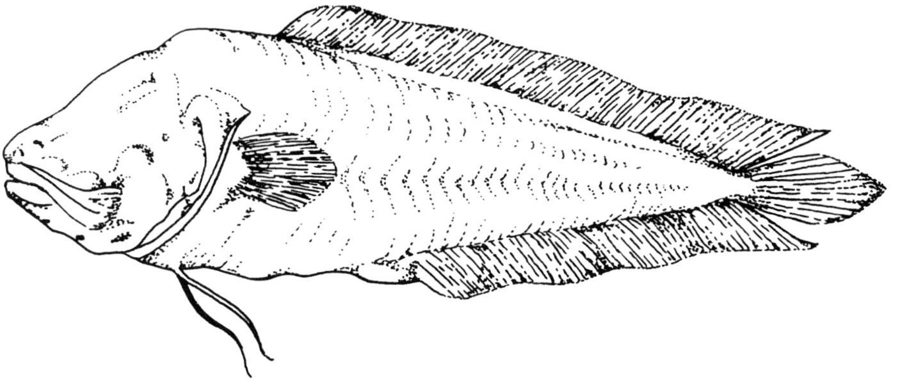

Common names: Sak kay (Mayan = white fish), dama blanca ciega (S). **Conservation status:** Endangered (IUCN), VU(D2) (IUCN). **Identification:** Total absence of eyes. Appearing white with pink posterior margins. Compressed and naked head with large, well developed sensory cavities and papillae. Longitudinal split at mouth edges, nose opening adjacent to superior lip. Scaled body. Long dorsal and anal fins not continuous with caudal. Males have claspers. Up to 95 mm SL. Drawing by Rhian Hicks, 90.5 mm SL. **Distribution:** Cenotes and groundwater on the Yucatán peninsula, México (20–21°N, 89–91°W). This species is sympatric in some locations with *Ophisternon infernale* (Synbranchidae) and *Rhamdia guatemalensis* (Pimelodidae) as well as with some endemic crustaceans (Chumba-Segura 1983). **Abundance:** As top predator in the anchialine community it is likely that the population is small (Pohlman et al. 1997). In freshwater habitats populations are also small, with few specimens observed in each cenote. It seems not widely distributed, recorded in at most 16% of caves. Underwater exploration should increase the number of recorded sites. **Habitat and ecology:** Lives in the freshwater aquifer. Temperature range is 22–27°C. Various crustaceans are the main food supply. **Reproduction:** Viviparous. Parturition from December to February. Young animals (22–34 mm SL) independent from birth. Two to 11 young depending on female size. **Threats:** The main threat is from water pollution, principally faecal coliforms, and excess nitrate (Cruickshank et al. 1980). Humans obtain their water from the aquifer and, until recently, disposed of wastewater directly back into the same aquifer, a few metres from where it was obtained. This situation is particularly acute under the largest city, Mérida, although some piped water is now available. In rural areas there is little or no sewage treatment and industrial and domestic waste is discarded into caves. The population of the peninsula is growing: 2.3×10^6 in 1990, 2.9×10^6 in 1995, and tourist numbers are also rising. This will place growing strain on the water resources of the area. For example, freshwater removal will lead to the incursion of saline waters into the freshwater zone. **Conservation action:** Wastewater in some parts of Mérida is now collected and pumped to treatment plants. The treated product is injected into saline groundwater 200 m below the surface. The injected water becomes saline and remains below the less dense freshwater lens. **Conservation recommendations:** Human water supply should be augmented by the collection (and subsequent treatment) of rainfall to supplement that obtained from the aquifer. All wastewater should be collected and properly treated. It should not be injected into the aquifer which contains a very delicate ecosystem. The best economic solution is probably to pipe wastewater out to sea although this will of course have its own (detrimental) effects. The continental shelf is wide and it will be expensive to dispose of wastewater far enough to avoid the immediate effect on the coastal systems. Detailed studies of the aquifer need to be made to monitor its health.

Chumba-Segura, L. 1983. Brotulidae: *Typhliasina pearsei*. Fauna de Los Cenotes de Yucatán No. 4. Departamento de Acuacultura y Biología Marina, Universidad de Yucatán. 9 pp.
Cruickshank, G., J. Aguirre, D. Kraemer & E. Craviota. 1980. Bacterial contamination of the limestone aquifer beneath Mérida, México. pp 341–345. *In*: Aquifer Contamination and Protection, UNESCO Publication No. 30.
Pohlman, J.W., T.M. Iliffe & L.A. Cifuentes. 1997. A stable isotope study of organic cycling and the ecology of an anchialine cave ecosystem. Mar. Ecol. Progr. Ser. 155: 17–27.

Environmental Biology of Fishes **62**: 215–222, 2001.
© 2001 *Kluwer Academic Publishers. Printed in the Netherlands.*

Status of the federally endangered Alabama cavefish, *Speoplatyrhinus poulsoni* (Amblyopsidae), in Key Cave and surrounding caves, Alabama

Bernard R. Kuhajda & Richard L. Mayden
Department of Biological Sciences, University of Alabama, Box 870345, Tuscaloosa, AL 35487-0345, U.S.A. (e-mail: bkuhajda@bama.ua.edu)

Received 1 March 2000 Accepted 14 December 2000

Key words: abundance, range, Tennessee River, southern cavefish, *Typhlichthys subterraneus*

Synopsis

Speoplatyrhinus poulsoni, the Alabama cavefish, is a federally endangered stygobitic fish that is restricted to Key Cave in northwestern Alabama. Due to the low abundance of this species and the continuous threats to the cave's recharge area, we surveyed Key Cave from 1992 to 1997. Alabama cavefish were present in five different pools and had numbers comparable to previous surveys performed in the 1970's and 1980's. Three different size classes were observed, indicating recruitment is occurring. These data suggest that the population is relatively stable, but future monitoring is recommended. The Alabama cavefish is an endangered species and is extremely difficult to capture, therefore any marking program to accurately estimate the population size is unlikely. A single specimen of *Typhlichthys subterraneus* was captured in Key Cave. This falsifies the previous assumption that these two species are allopatric. Because of this discovery, we resurveyed seven caves in the area around Key Cave for additional populations of *S. poulsoni*. These caves had either known populations of *T. subterraneus* or unidentified cavefish. No additional populations of *S. poulsoni* were discovered, but further investigation is warranted. The Key Cave National Wildlife Refuge was established in the high recharge area of Key Cave to protect this habitat, yet threats to the groundwater continue from encroaching urbanization.

Introduction

The Alabama cavefish, *Speoplatyrhinus poulsoni*, is a stygobitic fish distinguished from other members of the family Amblopsidae by several unique morphological features and is restricted to a single cave in northwestern Alabama. *Speoplatyrhinus poulsoni* was described by Cooper & Kuehne in 1974 from Key Cave, Lauderdale County, located southwest of Florence and just north of the Tennessee River (now Pickwick Reservoir) (Figure 1) within the Tuscumbia Limestone formation. Populations of the widespread southern cavefish, *Typhlichthys subterraneus*, are found in numerous caves in the vicinity of Key Cave, but the two species have not historically been shown to be sympatric. In contrast to *T. subterraneus* (and other amblyopsids), *S. poulsoni* possesses a long head with a laterally constricted and dorsally flattened snout, lacks bifurcate fin rays, and has incised fin membranes (Cooper & Kuehne 1974).

The U.S. Fish and Wildlife Service (USFWS) listed *Speoplatyrhinus poulsoni* as threatened in 1977. This listing was based on the likely restriction of *S. poulsoni* to Key Cave and its low abundance within the cave, estimated at less than one hundred individuals. This estimate was based on seven visits to Key Cave to collect the nine type specimens (Cooper & Kuehne 1974, Cooper[1]). Another concern was the possible

[1] Cooper, J.E. 1975. Order Percopsiformes, family Amblyopsidae, Alabama cave fish, *Speoplatyrhinus poulsoni* Cooper & Kuehne 1974. 2nd Symposium on Alabama endangered and threatened species, University of Alabama, Tuscaloosa. 6 pp.

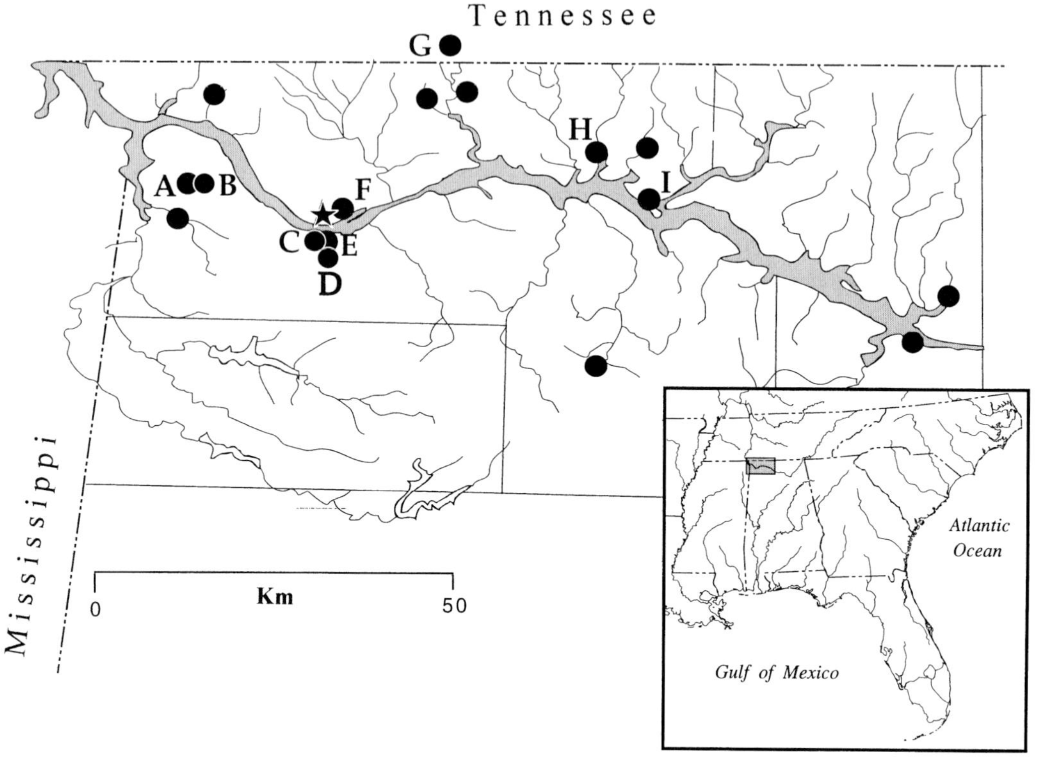

Figure 1. Star represents location of Key Cave. Circles represent distribution of *Typhlichthys subterraneus* in northwestern Alabama and extreme southern Tennessee. Caves with potential *Speoplatyrhinus poulsoni* populations: A = Persimmon Cave, B = Thomason Cave, C = Bell Cave, D = McKinney Pit Cave, E = Elbow Cave, F = Key Spring Cave, G = Hound Dog Drop Cave, H = Davis Bat Cave, I = Watkins Lake Cave.

disruption of the cave's ecosystem through disturbance of the annual maternal colony of gray bats, *Myotis grisescens*. Cooper & Kuehne (1974) considered the guano from these bats an important input of nutrients into the aquatic environment of Key Cave. Groundwater pesticide pollution and a proposed industrial park for the city of Florence were additional factors that contributed to this listing (USFWS 1977).

Surveys of Key Cave by USFWS personnel (James Stewart personal communication) and Richard M. Cobb (unpublished data) in the early and mid-1980's again suggested low population numbers for *S. poulsoni*, with a maximum of 10 specimens viewed during one trip. Cobb[2,3] also surveyed caves in the vicinity of Key Cave and failed to find any specimens of *S. poulsoni*, further supporting the restriction of the Alabama cavefish to Key Cave. A hydrological survey for a proposed solid-waste landfill for the city of Florence showed an existing sewage sludge disposal operation was likely to lie within the recharge area of Key Cave (Aley[4]). Additionally, herbicide and pesticide applications to abundant cotton fields in the recharge area were shown to have direct access to Key Cave through the numerous surface pool seeps. These factors lead the USFWS to reclassify *S. poulsoni* as endangered in 1988.

Despite this level of federal protection, urbanization and application of herbicides and pesticides to the recharge area of Key Cave continued into the 1990's. Additionally, only two pools located within the first 60 m of passage within Key Cave (A and B in Figure 2)

[2] Cobb, R.M. 1985. A reconnaissance of caves in Lauderdale and Colbert counties, Alabama for the Alabama cavefish, *Speoplatyrhinus poulsoni*. U.S. Fish and Wildlife Service Report, unpublished. 28 pp.

[3] Cobb, R.M. 1986. An attempt to collect specimens of cavefish at three cave sites in northwestern Alabama. U.S. Fish and Wildlife Service Report, unpublished. 8 pp.

[4] Aley, T. 1986. Hydrogeologic investigations for a proposed landfill near Florence, Alabama. A contract study and report for Waste Contractors, Inc., unpublished. 20 pp.

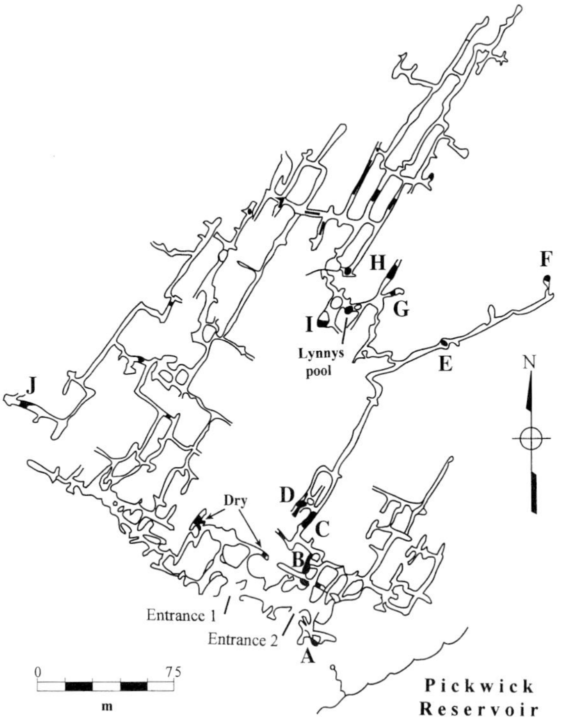

Figure 2. Map of Key Cave, Lauderdale County, Alabama, modified from map by Huntsville Grotto (National Speleological Society). Letters represent pools studied, see text for details.

had ever been repeatedly surveyed, leaving 3300 m of passage and numerous other pools poorly studied. Due to the continued threats of groundwater pollution, lack of information on the abundance and distribution of *S. poulsoni* throughout Key Cave, and a USFWS recovery objective[5] for continued monitoring of this species, we surveyed Key Cave for *S. poulsoni* between 1992 and 1997.

Cobb[2,3,6] considered 130 caves within the general area of Key Cave to search for additional populations of *S. poulsoni*. Twenty-eight caves had suitable conditions for possible populations of Alabama cavefish and were surveyed; no Alabama cavefish were found. Several of these caves had cavefish that Cobb could not positively identify. Additionally, Cobb excluded several caves with known populations of *Typhlichthys subterraneus* from his list of possible *S. poulsoni* habitat because the two species had historically been considered allopatric. During our survey of Key Cave, a specimen of *T. subterraneus* was captured from this cave (see details below). This discovery initiated our survey of caves in the area of Key Cave with known or presumed populations of *T. subterraneus*. We also resurveyed those caves where Cobb had reported unidentified cavefish or where muddy water had prevented observations, another objective in the recovery plan for the Alabama cavefish (USFWS[5]).

Key Cave survey

Methods

During each visit to Key Cave, observations were made in each of the two pools closest to Entrance 2 (A and B in Figure 2), which were always completely accessible, even during high water levels. Most of Pool C was also accessible during high water; pools farther back in the eastern half of the cave (D–F in Figure 2) were explored for additional specimens during low water conditions. When cavefish were observed, attempts were made to get as close as possible to the specimen for positive identification. To minimize disturbances of this protected species, all observations were made from the shoreline. The cave was closed to research from April through September due to the presence of a maternal colony of federally endangered gray bats, *Myotis grisescens*.

Habitats

Key Cave is a multi-level, fairly complex cave with most of the water and all cavefish observations in its eastern half (Figure 2). Pool A is closest to the cave entrance and is located at the bottom of a long steep slope of guano. The main channel is about 1.5 by 15 m, with water depth to 5 m and the water surface from 1 to 2 m below the rock ledge. During high water a slight current flows from the shallow end of the main pool, and a normally dry side pool exhibits some upwelling, but no current was discernible at lower water levels. We did not observe current in any other pools within Key Cave. Pool B is at the bottom of a long, tube-shaped passage. This pool is about 15 m long, 1–5 m wide,

[5] U.S. Fish & Wildlife Service. 1990. Alabama cavefish, *Speoplatyrhinus poulsoni* Cooper & Kuehne 1974 (second revision) Recovery Plan. Prepared by J.E. Cooper, North Carolina State Museum of Natural History, Revised by James H. Stewart, U.S. Fish & Wildlife Service, Jackson, MS, for U.S. Fish & Wildlife Service, Atlanta. 17 pp.

[6] Cobb, R.M. 1989. An attempt to secure a cavefish for identification in Bell Cave, Colbert County, Alabama. U.S. Fish and Wildlife Service Report, unpublished. 7 pp.

and 2–5 m deep. A ledge 1.5–3 m above the surface is only present around the narrower half of the pool. Pools A and B are the two pools that have historically been examined for *Speoplatyrhinus poulsoni*.

Pool C lies in a fairly narrow passage with a high ceiling, angled walls, and no ledges. Observations within this passage are limited to placing ones feet on a wall and hands and back on the other, and 'walking' down the passage above the pool (1 × 12 m). Pool depth is unknown due to the lack of visibility below the small openings where the walls nearly converge below the water's surface.

Access to waters farther into Key Cave is limited to low water conditions, as the far end of Pool C is impassable during high water. A total of five expeditions beyond this passage were undertaken during this study. The first of these remote pools (D in Figure 2) also has no ledges surrounding it. This pool averages 1.5 m deep and has a long section (2 × 10 m) off to one side that appears to continue underwater. The passage leading back to other pools has a low ceiling and the walls are very close together (1 m), making maneuvering over this section of the pool difficult.

An extensive stomach crawl through a 0.3 m tall passage with floors of mud and standing water leads past the historic pool E, which is currently a 5 m deep waterless hole in the passage floor. Pool F is a small (3 × 4 m), 0.5 m deep pool at the bottom of a 4.6 m pit at the end of this passage. As in other pools, there appears to be an underwater passage continuing off the back of this pool.

Pools G, H, and I and Lynnys Pool are also at the end of a very small and muddy passage. Pools G and I are small and have water that may be connected to other pools within Key Cave, however, no cavefish or cave crayfish have ever been observed in these pools. Pool H is very deep (depth unknown) and circular in shape, with a diameter of about 7 m and sheer walls around all sides; access to the far side of the pool is impossible without swimming and the use of climbing gear. Lynnys Pool is also in a sheer-sided circular pit of unknown depth. Pool H and Lynnys Pool appear to be isolated drip pools with no subterranean water connections. While we surveyed these two pools only once, the water in each was extremely clear and devoid of any visible aquatic life.

The western half of Key Cave is even more difficult to maneuver than the eastern half. Passages here are extremely small, angled, and muddy. Three pools were examined in the western half, the largest being Pool J. All pools lacked cavefish or cave crayfish, and the water was extremely clear at a time when waters known to contain cavefish were cloudy. We believe these arc isolated drip pools and are not connected to waters containing aquatic cave species. Several pools in the last one-third of the western half of Key Cave remain to be surveyed for potential habitat of Alabama cavefish.

Status determination

Pools A and B have been historically examined for the presence or absence of *Speoplatyrhinus poulsoni*. A total of 25 separate trips to one or both of these pools have occurred over the last 30 years (Table 1). Cooper[1] reported on seven trips during 1967–1970, when the type series of nine specimens was collected. Only pool A was sampled, and all visits occurred in March through May; except for one unproductive visit in July. The number of cavefish captured ranged from none to three individuals per visit. Cobb (unpublished data) reported on 12 visits to both pools A and B from November 1985 to March 1986, and again in September 1986, where he observed two to nine cavefish per visit. Observations were made from the shoreline for one hour total per visit. From 1992–1997, we surveyed both pools six times from October through February for an average of two hours per visit, and saw from a minimum of no cavefish to a maximum of three to six individuals per visit. The range of three to six cavefish results from the possibility of recounting the same individuals. We observed cavefish on two-thirds of our visits (4 of 6). The mean number of *S. poulsoni* observed per trip was highest in 1985–1986 with 4.6 individuals. Lower values were recorded in 1967–1970 (1.3) and 1992–1997 (1.7–2.7).

We visited pools A, B, and C from January through March 1995 for an average of three hours per visit, and observed no cavefish on one visit and only one cavefish during each of the other four trips (Table 1). The very low mean number of cavefish per trip (0.8) is most likely a reflection of the extremely high and cloudy water that was present during each of these trips.

Surveys of four or more pools in Key Cave have occurred once in November 1983 by USFWS personnel and five times by us from late September 1995 to February 1996 (Table 1). The 1983 expedition (James Stewart personal communication) observed a total of 10 *S. poulsoni* over six hours within five different pools (A–E in Figure 2); Pool E is no longer present, it is now a waterless hole in the cave floor. We surveyed four or

Table 1. Number of trips into Key Cave and number of *Speoplatyrhinus poulsoni* observed during various time periods. Data divided into number of pools visited to standardize comparisons.

Year	Number of trips	Number of trips with fish	Total number of fish	Range	Mean number of fish per trip
1 or 2 pools visited					
1967–1970 (Cooper)[1]	7	5	9	0–3	1.3
1985–1986 (Cobb)[1]	12	12	55	2–9	4.6
1992–1997	6	4	10(16)[2]	0–3(6)[2]	1.7 (2.7)[2]
3 pools visited – high water					
1995	5	4	4	0–1	0.8
At least 4 pools visited					
1983 (Stewart)[1]	1	1	10	—	10.0
1995–1996	5	5	39(42)[2]	5–10(12)[2]	7.8 (8.4)[2]

[1]Authors of data from previous surveys.
[2]Higher number in parentheses may include recount of the same specimens.

five pools five times and observed from 5 to 10–12 specimens with an average of 7.8–8.4 per trip. We averaged four hours per trip. Our data are comparable to the 1983 excursion.

It appears that recruitment within *S. poulsoni* is occurring, as we observed at least three size classes in Key Cave during our survey. The largest size class was approximately 40–50 mm standard length (SL), the middle size class included individuals around 30 mm SL, and the smallest individuals were approximately 12–20 mm SL.

Poulson (1963) found *Typhlichthys subterraneus* to be 9–10 mm SL at an age of 3 months and 10–23 mm SL at 5 months. We found very small specimens of *S. poulsoni* on 2 February 1995 (12 and 15 mm SL) and on 18 November 1996 (15 mm SL) in Pool F. Specimens about 20 mm SL were also observed during the above dates and on 25 January 1996 in pools A, D, and F. Assuming similar growth rates of young between these two species, spawning for *S. poulsoni* would have occurred during the summer months, which would contrast with the late winter and spring spawnings for other species of Amblyopsidae (Poulson 1963). The holotype of *S. poulsoni* is a female with ova that was collected in late May (Cooper & Kuehne 1974) which would lend support to a summer spawning. Unfortunately, we have no information on the presence of small *S. poulsoni* from April through September because of the resident maternal gray bat colony in Key Cave. Therefore spawning activity at other times of the year can not be ruled out.

The number of observed *S. poulsoni* in pools A and B was higher in the 1980's, but it is not known if these data reflect shortcomings of this method of population monitoring, natural population fluctuations, or better habitat for this time period. Even if these higher counts are accurate, all of the data indicate that the population appears to have been relatively stable and reproducing over the last three decades. This is the best determination that can be made with the extremely limited amount of data available for this highly inaccessible and federally protected fish. Surveys should continue in the future to ensure that there are no extreme decreases in the number of *S. poulsoni* observed in Key Cave.

Mark and recapture limitations

In any mark and recapture program, reliable collection of specimens is mandatory. We attempted to capture specimens of *S. poulsoni* during 14 trips, and were able to net only 3 specimens. The majority of habitat these cavefish occupy is inaccessible, the cavefish are often several meters from the potential collector, and the cavefish are very sensitive to any water movements created by a net. We baited minnow traps with pieces of crayfish, brine shrimp, and artificial fish attractants to collect Alabama cavefish. Although traps were quite effective for cave crayfish, no cavefish were ever trapped. Snorkeling and using a small aquarium dipnet would probably increase the number of specimens captured, but the difficulty in accessing most pools within Key Cave hinders any capture method.

Marking procedures used on other species of amblyopsids have been shown to cause no permanent disabilities (Smith & Welch 1978, Means & Johnson 1995,

Brown 1996). However, *S. poulsoni* is an endangered species with possibly one of the lowest population sizes of any North American freshwater fish. Therefore extensive testing on *T. subterraneus* would be necessary to demonstrate that any procedure would not be harmful to *S. poulsoni*. Because of the uncertainty of recapture, the preferred marking method would be a unique subcutaneous injection of dye for each specimen. Individuals could then be recognized from a distance and would not have the added stress of a recapture event. But until a reliable and safe procedure for marking specimens of *S. poulsoni* is established, the possibility of any accurate estimation of population size is very limited.

Sympatric cavefish species

The southern cavefish, *Typhlichthys subterraneus*, had never been observed in Key Cave until we collected a 60 mm SL specimen in January 1995. The specimen was netted in Pool B, a fin clip was removed, and it was released. We presumably observed the same individual over the last two years of the survey in Pool B on three other occasions and once in Pool A. This would indicate that this fish is resident in this area of Key Cave and that there is an underwater passage connecting pools A and B. We do not know if this specimen represents a unique recent invasion event, or if this species has a long history of co-inhabiting Key Cave with *S. poulsoni*. Regardless of the history, *S. poulsoni* and *T. subterraneus* do co-occur within Key Cave.

Survey of caves in the vicinity of Key Cave

Methods

Nine caves were selected as potentially harboring additional populations of *S. poulsoni* and ranged from 1.2 to 45 km from Key Cave (A–I, Figure 1; Table 2). Cobb[2] did not survey three caves to the south and west of Key Cave (Persimmon, Thomason, and McKinney caves) because previous biological surveys had shown the presence of *T. subterraneus*. With the discovery of both species of cavefish in Key Cave, we attempted to resurvey these caves. Key Spring Cave was too turbid on all three trips for Cobb[2] to determine if any aquatic cave organisms were present. Because of its close proximity to Key Cave (1.2 km), it was considered important to re-examine. The remaining caves (Bell, Elbow, Hound Dog Drop, Davis Bat, and Watkins Lake) had cavefish present that Cobb[2,3,6] could not positively identify.

Cave surveys included observing and identifying cavefish from either the shoreline of shallow or small pools or snorkeling in large deeper pools. We attempted to capture all observed cavefish with aquarium dipnets for photo-documentation, but in all cases positive identification was possible solely from in situ observations.

Results

We were successful in gaining access to seven of the nine caves in 1999. The owner of Persimmon and Thomason caves (A, B in Figure 1) denied us access based on unauthorized collecting of cave organisms

Table 2. Number of southern cavefish, *Typhlichthys subterraneus*, observed and captured during 1999 in nine caves with potential *Speoplatyrhinus poulsoni* habitat in northwestern Alabama and extreme southern Tennessee.

Cave	County/State	Distance from Key Cave (km)	Number of trips	Number observed	Number captured
Persimmon[1]	Colbert/AL	19	—	—	—
Thomason[1]	Colbert/AL	18	—	—	—
Bell	Colbert/AL	2	2	0	0
McKinney Pit	Colbert/AL	2.4	3	12	2
Elbow	Colbert/AL	2	3	1	0
Key Spring	Lauderdale/AL	1.2	2	0	0
Hound Dog Drop	Wayne/TN	37	1	2	0
Davis Bat	Lauderdale/AL	40	2	2	1
Watkins Lake	Lauderdale/AL	45	2	8	0

[1]Access denied by landowner.

by previous visitors to these caves. No cavefish were observed in two trips to Key Spring and Bell caves. Although Key Spring Cave (F, Figure 1) is extremely close to Key Cave, it has a different habitat; the entire cave is a huge slow-moving pool about 1 m deep and 155 m long. We experienced somewhat turbid waters within the cave, as did Cobb[2], and found the source to be several large common carp, *Cyprinus carpio*, stirring up the silt as they swam. We also observed a small sucker (Catostomidae) and numerous beaver cuttings. These epigean residents of Key Spring Cave indicate a direct connection to Pickwick Reservoir. Bell Cave (C, Figure 1) is located just across (south) the reservoir from Key Cave. Its only pool was very small due to the extremely low water conditions; cave crayfish were present, but no cavefish were observed.

Eleven trips to the remaining five caves resulted in the observation of 25 cavefish (Table 2). All cavefish were *T. subterraneus*; no *S. poulsoni* were observed. Of these 25 sightings, only three specimens were captured. Half of the observed specimens (12) were in Hound Dog Drop, Davis Bat, and Watkins Lake caves (G–I in Figure 1). These caves are relatively far from Key Cave (37–45 km) and they are located in a different geologic formation (Fort Payne Chert rather than Tuscumbia Limestone) (Cobb[2]). McKinney Pit and Elbow caves (D and E in Figure 1) are located just across (south) Pickwick Reservoir from Key Cave and have somewhat similar habitats to Key Cave. Twelve specimens of *T. subterraneus* were observed in McKinney Pit, which has over a dozen pools. One specimen was observed in the single, large pool at the rear of Elbow Cave.

Conclusions

No *Speoplatyrhinus poulsoni* were observed during our survey of seven caves in 1999. These data support the hypothesis that *Speoplatyrhinus poulsoni* is restricted to the Key Cave system. But given the small amount of total cavefish habitat accessible to us in each cave, these results do not eliminate the possibility of discovering a range extension for the Alabama cavefish in the future. Although McKinney Pit Cave had been 'surveyed for biological content' (Cobb[2]), we discovered a new species of cave shrimp during one trip. This is a prime example of how incomplete surveys of aquatic cave systems can be, given the limitations of sampling within this environment.

All pools in Key Cave have little to no flow; if flow is present it is restricted to water movement out of a submerged window. Key Spring Cave is an unlikely candidate for future surveys given its somewhat lotic habitat, the number of epigean species present in the cave, and the lack of a bat population. Hound Dog Drop Cave is also a flowing system, but a population of gray bats is present during the summer. This cave, along with Davis Bat and Watkins Lake may deserve continued surveys, but their distance from Key Cave and their presence in the Fort Payne Chert formation rather than Tuscumbia Limestone makes their candidacy suspect.

McKinney Pit, Elbow, and Bell caves are all located within 2.4 km of Key Cave and possess habitat that *S. poulsoni* is found in – deep pools with silt and rock bottoms and little or no flow. Additionally, McKinney Pit and Elbow caves have bat populations. Both caves are also likely to share the same aquifer; specimens of cave shrimp have been found in both and the rear of McKinney Pit is only 343 m from the back of Elbow Cave. These two caves have the greatest potential for discovering a new population of Alabama cavefish. The entrance to Bell Cave is only 30 m west from that of Elbow Cave and is probably in the same aquifer, but the pool in Bell Cave is small, especially under low water conditions, and it may not warrant intensive surveying. Finally, any unsurveyed caves in the vicinity of Key Cave that have pools should also be examined for populations of *S. poulsoni*.

Conservation status

Although the population of *S. poulsoni* in Key Cave appears to have remained relatively stable over the last three decades, disruptions within the recharge area could alter the status of this species. In January 1997 the USFWS purchased 1060 acres within the recharge area and established the Key Cave National Wildlife Refuge, which is managed by Wheeler National Wildlife Refuge personnel. This encompasses most of the high recharge area. All cotton production was immediately replaced with soybeans and corn, and chemical use was restricted. Although 80% of the refuge is currently agricultural land, it will be converted to upland forest and/or native grasslands within the next decade.

These measures should go far towards protecting the aquatic habitat in Key Cave, although urbanization of the recharge area continues. An industrial park for the city of Florence has again been proposed. Concerns include a lowering of the water table due to impermeable surfaces and runoff into the recharge area of

Key Cave from either the industrial site or from spills on the highways serving the park (USFWS personal communication). The absence of *S. poulsoni* in our survey of potential habitats outside of Key Cave further indicate that it is restricted to this single cave system, and vigilant protection of this habitat is paramount to the continued existence of this endangered species.

Acknowledgements

None of this research would have been possible without the caving expertise donated by Shane Stacy on both phases of this research, Randall Blackwood on the Key Cave survey, and Mark Grace on the survey of surrounding caves. Shane and Randall also collected some data for us in Key Cave. Kevin Roe (University of Alabama) and Jim Godwin (Alabama Natural Heritage Program) assisted in the survey of surrounding caves. Tennessee Valley Authority granted access to Key Cave, and Tuck Stone of the USFWS granted access to the Key Cave National Wildlife Refuge. Others who assisted in field work include Scott Atkins, Cesar Blanco, Herbert Boschung, John Brown, Jessica Doiron, Steve Francouer, Phil Harris, Keith Hudson, Dale Jones, Jonathan Keiser, Ron Larson, Ken Lister, Rhonda McClutcheon, Bob Meadows, Chance Mullins, David Neely, Keith Scott, Andrew Simons, Rex Strange, Larry Thompson, and Joe Tomelleri. The Hunstville Grotto of the National Speleological Society allowed us to use their map of Key Cave produced by Bill Torode, John Van Swearingen, Jim Hall, Eric Steenburn, and Lynny Byrd. Grants from the Alabama Department of Conservation and Natural Resources and the U.S. Fish and Wildlife Service, as well as funds from the University of Alabama Department of Biological Sciences supported this research.

References cited

Brown, J.Z. 1996. Population dynamics and growth of Ozark cavefish in Logan Cave National Wildlife Refuge, Benton County, Arkansas. M.Sc. Thesis, University of Arkansas, Fayetteville. 87 pp.

Cooper, J.C. & R.A. Kuehne. 1974. *Speoplatyrhinus poulsoni*, a new genus and species of subterranean fish from Alabama. Copeia 1974: 486–493.

Means, M.L. & J.E. Johnson. 1995. Movement of threatened Ozark cavefish in Logan Cave National Wildlife Refuge, Arkansas. Southwest. Nat. 40: 308–313.

Poulson, T.L. 1963. Cave adaptation in amblyopsid fishes. Amer. Midl. Nat. 70: 257–290.

Smith, P.W. & N.M. Welch. 1978. A summary of the life history and distribution of the spring cavefish, *Chologaster agassizi*, Putnam, with population estimates for the species in southern Illinois. Ill. Nat. Hist. Surv. Biological Notes (104): 1–8.

U.S. Fish & Wildlife Service. 9 September 1977. Final threatened status and critical habitat for five species of southeastern fishes. Federal Register 42 (175): 45526–45530.

U.S. Fish & Wildlife Service. 28 September 1988. Reclassification of the Alabama cavefish from threatened to endangered. Federal Register 53 (188): 37968–37970.

Part 3. Genetics and evolution

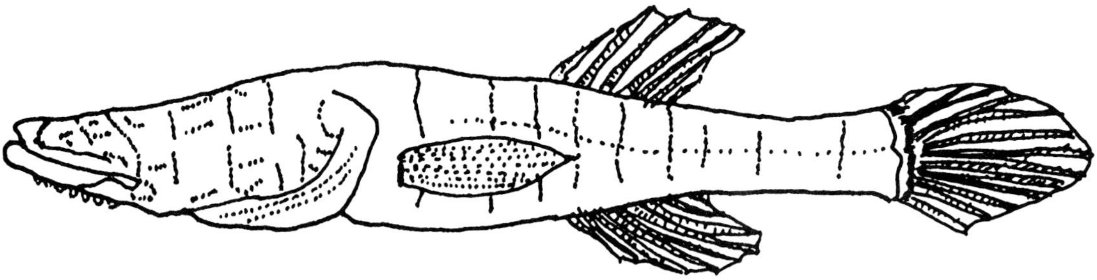

Environmental Biology of Fishes **62:** 225–231, 2001.
© 2001 *Kluwer Academic Publishers. Printed in the Netherlands.*

Genetic differentiation among populations of the cave fish *Schistura oedipus* (Cypriniformes: Balitoridae)

Richard B. Borowsky & Lisa Mertz
Department of Biology, New York University, Washington Square, NY, 10003, U.S.A.
(e-mail: rb4@scires.nyu.edu)

Received 6 March 2000 Accepted 8 December 2000

Key words: troglobitic fish, mtDNA, Thailand

Synopsis

DNA sequences were determined for portions of the mitochondrial control region and 16S rRNA genes in five populations of *Schistura oedipus,* a troglobitic fish from northwestern Thailand. Based on variation at seven sites out of 593 sequenced, the five populations fall into three groups, apparently isolated by altitude and distance. Nucleotide diversity within groups is on the order of 1.0×10^{-3} while the extra increment of diversity among groups is on the order of 5.0×10^{-3}. While the groups are distinct, the level of differentiation among them is low enough to accept provisionally that the five populations represent a single species. On the basis of 16S sequence, *Schistura oedipus* appears to be more closely related to *S. poculi* than to *S. reidi*, a proposed sister taxon.

Introduction

Thailand harbors a rich diversity of hypogean fishes, the extent of which is only gradually being revealed. Six species have been described (Weber et al. 1998), two newly discovered ones await description (unpublished data and Vidthayanon personal communication), and areas of under-explored karst promise new discoveries. Of the eight known cave fishes from Thailand, five are balitorids, two are cyprinids, and one is a silurid. These three families contain the largest numbers of freshwater fish species in Thailand: 62, 204, and 29 respectively, out of 570 in the most recent count (Vidthayanon et al. 1997).

The Thai cave fish species with the widest distribution is *S. oedipus* (Kottelat 1988). Populations of this species are known from five caves distributed over a linear range of 15 km in Mae Hong Son Province in Thailand's northwest (Figure 1). The five caves and their catalog numbers ('MH') (Dunkley 1995) are: Tham Huet (MH33), Tham Ban Louk Kow Larm (MH16), Tham Nong Pha Cham (MH91), Tham Nam Lang (MH57), and Tham Mae Lana (MH42). 'Tham' is Thai for 'cave'. One of us (RB) has visited each of these caves, and others in the surrounding area. Our tentative acceptance that this species is limited to these caves is based on the information of experienced local cavers, reviews of the caving literature on Northern Thailand (Dunkley 1995), and our field work.

Waters from these caves feed into the Nam Lang and Nam Khong rivers of the Salween river system. The heights above sea level at which cave fish occur, or at sinks, resurgences and river levels, were gleaned from topographic maps and local informants, and are given in Figure 1.

Present opportunities for gene flow among the five populations appear limited, as significant underground connections among the caves seem unlikely. In the past, such connections may have existed and, in fact, would be the most parsimonious explanation for the current distribution of the populations. The two populations in the east, in Tham Huet and Tham Nong Pha Cham are most clearly isolated by altitude. Although streams from both caves are tributary to the Nam Lang River, the caves are perched 155 and 370 m above river level, and feed to opposite shores. While cave fish in Tham Nam Lang (the type locality, Kottelat 1988) and Tham Mae Lana are found in internal tributaries only a few

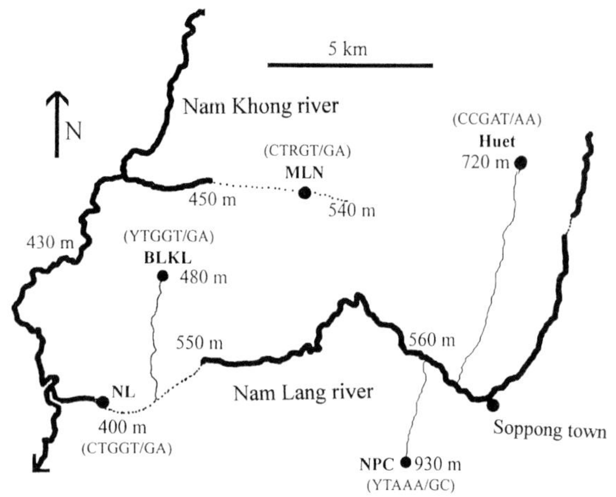

Figure 1. Distribution of known populations of *S. oedipus* in Mae Hong Son Province, Thailand. Each dot represents the location of a distinct cave fish population. Altitudes are heights above sea level in meters at cave fish localities or river base levels. Details of cave passage are not shown, although the main passages of Mae Lana (MLN) and Nam Lang (NL) caves were explored. Passages of Nong Pha Cham (NPC), Ban Louk Kow Larm (BLKL) and Huet caves have not been explored. The resurgence of Tham Huet is known, while those of NPC and BLKL caves are hypothesized. The seven letter sequences (NNNNN/NN) represent the bases found at the five variable sites in control region and the two variable sites in 16S rRNA.

meters above the main river levels, both caves have well defined resurgences that are confluent above ground, which would limit subterranean flow between them. Tham Ban Louk Kow Larm is located between Nam Lang and Mae Lana caves and is presumed to drain south to the waters of the Nam Lang, although this has not been proven.

Thus, *S. oedipus* is a species with five populations presently isolated from one another. Although gene flow among them might still be possible via surface waters, this route, if at all practical, must be minor. Under these circumstances, genetic divergence among the populations is expected, although its extent cannot be predicted in the absence of a detailed history of the erosion of the karst and the isolating events. If the isolation of the populations is very recent, divergence might be too small to detect. If of longer duration, its measured magnitude could provide clues to the chronology of isolation.

To test for isolation among these populations, we looked at the sequences of two mitochondrial DNA segments: control region (d-loop) and 16S rRNA. Control region evolves relatively rapidly while 16S rRNA evolves more slowly (Stepien & Kocher 1997), and the pair was chosen to provide a range of temporal and systematic resolution.

Material and methods

DNA for amplification was prepared from ethanol fixed fin clips obtained in the field using standard techniques (Sambrook et al. 1989). Tissue was digested in proteinase k and the digests were extracted twice with equal volumes of phenol, twice with phenol : chloroform : IAA (25 : 24 : 1) and twice with chloroform : IAA (24 : 1). DNA was ethanol precipitated from the extract. Fin clips were available from five individuals from Tham Huet, nine from Tham Nong Pha Cham, 13 from Tham Mae Lana, four from Tham Ban Louk Kow Larm and five from Tham Nam Lang, although full sequences were not obtained from all individuals for both gene regions. Sample sizes for the sequences are given in Table 4.

Species available as potential outgroups were *Schistura reidi, S. poculi* and *S. paucicinta.* All three are found in the Salween drainage in the Mae Hong Son and Tak regions, and *S. reidi* has been suggested as a likely sister to *S. oedipus* (Kottelat 1988, 1990).

Initial primer sequences were obtained from the literature (control region – Meyer et al. 1990, Mugue 1996; 16S rRNA – Parker & Kornfield 1996) and additional primers for sequencing were designed from *Crossostoma lacustre* sequence (Tzeng et al. 1992) and preliminary sequence data. The initial primers (5′–3′) were: LProf (AACTCTCACCCCTAGCTCCCAAAG), Hl6498+3 (CCTGAAGTAGGAACCAGATGCAG), 16SvF (TACATAACACGAGAAGACC), and 16SvR (GTGATTGCGCTGTTAATCC). In *S. oedipus*, these primer sets amplified fragments of 366 and 227 bases from the control region and 16S rRNA regions, respectively. The amplified 16S fragments correspond to bases 3137 through 3364 and the control region fragments to bases 16531 through 327 in the *Crossostoma lacustre* mtDNA map (Tzeng et al. 1992).

Amplified fragments were purified for sequencing using Wizard PCR preps (Promega) and sequenced using the ABI ‘Big Dye’ terminator sequencing kit. Termination products were run on an automated sequencer (ABI 377), and sequences were assembled and aligned using associated software (Factura 2.2 and AutoAssembler 1.3 – PE Cetus and DNASIS 2.0 – Hitachi).

The sequence data were used to calculate nucleotide diversities (π) for both gene regions, using the formulae of Nei & Li (1979). For calculations of π, all 366 control region and 227 16S rRNA bases were used. In addition, pair-wise distances (uncorrected 'p') were calculated for each gene region separately among populations and species using PAUP* (ver. 4.0b4a, Swofford[1]). For phylogenetic analyses, only a shorter segment of control region (282 bases) could be aligned because the primers did not work well in potential outgroup species. In aligning 16S rRNA sequence of *S. oedipus* with the *S. poculi* outgroup sequence, an insertion/deletion increased the length from 227 to 228 positions.

Phylogenetic analyses were done using PAUP*. Parsimony analysis was performed using full heuristic searches of bootstrapped data (1000 replications) treating characters as unordered. Distance analysis (uncorrected 'p') was performed by neighbor-joining of bootstrapped data (1000 replicates). Trees were obtained for both data sets separately and for a combined data set including sequence from both genes. All three outgroup species were used in the analysis of 16S rRNA data. Control region was more difficult to sequence in the outgroup species and only data from *Schistura poculi* was available for analysis of the control region and combined data set.

Results

Both primer sets amplified appropriate mtDNA regions for sequencing. The amplified fragments, exclusive of primers, were 366 bases in length (control region) and 227 bases in length (16S rRNA). A search of the GenBank database using our putative control region sequence from *S. oedipus* gave numerous hits, the first 90 (= all) of which were from teleost d-loop. A similar search using putative 16S rRNA sequence gave 225 hits, the vast majority of which were teleost 16S rRNA. In both searches, the closest matches were from cobitoids. We conclude that we amplified and sequenced what the primers were designed to amplify. The sequence will be submitted to GenBank.

Five sites were variable among populations in control region and two in the 16S rRNA gene (Tables 1, 2). The bases at these sites are shown in Figure 1 in the context of their geographic distribution.

The relationships among populations were clarified by phylogenetic analysis. The data from both regions were combined for analysis and bootstrapped (with resampling, 1000 replicates) for parsimony analysis (Figure 2). The three populations at or close to river level form one significant cluster (bootstrap value = 0.65). This group then clusters with Tham Huet (bootstrap = 0.73). Tham Nong Pha Cham is basal in the tree. Distance analyses revealed essentially the same result, with the only difference being an apparent cluster of Tham Ban Louk Kow Larm and Tham Nam Lang.

Genetic distances among the populations of *S. oedipus* are low relative to those among species (uncorrected 'p', Table 3). For both gene regions, interpopulational distances average 0.003–0.004 while interspecies distances are an order of magnitude higher, ranging from 0.04 to 0.16. For 16S rRNA, the distance between *S. oedipus* and *S. reidi* (0.13) is about twice that between *S. oedipus* and *S. paucicinta* (0.07) or *S. poculi* (0.09).

Multiple individuals from each population of cave fish (range 2–9, mean = 4.8) were sequenced for each gene, so the data were suitable for estimates of nucleotide diversity (π) within and between populations. As expected, π was greater for control region sequence than for 16S rRNA (Table 4). Sequence variation within populations averaged less than 10^{-3} for control region and was not detected for 16S rRNA. The live populations seemed to fall into three sets on the basis of sequence variation. One set was composed of Nam Lang, Mae Lana and Ban Louk Kow Larm caves, while Tham Nong Pha Cham and Tham Huet were each unique. In essence, these are the same relationships revealed by parsimony analysis.

Nucleotide diversity was low within sets and was higher among sets. For control region, diversities averaged 1.2×10^{-3} within sets and 7.8×10^{-3} among. For 16S rRNA, diversities averaged 0.00×10^{-3} within sets and 5.0×10^{-3} among ('within set' includes 'within population' figures). Taking the within set averages as baselines, the extra diversity attributable to population divergence was on the order of 5.0×10^{-3}.

Discussion

The five populations of *S. oedipus* exhibit a consistent genetic divergence from one another. Among

[1] Swofford, D.L. 2000. PAUP*. Phylogenetic analysis using parsimony (* and other methods), Version 4. Sinauer Associates, Sunderland.

Table 1. 16S rRNA sequence data for the five cave populations of *S. oedipus*, *S. paucicinta*, *S. poculi* and *S. reidi*. Bases 1 and 227 correspond to positions 3137 and 3364 in the Crossostoma map (Tzeng et al. 1992). See caption to Figure 1 for population abbreviations.

	10 20 30 40 50	
NPC	TTTGGAGCTTAAGGTACAAATCCACCTACGTTAAACAGCTTC-CTAAACA	49
HUET	....-.......	49
MLN	....-.......	49
BLKL	....-.......	49
NL	....-.......	49
pauci	....................C.....C.T.............--...G..	48
pocul	....................C.....C.T..C.........A-....T..	49
reidi	.....................T...............A...AC..T..T.	50
	60 70 80 90 100	
NPC	AGCACAAACCTAATAGAATATGAAATTTTACCTTCGGTTGGGGCGACCGC	99
HUET	....	99
MLN	....	99
BLKL	....	99
NL	....	99
pauci	...G......CG..........G.G.......................A.	98
pocul	...G.............G....G.G.......................A.	99
reidi	..TGT.............A...G.........................AT	100
	110 120 130 140 150	
NPC	GGAGAAC-AAAAATCCTCCAAGTGGATTGGGGACTTCCCTAAAACCAAG-	147
HUET	.......-..	147
MLN	.......-..	147
BLKL	.......-..	147
NL	.......-..	147
pauci	.......A.G................C....A.........G.......A	148
pocul	.......A.G................C....A.........G.......-	148
reidi	.......A.......................ACAAA............A-	149
	160 170 180 190 200	
NPC	AAAGACATTTCCAAGTCACAGAAAATCTGACCAAATATGATCCGACCACC	197
HUET	..............A...................................	197
HUET	....	197
MLN	....	197
BLKL	....	197
NL	....	197
pauci	.W..	198
pouel	....G.....	198
reidi	..TA...CC..T............T.........T.........G..-T.	198
	210 220 230	
NPC	CAGATCGATCAACGAACCAAGTTACCCTAG	227
HUET	A.............................	227
MLN	A.............................	227
BLKL	A.............................	227
NL	A.............................	227
pauci	.G.C..........................	228
pocul	.G.GC.........................	228
reidi	AC.GC.........................	228

Table 2. Control region sequence data for the five cave populations of *S. oedipus* and *S. poculi.* Bases 1 and 366 correspond to positions 16531 and 327 in the Crossostoma map (Tzeng et al. 1992). See caption to Figure 1 for population abbreviations.

	10 20 30 40 50	
NPC	CCAGAATTYTAAATTAAACTATTTTCTGTACCTTATGGTATAGTACATTA	50
HUET	........C...	50
MLN	........C...	50
BLKL	........C...	50
NL	........C...	50
pocul	........C......................TA.................	50
	60 70 80 90 100	
NPC	TATGCATAATATTACAT-AATGTGCTAAATACATTTATGTATAATCACCT	99
HUET	.................-................C...............	99
MLN	.................-................................	99
BLKL	.................-................................	99
NL	.................-................................	99
pocul	.................T....C.T.........G...........C..-	99
	110 120 130 140 150	
NPC	ATTGAAGAATTAGACCATAAAGCAAGTACTAAAAATTAAGACGTACATAA	149
HUET	....	149
MLN	....	149
BLKL	....	149
NL	....	149
pocul	...AG.....T.C.............C....TT......GTA........	149
	160 170 180 190 200	
NPC	CACATACTGTCAATACTCAATATTAAGTTATATAAAACATGGAAAATCCA	199
HUET	........................G.........................	199
MLN	........................R.........................	199
BLKL	........................G.........................	199
NL	........................G.........................	199
pocul	T....C.............GA...RR............-G.....T.T..	198
	210 220 230 240 250	
NPC	TAATACCCCTAAAAGTCCATCATAAATGTTTCCTTGTATGACTCAACCCA	249
HUET	....	249
MLN	.G..	249
BLKL	.G..	249
NL	.G..	249
pocul	....C.............T.G.A..CATC.-.....C.T..C........	247
	260 270 280 290 300	
NPC	CATCTATCAGACAATTCTTAATGTAGTAAGAACCACCAACCAGTTTACAT	299
HUET	....	299
MLN	....	299
BLKL	....	299
NL	....	299
pocul	...T..C.T.......................??????????????????	297
	310 320 330 340 350	
NPC	AATTGTTAAAAATCCATGATAGAATCAGGGACTAAACTTGGAAATACGGT	349

Table 2. (Continued)

```
HUET    ..................................................   349
MLN     ..................................................   349
BLKL    ..................................................   349
NL      ..................................................   349
pocul   ??????????????????????????????????????????????????   347

                  360
                   .
NPC     ATTTAGTGAATTATTTC   366
HUET    .................   366
MLN     .................   366
BLKL    .................   366
NL      ..A..............   366
pocul   ?????????????????   364
```

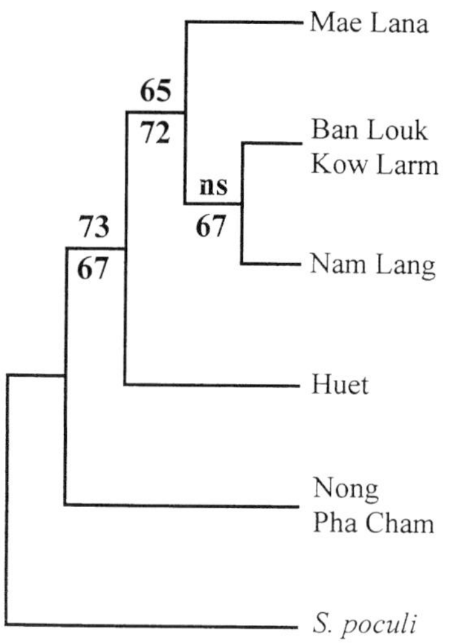

Figure 2. Relationships among the five cave populations of *S. oedipus*. Bootstrap values for the parsimony analysis are above the branches. Bootstrap values for the neighbor joining analysis are below the branches. The outgroup in both analyses was *S. poculi.*

Table 3. Genetic distances ($\times 10^{-3}$) among populations of *Schistura* species. Figures above the diagonal are for control region and below are for 16S rRNA. Population symbols for *S. oedipus* as in Figure 1. Other species are *S. paucicinta*, *S. poculi* and *S. reidi.*

	Huet	NPC	MLN	BLKL	NL	*S. pauci*	*S. pocul*
Huet	—	4	4	4	4	NA	134
NPC	9	—	4	7	7	NA	134
MLN	4	4	—	0	0	NA	137
BLKL	4	4	0	—	0	NA	138
NL	4	4	0	0	—	NA	137
S. pauci	79	70	75	75	75	—	NA
S. pocul	92	84	88	88	88	40	—
S. reidi	133	133	128	128	128	158	141

Table 4. Nucleotide diversity ($\times 10^{-3}$) within and among *S. oedipus* populations. Figures in bold face are for control region and in plain type are for 16S rRNA. Population symbols as in Figure 1. The numbers sequenced in each population were: Huet (4/5 for CR/16S), NPC (7/2), MLN (7/9), BLKL (3/3), NL (4/4).

	Huet	NPC	MLN	BLKL	NL
Huet	0.0/**0.0**	**9.3**	**7.4**	**6.4**	**5.5**
NPC	8.8	0.0/**1.3**	**7.3**	**9.5**	**9.3**
MLN	4.4	4.4	0.0/**1.1**	**2.9**	**2.0**
BLKL	4.4	4.4	0.0	0.0/**1.2**	**0.9**
NL	4.4	4.4	0.0	0.0	0.0/**0.0**

population nucleotide diversities are an order of magnitude greater than within population figures. Also, the three population clusters in the species each had at least one diagnostic base that distinguished it: Tham Huet and Tham Nong Pha Cham populations both had two, and the cluster of Tham Mae Lana, Tham Ban Louk Kow Larm and Tham Nam Lang had one. Fixed differences among populations are strong evidence of restricted gene flow.

Nevertheless, the overall differentiation among populations of *S. oedipus* was low compared to differences among species, again by an order of magnitude (Tables 3, 4). This suggests that the five populations be treated, at least for the present, as a single species, in spite of the apparently fixed diagnostic differences. A determination of whether the three groups merit taxonomic distinction is best left to the future, when more is known about phenotypic variation among populations.

Altitude may be the most significant barrier to gene flow among the cave populations. The three populations to the west, at or just above base river levels, form one of the groups. Each of the two populations to the east, both in perched waters 155 and 370 m above river level, comprises its own isolated group.

It is unknown whether the populations came from separate subterranean invasions of a common ancestor, or from a unique invasion with subsequent isolation and differentiation. The latter is more likely because of the geographically restricted distribution of the five populations, but these hypotheses await identification of the epigean sister species for critical testing.

Kottelat (1988) discussed the possible relationships of *S. oedipus*, but conceded that these would remain unclear until the phylogenetic relationships of the surface balitorids of the region were better understood. Among possible sister taxa he mentioned *S. poculi* and *S. reidi*, favoring the latter because of its widespread occurrence in Nam Mae Pai and Mae Nam Yuam basins (Salween drainage). The data on 16S rRNA presented here, as well as unpublished data (Borowsky, Kottelat & Vidthayanon in preparation) on Cytochrome b sequence suggest that *S. oedipus* is only distantly related to *S. reidi*, and much more closely related to a cluster of species containing *S. poculi*. *Schistura poculi* is also very common in the same regions as *S. oedipus* and *S. reidi*.

The degree of genetic divergence among the three groups of *S. oedipus* allows a rough estimate of the time of isolation. The nucleotide diversity distance among populations (above the base levels within) is on the order of $0.5 \pm 0.14\%$. This is based on approximately equal lengths of sequences generally accepted to be quickly and slowly evolving. Mitochondrial DNA sequences evolve at a rate of approximately 2–4% per million years (Li & Graur 1991). Thus, a first approximation to the time since separation of the groups is about 1.3 to 2.5 ± 0.4 to 0.7×10^5 y. The standard errors of estimate are based only on the expected errors in measuring divergence and do not include errors in clock calibration or consistency. As such, they overestimate the precision of the estimate by an unknown, but probably significant factor (Hillis et al. 1996). Thus, the estimated divergence times, on the order of 1×10^5 y, are useful as a guide, but should not be considered reliable to more than order of magnitude.

Acknowledgements

This work was submitted in partial fullfillment of requirements for a MS degree from NYU (LM) and was supported by grants from the National Science Foundation (INT 9605200) and the NYU Research Challenge Fund to KB. Thanks to Chavalit Vidthayanon (Thai Fisheries Department), Dean Smart (Thai Forestry Department), and John Spies (Mae Hong Son) for aid in the field and sharing their deep knowledge of the fishes, caves, and topography. Gerry Campbell, Nikolai Mugue, Steve Smith and Luis Espinasa were especially generous with time and effort in the field.

References cited

Dunkley, J. 1995. The caves of Thailand. Speleological Research Council, Ltd., Sydney. 124 pp.

Hillis, D.M., B.K. Mable & C. Moritz. 1996. Applications of molecular systematics: the state of the field and a look to the future. pp. 515–541. *In*: D.M. Hillis, C.M. Moritz & B.K. Mable (ed.) Molecular Systematics, Sinauer, Sunderland.

Kottelat, M. 1988. Two species of cavefishes from northern Thailand: description of a new blind homaloptera and a new *Nemacheilus* with degenerated eyes (Osteichthyes: Homalopteridae). Rec. Austr. Mus. 40: 225–231.

Kottelat, M. 1990. Indochinese nemacheilenes: a revision of nemacheilene loaches (Pisces: Cypriniformes) of Thailand, Burma, Laos, Cambodia and southern Viet Nam. Verlag Dr. Friedrich Pfeil, Munich. 262 pp.

Li, W-H. & D. Graur. 1991. Fundamentals of molecular evolution. Sinauer, Sunderland. 284 pp.

Meyer, A., T.D. Kocher, P. Basasibwaki & A.C. Wilson, 1990. Monophyletic origin of Lake Victoria cichlid fishes suggested by mitochondrial DNA evidence. Nature 347: 550–553.

Mugue, N. 1996. Population genetics of *Fundulus heteroclitus* (L.) in New York/New Jersey harbor estuary. Ph. D. Thesis, Rutgers, the State University of New Jersey, New Brunswick. 118 pp.

Nei, M. & W-H Li. 1979. Mathematical model for studying genetic variation in terms of restriction endonucleases. Proc. Nat. Acad. Sci. USA 76: 5269–5273.

Parker, A. & I. Kornfield. 1996. An improved amplification and sequencing strategy for phylogenetic studies using the mitochondrial large subunit rRNA gene. Genome 39: 793–797.

Sambrook, J., E.F. Fritsch & T. Maniatis. 1989. Molecular cloning. Cold Spring Harbor Laboratory Press, Cold Spring Harbor. 1659 pp.

Stepien, C.A. & T.D. Kocher. 1997. Molecules and morphology in studies of fish evolution. pp. 1–11. *In*: T.D. Kocher & C.A. Stepien (ed.) Molecular Systematics of Fishes, Academic Press, New York.

Tzeng, C.S., C.F. Hui, S.C. Shen & P.C. Huang. 1992. The complete nucleotide sequence of the *Crossostoma lacustre* mitochondrial genome. Nucl. Acids Res. 20: 4853–4858.

Vidthayanon, C., J. Karnasuta & J. Nabhitabhata. 1997. Diversity of freshwater fishes in Thailand. Office of Environmental Policy and Planning, Bangkok. 102 pp.

Weber, A., G.S. Proudlove & T.T. Nalbant. 1998. Morphology, systematic diversity, distribution, and ecology of stygobitic fishes. pp. 1179–1190. *In*: C. Juberthie & V. Decu (ed.) Encyclopaedia Biospeologica (Tome II), Societe de Biospeologie, Moulis-Bucarest.

Environmental Biology of Fishes **62:** 233–237, 2001.
© 2001 *Kluwer Academic Publishers. Printed in the Netherlands.*

Origins and relationship of cave populations of the blind Mexican tetra, *Astyanax fasciatus*, in the Sierra de El Abra

Luis Espinasa[a] & Richard B. Borowsky[b]
[a]*CEAMISH Universidad Autónoma del Estado de Morelos, Ave. Universidad # 1001, Colonia Chamilpa, Cuernavaca, Morelos 62210, México (e-mail: espinasl@buzon.uaem.mx)*
[b]*Department of Biology, New York University, 1009 Main Building, Washington Square, NY 10003, U.S.A.*

Received 20 July 2000 Accepted 11 December 2000

Key words: troglobite, RAPD, AP-PCR, cave colonization

Synopsis

The blind morph of *Astyanax fasciatus* (Pisces: Characidae) has been more thoroughly studied than any other cave inhabiting organism. Most studies of *A. fasciatus* have used individuals from different caves of the Sierra de El Abra, Mexico, and have assumed that each population independently evolved to live in the cave environment. We analyzed the relationships among several cave populations that delineate the Sierra de El Abra using RAPD markers. The results indicate that all cave populations are more closely related to one another than they are to the surface populations. This suggests that present day cave populations derived from a common ancestral stock, most likely due to a single colonization event, or alternatively, that strong gene flow among cave populations has occurred, raising precaution against considering each cave population as independent.

Introduction

A recurring problem in the study of cave organisms is how to interpret the distribution of a single species which is found in more than one cave. There are two alternative explanations for this phenomenon: (1) a single colonization event was followed by underground dispersion to the other caves, or (2) multiple colonization and subsequent convergence occurred independently in each population. In this study we address this question in the Mexican characid fish *Astyanax fasciatus* (Cuvier, 1819), *A. mexicanus* (Filippi, 1853).

A. fasciatus is divided into two morphs: the hypogean, a blind, cave-inhabiting troglobite and the epigean, a surface dwelling, eyed and pigmented morph. While the epigean morph is more or less cosmopolitan within Mexico, the hypogean morph is restricted to a 150 km portion of the Sierra Madre Oriental in northeast Mexico, in the limestone ridges of Micos, Sierra de Nicolás Pérez, Sierra de Guatemala, and Sierra de El Abra (Figure 1).

Ever since the first description of a blind, troglobitic morph of *A. fasciatus* (Hubbs & Innes 1936), scores of papers have been written on the species (reviewed in Reddell 1981, Wilkens 1988, Romero 2001). Essentially every aspect of its morphology, behavior, and genetics has been examined at least once. Despite this intensive examination however, considerable confusion remains about the origin and relationship of the 29 cave populations described thus far, of which the great majority are found in the Sierra de El Abra. Most research on *A. fasciatus* has also been conducted on El Abra populations and in these studies, both monophyly and polyphyly of the cave populations have been assumed and no clear consensus has been reached (reviewed by Mitchell et al. 1977). Gene flow among caves has sometimes been invoked, but not backed with genetic data.

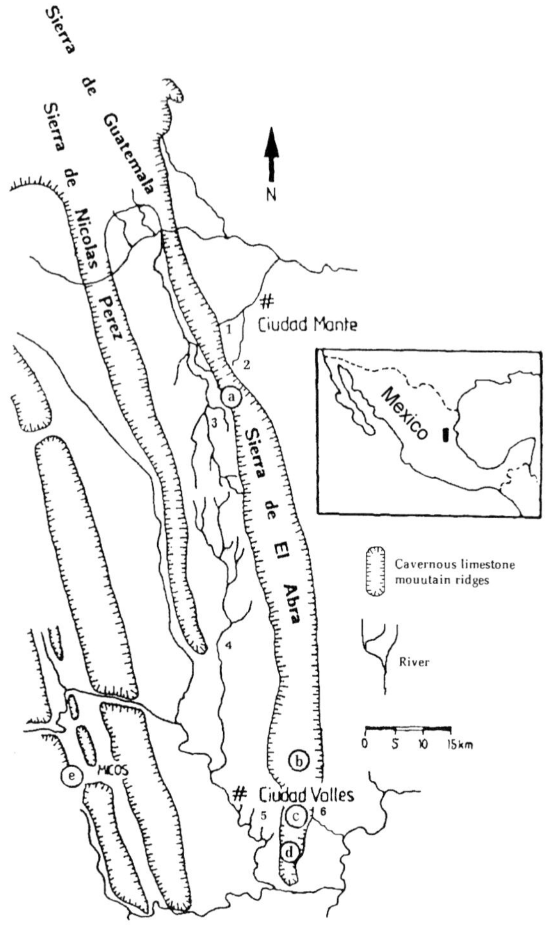

Figure 1. Geographical map of the Huastecan province inhabited by troglomorphic *Astyanax*. Cave localities used in this paper are represented by letters (a) = Pachón, (b) = Tinaja, (c) Curva, (d) = La Cueva Chica, (e) = Río Subterráneo, and surface localities by numbers (1) = Mante, (2) = Santa Clara, (3) = Arroyo B, (4) = Yerbaniz, (5) = Valles, (6) = Choy.

In order to elucidate the origins and relationships of the blind morph of *A. fasciatus* we analyzed the relationships of several cave and surface populations of this species distributed throughout most of the Sierra de El Abra using arbitrarily primed polymerase chain reaction (AP-PCR or RAPD; Welsh & McClelland 1990, Williams et al. 1990).

The localities for this study were selected to delineate all the Sierra de El Abra as follows (Figure 1): Pachón is the northernmost cave in the Sierra with fish populations, La Cueva Chica cave is in the southernmost tip, and Tinaja and Curva caves are located in-between. Surface collections were made in various locations along the perimeter of the Sierra; Mante and Santa Clara in the north and northeast; Arroyo B, Valles and Yerbaniz in the west; and Choy in the southeast. A population in the Micos area from Río Subterráneo cave, which is known to have an origin independent from any Sierra de El Abra cave populations (Wilkens 1988), was also included. As the outgroup, *A. fasciatus* individuals collected in southern México from the Granadas River were selected. The Granadas River is part of the Balsas drainage system which flows into the Pacific Ocean. Obregon-Barbosa et al. (1994) has argued that this population should be classified as a different species from the populations in northern drainage systems of the Gulf of Mexico, to which all Sierra de El Abra localities belong.

Materials and methods

Fish samples were collected from the Sierra de El Abra (Pachón, Tinaja, Curva and La Cueva Chica caves), Micos region (Río Subterráneo cave) and from the surface rivers that surround the Sierra de El Abra (Mante, Santa Clara, Arroyo B, Valles, Yerbaniz and Choy). Cave and surface populations were as designated by Mitchell et al. (1977). Specimens collected from the Granadas River, a tributary to the Amacuzac River, northeast of Taxco, Guerrero, México were used as an outgroup. Samples were either preserved in 95% ethanol or kept alive in the laboratory. Specimens were deposited at New York University, in the laboratory of R. Borowsky. DNA was isolated from tail fin tissue by phenol/chloroform extraction and AP-PCR was performed using procedures similar to those previously described (Welsh et al. 1991, Borowsky et al. 1995). Two individuals per locality were analyzed. Samples were amplified in three conditions. Each 20 μl reaction contained 2.5, 10.0, or 50.0 ng of DNA, in 10 mM Tris (pH 8.3), 4 mM $MgCl_2$, 0.2 mM each dNTP, 0.5 U Taq polymerase (Amplitaq, Perkin Elmer–Cetus) and 1 mM oligonucleotide primer. Four reactions were done, each with a single primer. Primers used were: Cp1=5'gatgagttcgtgtccgtacaactgg3'; 3138=5'gtcttgttggagatgcacgtgccccttgc3'; 3139=5'ggagaaattatggagggaaat3', 3140=5'gttcctcaggatcaaagtatgtac3'.

Reactions were cycled five times through the following temperature profile in a Hybaid thermocycler: 94°C for 70 s to denature the template, 40°C for 5 min to allow arbitrary priming, and 72°C for 3 min to allow primer extension. These low stringency cycles were followed by 35 higher stringency cycles: 94°C, 50°C for

1 min, and 72°C for 90 s. Reaction products were run on agarose gels and of the three reactions per individual with different DNA concentrations, the two with highest intensity and clarity of bands were selected. These were then run in 45 cm long, 6% polyacrylamide gels (29 : 1, acrylamide : N,N´-methylene-bis-acrylamide, w/w) and silver stained as in Espinasa & Borowsky (1998). These highly sensitive methods allow visualization of DNA bands of low concentration and thus more bands are observed per gel. All products of a single primer, representing two individuals, with two concentrations, from each locality were run in a single gel, such that four gels were obtained. The fragment distributions thus visualized (AP-PCR fingerprints) were compared among populations using a size match criterion within the gel, and each uniquely sized band was taken to be a character and the character states were scored as 'present' or 'absent', as in Borowsky et al. (1995). Bands that showed variability within the different concentration reactions of the same individual were eliminated from the analysis. Phylogenetic trees were obtained by parsimony methods in which variation between the two individuals within a population was set as polymorphic characters using PAUP 3.1.1.[1] For bootstrap analyses, 1000 replicates were performed (Felsenstein 1985). The minimum number of changes (inferred synapomorphies) were calculated with the aid of MacClade 3.08.[2]

Results

Each gel showed approximately 22 bands (Figure 2) and a total of ninety bands were scored. The number of bands from any individual ranged from 39 to 47. All of the most parsimonious trees obtained show two groups, one formed by the Sierra de El Abra cave populations, and the other by the surface and Micos populations. There are small variations in the branching pattern within the surface/Micos group, but the two groups always fall out together. The number of steps (character changes) in these trees is 185. All branch rearrangements in which Sierra de El Abra hypogean morphs are polyphyletic increases the tree length by at least four steps. That the Sierra de El Abra cave populations are monophyletic as compared to the surface populations is supported by a 76% bootstrap (Figure 3) and by a minimum of 7 changes (inferred synapomorphies).

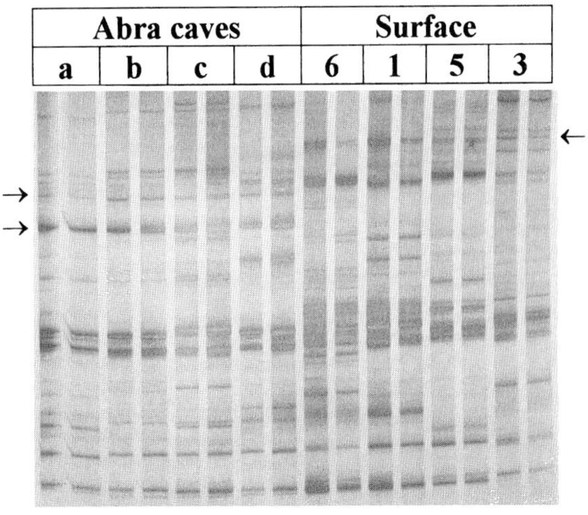

Figure 2. AP-PCR fingerprint fragment distribution. Two amplification products of a single individual per locality are shown. Marked with arrows on the left are two bands (inferred synapomorphy) shared by the Sierra de El Abra cave populations. Marked with an arrow on the right is a band shared by surface populations. Cave localities used are represented by letters (a) = Pachón, (b) = Tinaja, (c) = Curva, (d) = La Cueva Chica, (e) = Río Subterráneo, and surface localities by numbers (1) = Mante, (2) = Santa Clara, (3) = Arroyo B, (4) = Yerbaniz, (5) = Valles, (6) = Choy.

Discussion

Cave populations of *A. fasciatus* within the Sierra de El Abra are more closely related to one another than they are to the surface populations. This suggests that all Sierra de El Abra cave populations derive from a common ancestral stock, most likely due to a single colonization event of the cave environment in the Sierra de El Abra. Alternatively, it can be proposed that at some point in time, there was a strong subterranean gene flow among the caves. Our RAPD data also support Peters et al. (1975) enzyme experiments, which showed that the Micos morphs are of separate origin and phylogenetically younger than those of Sierra de El Abra.

Regardless of which scenario best explains the genetic relationship of Sierra de El Abra cave populations, it is likely that cave connections throughout the Sierra had been extensive in the past because cave populations (Pachón and La Cueva Chica) are located as far away as 80 km from each other. Since it is unlikely that

[1] Swofford, D.L. 1993. PAUP: phylogenetic analysis using parsimony, Version 5.3.3. Illinois Natural Survey, Champaign.

[2] Maddison, W.P. & D.R. Maddison. 1999. MacClade: analysis of phylogeny and character evolution, Version 3.08. Sinauer Associates, Sunderland.

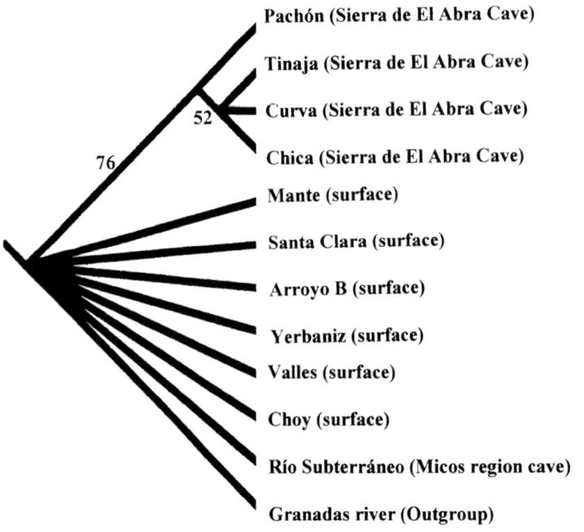

Figure 3. Fifty percent bootstrap tree from the parsimony analysis of the RAPD data from cave populations from Sierra de El Abra caves (Pachón, Tinaja, Curva and La Cueva Chica), Micos region cave (Río Subterráneo), and surface river populations (Mante, Santa Clara, Arroyo B, Yerbaniz, Valles and Choy). *Astyanax* from the Balsas drainage system (Granadas River) is used as an outgroup.

a single cave spanned the whole 80 km of the range at any one time, underground dispersal probably occurred progressively through time via the formation and erosion of the cave systems in this area. Such movements of drainage divides, with the subsequent capture of one population from one basin to another, have been recorded in cavernicole *Gammarus minus* by Culver et al. (1995) in West Virginia, and are equally feasible for *A. fasciatus* in the Sierra de El Abra.

Finally and as a precautionary note, most studies on *A. fasciatus* have used individuals from Pachón and La Cueva Chica caves of the Sierra de El Abra and have assumed them to be populations which independently evolved to live in the cave environment. Publications dealing with rates of troglobitic evolution have often assumed that the Pachón population was the first to colonize the cave environment (e.g. Wilkens 1988). This interpretation is not supported by our data. A more consistent conclusion is that different troglobitic morphologies result from different levels of gene flow from surface populations (Mitchell et al. 1977), and from the independent acquisition of troglobitic characters through natural selection or the accumulation of neutral mutations. This would also explain why the genetic basis for eye loss is different in the Pachón and La Cueva Chica cave populations (Wilkens 1971). Finally, it also should be taken into consideration the fact that gene flow and even introgressive hybridization between surface and hypogean populations may have occurred after the original cave population became isolated (Romero 1983, Langecker et al. 1991).

Acknowledgements

We thank the Mexican government and SEDESOL for providing the necessary permit for collections (Permit # A0070200347). This work was funded by an NYU RCF award and an Axelrod Foundation grant to RB. We thank Oscar Dorado, for general support, as well as two anonymous reviewers and Aldemaro Romero for their valuable suggestions.

References cited

Borowsky, R., M. McClelland, R. Cheng & J. Welsh. 1995. Arbitrarily primed DNA fingerprinting for phylogenetic reconstruction in vertebrates: the *Xiphophorus* model. Mol. Biol. Evol. 12: 1022–1032.

Culver, D.C., T.C. Kane & D.W. Fong. 1995. Adaptation and natural selection in caves. The evolution of *Gammarus minus*. Harvard University Press, Cambridge. 233 pp.

Espinasa, L. & R. Borowsky. 1998. Evolutionary divergence of AP-PCR (RAPD) patterns. Mol. Biol. Evol. 15: 408–414.

Felsenstein, J. 1985. Confidence intervals on phylogenies: an approach using the bootstrap. Evolution 5: 359–367.

Hubbs, C.L. & W.T. Innes. 1936. The first known blind fish of the family Characidae: a new genus from México. Occ. Pap. Mus. Zool. Univ. Mich. 342: 1–7.

Langecker, T.G., H. Wilkens & P. Junge. 1991. Introgressive hybridization in the Pachon Cave population of *Astyanax fasciatus* (Teleostei Characidae). Ichth. Expl. Freshw. 2: 209–212.

Mitchell, R.W., W.H. Russell & W.R. Elliot. 1977. Mexican eyeless characin fishes, genus *Astyanax*: environment, distribution, and evolution. Sp. Publ. Mus. Texas Tech Univ. 12: 1–89.

Obregon-Barboza, H., S. Contreras-Balderas & M. Lozano-Vilano. 1994. The fishes of northern and central Veracruz, México. Hydrobiologia 286: 79–95.

Peters, N., A. Scholl & A. Wilkens. 1975. Der Micos-fisch Hohlenfisch in Statu Nascendi oder Bastard? Ein Beitrag zur Evolution der Höhlentiere. Z. zoöl. Syst. Evolutionsf. 13: 110–124.

Reddell, J.R. 1981. A review of the cavernicole fauna of México, Guatemala and Belize. Mem. Mus. Univ. Texas Austin Bull. 27: 1–327.

Romero, A. 1983. Introgressive hybridization in a population of *Astyanax fasciatus* (Pisces: Characidae) at La Cueva Chica. Nat. Speleol. Soc. Bull. 45: 81–85.

Romero, A. 2001. It's a wonderful hypogean life: a guide to the troglomorphic fishes of the world. Env. Biol. Fish. 62: 13–41 (this volume).

Welsh, J. & M. Mcclelland. 1990. Fingerprinting genomes using PCR with arbitrary primers. Nucl. Acids Res. 18: 7213–7218.

Welsh, J., M. Mcclelland, R.J. Honeycutt & B.W.S. Sobral. 1991. Parentage determination in maize hybrids using arbitrarily primed PCR. Theoretical and Applied Genetics 82: 473–476.

Wilkens, H. 1971. Genetic interpretation of regressive evolutionary processes: studies on hybrid eyes of two *Astyanax* cave populations (Characidae, Pisces). Evolution 25: 530–544.

Wilkens, H. 1988. Evolution and genetics of epigean and cave *Astyanax fasciatus* (Characidae, Pisces). Support for the neutral mutation theory. Evol. Biol. 23: 271–367.

Williams, J.G., A.R. Kubelik, K.J. Livak, J.A. Rafalski & S.V. Tingey. 1990. DNA polymorphisms amplified by arbitrary primers are useful as genetic markers. Nucl. Acids Res. 18: 6531–6535.

Environmental Biology of Fishes **62:** 238, 2001.
© 2001 *Kluwer Academic Publishers. Printed in the Netherlands.*

Threatened fishes of the world: *Caecobarbus geertsii* Boulenger, 1921 (Cyprinidae)

Graham S. Proudlove[a] & Aldemaro Romero[b]
[a]*Department of Zoology, Manchester Museum, Manchester University, Manchester, M13 9PL, U.K. (e-mail: g.proudlove@umist.ac.uk)*
[b]*Environmental Studies Program and Department of Biology, Macalester College, 1600 Grand Ave., St. Paul, MN 55105-1899, U.S.A.*

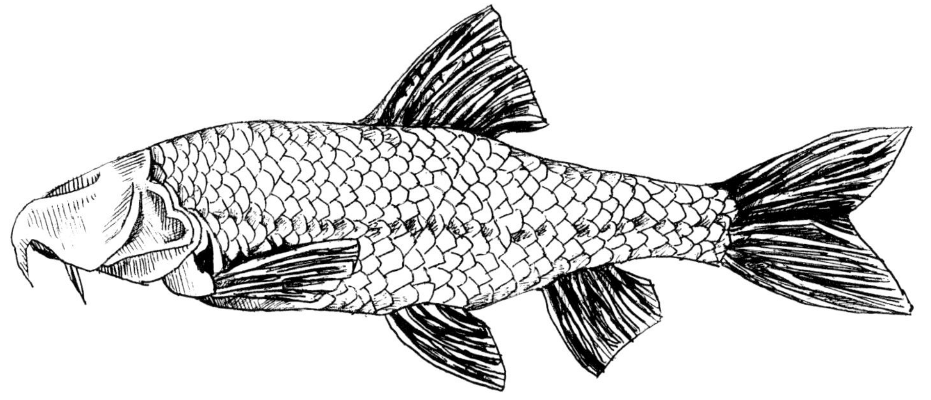

Common name: Congo blind barb (E), barbu aveugle (F). **Conservation status:** Vulnerable (VU A1e, B1 + 2e, D2), World Conservation Monitoring Centre, CITES Appendix II. **Identification:** It is very similar to primitive species of the genus *Barbus*, except for the lack of externally visible eyes and pigmentation. Laterally compressed body. Snout rounded and depressed by the mouth. D II/7-8, A III/5, P I/11-15, V I/7-8. Upper lobe of the caudal fin slightly longer than the lower one. Vestigial eyes present. Maximum length: 110 mm. It has very thin ctenoid scales that on the lateral line number 28–29. Head scaleless (Boulenger 1921). **Distribution:** In lower Congo near Kanka (formerly Thysville) near Mbanza-Ngungu (5° 18′ S, 14° 50′ E), Congo River system (Heuts & Leleup 1954). **Abundance:** The seven caves where this fish has been found may total 7000 individuals. **Habitat and ecology:** They live in caves characterized by cyclic food scarcity (Heuts 1951). **Reproduction:** They spawn in a side to side fashion and produce demersal adhesive eggs. Young animals are observed only after the rainy season. Longevity may surpass 15 years. **Threats:** It is a species commercially traded as an aquarium novelty. Large numbers have been exported in the past to industrialized nations. In 1993 alone a CITES certificate was issued to import 1500 individuals into the Unites States. Since they have never been bred in captivity, demand is constant. **Conservation action:** This is the only hypogean fish whose trade is regulated by CITES. **Conservation recommendations:** Collection should be limited for scientific purposes only and for good reasons. Protection of its habitat should be carried out. Yet, probably the best way to protect this species would be by developing a protocol for its captive breeding which will eliminate the demand for capturing it in its natural habitat. **Remarks:** It tends to aggregate based on conspecific odors, is scotophilic (moves away from light), and responds to the alarm substance (Jankowska & Thinès 1982). This species was the first troglomorphic species described outside the American continent. A great deal of experimental work has been carried out on the behavior and physiology of this species (Romero & Benz 2000).

Boulenger, G.A. 1921. Description d'un poisson aveugle decouvert par M.G. Geerts dans la grotte de Thysville (Bas-Congo). Rev. Zool. Afr. 9: 252–253

Heuts, M.J. 1951. Ecology, variation and adaptation of the blind African cave fish *Caecobarbus geertsii* Blgr. Ann. Soc. Roy. Belg. 82: 155–230.

Heuts, M.J. & N. Leleup. 1954. La géographie et l'ecologie des grottes du Bas-Congo. Les habitats de *Caecobarbus geertsi* Blgr. Ann. Mus. Roy. Congo Belge 35: 10–71.

Jankowska, M. & G. Thinès. 1982. A comparative study of group density in cave and epigean fishes. Behav. Proc. 7: 281–294.

Romero, A. & K. Benz. 2000. The unsung heroes of speleology. Nat. Speleol. Soc. News 58: 106, 126.

Environmental Biology of Fishes **62:** 239–249, 2001.
© 2001 *Kluwer Academic Publishers. Printed in the Netherlands.*

Morphological and physiological correlates of evolutionary reduction of metabolic rate among amblyopsid cave fishes

Thomas L. Poulson[a]
Department of Biological Sciences, University of Illinois-Chicago, Chicago, IL 60607-7060, U.S.A.
[a]*Current address: 318 Marlberry Circle, Jupiter, FL 33458-2850, U.S.A.*
(e-mail: tomandliz@adelphia.net)

Received 19 July 2000 Accepted 15 December 2000

Key words: cave adaptation, gill area, ventilation rate and amplitude, brain and muscle tissue metabolic rate, thyroid activity estimates, *Chologaster cornuta*, *Chologaster agassizi*, *Typhlichthys subterraneus*, *Amblyopsis spelaea*, *Amblyopsis rosae*, deep-sea fishes, convergent and parallel evolution

Synopsis

I report on multiple correlates of reduction in metabolic rate during evolutionary adaptation to caves in the fish family Amblyopsidae. The family includes six species with one surface-dweller, *Chologaster cornuta*, a facultative cave dweller, *Chologaster agassizi*, and four obligate cave-dwellers that, on the basis of eye degeneration, have been isolated in caves for increasing times: *Typhlichthys subterraneus* < *Amblyopsis spelaea* < *Amblyopsis rosae* < *Speoplatyrhinus poulsoni*. Of those traits I examined, the strongest correlates of reduction in whole fishes metabolic rate were reduction in ventilation frequency and volume > brain metabolic rate > gill surface area. Relative amplitude declined and the estimated ventilation minute-volume decreased 5.9 fold from a value of 1.12 to 0.19 ml. Log brain rate of oxygen consumption was directly related to log body mass, $b = +0.75$, and decreased from 1.5 to 0.06 μl per mg dry mass per hour among species. No single gill trait accounted for the reduction in total lamellar surface area from 228 to 103 $mm^2\ g^{-1}$. There was no reduction in muscle metabolic rate or histological indices of thyroid activity. Log muscle rate of oxygen consumption was inversely related to log body mass, $b = -0.60$, but did not differ among species. The mean total thyroid follicle volume among species, from 0.010 to 0.020 $mm^3\ g^{-1}$ of adult fishes, showed no relation to interspecific differences in whole fish metabolic rate. In the general discussion I show that no trait that could contribute to a lowered metabolic rate is in the same rank order as metabolic rate, though collectively the reductions were in the same order as metabolic rates. I explain that this is as expected with convergent evolution of complex traits. Finally I discuss the literature about central nervous system control of metabolic rate.

Introduction

It has been argued that caves are energy-limited and this explains why many obligate cave species (troglobites) have reduced metabolic rates (Poulson 1964). Direct measures of reduced metabolic rate were oxygen consumption and indirect measures were ventilation frequency and weight loss during starvation. To explain possible physiological and morphological bases for this energy conservation in cave animals, I here report on a family of fishes, the Amblyopsidae, that have evolutionary reduction metabolic rates (Poulson 1963, 1985) from surface, to facultative cave (troglophilic), and to obligate cave species (troglobitic). The surface species is the swamp fish *Chologaster cornuta*, the troglophile is the spring-cave fish *Chologaster agassizi*, and the three troglobites, that have been isolated in caves for increasing evolutionary time based on increasing eye degeneration, are *Typhlichthys subterraneus*, *Amblyopsis spelaea*, *Amblyopsis rosae*, and *Speoplatyrhinus poulsoni*. The reduction in standard, routine, and active metabolic rates follows this same order (Poulson 1963, 1985). In this paper I present data on morphological and physiological traits that each may contribute to reduction in these whole animal metabolic rates. I consider gill surface area, ventilation frequency and relative amplitude, muscle and brain metabolic rate, and histological indices of thyroid activity. And, I discuss how my data relate to convergence and parallelism in cave fishes.

The Amblyopsidae have been studied from a number of points of view. Rosen (1962) analyzed their phylogenetic relationships and Woods & Inger (1957) did a systematic revision. I have extended Eigenmann's pioneering work on eye structure (1909) to all aspects of their biology (Poulson 1961, 1963, 1985). In my initial work I predicted the existence of a species with still further paedogenetic cave adaptation and Cooper & Kuehne (1974) found and described a new species that fit my prediction, *Speoplatyrhinus poulsoni*.

Materials and methods

Source and care of fish

Fishes used for the present study came from several localities. The *Chologaster cornuta* came from swamps in Bertie County, North Carolina. All *Chologaster agassizi* came from cave springs in Union County, Illinois and all *Typhlichthys subterraneus* came from caves in Madison County, Alabama. The *Amblyopsis spelaea* came from caves in Lawrence County, Indiana and Breckinridge County, Kentucky. The *Amblyopsis rosae* came from caves in Newton County, Missouri and Benton County, Arkansas. As I have documented (Poulson 1961), the live food supplies for the *Chologaster* species were three to four orders of magnitude greater than for any of the cave species. Among caves the food supplies were very similar, though those in Arkansas gray bat caves were an order of magnitude higher than any other study caves during the growing season.

All spring-cave and cave fishes were maintained in aquaria in the dark except when fed waterfleas (*Daphnia magna*) and examined for health in red light. Aquarium water had similar chemistry and temperature to cave waters with pH = 7.5, total carbonate hardness ~175, and temperature 13°C. Other details about the natural habitats and laboratory care can be found in Poulson (1961).

Gill surface area

The respiratory area of the gills was determined for two 0.9–1.0 g individual of each species. The weeks of work required to make measurements precluded larger sample sizes but the total intraspecific range of variation for different aspects of gill morphology was very low just as Gray (1954) found in an extremely detailed study of many marine species. In each specimen, all eight gill bars were removed. One side was chosen for all measurements and the calculated surface area was doubled. Measurement of the length of each gill bar, the number and length of gill filaments, and number of lamellae per 1 mm of filament allowed estimation of total numbers of lamellae. All measurements were made on fresh tissue using a calibrated grid at 100×. Numbers of filaments were per 1 mm of gill bar measured for the entire length of both the proximal and distal hemibranch. Length of the gill bar was defined as the length of the arc along the bases of the filaments including the short ones at either end of each bar. Average filament length was obtained by measuring two filaments just anterior to the start of the caudal acute hook of each bar (location of longest filaments), two middle filaments, and two unstunted filaments from the anterior end of each bar and then interpolating to get an average. The number of lamellae per 1 mm of filament was obtained by counting the longest series of lamellae present on each of the six filaments studied. There was no change of density of lamellae along a filament. Surface area of a lamella was obtained by measuring length and width of 100 lamellae at 430×, on the same 8 μ sections of the head used for study of thyroid histology and doubling the area for the two sides of the lamella.

Frequency and relative amplitude of ventilation

Ventilation frequency and relative amplitude were measured by recording electrical activity associated with contraction of the branchial musculature. Fishes were placed in a stoppered glass tube submerged in an aquarium at 13 ± 0.3°C. The stopper in each end of the tube had a brass screen electrode separated from the fishes by a lucite screen. Leads from the electrodes fed through an EMG preamplifier to a Gilson Polygraph. My observations confirmed that the frequency of recorded waves and their relative amplitudes corresponded to the actual ventilation frequency and relative amplitude. During recording, water from the aquarium circulated through the 60 ml tube at a rate of 20 ml min^{-1}. Data were recorded continuously for 12–24 h. Swimming activity was detectable as gross disturbance of the ventilation pattern. The only ventilation data reported are for resting fishes that had been in the apparatus for at least two hours. Five 0.9–1.1 g fish were studied for each species.

Ventilation volumes for resting fish of 1 g (Table 4) were calculated from measured ventilation frequency and relative amplitude plus estimated branchial volume. I measured the inside widths, lengths, and heights of the branchial cavity of each species to calculate volumes in cubic mm and converted this to ml. Then I multiplied this volume in turn by the ventilation frequency in numbers per minute and the relative amplitude of ventilation to obtain ventilation volume in ml per minute.

Brain and muscle oxygen consumption

Oxygen consumptions of brain and muscle were studied at 15°C using standard Warburg respirometry. Methods used followed Ekberg (1958) exactly. The tissue slurries were made up to a volume of 2 ml with cold freshwater teleost Ringers and placed in the bottoms of mercury calibrated Warburg vessels. Each vessel contained 0.2 ml of 10% KOH on fluted filter paper in the greased center well, to absorb respired carbon dioxide, and 0.5 ml of succinate substrate in the side arm. Two thermobarometers were used for each run.

Each species measured represented, as far as possible, its natural weight range (neither *Chologaster cornuta* nor *Amblyopsis rosae* were available when these studies were done). This was important since rates almost always vary allometrically with body weight. Goldfish, *Carassius auratus*, of 1–30 g were used as a non-cave species and used as a control on methodology since Ekberg (1958) had done extensive studies on their tissue metabolic rates. Each fish was removed from its aquarium, blotted, weighed, and killed by decapitation. The body was kept in cold Ringers while the brain was removed and cleared of meninges. The entire brain (20–70 mg) was placed in a depression slide with 0.5 ml of cold Ringers and gently ground with the bottom of a test tube until the slurry appeared homogeneous. The brain slurry was then placed in a Warburg vessel. The muscle was dissected from the epaxial area of the body, minced with fine scissors, and a slurry was prepared as with the brain. In smaller fishes the entire epaxial mass was needed to make up the approximately 200 mg of muscle slurry added to the Warburg vessel. Sample size of amblyopsids was 5–16 fish for muscle and 4–5 fish for brain oxygen consumption studies.

Great care in procedures was necessary because tissue samples, particularly of brain, were so small. Pooled samples were not used because I wanted to minimize the time between death and first measurement and wanted to avoid scatter due to expected allometric differences in weight-relative oxygen consumption from fishes of different size. Graphs of oxygen consumption were plotted for each vessel and the linear part of the curve, usually from about 15 to 300 min, was used as the experimental value.

To begin oxygen consumption runs, Warburg vessels were equilibrated in the water bath at 15°C for 10 min before tipping the substrate from the vessel side arms. Readings were taken every 15 min as long as steady oxygen consumption was observed. When the run was terminated, the fluted filter paper was removed and the center well was cleaned. Then the tissue and substrate were rinsed with distilled water into tarred aluminum pans and dried about 40 h at 90°C in order to determine tissue weights. Corrections were made for the chemical weight in the experimental vessels using the thermobarometers as blanks.

Histological indices of thyroid activity

As discussed in any detailed treatise on thyroid endocrinology, thyroid activity as hormone turnover can be measured radiochemically or a relative value can be estimated from histological sections. Histologically, hormone turnover depends on a combination of measures. Follicle cell height and width is an index of cell volume and thus volume of secreted hormone precursors. Vacuolization of colloid is an index of mobilization of hormone from follicle lumens. In fishes, that all have a diffuse thyroid, total volume of all follicles is an index of the total capacity to produce hormone.

Indices of thyroid activity were calculated from histological measurements in fishes collected in the non-growing season, November and January, and in the growing season, July and August. Fishes were 35–55 mm SL, adult, and both males and females were included (sex determined by dissection). Fishes were fixed in alcoholic Bouins immediately after collection. One of each species was kept in the laboratory from summer through fall before fixing as a check for any differences in thyroid activity due to differences in iodine content of their natural cave waters. After fixation the lower jaw, heart, and head kidney area were removed, decalcified, embedded in paraffin, serially sectioned at 8 μ, and stained with Harris' haemotoxylin and eosin.

One histological index of thyroid activity is total thyroid volume. In 2–3 individuals of ~1 g fish of each

Table 1. Number of gill filament lamellae contributed most importantly to total gill surface area which, in turn, was closely correlated with reduction of whole fish standard metabolic rate among amblyopsid fish species. See Figures 1 and 2 for other aspects of gill morphology that were correlated with decline in standard metabolic rate with increasing cave adaptation. N = 2 for each species with weights ranging from 0.91 to 1.05 g.

Species	Lamellae on gill filaments		Gill area (mm^2)	Metabolic rate (O_2 ml g^{-1} h^{-1})
	Number	Area (mm^2)		
Chologaster cornuta surface	43 500	0.00524	227.8	0.065
Chologaster agassizi spring-cave	32 300	0.00534	172.4	0.028
Typhlichthys subterraneus cave 1	28 600	0.00592	169.1	0.016
Amblyopsis spelaea cave 2	25 400	0.00638	162.1	0.018
Amblyopsis rosae cave 3	18 200	0.00568	103.4	0.011

species, I measured total volume of follicles from serial sections examined at 430×. I measured the area of the follicles and multiplied it by 0.008 μm (8 μ sections) to get mm^3. This was so time and labor intensive that I estimated thyroid volumes in the same 2–3 individuals where every section was measured and for four other fishes by measuring every 10th section at 100×. The estimated volume for the 2–3 fishes was 10–15% lower than the total volume for the same individuals so the estimated volumes for the four other fishes were corrected accordingly.

Another index of thyroid activity is a combination of follicle cell height and vacuolization of the colloid secreted by the follicle cells. This was determined for each of 6–7 fishes (half in the growing season and half in the non-growing season) in 50 follicles chosen randomly. At the greatest diameter of each follicle, cell height was measured at 970× (oil immersion) by averaging the heights of cells separated by 90° along the circumference of each follicle. Vacuolization of colloid was scored from 1, no vacuolization and intense red staining with eosin, through 5, vacuolization of the entire colloid and light pink stain; a value of 6 was assigned if there was no colloid visible.

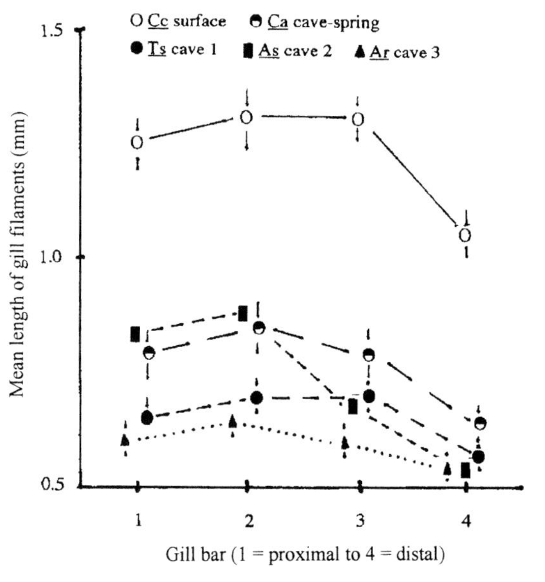

Figure 1. The mean length of gill filaments among amblyopsid fishes with increasing cave adaptation from the surface-dwelling swampfish, *Chologaster cornuta*, to the spring-cave fish, *Chologaster agassizi*, and with increasing evolutionary time isolated in caves *Typhlichthys subterraneus* < *Amblyopsis spelaea* < *Amblyopsis rosae*. N = 2 for each species with the total range of filament lengths shown by the bars. Only the swampfish showed clearly higher and different values than for the other species.

Results and interpretation

Gill surface area

There was no single aspect of gill morphology that explains the overall decrease in total gill surface area from surface-dwelling, to facultative, through obligate species of amblyopsid fishes that have decreasing whole animal metabolic rates (Table 1, Figures 1, 2). Given the total range of variation shown in Figures 1 and 2 only the surface species and the most cave adapted species were clearly different from the intermediate species. *Chologaster cornuta* had very long gill filaments (Figure 1) with very high calculated total number of lamellae (Table 1) and so it had by far the highest calculated total gill respiratory area; this fit with its highest metabolic rate. *Amblyopsis rosae* had

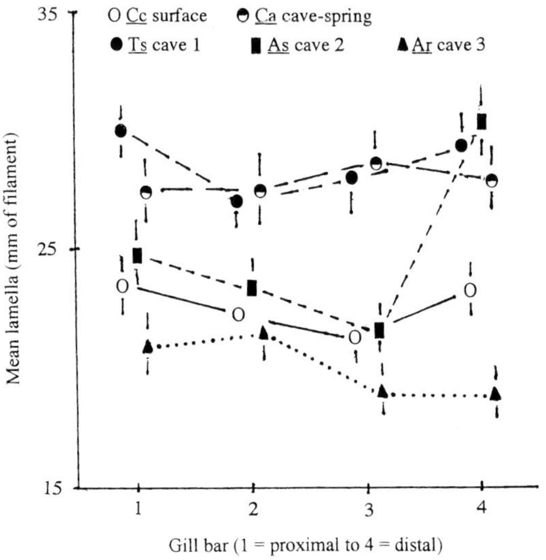

Figure 2. The mean number of lamellae per 1 mm of gill filament was least (range of all values shown by bars) for the most cave adapted amblyopsid, *Amblyopsis rosae*. Overall there was no pattern with cave adaptation for the other species. Hundreds of lamellae were measured for two individuals of each species.

consistently the lowest values for traits contributing to gill surface area; it use was probably not significantly different from intermediate species but it had by far the lowest total calculated respiratory gill area and this fit with its lowest metabolic rate. But for all the intermediate species there were only slight differences in traits contributing to total gill area and these traits were not in the same rank order as metabolic rates.

My overall data cast doubt on the validity of predicting weight-relative gill surface area either from the density of gill lamellae alone, the length of gill filaments alone, or from activity level or metabolic rate of a species alone (Gray 1954). For example, the amblyopsid with the greatest gill area, *Chologaster cornuta*, is inactive during the day, does not swim actively at night, had a moderate to low density of lamellae, and the lowest surface area of lamellae. And, *Amblyopsis spelaea* had much lower gill area and metabolic rate but swims all the time, had a moderate density of lamellae, and the highest surface area of lamellae.

Ventilation frequency and relative amplitude

Table 2 shows the ventilation rates per minute and relative amplitudes for resting spring-cave and cave amblyopsid fishes. Both aspects of ventilation declined in the same order as cave adaptation and metabolic rate; *Chologaster agassizi* had the highest values and *Amblyopsis rosae* the lowest (Tables 1, 2). *C. agassizi* had the most irregular ventilation, with frequent bouts of apnea, and *A. rosae* had the most regular ventilation, with almost no bouts of apnea in the 24 h record. On the other hand, the other two cave species, with very similar metabolic rates, showed ventilation patterns that were different. *Typhlichthys subterraneus* had a slightly lower metabolic rate than *Amblyopsis spelaea* but its adjusted ventilation pattern suggests that it should have a higher metabolic rate. Estimated ventilation volumes in ml min^{-1} were loosely related to whole animal metabolic rate (Table 4). Generally the calculated branchial volumes increased with increasing cave adaptation as relative head size increased from spring-cave to cave species (Poulson 1963). Large heads are associated with more numerous surface lateral line organs that are used to detect moving prey and obstacles (Poulson 1963). The surface species with by far the highest metabolic rate showed the second highest ventilation volume, and the spring-cave species with the next highest metabolic rate had the highest ventilation volume, and both were much higher than for any of the cave species. There was greater decreased frequency and relative volume of ventilation than increase in branchial volume. So, among the cave species, *Amblyopsis rosae* clearly had the lowest ventilation volume and metabolic rate. The other two cave species were similar in metabolic rate but ventilation volume of the species with shortest evolutionary time in caves, *Typhlichthys subterraneus*, was clearly more than for the species with the next longest evolutionary time in caves, *Amblyopsis spelaea*.

Brain and muscle oxygen consumption

Despite the relatively small sample sizes, I consider the amblyopsid data shown in Figures 3 and 4 to be reliable because my values for goldfish were taken at the same time and they overlap those taken on goldfish by Ekberg (1958). Figure 3 shows the usual inverse allometric relation between log body weight and log muscle rate of oxygen consumption (Vernberg 1954) in all species of amblyopsid tested and in the goldfish with b −0.60. The muscle rates of oxygen consumption for goldfish were higher than, and did not overlap, those for amblyopsids that did not differ among species. I expect that, if data were available, the surface-dwelling amblyopsid, *Chologaster cornuta*, would show similar

Table 2. Resting ventilation frequency in number per minute and relative ventilation amplitude decreased with cave adaptation, estimated branchial volume (see methods), and standard metabolic rate (Poulson 1963) among species of amblyopsid fishes. Ventilation rate was adjusted to lg fishes based on a b = −0.64 of metabolic rate vs. mass (Poulson 1963) and relative ventilation amplitude was adjusted to lg fishes assuming that the estimated branchial volume is related to ventilation amplitude as the 2/3 power. N = 4 for 0.7 − 1.5 g fish of each species.

Species	Branchial volume (mm^3)	Ventilation		Mass-adjusted ventilation	
		Frequency	Amplitude	Frequency	Amplitude
Chologaster agassizi cave-spring	25	36 (29–44)	1.5 (0–2.8)	42	1.3
Typhlichthys subterraneus cave 1	50	19 (15–21)	0.8 (0.4–1.0)	23	0.6
Amblyopsis spelaea cave 2	84	9 (4–11)	0.6 (0.4–1.2)	9	0.4
Amblyopsis rosae cave 3	81	7 (5–9)	0.3 (0.2–0.5)	6	0.4

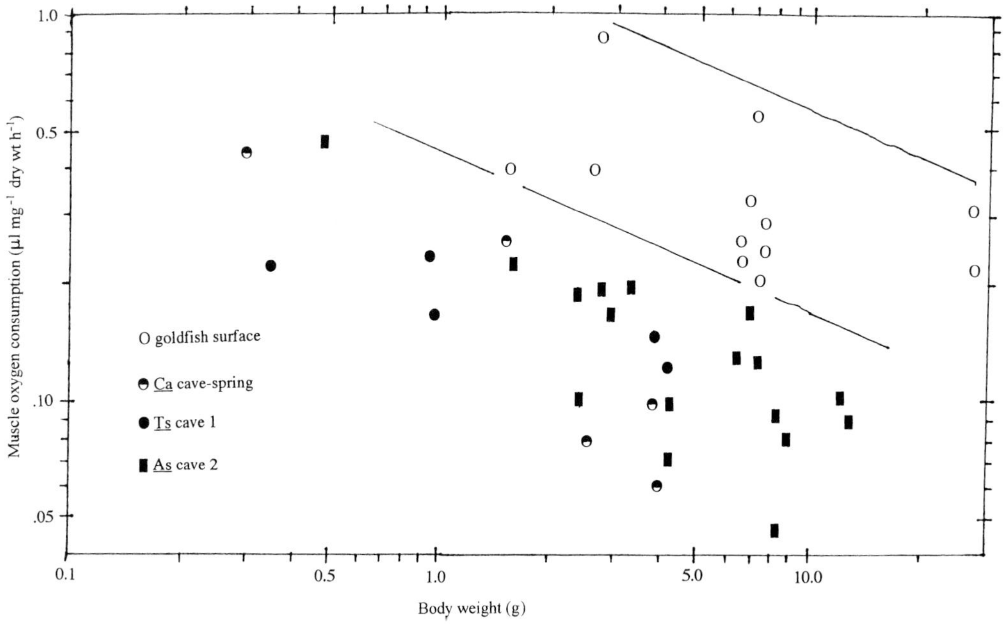

Figure 3. Muscle rates of oxygen consumption were clearly higher for a surface species, the goldfish *Carassius auratus*, than for the springcave and two cave amblyopsid species. However the amblyopsid values overlap and there was no pattern in reduction in rate from the springcave fish, *Chologaster agassizi*, to the least cave adapted cave species, *Typhlichthys subterraneus*, to the more cave adapted, *Amblyopsis spelaea*. The slopes of the regression lines were ∼ −0.75 for all species. Each point represents a different individual.

muscle metabolic rates to the goldfish which is also a surface-dweller.

The brain oxygen consumptions showed a very different pattern (Figure 4). For goldfish there was a hint of an inverse relation between rate of oxygen consumption and body weight but the b of −0.20 was not significant. For amblyopsids there was a significant direct relation with a b ∼ +0.75. As a result the values for goldfish and amblyopsids overlapped at some body sizes. These results for amblyopsids are

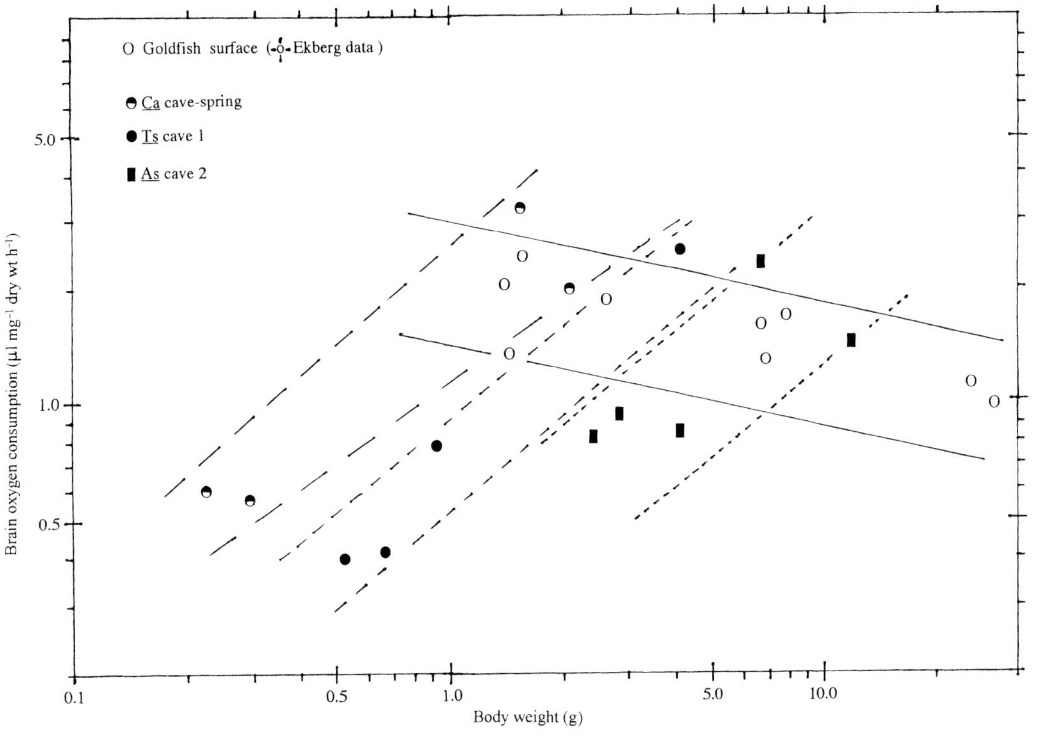

Figure 4. Brain rates of oxygen consumption showed different patterns with weight in goldfish *Carassius auratus* (slope not significantly different from zero) and three species of amblyopsid (slopes + 0.60 to 0.75). Among amblyopsids, the cave-spring species had the highest brain metabolic rate followed by the least cave adapted *Typhlichthys subterraneus*, and the more cave adapted *Amblyopsis spelaea*. Each symbol is for a different individual. There were not enough individuals to compare the regressions statistically. The values for each species of amblyopsid overlapped those for the goldfish at intermediate weights.

at variance with those of Vernberg and Grey (1953) who concluded that brain rates of oxygen consumption were independent of body weight in a series of marine teleost fishes species. Among the amblyopsids tested, *Chologaster agassizi* had the highest brain oxygen consumption and *Amblyopsis spelaea* had the lowest with *Typhlichthys* intermediate, just as with whole animal metabolic rates.

Histological indices of thyroid activity

There were few and only slight differences in indices of thyroid activity for amblyopsids either intraspecifically or interspecifically. There were no differences intraspecifically by sex or maturity. There were only slight or no increases in the growing season except for *Amblyopsis rosae* in Arkansas caves where the food supply is much higher under maternity roosts of gray bats (Table 3, Figure 5). There were only slight interspecific differences in indices of thyroid activity among amblyopsid species and they were not in the same order as increasing cave adaptation among the evolutionary series of obligate surface-dwelling, facultative cave, and obligate cave-dwelling amblyopsids (Tables 3, 4). Subtle interspecific differences among amblyopsids were much more striking for total thyroid volume than for either cell height or colloid vacuolization. This is consistent with laboratory results that showed that guppies, *Lebistes reticulatus*, fed high food rations had higher growth rates and thyroid follicle numbers than guppies fed low food rations (Woodhead & Ellet 1967).

The subtle pattern or lack of pattern I found is not due either to differences in iodine availability among caves or to inaccuracy of the techniques. There were no differences in iodine availability in water among caves because the patterns for each species remained in individuals held in the same water in the laboratory for

Table 3. Histological indices showed that for adult fishes (0.6–1.5 g) in the growing season (spring–summer), surface-dwelling guppies had higher thyroid activity than amblyopsid fishes but that there was no relation of thyroid activity to metabolic rate or growth rates among amblyopsid species from cave-spring to cave species. Even within the most cave adapted species, *Amblyopsis rosae*, there was as much difference in thyroid indices between caves with very different food supplies as among species that live in caves with similar food supplies (see text). N = 6–7 for each species.

Species	Histological indices of thyroid activity		
	Follicle volume (mm^3)	Follicle height (μ)	Colloid vacuolization
Lebistes reticulatus surface	0.044	1.9	1.5
Chologaster agassizi cave-spring	0.0036	3.8	2.7
Typhlichthys subterraneus cave 1	0.0028	3.1	1.8
Amblyopsis spelaea cave 2	0.0016	5.3	2.1
Amblyopsis rosae, bat guano cave 3	0.0033	5.1	4.0
Amblyopsis rosae, no bat guano cave 3	0.0018	6.2	4.3

six months. My indices accurately reflect real thyroid activity. Thus all species showed subtle increases in follicle cell height and colloid vacuolization in the growing season (Figure 5), and the *Amblyopsis rosae* from Benton County Arkansas gray bat caves and a high food supply (Poulson 1961), had a higher growth rate and higher histological indices of thyroid activity than fishes from Newton County, Missouri caves without bats and a low food supply comparable to that for the other cave species (Poulson 1961).

The lack of clear interspecific differences in indices of thyroid activity among amblyopsid fishes is not consistent with earlier studies of cave fishes. Oliverau & Francotte-Henry (1956) and Oliverau (1960a,b) had claimed that there is a causal relation between low follicle cell heights and low vacuolization of colloid and low rates of metabolism and growth in several species of cave fishes but they did not measure the more accurate index of total thyroid volume and did not have measures of metabolic rate for some of the species. Also the amblyopsid species with the lowest metabolic rate, *Amblyopsis rosae*, had the highest follicle cell heights and vacuolization of colloid (Table 3) and an intermediate total thyroid volume (Table 4).

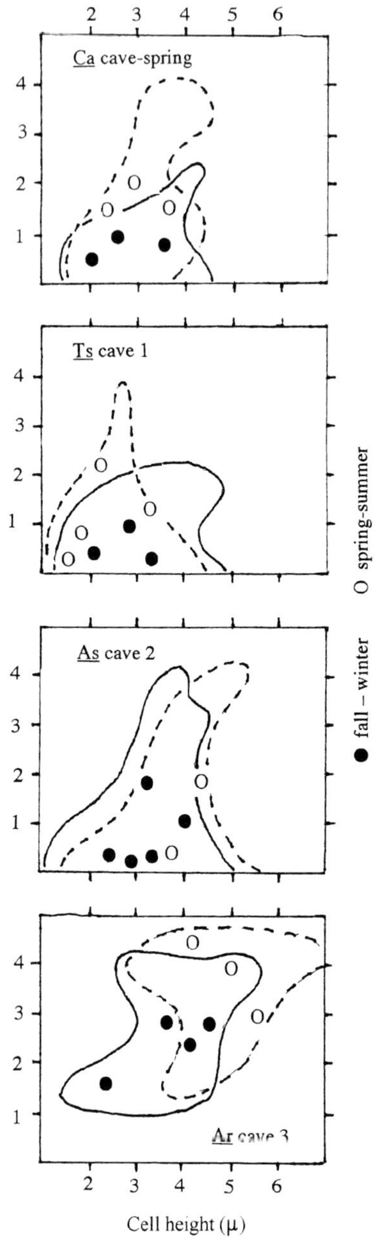

Figure 5. Among amblyopsid species there were no differences with cave adaptation in histological indices of thyroid activity in adults. There were subtle increases in thyroid follicle cell height and/or colloid vacuolization from the non-growing season in fall-winter to the growing season in spring–summer, most clearly for *Amblyopsis rosae*. The lines enclose 95% of the 50–100 follicles measured for each species and the symbols are the grand means for each individual.

Table 4. Summary of traits that could contribute to whole fishes metabolic rate in swamp, spring-cave, and cave species (similar food supplies) of Amblyopsidae. All rates adjusted to 1 g fishes, based on measured regressions of rate vs. body weight, and taken at 13–15°C. (*Chologaster cornuta* data taken at 20°C and adjusted to 15°C assuming a Q10 of 2.0.) Where ties or near ties in rank of metabolic rate vs. trait among amblyopsid species occurred the ranks were switched and both probabilities for Spearman's r are given. The columns are as follows: 1 = species with Cc *Chologaster cornuta*, Ca *Chologaster agassizi*, Ts *Typhlichthys subterraneus*, As *Amblyopsis spelaea*, and Ar *Amblyopsis rosae* (cave species evolutionary time in caves ranked 1, 2, and 3); column 2 = standard metabolic rate in ml O_2 g^{-1} h^{-1}; column 3 = total gill surface area in mm; column 4 = brain oxygen consumption in ml mg^{-1} dry weight h^{-1}; column 5 = muscle oxygen consumption; column 6 = estimated ventilation volume in ml per minute; and column 7 = total thyroid follicle volume in mm^3.

Species	Metabolic rate	Gill area	Tissue metabolic rate		Ventilation volume	Thyroid volume
			brain	muscle		
Cc swamp	0.065	228	—	—	1.12	—
Ca spring	0.028	172	1.50	0.25	1.37	0.020
Ts cave 1	0.016	169	0.30	0.28	0.46	0.010
As cave 2	0.018	162	0.060	0.21	0.69	0.016
Ar cave 3	0.010	103	—	—	0.19	0.014
Sr p =		0.04 and 0.46	0.17 and 0.21	0.17 and 0.83	0.04 and 0.07	0.46 and 0.79

Discussion

Traits that contribute to the decline in metabolic rate among amblyopsid fishes

Table 4 shows that no one physiological or morphological trait was in the same rank order as the decline of metabolic rate observed from a surface-dwelling amblyopsid species *Chologaster cornuta*, to a facultative amblyopsid cave dweller *Chologaster agassizi*, and to the obligate amblyopsid cave species isolated in caves for increasing evolutionary time *Typhlichthys subterraneus*, *Amblyopsis spelaea*, and *Amblyopsis rosae*. To evaluate these data I used Spearman's rank correlation because there was a mix of continuous variables, composite indices, and relative variables. I also calculated probability pairs for switches of rank in cases where there were ties or close values. The order of significance for possible contributions to whole fish standard metabolic rate was as follows: ventilation frequency ($p = 0.04$ and 10) > relative ventilation volume ($p = 0.04$ and 17) > brain oxygen consumption ($p = 0.17$ and 0.2 1) > gill surface area ($p = 0.04$ and 0.46) > muscle oxygen consumption ($p = 17$ and 0.83) > thyroid follicle cell height ($p = 0.21$ and 0.79) > total thyroid follicle volume ($p = 0.46$ and 0.79 > thyroid colloid vacuolization ($p = 0.79$ and 0.96). Among the cave amblyopsids, *Typhlichthys* had a slightly lower metabolic rate than *Amblyopsis spelaea* but had much higher values for brain oxygen consumption, ventilation frequency, and relative ventilation volume. If *A. spelaea* had a slightly lower metabolic rate than *Typhlichthys* then the correlations with metabolic rate, in Table 4, of ventilation volume, brain oxygen consumption, and thyroid volume would be much improved; other correlations would not be changed appreciably.

Central nervous system control of metabolic rate

My data add to the body of literature that implicates the central nervous system in the control of metabolic rate in fishes. Vernberg and Gray (1953) found a positive relation of brain oxygen consumption and 'activity level' in a series of marine teleosts and they went so far as to say '... it might seem possible to suggest that an integral part of the coordinating system of the body, the brain, is extremely important in maintaining the basal metabolic rate of the entire organism'. On the one hand, their suggestion is weakened by their equally good correlation between whole animal metabolic rate and gill surface area. This is further weakened by the fact that goldfish had higher brain oxygen consumptions than amblyopsid cave fishes at some body sizes. On the other hand, their suggestion is strengthened by many more facts. First, the brain is a very small percent of body mass and yet the rank order of brain and whole fish metabolic rates among amblyopsids were similar. Second, McFarland's (1959) 'dissection' of maintenance metabolic rate using progressive anesthesia for several non-cave species found that 50% of maintenance metabolic rate was related to functions controlled by the brain, namely equilibrium and muscle tonus. The remaining 50% was split between systemic functions, such as ventilation and circulation, and tissue metabolic rate, the bulk of which must be muscle and viscera. Third, the ingenious experiment of acclimating the head and body of an eel to different temperatures (Jankowsky 1966, in Prosser 1967) showed that whole organism metabolic rate is a function only of the head's acclimation temperature. Further, this conclusion of control of acclimation by the brain is consistent with the type of experiment which show that there is a progressively narrower temperature range of acclimation possible for simple reflexes, avoidance reactions, and maze learning (Prosser 1967). Finally considerable data show a decrease in brain mass with extreme metabolic economy in bathypelagic fishes (Marshall 1960, Poulson 1971).

Unlike many bathypelagic fishes (Marshall 1960) and some ophidiid bathybenthic fishes (Horn et al. 1978), amblyopsid cave fishes do not have reduced brain sizes (Poulson 1961). The very small brain sizes of deep sea fishes is correlated with reduced muscle mass and high water content that function to reduce metabolic cost and increase static lift in the absence of a swimbladder. These bathypelagic fishes have hypertrophied sensory systems, such as lateral line and olfactory systems, but the relative increase in brain centers associated with these senses has not balanced the loss in brain volume with decreased visual systems and cerebellar muscle tonus centers associated with extreme loss of steady and burst swimming ability. In contrast, cave amblyopsids retain burst swimming ability, have normally developed muscles and swim bladders, and the decrease in volume of brain optical centers is compensated by an increase in lateral line and tactile input and information processing centers (Poulson 1963). This is in contrast to a fish that uses weak electrical output to navigate and feed and has a very large brain associated with the complexity of information processing (Nilsson 1996). The convergence of relatively increased head size, due to paedogenesis, results in an incredibly increased lateral line system that all the amblyopsid cave species use for detection of obstacles and prey (Poulson 1963).

Genetic bases for evolution of reduced metabolic rates in amblyopsid cave fishes

Overall, my data show that there were multiple morphological and physiological ways to reduce metabolic rate and no one amblyopsid cave species showed them all. I expected this with convergent evolution of metabolic economy because of previous data on convergent eye reduction (Eigenmann 1909, Poulson 1963, 1985). Thus with the convergent reduction in eye size and number of structures lost among amblyopsid cave species there were interspecific differences in which eye structures showed reduction or loss.

My data on reduction in metabolic rate among amblyopsid cave fishes bear on the controversy about the genetic mechanism for convergent reduction in traits (Wilkens 1985, Poulson 1985). If reduction in metabolic rate were simply due to more of the same multiple quantitative genes, each with a similar additive effect (e.g. Wilkens 1985), then I should have found a parallel and progressive reduction in the same traits that influence metabolic rate; I did not see this. Instead different traits, even within a structure such as gills (Figures 1, 2), accounted for the overall reductions which collectively contributed to lowered metabolic rate. This is the same argument I have made for reduction in eyes among cave amblyopsids (Poulson 1963, 1985) where overall eye size has decreased evolutionarily but the order in which structures degenerate or disappear is different among species. Thus there must be both quantitative genes, such as eye size (Wilkens 1985), and structural genes, such as for eye muscles or choroid pigment or crystalline lens protein, that contribute to evolutionary reduction in eye size and complexity. By extension, there must be both quantitative and structural genes for traits such as gills and for ventilation.

Of course the real test of the relative importance of convergence and parallelism in cave animals has come from molecular evidence. Thus, Culver et al. (1995) have shown that different populations of the same amphipod species have independently invaded caves even though they are morphologically quite similar in reduction of eyes and hypertrophy of antennal sense organs. On this basis I predict that different populations of what is now recognized as one species, especially *Typhlichthys* from different geomorphic areas in Kentucky to Alabama and in disjunct populations in Missouri, will be shown, by molecular data, to have independently been isolated in caves. And, the relative contribution of morphological and physiological traits to decreased metabolic rates will be found to be different among these same populations.

Acknowledgements

This work could not have been done without the superb technical assistance of an NSFURP student Stanley Drezek. His care to detailed documentation and standardization of protocols was exemplary. Comments on a much earlier draft, by Grace Pickford, helped to improve my presentation especially about the histological indices of thyroid activity. These studies were partially supported by a USPHS – Neurological Diseases and Blindness Grant to T.L. Poulson (1963–1968).

References cited

Cooper, J.E. & R.A. Kuehne. 1974. *Speoplatyrhinus poulsoni*, a new genus and species of subterranean fishes from Alabama. Copeia 1974: 486–493.

Culver, D.C., T.C. Kane & D.W. Fong. 1995. Adaptation and natural selection in caves: the evolution of *Gammarus minus*. Harvard University Press, Cambridge, 223 pp.

Eigenmann, C.H. 1909. Cave vertebrates of America, a study in degenerative evolution. Carnegie Inst. Wash. Publ. (104): 1–241.

Ekberg, D.R. 1958. Respiration in tissues of goldfishes adapted to high and low temperatures. Biol. Bull. 114: 308–316.

Gray, I.E. 1954. Comparative study of the gill area of marine fishes. Biol. Bull. 107: 219–225.

Horn, M.H., P. W. Grimes, C.F. Phleger & L.L. McClanahan. 1978. Buoyancy function of the enlarged fluid-filled cranium in the deep-sea ophidiid fish *Acanthonus armatus*. Mar. Biol. 46: 335–339.

McFarland, W.N. 1959. A study of the effects of anesthetics on the behavior and physiology of fishes. Bull. Inst. Mar. Sci. 4: 23–55

Marshall, N.B. 1960. Swimbladder structure of deep-sea fishes in relation to their systematics and biology. Discov. Rep. 31: 1–122.

Nilsson, G.E. 1996. Brain and body oxygen requirements of *Gnathonemus petersii*, a fish with an exceptionally large brain. J. Exp. Biol. 199: 603–607.

Olivereau, M. 1960a. Etude anatomique et histologique de la glande thyroide *d' Uetiglanis zammaranoi*, poisson aveugle et cavernicole, comparison avec un Clariidae voison, *Clarias bathypogon*. Ann. Soc. Roy. Zool. Belg. 90: 99–116.

Olivereau, M. 1960b. Hyperplasie thyroideienne et presence de follicules thyroidiens intra-renaux, chez un exemplaire de *Typhlogarra widdowsoni*, poisson aveugle et cavernicole de I'lrak. Ann. Soc. Roy. Zool. Belg. 90: 117–125.

Olivereau, M. & M. Francotte-Henry. 1956. Etude histologique de la glande thyroide de *Caecobarbus geertsi*. Ann. Soc. Roy. Zool. Belg. 86: 129–150

Poulson, T.L. 1961. Cave adaptation in amblyopsid fishes. Ph.D. Dissertation, U. Michigan Microfilms 61–2787. 185 pp.

Poulson, T.L. 1963. Cave adaptation in amblyopsid fishes. Amer. Midl. Nat. 70: 257–290.

Poulson, T.L. 1964. Animals in aquatic environments: animals in caves. pp. 749–771. *In:* D.B. Dill (ed.) Handbook of Physiology: Environment, Amer. Physiol. Soc., Washington, D.C.

Poulson, T.L. 1971. Biology of cave and deep-sea organisms: a comparison. Bull. Nat. Speleol. Soc. 33: 51–61.

Poulson, T.L. 1985. Evolutionary reduction by neutral mutations: plausibility arguments and data from amblyopsid fishes and linyphiid spiders. Bull. Nat. Speleol. Soc. 47: 109–117.

Prosser, C.L. 1967. Mechanisms of temperature acclimation in fishes. pp. 375–409 *In:* A.S. Troshin (ed.) Cell and Environmental Temperature, Pergamon Press, London.

Rosen, D.E. 1962. Comments on the relationships of the North American cave fishes of the family Amblyopsidae. Amer. Mus. Novitates No. 2109: 1–35.

Vernberg, I.F. 1954. The respiratory metabolism of tissues of marine teleosts in relation to activity and body size. Biol. Bull. 106: 360–370.

Vernberg, I.F. & I.E. Gray. 1953. A comparative study of the respiratory metabolism of excised brain tissue of marine teleosts. Biol. Bull. 104: 445–449.

Wilkens, H. 1985. The evolution of polygenic systems, studied on epigean and cave populations of *Astyanax fasciatus* (Characidae, Pisces). Bull. Nat. Speleol. Soc. 47: 101–108.

Woodhead, A.D. & S. Ellet. 1967. Effects of food-restriction on thyroid activity in the guppy, *Lebistes reticulatus* (Peters). Exp. Geront. 2: 49–56.

Woods, L.P. & R.F. Inger. 1957. The cave, spring, and swamp fishes of the family Amblyopsidae of central and eastern United States. Amer. Midl. Nat. 58: 232–256.

Environmental Biology of Fishes **62:** 251–261, 2001.
© 2001 *Kluwer Academic Publishers. Printed in the Netherlands.*

Convergent adaptations to cave life in the *Rhamdia laticauda* catfish group (Pimelodidae, Teleostei)

Horst Wilkens
Zoologisches Institut und Zoologisches Museum, Universität Hamburg, Martin-Luther-King-Platz 3, 20146 Hamburg, Germany (e-mail: wilkens@zoologie.uni-hamburg.de)

Received 5 June 2000 Accepted 11 January 2001

Key words: troglobitic evolution, eye reduction, ontogenetic regression, phylogenetic age

Synopsis

The nocturnal Mexican catfish *Rhamdia laticauda* (Pimelodidae, Teleostei) is the surface sister species of a number of cave species. Comparison between two of them, *R. zongolicensis* and *R. reddelli*, reveals an intermediate state of reduction of the biologically functionless eyes, melanin pigmentation and the negative phototactic behavior. The surface species is perfectly adapted to life in darkness. Therefore only few constructive adaptations are developed in the cave species. For survival under conditions of low food supply in the caves the barbels are elongated to improve the senses of taste and touch and more fat can be deposited in the cave fish tissue. Due to convergent evolution the two cave species are morphologically much alike and show only minor diagnostic meristic differences. From geological data it can be concluded that the two cave species started troglobitic evolution at the end of Pleistocene.

Introduction

The evolution of troglobites is very often characterized by the existence of several cave forms which are closely related to one single and still existing surface sister species. This makes possible the comparative morphological study of the convergent adaptational processes as well as the question of which features are prerequisite of cave colonization. Relationships between the different cave and surface species may be so close that the genetics of the adaptation process can be analyzed by hybridization.

The neotropical catfish genus *Rhamdia* started colonization of the North American continent at the end of Pliocene, when the Middle-American landbridge closed (Myers 1966). Today its northern distribution limit is not far north of the Isthmus of Tehuantepec in the state of Veracruz in Mexico (Miller 1984, Weber & Wilkens 1998). According to Miller (1984), the genus is systematically subdivided into the *Rhamdia guatemalensis*-group consisting only of the species *R. guatemalensis* (Günther 1864) and the *R. laticauda*-group. The latter contains the surface *R. laticauda* (Heckel 1858) and several cave species, some of which are still to be described (Sbordoni et al. 1977, Mosier 1984, Weber 1996). Here I present a comparative analysis of the cave species *R. reddelli* Miller, 1984 and *R. zongolicensis* Wilkens, 1993 with their surface sister species *R. laticauda* and the closely related *R. guatemalensis*.

Material and methods

Specimens of *R. zongolicensis* were caught in the Cueva del Túnel (Comalapa, Veracruz) (Robertson 1983a) and specimens of *R. reddelli* in the Cueva del Nacimiento del Río Subterraneo (Cañada San Antonio, Oaxaca) (Miller 1984, Wilkens et al. 1991) in March 1982. The fish were kept and bred in the laboratory. Spawning was usually induced by the injection of grinded carp pituitary solution into the dorsal musculature (0.1–0.2 ml of one grinded pituitary dissolved in 5 ml NaCl-solution). Eggs and sperm were striped into a glass bowl about 16 h later, where they were mixed for fertilization. *R. laticauda* was caught in the Río Metlac (Fortín, Veracruz) in 1989, *R. guatemalensis* were derived from a pool at the entrance of the Cueva del Nacimiento del Río San Antonio (Wilkens et al. 1991) in 1988. Both surface species were kept and bred in the laboratory. The keeping and breeding of all individuals of all species was generally easy. Only *R. reddelli* presented difficulties, probably because this fish is not adapted to life in stagnant water.

For eye histology in adult fish paraffin sections (10 μm) of formaline preserved specimens were prepared and stained with Pasini dye. For the study of larva and juvenile eyes semi-thin sections (0.05–1 μm) were stained with toluidineblue (1–2 min, 60°C). Melanophores were counted after melanin contraction was induced by application of L-noradrenaline (1 : 1000) on the body surface. Barbels were measured in specimens which had been kept separated for a longer time period to prevent loss or damage by biting. The maximum yolk diameter was measured before the yolk was covered with the germ.

For the study of the phototactic behavior the fish were placed into 25 l tanks divided into a dark and a light compartments. After seven days of acclimation under bright light (620 lux at the bottom of the lighted half) in a light-dark cycle of 12 : 12 h, the time the fish spent in the lighted compartment was recorded during a time period of 5 min three times a day over five days. Subsequently, the same experiments were performed using lower light intensities (50 lux and 3 lux at the bottom of the lighted half).

The feeding experiments were performed with adult specimens kept separately in single 25 l tanks in darkness. No food was given for about seven months. Feeding before and after this phase was ad lib.

Distribution and habitat

R. laticauda

The surface fish *R. laticauda* (Figure 1) lives in the cool (22°C), rapidly flowing upper sections of the rivers of the Atlantic slope of the Sierra Madre Oriental with the Río Jamapa and Atoyac (Veracruz) as northern distribution limits (Miller 1984). It is closely related to the cave species described below and several other troglobitic populations in Mexico and Belize (Greenfield et al. 1982, Moisier 1984, Weber & Wilkens 1998), all of which derive from the same ancestral species.

R. zongolicensis

There are four caves in the karst region of the Sierra de Zongolica (Veracruz) from which *Rhamdia* cave fish are reported (Robertson 1983a,b, Mosier 1984). *R. zongolicensis* (Figure 2) was described from the Cueva del Túnel (Cueva del Ostoc) (Wilkens 1993). The cave fish here live in stagnant pools where there is no other aquatic cave fauna present. The cave is situated 300 m above sea level and the surrounding Sierra reaches altitudes of 2750 m above sea level with a mean annual temperature of 17.4°C at 1000 m altitude. Consequently, water temperature is low in the cave (around 20°C). The connection to the groundwater base level seems only to exist during the rainy season, when the water is accumulated. The fish caught in this cave showed severe signs of malnutrition in some sampling years. This malnutrition can only be explained by the food supply being extremely limited for longer periods of time and may have a correlation with the amount of rainfall (Wilkens 1993).

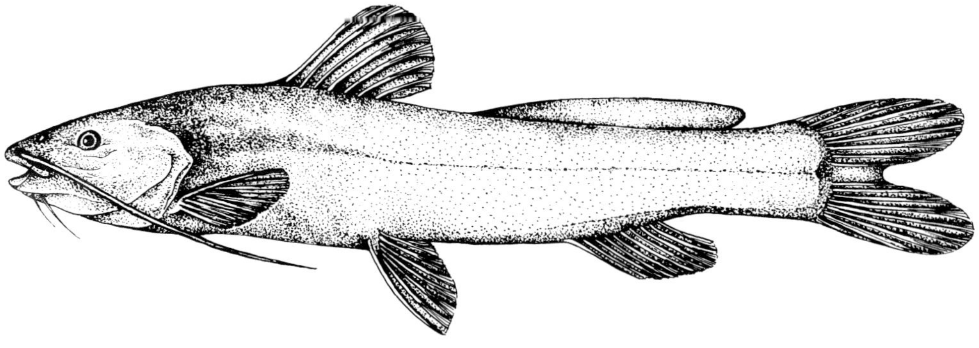

Figure 1. Surface *R. laticauda* (140 mm standard length).

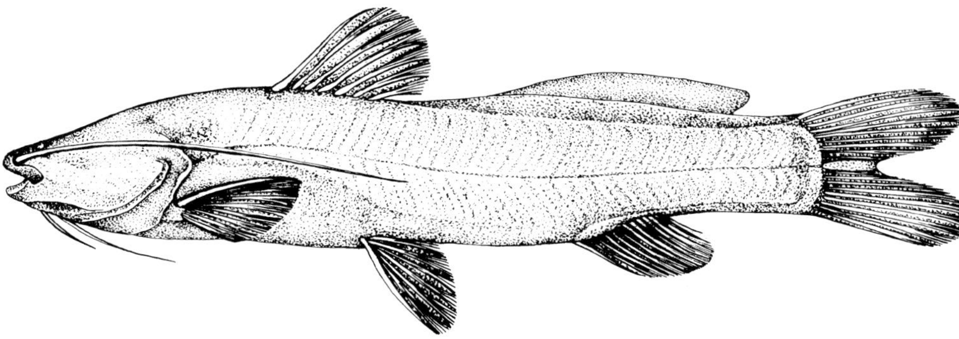

Figure 2. Cave *R. zongolicensis.*

The Sierra de Zongolica hydrographically drains into the Río Blanco basin in the north and into the Río Tonto in the south. It cannot be excluded therefore, that in addition to *R. zongolicensis* there may be other troglomorphic *Rhamdia* populations or species in caves of the Río Blanco affluents in the Sierra de Zongolica. The geographically close existence of cave populations or species being derived from one single ancestor is not unusual. It has also been described for the characid *Astyanax fasciatus* (= *A. mexicanus*) (Mitchell et al. 1977, Wilkens 1988).

R. reddelli

The cave fish *R. reddelli* is found in a cave called Cueva del Nacimiento del Río San Antonio. It lies at a distance of about 30 km away from the location of *R. zongolicensis* (Miller 1984, Wilkens et al. 1991). Both caves belong to the Río Tonto drainage. The San Antonio cave is situated at 100 m above sea level within a mountain ridge of up to 600 m altitude extending southward of Acatlán (Oaxaca). The mountain ridge is surrounded by flat plains and thus topographically isolated from the Sierra de Zongolica. The cave contains one of the creeks draining the mountain ridge and gathering in the Río San Antonio, which flows at the foot of the Sierra. The waterflow in the cave may fluctuate considerably depending on the amount of rainfall in the Sierra. The cave waters have a temperature of 22°C. In addition to *R. reddelli* troglobitic aquatic cave life consists of four crustacean species. In the Río San Antonio outside the cave the ichthyofauna is composed of species reflecting the thermophilic community typical of the Gulf Coast Plain including *R. guatemalensis* (Miller 1984, Wilkens et al. 1991).

R. guatemalensis

Unlike *R. laticauda*, the surface species *R. guatemalensis* inhabits the warm Gulf Coastal lowland waters (Miller & Smith 1986). It has also colonized the groundwater system and the cenotes of the Yucatan peninsula (Wilkens 1982). The populations from the Yucatan diverge from those of the other regions by a darker body pigmentation. These darker body pigmentation cannot be influenced by physiological colour change (unpublished data). *R . guatemalensis* is found in the plains surrounding the mountain ridge in which *R. reddelli* is found. It penetrates into light-poor cave entrance areas which exist at several places at the outer margin of the Sierra.

Regressive traits

Eyes

The surface *R. laticauda* possess the well-developed, comparatively small sized eyes typical of most catfishes. The retina is much thinner than in *A. fasciatus* (Characidae) (Wilkens 1988). This is probably caused by the lower number of visual cells developed in this nocturnal species. As is typical in catfishes, optic papillae are developed (Ali & Anctil 1976) with the optic nerve penetrating the retina in five separate portions. The eyes of the *Rhamdia* cave forms are reduced. In most adult specimens they are not visible externally, because the scleral and the epidermal parts of the cornea have separated and lost their tight contact. Due to this fact, the eyes may completely sink beneath the body surface (Figure 3c). In individuals with larger eye rudiments, this contact may persist or

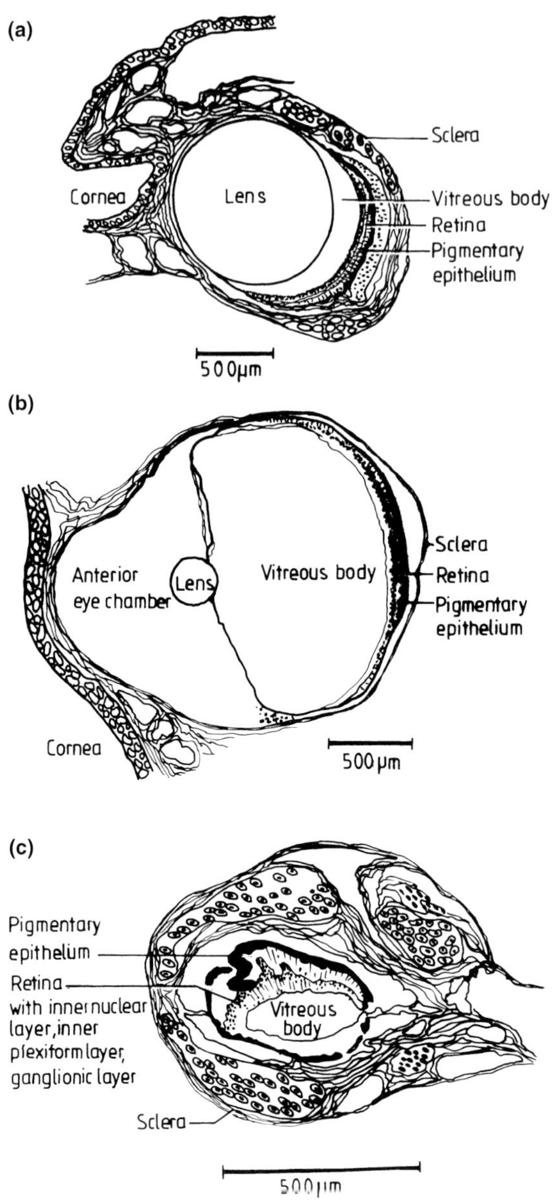

Figure 3. Variability of eye reduction in three adult specimens of *R. reddelli*: a – eye with well-developed retina and large lens in place of the vitreous body, b – large eye with well-developed retina and small lens, c – extremely reduced eye without lens and sensory cells.

cause a pit-like depression on the body surface, with the eyes still visible (Figures 3a,b).

The histological-anatomical differentiation of the adult eyes is dependent on their overall size. The larger the eye, the more it is differentiated (Figures 3a,b,c). The following structural characteristics can be observed in the eyes of adult specimens: retinal reduction proceeds by thinning of the specific layers and has reached its maximal degree with the loss of the visual cells and outer nuclear and plexiform layers in the smallest eyes; the smallest rudiments still contain ganglionic, inner nuclear, and plexiform layers and develop Müllerian cells and inner and outer limiting membranes (Figure 3c); the pigmentary epithelium flattens when the visual cells are reduced. Lens reduction is also correlated to that of the other single structures. However, as already observed in crosses of *A. fasciatus* (Wilkens 1988), the correlation of size and differentiation between the lens and the other single structures of the eye is rather loose; in some eyes the lens may be so large that this organ completely fills the vitreous body and even penetrates into the anterior eye chamber (Figure 3a). Conversely, large eyes with well-developed retinas may contain extremely small lens rudiments (Figure 3b). All rudiments have an optical nerve, the size of which is dependent on the degree of retina differentiation. Optic papillae are existent but have a rudimentary character. Only two instead of five retinal penetrations are developed.

The eye size of adult cave fishes shows an extraordinary variability (Figure 4). The largest ones are only slightly smaller than those of the surface fish, the smallest ones are reduced to tiny rudiments. Although the ranges overlap, the eyes of *R. reddelli* on average are smaller than those of *R. zongolicensis* (Figure 5a). In *R. zongolicensis* a considerable discrepancy of eye size can be observed in the same specimen (left-right-asymmetry) (Figure 4). More than 50% of all specimens exhibit a size difference of more than 10% (in percent of larger eye); differences of 70% occur. Variability develops during ontogeny and is completed when maturity is reached. This development is correlated to the process of ontogenetic regression being typical of cave-fish eyes in general. The growth of the cave-fish eye shows negative allometry ($y = 0.918x + 32.113$, $r = 0.802$ vs. $y = 1.19x + 0.778$, $r = 0.978$ in the surface form).

The eye development of surface *R. laticauda* proceeds much slower than in *A. fasciatus*. The primary lens fibers in *R. laticauda* are formed at an age of about 144 h and the lens nucleus starts building crystallines after 216 h. In *A. fasciatus* this takes place after 30 and 35 hours respectively (Anderßon 1992, Langecker et al. 1993). The eyes of cave *Rhamdia* are smaller than those of the surface *Rhamdia* from the very beginning.

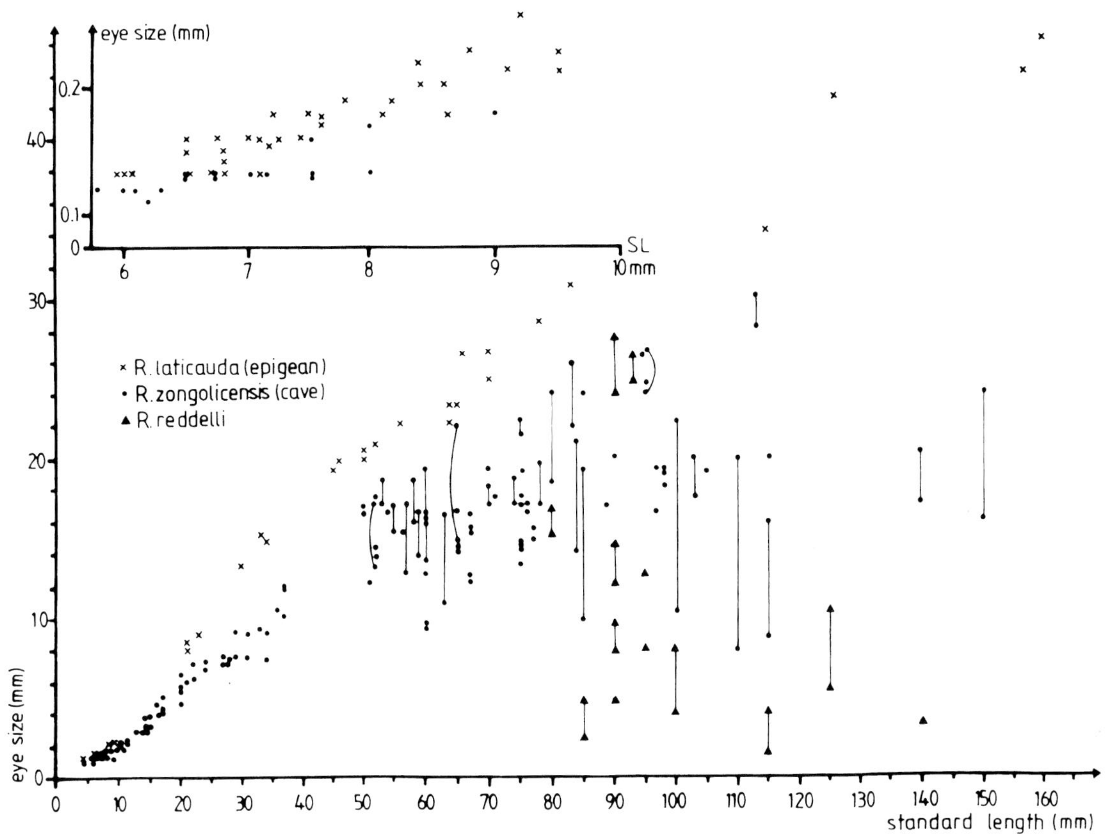

Figure 4. Correlation of eye size and body length in *R. laticauda*, *R. zongolicensis* and *R. reddelli*. Connecting lines indicate extremes of left-right asymmetry found in one specimen.

On an average the diameter of the first anlage deviates by 0.03 mm from the surface eye. Due to the process of ontogenetic regression this discrepancy increases during growth (Figure 4). Almost all single structures in the cave fish eye are developed at a smaller size. Although the retina as a whole is thinner, all layers and the typical single structures like Muellerian cells are developed; however, gross differences are observable in the lenses. After the first lens anlage and the development of the primary lens fibers has been completed, divergences between surface and cave fishes emerge. The rate of cell division in the lens nucleus is reduced and necrotic processes arise here. The lens epithelium is developed, but the secondary lens fibers derived from it do not elongate properly. The production of crystallins, which in surface eyes start in the lens nucleus and then proceed into primary and secondary lens fibers, is subjected to variability. In very small cave-fish lenses produce no crystallin granula and in medium sized lenses only few crystallin granula are developed, whereas in large lenses a rather normal development takes place (Figure 6).

Pigmentation

In absolute darkness individuals of both cave species are pale. When kept for some months under daylight conditions they will develop greyish pigmentation. This is caused by the light induced development of melanin in the melanophores; however, in comparison to the surface species the number of melanophores, which still are able to perform physiological color change, is reduced by about 50% (Figure 5b).

Phototactic behavior

R. laticauda is a nocturnal fish, showing strong scotophilia. The sensitivity towards light is such that the

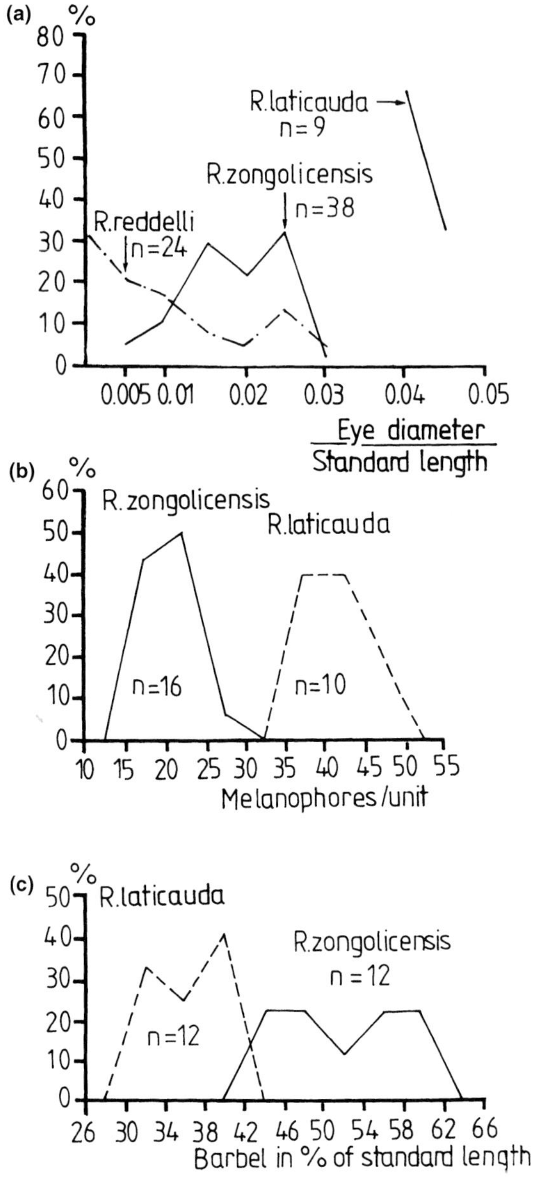

Figure 5. Comparison of a – relative eye size, b – number of melanophores, and c – barbel length in *R. laticauda*, *R. zongolicensis* and *R. reddelli*.

response at 3 lux is not less pronounced than at 650 lux. Both cave species also display scotophilia although to a lesser extent. *R. reddelli* displays less scotophilia than the surface *R. laticauda* at all light intensities measured. In *R. zongolicensis* scotophilia is not uniform. It decreases continuously from 6% at 620 lux to 18% (50 lux) and 35% at 3 lux (Figure 7).

The surface *R. guatemalensis* also shows scotophilia. This fish is also nocturnal and begins to be active at twilight; however, several populations living in cenotes in the Yucatan are even day-active. It was experimentally shown that the degree of negative phototactic behavior in *R. guatemalensis* is less intensive, than in *R. laticauda* (Figure 7) (Wilkens et al. 1991, Langecker & Wilkens 1992).

Constructive traits

Barbels

Pimelodid catfishes possess three pairs of barbels, the longest of which are inserted on the upper jaw. They are covered by taste buds and used like antennae for tasting and touching. In adult surface *R. laticauda* these barbels are about 40% of the body length and are shorter than in *R. guatemalensis*. In the cave fish the barbels are considerably elongated reaching 50% of body length. Barbel enlargement already starts in juvenile cave fish. Growth is positively allometric ($y = 1.26x - 0.78$ vs. $y = 1.16x - 0.68$ in the surface fish) (Figures 5c, 8).

Survival and recovery from conditions of starvation

Individuals from the La Cueva del Túnel were observed to be extremely undernourished in some of the sampling years. Comparisons between these and those kept in the laboratory lead us to believe that the ones in natural conditions may have starved for more than one year (Wilkens 1993). In the laboratory these individuals have been able to recover from starvation. Our observations suggest that the cave fish are much better adapted to lack of food than the surface fish (Figure 9). Three of four individuals of the surface *R. laticauda* died after four to six months of starvation. Only one individual survived a seven-month starvation period; however, all four specimens of *R. zongolicensis* survived when subjected to the same conditions. Loss of weight in the surface fish was more rapid than in the cave fish. After a time period of two months the surface fish had lost twice as much of their original body weight in comparison with the cave fish. After seven months, slightly more than 50% of the body weight was lost in the surface fish in contrast to 25% (2 individuals), 40% (1 individual) and 45% (1 individual) in the cave fish. Furthermore, the time span of recovery

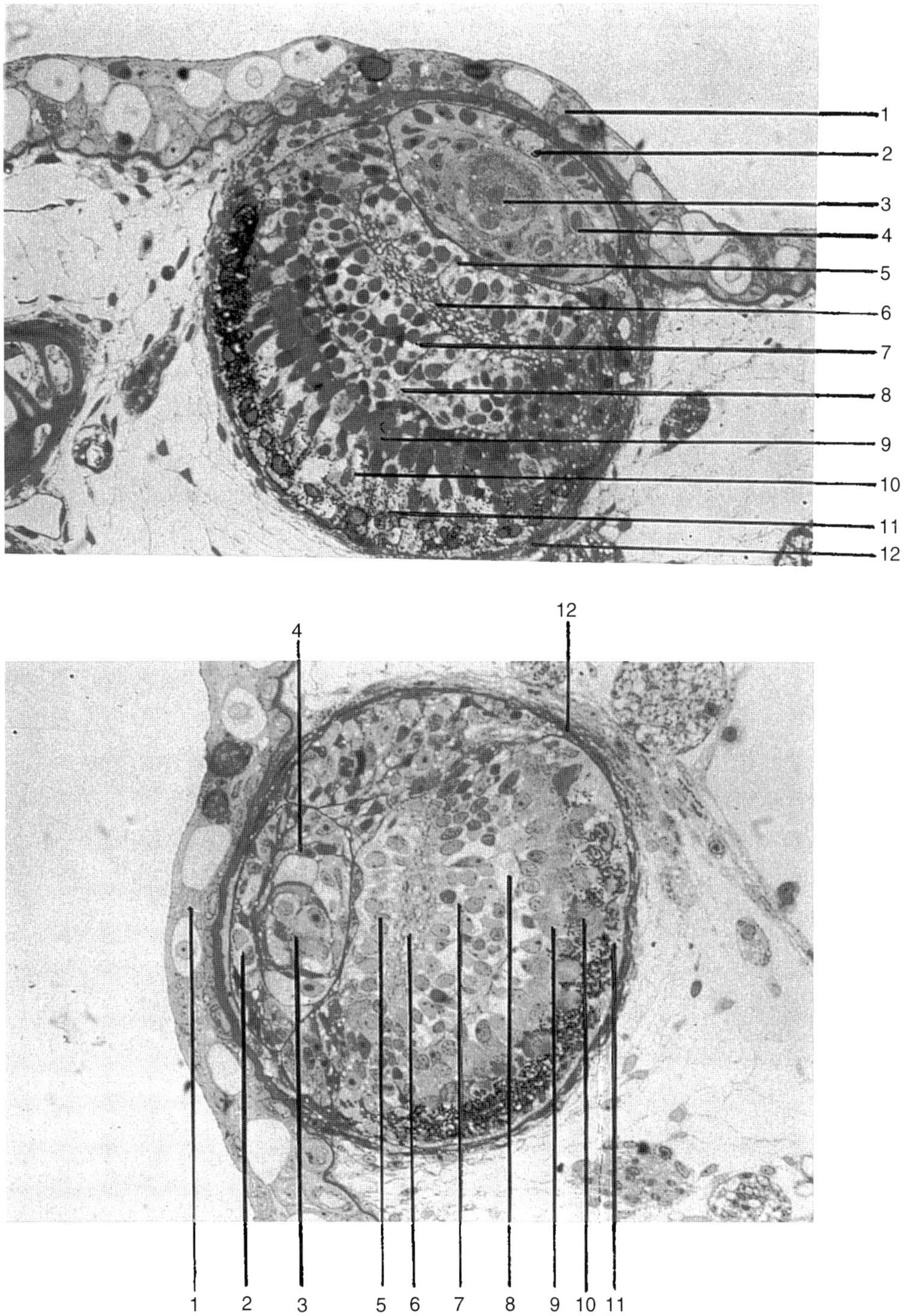

Figure 6. Cross sections of larva eyes of *R. laticauda* (above) (horizontal eye diameter = 0.18 mm; standard length 7.2 mm) and *R. zongolicensis* (below) (horizontal eye diameter = 0.17 mm; standard length 8.9 mm). 1 = cornea, 2 = lens epithelium, 3 = lens nucleus, 4 = lens fibre, 5 = ganglionic layer, 6 = inner plexiform layer, 7 = inner nuclear layer, 8 = outer pexiform layer, 9 = outer nuclear layer, 10 = sensory cells, 11 = pigmentary epithelium, 12 = sclera.

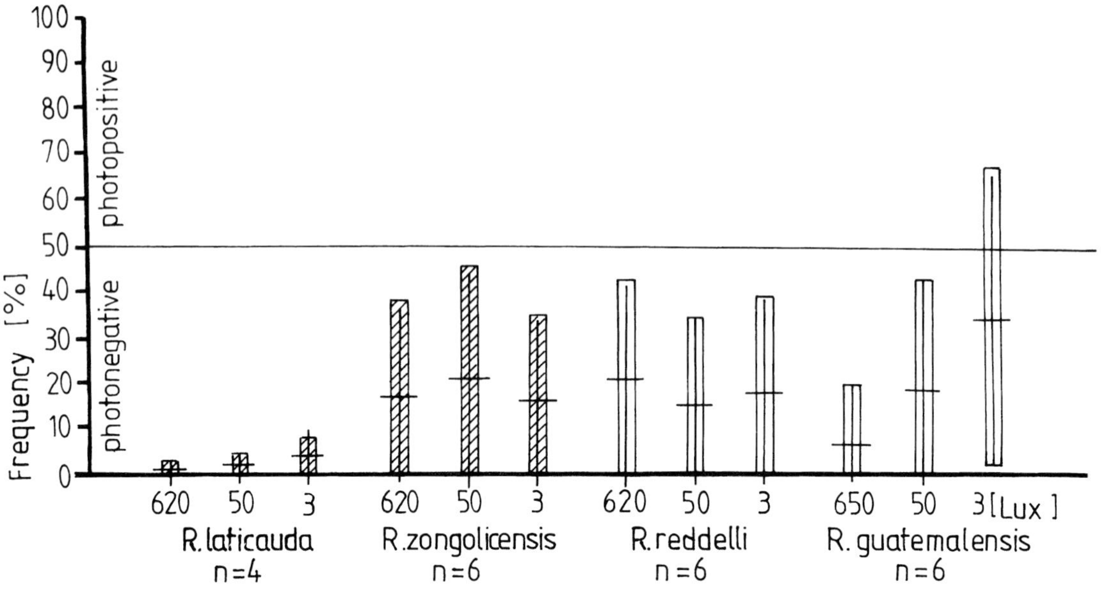

Figure 7. Phototactic behavior of two surface and two cave *Rhamdia* species.

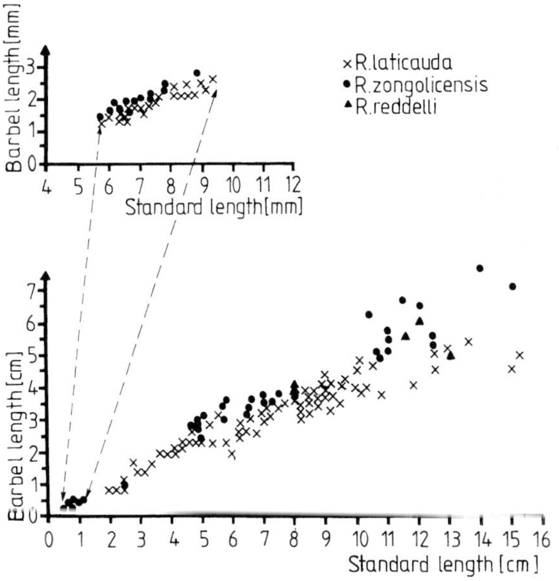

Figure 8. Correlation of barbel and body length during ontogenetic growth in *R. laticauda*, *R. zongolicensis* and *R. reddelli*.

from starvation was much shorter in the cave fish. After two to five months of feeding, individuals of the latter regained their original weight. The only remaining surface fish had not even achieved its original condition after six months.

Preadaptive traits

Egg yolk content

Comparative measurements showed that the species of the *R. laticauda*-group have yolk rich eggs. The yolk content is double of that of *R. guatemalensis*. There are no differences between the surface *R. laticauda* and the two closely related cave species (Figure 10).

Discussion

As is common in troglobites, all light dependent structures in cave *Rhamdia* are in the process of structural reduction; however, troglomorphism in the *Rhamdia* cave fishes is not as complete as in other troglobites, but rather is in an intermediate interval. The eye size is highly variable between surface and cave species and the genetic basis of melanin pigmentation also seems to be only slightly changed. The intense scotophilia of the surface fish is reduced to a certain degree in the cave fishes. Like in *A. fasciatus*, the pineal organ is structurally almost completely intact in *R. zongolicensis* (Langecker & Wilkens 1992).

Comparison of the state of regression between the two cave species reveals slight differences. *R. zongolicensis* displays a lower degree of both eye

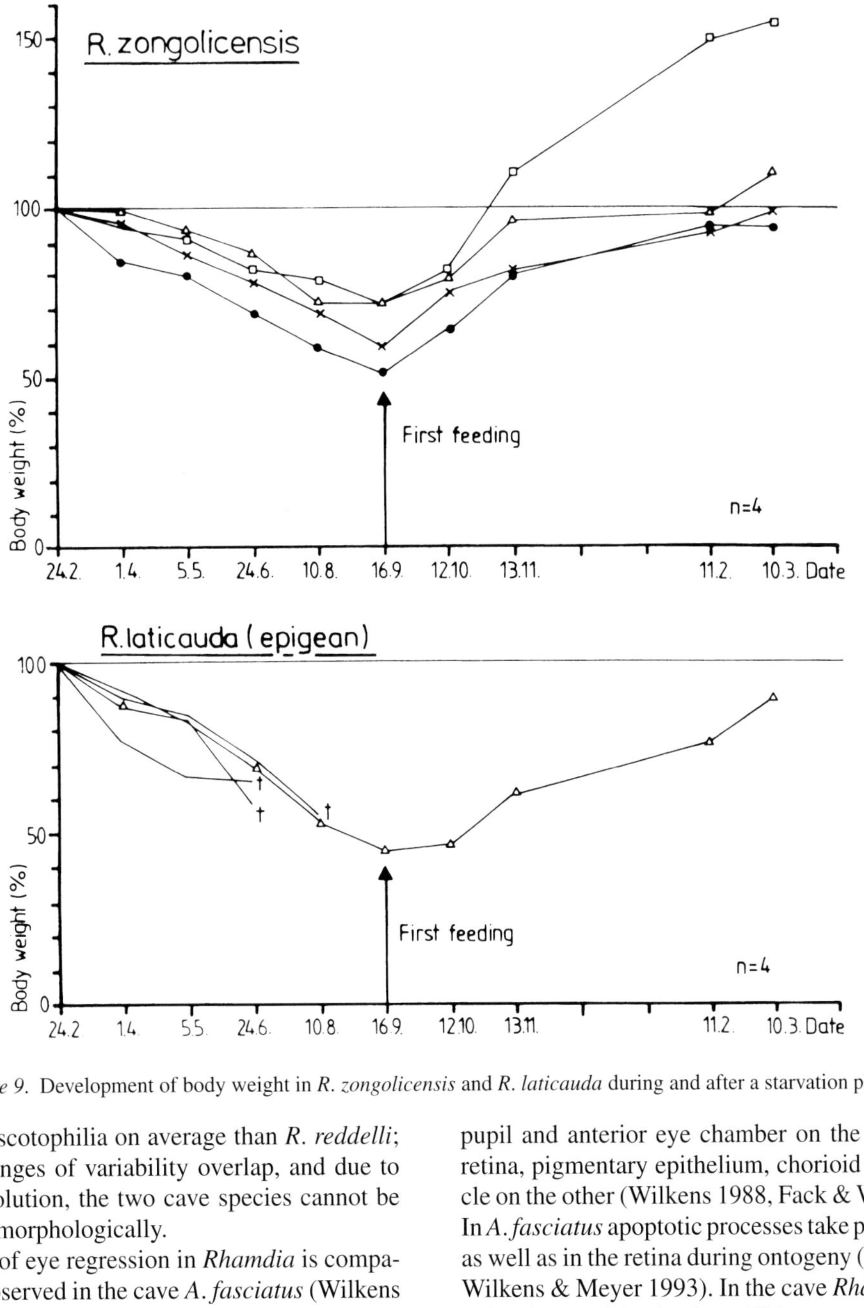

Figure 9. Development of body weight in *R. zongolicensis* and *R. laticauda* during and after a starvation phase.

reduction and scotophilia on average than *R. reddelli*; however, as ranges of variability overlap, and due to convergent evolution, the two cave species cannot be discriminated morphologically.

The pattern of eye regression in *Rhamdia* is comparable to that observed in the cave *A. fasciatus* (Wilkens 1988). The degree of eye differentiation is correlated to eye size, with the complete loss of single structures like lens or sensory cells to be observed only in very small eyes. The combination of large as well as small lenses with fully differentiated retinae (Figures 3a,b) confirms observations in *A. fasciatus* showing that – though inductively correlated – two rather independent inductive systems of eye development exist in the eye: lens, pupil and anterior eye chamber on the one hand and retina, pigmentary epithelium, chorioid and lens muscle on the other (Wilkens 1988, Fack & Wilkens 1989). In *A. fasciatus* apoptotic processes take place in the lens as well as in the retina during ontogeny (Wilkens 1988, Wilkens & Meyer 1993). In the cave *Rhamdia* this was only observed in the lens. This seems to be caused by the lesser degree of reduction in *Rhamdia* eyes, which still enables the formation of a thinner, though structurally complete, retina with a lower number of sensory cells in most specimens. The important role of lens development in cave fish eye degeneration has also been demonstrated for *A. fasciatus* (Yamamoto & Jeffery 2000).

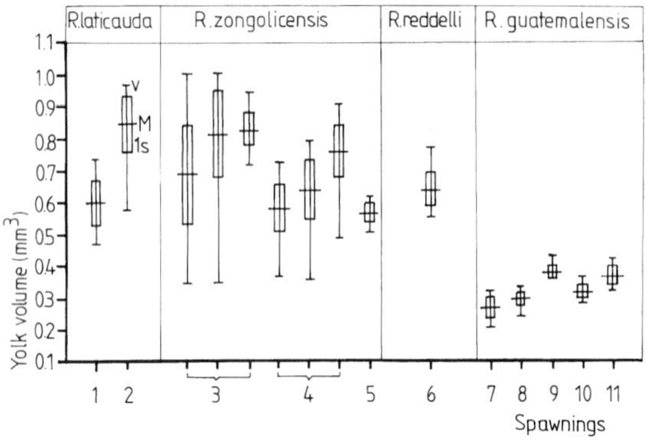

Figure 10. Comparison of yolk volume in two surface and two cave *Rhamdia* species (each number indicates different female specimens of which 45 eggs were measured in every clutch; v = variation, m = mean, s = standard deviation).

Because of the early ontogenetic existence of all single structures in *Rhamdia* and the extreme structural left–right asymmetry between the eyes of many adult individuals, it is quite possible that eye reduction is due mainly to the mutation of developmental genes, not structural ones. This has also been demonstrated for the eye of *A. fasciatus* (Wilkens 1998, Culver & Wilkens 2000).

The ancestor of the cave species was possibly a troglophile already perfectly adapted to life in darkness, similar to *R. laticauda*. As a consequence, only few newly developed morphological adaptations to cave life can be found in the cave fishes. There is neither an improvement of the lateral line sensory system nor of structure and quantity of the electro-sensitive organs (Weber 1995). The main adaptations focus on prerequisites for survival under food scarcity, such as the detection of food by the elongation of barbels and increased fat storage.

Silfvergrip (1996) has changed the taxonomic status of the two cave species *R. reddelli* and *R. zongolicensis* into two different subterranean populations of *R. laticauda* based on their morphological similarity. However, it can be shown that considerable genetic divergences exist between them. All F1-crosses between the two cave and the surface species *R. laticauda* consist exclusively of females (Haldane's Rule). This shows that the genetic basis of male sex determination has divergently evolved in the three species (Wilkens 1993) and postzygotic isolating mechanisms have developed. Furthermore, the evolution of separate gene pools is proven by mtDNA-analysis (unpublished data). Therefore it can be concluded that the morphological similarity is the result of convergent evolution and the cave catfishes should be treated as separate species.

Whereas in the *R. laticauda*-group several cave living species and one subspecies have been described, no true troglobite has yet been found in the *R. guatemalensis*-group. Presumably *R. guatemalensis* is not able to compete with specialized cave fishes. In the Yucatan peninsula, for example, *R. guatemalensis* is widely distributed in the groundwater system (Wilkens 1982); however, the majority of specimens concentrate in the influence area of the cenotes, where the trophic influx is more intense (Pohlman et al. 2000). Because its energy demands are higher than that of the better adapted cave fishes *Ogilbia pearsei* (Hubbs 1936) and *Ophisternon infernale* (Hubbs 1936), which also occur in the underground waters of Yucatan, *R. guatemalensis* has not developed troglomorphic characters. A similar situation seems to occur between *R. guatemalensis* and the cave fish of the *R. laticauda*-group where they are found in close association. Furthermore, the egg yolk content in *R. laticauda*, which is double of that of *R. guatemalensis*, may play an important role in favoring the development of cave forms in the *R. laticauda*-group where larger larvae with longer barbels are found (Junge 1993).

R. zongolicensis and *R. reddelli* are characterized by a comparatively low degree of eye and pigment reduction. This suggests a rather recent origin of these cave fishes. This observation is congruent with the biogeographical development in Northern Mexico during the Pleistocene. Due to climatic cooling the distribution limits of the thermophilic fish species were shifted from north to south and from higher to lower altitudes (Rosen 1998). Consequently, the surface *R. laticauda* was presumably distributed in the Gulf Coast Plain during the last glaciation, whereas *R. guatemalensis* had retreated further south. It was at that time the ancestor of *R. reddelli* probably invaded the San Antonio Cave and became isolated when, at the beginning of the Holocene, the external temperatures rose and the areas surrounding the mountain ridge became too warm for *R. laticauda* and were colonized by *R. guatemalensis*. Thus the chronological age of the cave species can be calculated as no more than 8000 to 10 000 years.

References cited

Ali, M.A. & M. Anctil. 1976. Retinas of fishes. Springer-Verlag, Berlin. 284 pp.

Anderßon, O. 1992. Vergleichende Untersuchung der Augenentwicklung der oberirdischen und der Höhlenformen der Gattung *Rhamdia* (Pimelodidae, Pisces). Diploma Thesis, University of Hamburg, Hamburg. 60 pp.

Culver, D. & H. Wilkens. 2000. Critical review of the relevant theories of the evolution of subterranean animals. pp. 389–407. *In:* H. Wilkens, D. Culver & W. Humphreys (ed.) Ecosystems of the World, Volume 30, Subterranean Ecosystems, Elsevier, Amsterdam.

Fack, H. & H. Wilkens. 1989. Eye reduction in hybrids and a naturally variable cave form of *Astyanax fasciatus* (Characidae, Pisces). Mém. Biospéol. 16: 157–161.

Greenfield, D.W., T.A. Greenfield & R.L. Woods. 1982. A new subspecies of cave-dwelling pimelodid catfish, *Rhamdia laticauda typhla* from Belize, Central America. Brenesia (19/20): 563–576.

Junge, P. 1993. Egg size and larval development of epigean and cave forms of *Rhamdia* (Pimelodidae, Pisces). pp 165–173. *In:* H. Schröder, J. Bauer & M. Schartl (ed.) Trends in Ichthyology, Blackwell Scientific Publications, Oxford.

Langecker, T.G. & H. Wilkens. 1992. Comparative ultrastructural studies on the pineal organ of the Mexican catfish *Rhamdia laticauda* (Heckel 1858) and one of its cave-dwelling derivatives (Pimelodidae, Teleostei). Acta Zool. 73: 247–253.

Langecker, T.G., H. Schmale & H. Wilkens. 1993. Transcription of the opsin gene in degenerate eyes of cave-dwelling *Astyanax fasciatus* (Teleostei, Characidae) and of its conspecific epigean ancestor during early ontogeny. Cell Tiss. Res. 273: 183–192.

Miller, R.R. 1984. *Rhamdia reddelli*, new species, the first blind pimelodid catfish from Middle America, with a key to the Mexican species. Trans. San Diego Soc. Nat. Hist. 20: 135–144.

Miller, R.R. 1986. Origin and geography of the fishes of Central Mexico. pp. 487–517. *In:* C.H. Hocutt & E.O. Wiley (ed.) The Zoogeography of North American Freshwater Fishes, John Wiley & Sons, New York.

Mitchell, R.M., W.H. Russell & W.R. Elliott. 1977. Mexican eyeless characin fishes, genus *Astyanax*: environment, distribution, and evolution. Sp. Publ. Mus. Tex. Tech Univ. (12): 1–89.

Mosier, D. 1984. Cave dwelling populations of *Rhamdia* (Pimelodidae). Assoc. Mex. Cave Stud. Act. Newsl. 14: 40–44.

Myers, G.S. 1966. Derivation of the freshwater fish fauna of Central America. Copeia 1966: 766–773.

Pohlman, J. W., L.A. Cifuentes & T.M. Iliffe. 2000. Food web dynamics and biogeochemistry of anchialine caves: a stable isotope approach. pp. 351–363. *In:* H. Wilkens, D. Culver & W. Humphreys (ed.) Ecosystems of the World, Volume 30, Subterranean Ecosystems, Elsevier, Amsterdam.

Robertson, S. 1983a. Zongolica project. 1983. Assoc. Mex. Cave Stud. Act. Newsl. 13: 36–41.

Robertson, S. 1983b. Cueva del Túnel, Veracruz. Assoc. Mex. Cave Stud. Act. Newsl. 13: 47–49.

Rosen, D.E. 1978. Vicariant patterns and historical explanation in biogeography. Syst. Zool. 27: 159–188.

Sbordoni, V., R. Argano, V. Vomero & A. Zullini. 1977. Ricerche sulla fauna cavernicola del Chiapas (Messico) e delle regioni limitrofe: grotte esplorate nel 1973 e nel 1975. Criteri per una classificazione biospeleologica delle grotte. Quaderni Accad. Naz. Lincei (171, Part 3): 5–74.

Silfvergrip, A.M.C. 1996. A systematic revision of the neotropical catfish genus *Rhamdia* (Teleostei, Pimelodidae). Swedish Museum of Natural History, Stockholm. 156 pp.

Weber, A. 1995. The lateral line system of epigean and cave dwelling catfishes of the genus *Rhamdia* (Pimelodidae, Teleostei) in Mexico. Mém. Biospéol. 22: 215–225.

Weber, A. 1996. Cave dwelling catfish populations of the genus *Rhamdia* (Pimelodidae, Siluroidei, Teleostei) in Mexico. Mém. Biospéol. 23: 73–85.

Weber, A. & H. Wilkens. 1998. *Rhamdia macuspanensis*: a new species of troglobitic pimelodid catfish (Siluriformes; Pimelodidae) from a cave in Tabasco, Mexico. Copeia 1998: 998–1004.

Wilkens, H. 1982. Regressive evolution and phylogenetic age: the history of colonization of freshwaters of Yucatan by fish and crustacea. Assoc. Mex. Cave Stud. Bull. 8: 237–243.

Wilkens, H. 1988. Evolution and genetics of epigean and cave *Astyanax fasciatus* (Characidae, Pisces). Support for the Neutral Mutation Theory. Evol. Biol. 23: 271–367.

Wilkens, H. 1993. A new species of *Rhamdia* (Pisces: Pimelodidae) from a cave in the Sierra de Zongolica (Veracruz, Mexico). Mitt. Hamb. Zool. Mus. Inst. 90: 375–378.

Wilkens, H. 1998. Genetics of cave fishes. pp. 1201–1213. *In:* C. Juberthie & V. Decu (ed.) Encyclopaedia Biospeologica, Tome II, Société de Biospéologie, Moulis.

Wilkens, H., P. Junge & T. G. Langecker. 1991. Speciation in troglobites: studies in the San Antonio Cave (Oaxaca, Mexico). Int. J. Speleol. 20: 1–13.

Wilkens, H. & M. Meyer. 1993. Eye formation and regression during early ontogeny in cave fish. pp. 185–192. *In:* H. Schröder, J. Bauer & M. Schartl (ed.) Trends in Ichthyology, Blackwell Scientific Publications, Oxford.

Yamamoto, Y. & W. R. Jeffery. 2000. Central role for the lens in cave fish eye degeneration. Science 289: 631–633.

Environmental Biology of Fishes **62:** 263–275, 2001.
© 2001 *Kluwer Academic Publishers. Printed in the Netherlands.*

A review of morphological and behavioural changes in the cave molly, *Poecilia mexicana*, from Tabasco, Mexico

Jakob Parzefall
Zoologisches Institut und Zoologisches Museum der Universität, Martin Luther King Platz 3, D 20146 Hamburg, Germany (e-mail: parzefall@zoologie.uni-hamburg.de)

Received 19 June 2000 Accepted 23 January 2001

Key words: regressive evolution, behaviour, biospeology, fish ecology

Synopsis

The poeciliid fish *Poecilia mexicana* successfully colonised a sulfur cave in Tabasco, Mexico. The eye size, melanin pigmentation and scale tuberances as well as aggressive and schooling behaviour are reduced in the cave fish. Besides these regressive characters some constructive ones have been found. The genital region of the female is enlarged to a genital pad and in both sexes the number of taste buds is considerably increased on the upper side of the head. The cephalic lateral line shows a hyperdevelopment with an incomplete covering of some channels. Nearly all of these characters studied form a genetically based gradient from the beginning to the end of the cave, which seems to be the result of gene flow from the outside to the inner part of the cave. A special behavioural adaptation was found among females which prefer bigger males during courtship. In darkness, only the cave females show this preference.

Introduction

The first and only known cavernicolous population of a poeciliid fish *Poecilia mexicana* (Steindachner 1863) was described as *P. sphenops* nearly 30 years ago (Gordon & Rosen 1962). They provided a detailed description of the Cueva del Azufre (also refered to as Cueva de la Sardina, Cueva de las Sardinas, Cueva de Villa Luz) and its organisms. The cave is about 300 m long and the principal chambers (Figure 1) have vaulted ceilings that are up to 12 m high. There are more than 20 skylights in the roof which give a variable amount of light in some parts. From chamber VI on the cave is essentially completely dark. For example, light intensity in chambers X and XIII is 0.0005 and 0.0005 lumen per squarefoot, which is basically dark (Schlupp & Parzefall unpublished). Recently a more detailed map and description of the cave has been published (Hose & Pisarowicz 1999). The authors created many new names for the different parts of the cave, without referring to the already published map (Gordon & Rosen 1962). This map is used by all authors cited in the present review. Therefore, I decided to use the terminology erected by Gordon & Rosen (1962) to avoid unnecessary confusion.

The first laboratory studies on the behaviour of *P. mexicana* were carried out shortly after the discovery of this cave populations (Walters & Walters 1965). They pointed out that the courtship in epigean and cave fishes under light conditions is optically mediated, while in darkness it involves physical contact. They also confirmed the findings of Gordon & Rosen (1962) that morphological characters vary within the cave population in correlation with distance from the cave entrance. These continuous morphological gradients are independent of the light in the cave, which does not show a gradient.

In 1964 C. Kosswig, N. Peters and C.D. Zander visited the cave and brought the first live fish to our laboratory. Since that time and with material collected in 1970, 1975, 1982 and 1996 our group at the University of Hamburg has bred and studied this fish in our laboratory. We compared our laboratory findings with in situ observations that took place during

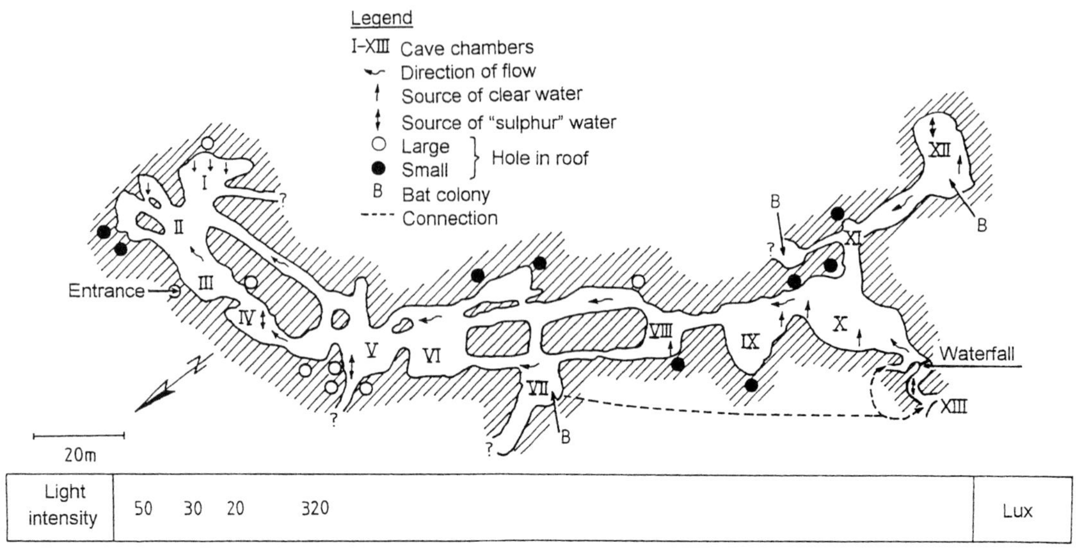

Figure 1. Map of the Cueva del Azufre, near Tapijulapa, Tabasco, Mexico. Light intensities measured in some chambers only (after Gordon & Rosen 1962, Peters & Peters 1968, Parzefall 1979 and original data).

10 expeditions between 1970 and 2000. As a result the cave molly is one of best characterised cave fishes besides the cave populations of *Astyanax fasciatus* (Wilkens 1988). This paper summarises our findings.

Material and methods

Fish from chamber IV and X have been kept and bred under artificial daylight with a 12 : 12 light/dark cycle and at a temperature of 26°C (±2°). We divided the stocks from chamber XIII into two lineages: a daylight stock (PSXIII L) and one kept in darkness (PSXIII D). The epigean individuals come from populations from the Rio Teapao near Teapa (PST), the Rio Oxolatan near Tapijulapa (PSOx) and the milky Arroyo coming out of the cave (PSO). In one of the early studies an aquarium stock (Liberty molly (PSLib)) were used. For comparison, aquarium stocks of *P. velifera* and *Xiphophorus helleri guentheri* from British Honduras were used. Fish were fed twice a day with commercially available flake food supplemented by cladocerous, tubifex worms and *Artemia* naupliae.

The fish were tested in visible light of about 900 lux and also in darkness with infrared light (800 nm) at 540 lux. We tested males for aggression with other males of the same size and in a second test adding a female. In choice tests we offered a female to two males of different size in perforated cylinders of polyacryl in opposite sides of a 25 l (25 × 25 × 50 cm) tank and we controlled for side biases. For the schooling behaviour the test fish had the choice between an empty compartment on one side and the school of conspecifics on the other side of a 300 l tank. All tests were controlled for side biases. To avoid sexual attraction only large juvenile individuals have been tested in visible light or in darkness with an infrared camera (Parzefall 1993). This was the only exception where we did not use adults.

For morphological studies I used a stereo microscope. Tissues and whole fish prepared for sectioning were fixed in formalin or Bouin's picroformol. For the study of dermal sense organs a modified (Schemmel 1967) chromatin fixation was used. Measurements of the genital region were made after staining with a 10% solution of methylene blue in 70% alcohol.

For the statistics parametric and non parametric tests have been used. The exact test for the traits studied is given together with the results.

Results

Morphological changes

Regressive characters. The eye size of the cave fish is significantly reduced and a gradient has been observed (Gordon & Rosen 1962, Peters & Peters 1968). This gradient has been stable for many years (Figure 2). But even in the fish having the smallest eyes (chamber XIII, Peters & Peters 1968) the eyes are still functional

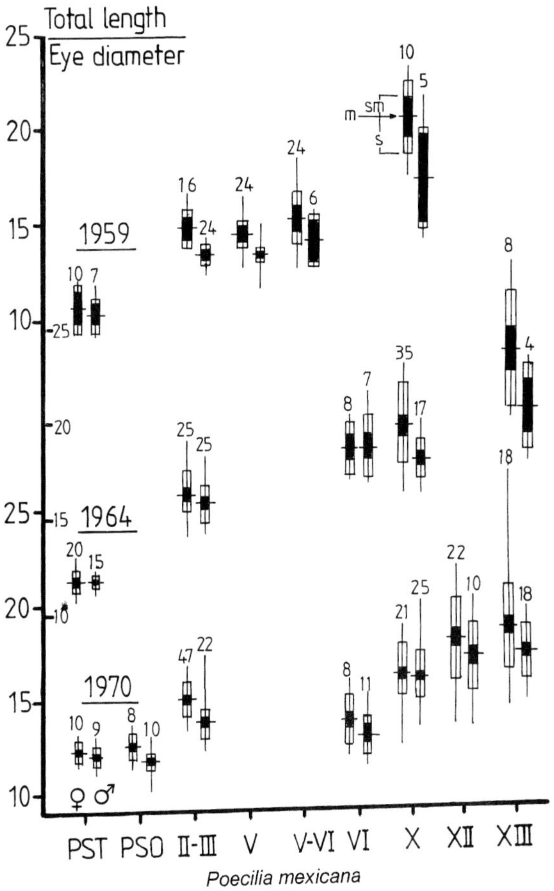

Figure 2. Relative eye size (total length of fish/eye diameter) in males and females of *Poecilia mexicana* collected in different cave chambers (II–XIII) and outside (PSO = near the cave entrance, PST = Rio Teapao near Teapa). Sample sizes are the numbers above each column, m = mean, sm = standard error of the mean, sd = standard deviation (after Peters et al. 1973).

despite some changes in the retina. A study of the spectral sensitivity resulted in slight differences between populations: the absorption-maxima of the four types of cones found are variable in the cave population of chamber XIII (Körner 1999). It could be demonstrated by cross breeding that the eye reduction is genetically based (Peters & Peters 1968, Wilkens unpublished data).

Besides the eyes, the melanin pigmentation is also reduced in the cave fish. This character is also genetically based (Peters et al. 1973, Wilkens unpublished data). Some of the fish collected in the cave and many of the stock cave fishes are yellow. They seem to have accumulated carotenoid pigments. Cross breeding experiments revealed a recessive, monogen inheritance

Table 1. Scale protuberances on the head in males.

Species/population	n	degree of development of scale protuberances				
		very good	good	weak	a few	absent
X. helleri guentheri	20	0	0	0	0	0
Poecilia velifera	20	1	4	3	11	1
P. mexicana[1] Liberty	20	9	7	3	1	0
P. mexicana[1] cave II–III	20	1	7	11	1	0
P. mexicana[1] cave VI	7	0	1	1	4	1
P. mexicana[1] cave X	18	1	3	5	5	4
P. mexicana[1] cave XIII	20	0	2	5	1	12

[1]Original material, all others laboratory reared fishes.

of this coloration (Parzefall unpublished). The scale protuberances, which the males of many poeciliids develop on the upper side of the snout (Figure 3) are more or less reduced in the cave with the highest degree of reduction in chamber XIII (Table 1).

Constructive characters. The genital region of the cave fish has been described by Walters & Walters (1965) as a large fleshy genital pad enveloping the bases of anal rays 1 and 2. We have measured the size by using a grid (Figure 5a), counting the squares covered by the genital region in females collected from four cave pools (Parzefall 1970). The relative size (TL/pad size) seems to increase from chamber II–III to chamber XIII (Figure 4). The epigean fish has no prominent genital pad.

Anatomical observations show that the pad develops many folds (Figure 5b). On the basis of visual inspection in these folds we found a great concentration of mucous cells common and frequent in the epidermis of fish. In the genital region of the epigean fish only very small folds can be found and the mucous cells are spread over the whole region (Parzefall 1970).

Zeiske (1968) reported that epigean males of *P. mexicana* can identify females by nipping in the genital region even when the nostrils are blocked. Therefore, we studied the taste organs in closely related species and different *P. mexicana* populations. All the species studied (*P. velifera*, *P. mexicana*, *Xiphophorus helleri*) have taste buds on the body surface with a concentration on the upper side of the head (Figure 6) in both sexes that enables them to receive a taste signal during the nipping behaviour. The number of these organs is considerably large (Figure 7) in the cave fish of chamber XIII in comparison to an epigean aquarium stock (Parzefall 1970). The taste organs in specimens of the other cave chambers have yet to be studied.

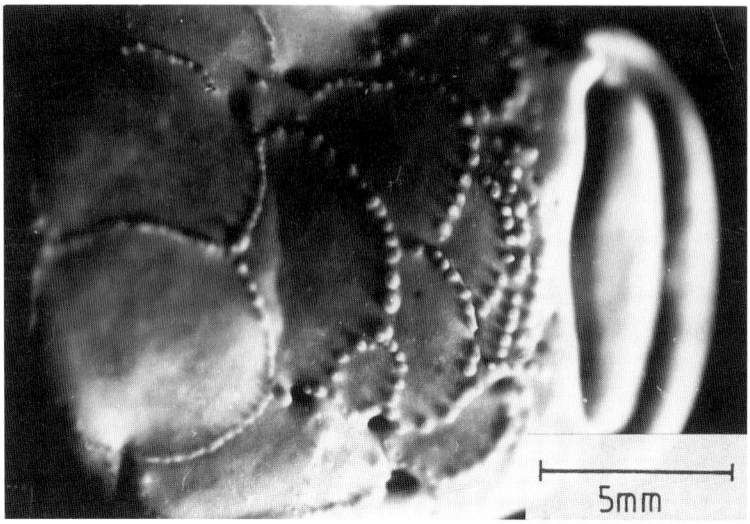

Figure 3. Upperside of the head in an epigean *P. mexicana* male with typical scale protuberances (after Parzefall 1970).

The cephalic lateral line system has already been characterized by Walters & Walters (1965) as hyper developed with an incomplete covering of some channels and a consequent exposure of the channel neuromasts. We subsequently confirmed these findings (Figure 6). In addition the head and trunk of the cave fish are dotted with large isolated neuromasts.

Behavioural and physiological changes

Regressive characters

Adult epigean *P. mexicana* can be found in mixed schools in deeper parts of rivers and ponds while the young prefer shallow areas with plant cover. Within groups a size dependent rank-order exists that is normally independent for the two sexes. Males compete mainly for females and the females for food. Aggressive behaviour consists of seven different patterns: aggressive fin-erection, tail-lifting, s-position, tail-beating, ramming, biting and circling (Parzefall 1969). In the cave form all aggressive patterns are reduced. This reduction increases gradually from the entrance to the inner end of the cave (Parzefall 1974). Figure 8 shows changes in these aggressive behaviours depending upon origin of subjects studied. The reduction in aggressive patterns is independent of light intensity measured in the different cave chambers. After testing male cave individuals with good visual orientation with males of the epigean population the reduction patterns remain unchanged (Figure 9). The mean for the number of aggressive movements in the cave fish is about zero. In the mixed tests the majority of the cave males answered the initial attacks of the epigean males with sexual behaviour, only a few responded defensively for a short period of time. No cave males ever showed aggression. When we tested the epigean males in darkness under infrared conditions the fish changed their normal swimming behaviour to a slow gliding. Coming into contact with another male they performed tail-beating as the only aggressive pattern. In addition, males of both an epigean and hypogean population have been tested for aggression in dyadic contests (Parzefall et al. 1997). We tested three different groups of adults: epigean fish, cave fish reared under light conditions and cave fish reared in darkness. We measured the optomotoric response to make sure that all tested males had comparable visual abilities. A significant difference was found between epigean fish (PSO) and dark reared cave fish (PSXIII D) ($p < 0.05$, $n = 10$, sign test and G-test). In subsequent aggression tests with light of various intensities (0.01; 0.1; 0.5; 1; 5; 50; 300; 1000 lux) different light levels were required to release single aggressive patterns. In total darkness only tail-beating was observed. Fights occurred at first at 5 lux. Some patterns increased with light intensity. The cave stock maintained in light for about 80 generations has an aggression level comparable to the one of their epigean relatives. The data reveal the importance of light for aggressive behaviour in *P. mexicana*. It seems that the epigean fish when colonising the dark could no

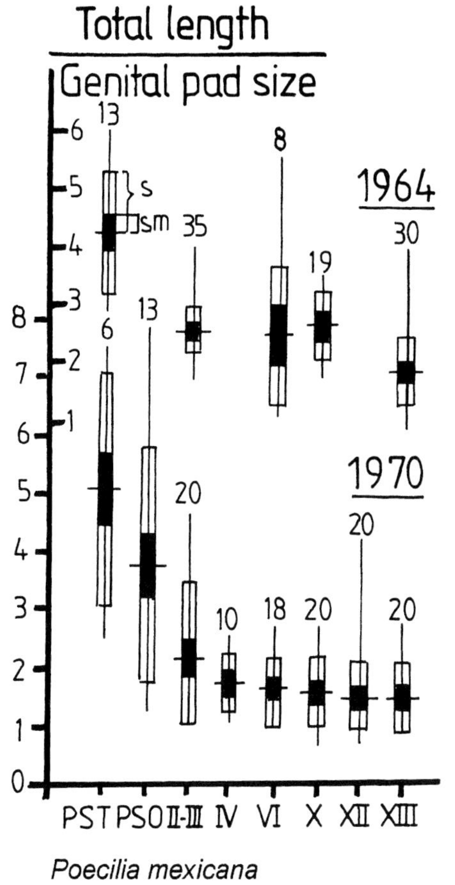

Figure 4. Relative size (TL/area) of the genital pad for samples of two years in *P. mexicana* females. For explanations see Figure 2 (after Parzefall 1970, Peters et al. 1973).

longer compete for females using the complete set of aggressive patterns.

When we tested hybrids (Parzefall 1979, 1982) between cave and epigean fish the frequency distribution for aggressive fin-erection and s-position revealed for all generations tested a genetically based reduction of the aggressive behaviour within the cave (Figure 10). The frequency distribution for the two patterns presented show a clear separation of the parental generations. The F_1-generation has an intermediate value for the average and the variability is practically halved in comparison to the epigean forms. The frequency distributions for the F_2-generations and the back-crosses are very similar. Normally one should expect a separation of the two back-crosses as it is expressed for the aggressive fin-erection. We can state that the reduction for aggression is genetically based and controlled by a polygenic genetic system. We also controlled our F_2 data for possible correlations: there is no significant correlation between the eye diameter and aggressiveness in the F_2 males tested, but despite free recombination a positive correlation was found for several aggressive patterns as in the epigean form. In comparison with the epigean stock, the reduction of all aggressive patterns is similar when expressed percentually (Parzefall 1979). From this and from visual inspection of the frequency distributions we can conclude that a closely linked polygenic system exhibiting additive gene interaction exists and controls the aggressive behaviour in *P. mexicana*. The absence of a correlation between the regressive changes in aggression and the reduction of the eyes in the F_2 males tested suggests that regression in these characters is under independent genetic control. As already mentioned, the epigean fish show schooling in their habitat. Schooling behaviour is widely distributed among fish and the important role of visual orientation is well known from various studies on schooling (Partrige & Pitcher 1980). In the absence of visual orientation the fish studied do not school. The cave population of *P. mexicana* does not school at all. To answer the question whether this difference results from the absence of visual orientation or a genetically based reduction on the behaviour level we have tested two epigean stocks (PST, PSO), cave fish from chamber XIII (PSXIII L), and hybrids (Parzefall 1993). To avoid sexual attraction only larger juvenile fishes have been tested. The epigean fish of the Rio Teapao (PST) oriented itself in visible light to the school. But the tendency to follow the school of conspecifics is already reduced in the epigean stock living near the cave entrance (PSO) where the water appears milky from suspended sulfur (Figure 11) ($p < 0.05$, U-test). There is no difference in the schooling tendency between PSO and the cave fish. In both of these stocks there is strikingly more variability than in the epigean fish (PST). However, in darkness no preference for the school has been recorded. The F_1 hybrids of PST $\times$ PSXIII are at the same level as the epigean parental generation and significantly different from the cave fish ($p < 0.002$, U-test). The F_2 generation differs significantly from the putative epigean ancestors ($p < 0.01$, U-test) in the mean and in the high variability, too. The backcrosses can not be separated. There is only a significant difference in the cave fish backcross from the epigean and the F_1 generation ($p < 0.002$; $p < 0.01$, U-test). These data confirm the important role of the visual system for schooling behaviour in

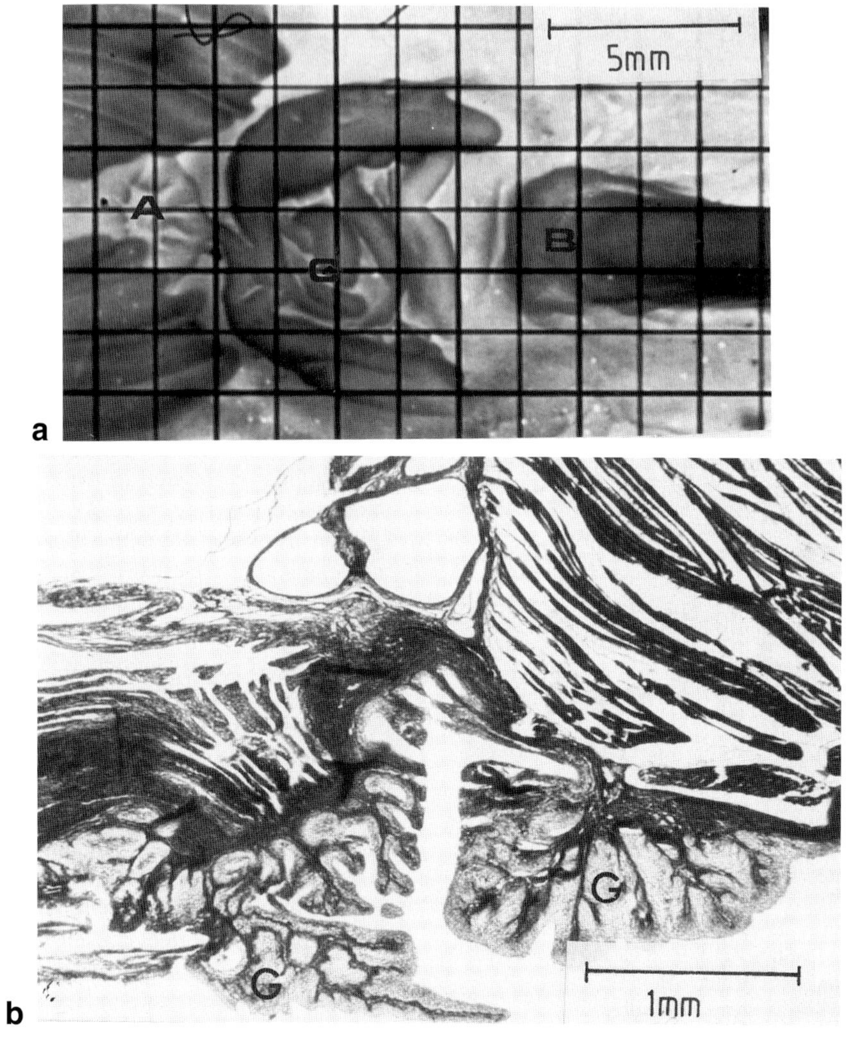

Figure 5. a – Genital pad of a cave fish female with the grid for size measuring, b – cross-section. The head of the fish is to the left. A = anus, G = genital pad, B = basis of the anal fin (after Parzefall 1970).

P. mexicana and a genetically fixed reduction which already takes place in the milky Arroyo del Azufre outside of the cave.

During courtship the males position themselves before or besides a female with their coloured fins erected. A male follows a female and tries to nip and copulate. During the nipping behaviour it receives information about the sexual status of the female. Females have a cycle of about 28 days and are very attractive for about 3 days typically after giving birth (Parzefall 1973). Contrary to other Poeciliids like *P. velifera* or *P. latipinna* a striking visual display with erected fins (Parzefall 1969) is absent in *P. mexicana*. In this respect, this species seems already preadapted for the successful colonisation of caves. The only difference is that the cave fish follow the female directly when coming into contact and try not to loose contact.

Constructive characters

As already noted by Gordon & Rosen (1962) and shown by our chemical studies (Peters et al. 1973) the cave water contains toxic hydrogen sulfide. When we placed cave fish of different size in perforated plastic boxes at the bottom in the cave stream for some days the fishes survived. Therefore we tested in the laboratory cave and epigean fish of *P. mexicana*, *Astyanax fasciatus* and

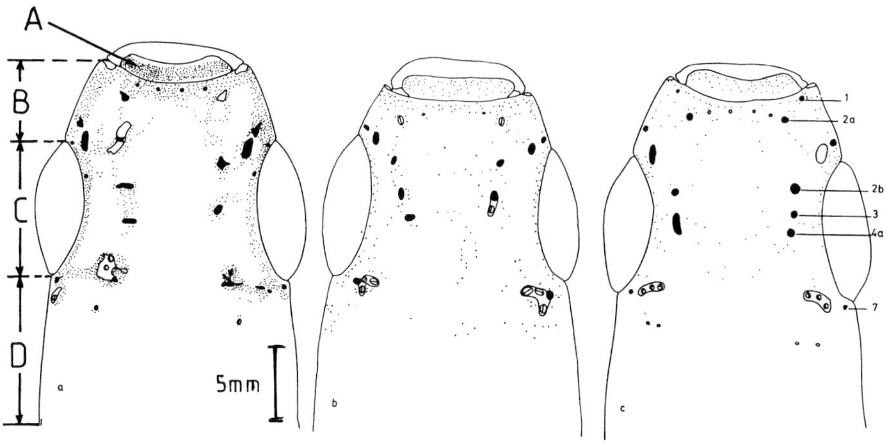

Figure 6. Distribution of the taste buds and lateral line system in four areas on the upper side of three specimens: a – *P. mexicana* cave female chamber XIII (PSXIII L), b – epigean female (PSLib), c – *Xiphophorus helleri guentheri* female from British Honduras. Black parts (1–7) are openings of the channel-system. Some channels are not closed and neuromasts therefore visible (2a, 3–4a, 7) (after Parzefall 1970).

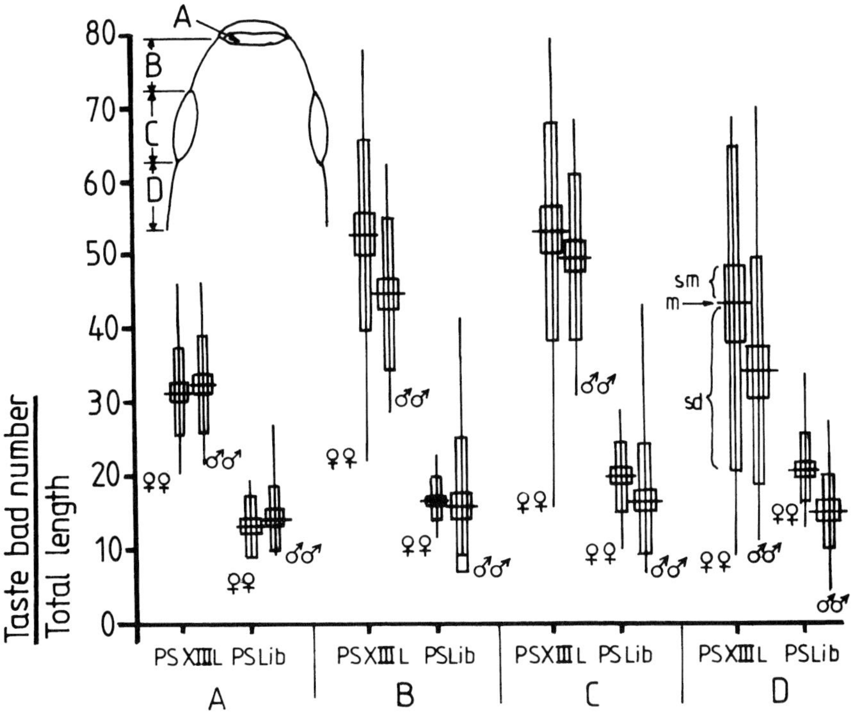

Figure 7. Relative density of taste buds in four different areas on the upper side of the head. A lips, B lips to beginning of the eyes, C between the eyes, D to the pectoral fin (n = 20 for each stock). Further explanations see Figures 2 and 6 (after Parzefall 1970).

a *Xiphophorus helleri* population from British Honduras for a possible hydrogen sulfide tolerance in an artificial stream of a closed polyacryl chamber and in an open tank using a hydrogen sulfid content of 1.4 to 4.6 mg l^{-1}. In the open tank nearly all fish tried to jump out of the water. In the closed chamber the fishfirst swam against the stream, tried to escape and finally unable to swim turned ventral side up. When we

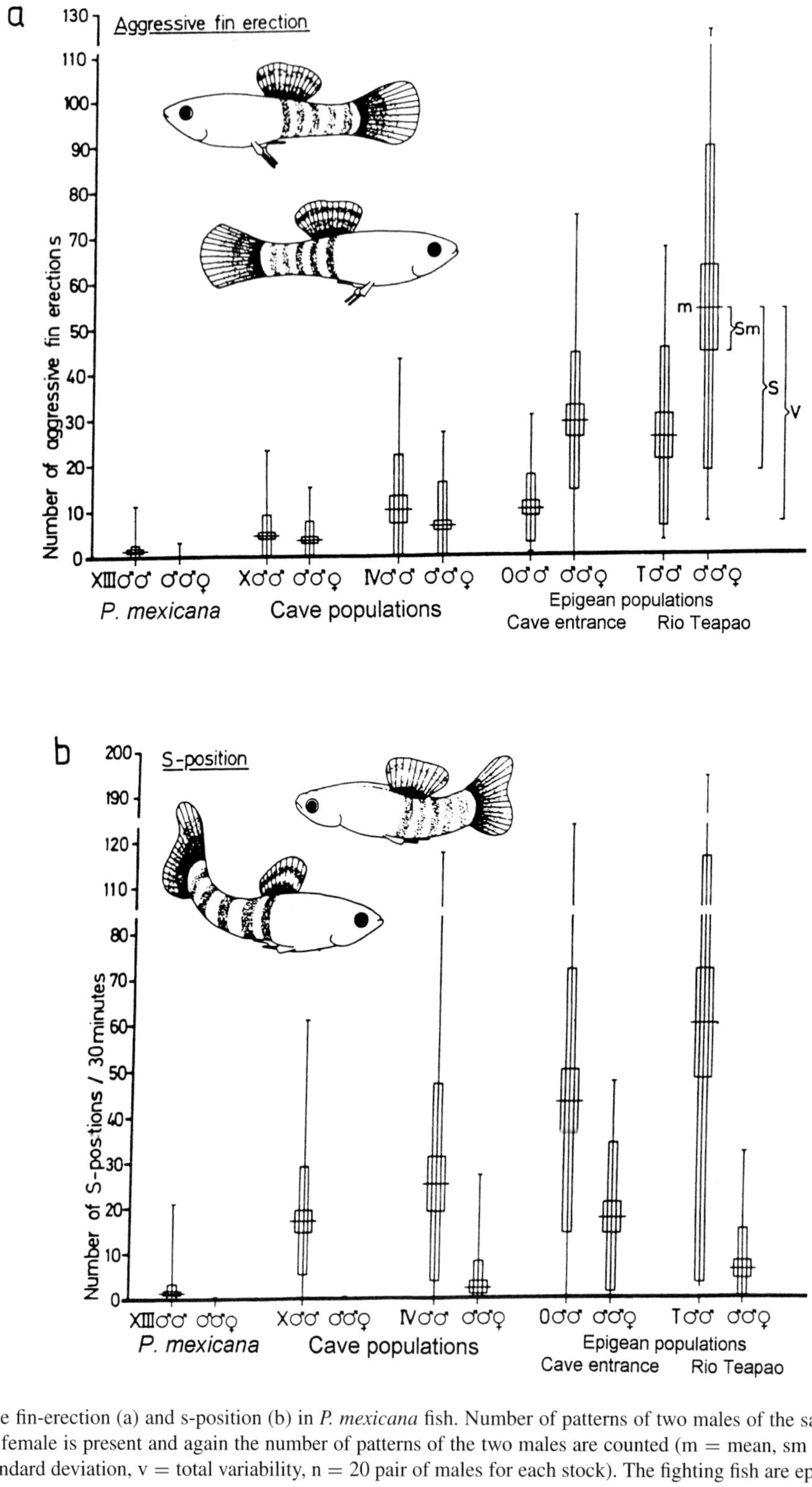

Figure 8. Aggressive fin-erection (a) and s-position (b) in *P. mexicana* fish. Number of patterns of two males of the same size per test. In the second part a female is present and again the number of patterns of the two males are counted (m = mean, sm = standard error of the mean, s = standard deviation, v = total variability, n = 20 pair of males for each stock). The fighting fish are epigean ones (after Parzefall 1982).

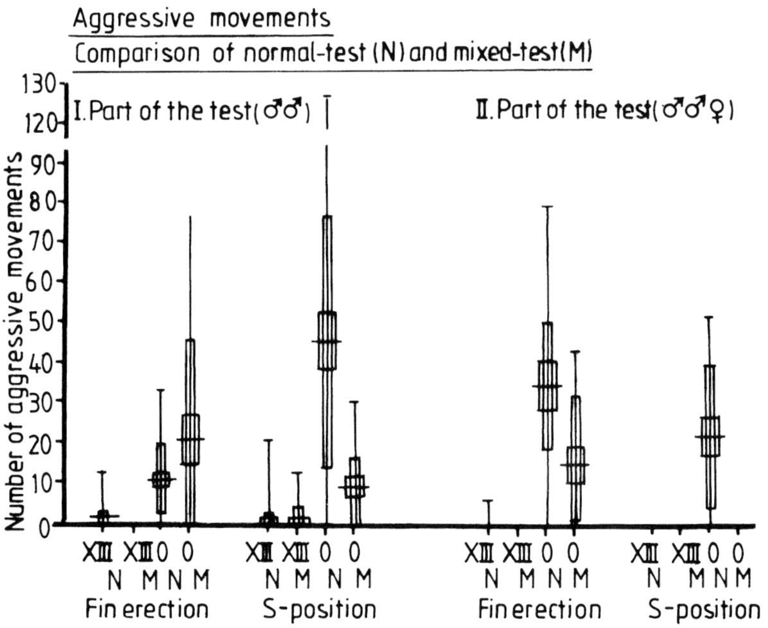

Figure 9. Aggressive fin-erection and s-position in *P. mexicana* stocks in different test situations. N = two equal sized males of the same stock tested, M = cave (chamber XIII light reared) against epigean male (PSO) of the same size tested, n = 20 males (after Parzefall 1982).

immediately brought this fish in clean water they recovered. Limited data (Peters et al. 1973) suggested that *X. helleri* and *A. fasciatus* always stopped swimming when the *P. mexicana* were still in good condition. In all the tests they resisted for a longer period than the two other species. However, our sample size was too small to separate cave and epigean fish of *P. mexicana*. In general we can state that there is a certain hydrogen sulfide tolerance in *P. mexicana*.

The epigean fish can mainly be observed in the free water and between plants. The slowly swimming schooling fish fed on algae and plant material. The cave fish tend to remain close to the bottom or the water surface during feeding. These behavioural differences exist in the natural habitats as well as in the laboratory (Parzefall 1979). It seems that the cave ecosystem does not primarily depend on energy input of epigean origin but is principally based on chemo-lithoautotrophic primary production of bacteria of the genus *Beggiatoa* that occur inside the cave. The water of the cave is saturated with hydrogen sulfide, which obviously is the energy source for chemoautotrophic, sulfur oxidising bacteria. The bacteria are the major food item of a large population of chironomid larvae and of the cave fish (Langecker et al. 1996). The fish feed on bacteria and chironomid larvae at the bottom and adult chironomids at the water surface. The bat colonies occurring inside the cave take a subordinate part in the energy input of the cave.

The female choice behaviour in many of the species studied reveals a preference for bigger males (Taborsky 1994). In the poeciliids studied there also exists such a preference (Bisazza & Pilastro 1997). In visible light the females of the epigean populations of the Rio Oxolatan and the cave population of *P. mexicana* prefer the compartment near bigger males separated by a glass pane. The females of the epigean fish of the milky Arroyo unexpectedly prefer smaller males (Körner 1999). During experiments in darkness with infrared light where the fishes are separated from the male by a grid, the cave fish is able to differentiate only between males of different size and prefers the bigger ones (Figure 12).

Discussion

The distinctive characteristic of the cavernicolous *Poecilia mexicana* is the genetically based gradient of the regressive and constructive characters studied. This

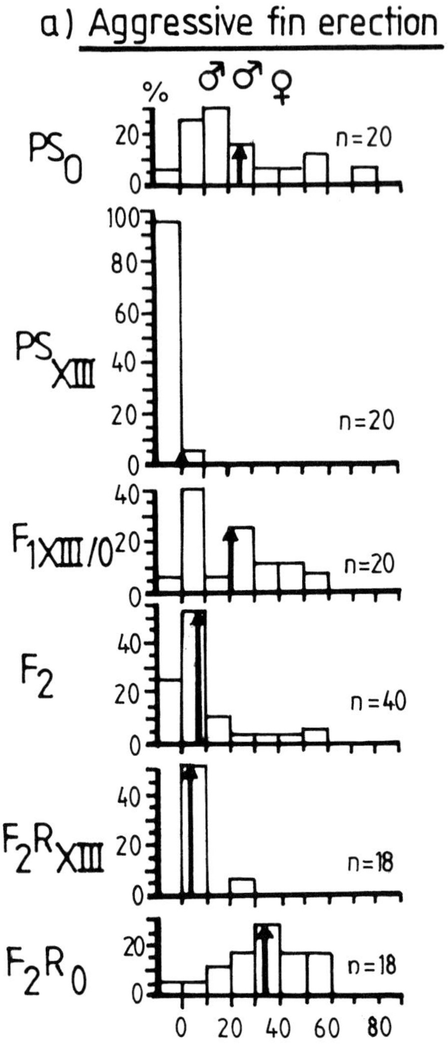

Figure 10a.

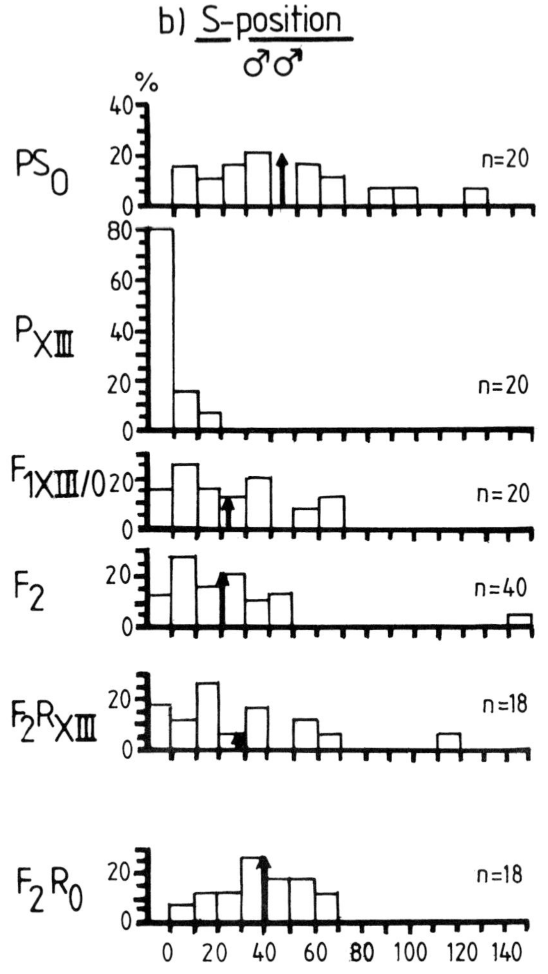

Figure 10b. Aggressive fin-erection (a) and s-position (b) in *P. mexicana* hybrids. Frequency distribution for PS_O = epigean stock near the cave entrance, PS_{XIII} = cave fish from chamber XIII, F_1, F_2, F_R = hybrid generations. The mean is shown by the arrow (after Parzefall 1979).

gradient which has been observed for many years can not be directly correlated to light conditions or any other ecological factor in the cave. Gordon & Rosen (1962) suggested that an undiscovered, truly cavernicolous upstream stock of mollies exists and causes the gradient by gene flow. For years our research group at the University of Hamburg and others (e.g., Hose & Pisarowicz 1999) have been studying this cave but have been unable to find such a stock. However, we know since our first visit in 1970 that the local Zoque Indians poison the fish in the main stream of the cave once a year during Easter. After that, we have observed cave mollies only in the shallow areas (Peters et al. 1973, Parzefall unpublished). This ceremony could possibly stabilise the gradient, and a founder stock does not seem necessary. The differences between epigean and cave fish so far studied by laboratory hybridisation are genetically based and gene flow from outside can cause the gradient. Restricted mobility of fish between the chambers seems to stabilise the gradient.

Regressive characters

Reduction of eye and pigmentation are a common phenomenon in cave dwellers (Wilkens 1986, 1987, 1988). In *P. mexicana* the degree of reduction is limited with eyes still functional. This can be explained by a recent

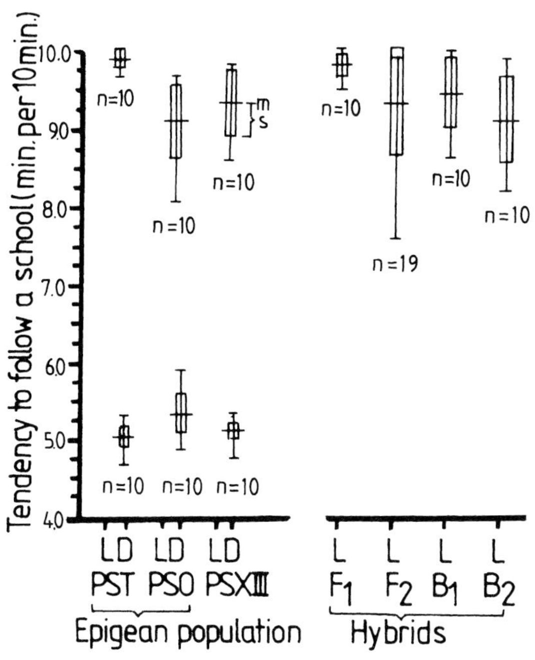

Figure 11. Tendency to follow the school during 10 min test in three different stocks of *P. mexicana* and the hybrids between PST = epigean fish of the Rio Teapao, PSXIII = cave fish from the cave chamber XIII light reared, PSO = epigean fish near the cave entrance, F_1, F_2 = hybrid generations, B_1 = back gross to PST, B_2 = back gross to PSXIII, L = tested in light, D = tested in darkness with infrared, m = mean, SD = standard deviation (after Parzefall 1993).

colonisation of the cave or by the fact that these characters cannot be reduced completely because of some conditions in the cave. The meaning of the reduction of the scale protuberances in the cave population remains unclear. If they are used to detect tactile stimuli during nipping one should expect a progressive development and not a reduction. The reduction of aggressive behaviour is up to now the only example in cave dwellers. From the data in the different taxa we can conclude that aggressive behaviour serves to stabilise a rank-order or establish a territory exists in general in dark conditions (Parzefall 2000). But there are some differences to the aggression in epigean habitats: the territories for food and reproduction seem to be smaller in caves, probably caused by the fact that non-visual signals like waterwaves are effective in shorter distances only. During fights in darkness the animals act in short distances of a few centimetres which allow the perception of waterwaves caused by tail-strokes or the animals have direct contact with biting, head-butting or tail-beating (Ercolini et al. 1981, Hausberg 1995, Trajano 1991). Submissive positions seem to be rare in dark habitats which facilitates flight and hiding. Freeze has been described as submissive behaviour only in the cave amblyopsids (Bechler 1983). It is possible that *P. mexicana* does not need to display aggression in the cave habitat, but in comparison to other cave dwellers one can also argue that the phylogentically young cave fish *P. mexicana* may not have completely evolved. For example, *A. fasciatus* reduced the aggressive pattern of the epigean ancestor and replaced it by a new type of aggressive behaviour (Hausberg 1995). The reduction of aggressive patterns in the cave population of *Poecilia mexicana* could also be caused by the absence of optical releasers for some aggressive pattern which are connected with striking dark body coloration. This means that these elements are not effective in darkness and therefore not controlled by stabilising selection any more. It is also possible to argue that fights in darkness are too costly and of disadvantage resulting in directional selection against aggression. Thus males should be better off keeping contact with a female they met in darkness than to fight with an opponent. Following this argument the final outcome could be a complete reduction of the aggressive behaviour in *P. mexicana*. The absence of stabilising selection seems to be a plausible explanation for regression of schooling behaviour that already took place in the milky river outside of the cave.

Constructive characters

It seems that many constructive characters have evolved to improve communication between males and females in darkness: the genital pad can guide the nipping male to the genital region for the intromission of the gonopodium. The mucous cells in the folds of the pad probably produce chemical signals. We have demonstrated that a direct body contact during nipping is necessary for the male to identify a sexually attractive female (Parzefall 1973) with his gustatory equipment (Zeiske 1968). The enlarged area with an increased number of taste buds on the upper side of the head seems to be an adaptation for better female identification mechanism. The fact that males and females have more taste buds is an indication that food detection is probably also improved in connection with the open head channels of the lateral line system. This requires further studies. In the cave *A. fasciatus* Schemmel (1967, 1980) found an improvement of the gustatory

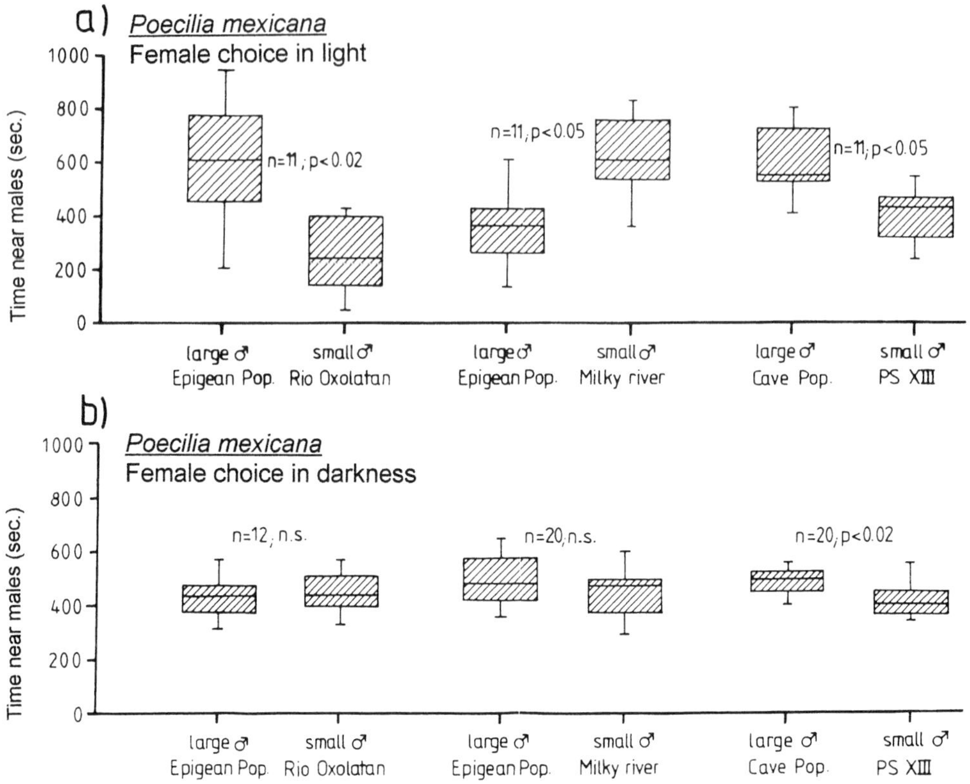

Figure 12. Female choice in different populations of *Poecilia mexicana* in light (a) and darkness (b). The female has the choice to swim to the compartment of a big or a small mature male. The time spent in a male compartment is measured. A clear partition prevents a direct contact. The middle line in the box plots represents the median, the upper end of the box the 75% value, the lower end of the box the 25% value. The wiskers represent the 90% and 10% values, respectively (after Körner 1999).

equipment on the ventral parts of the head and changes in the feeding position. Both traits are genetically independent in the cave *A. fasciatus*. In competition experiments with small pieces of meat in total darkness 80% of the food was found by Pachón cave fish but only 20% by epigean individuals (Hüppop 1987).

The most interesting result in *P. mexicana* is the successful female choice in the cave fish in darkness. The cave fish must have switched its sensory mechanism from a visual one to the lateral line system for the detection of water waves or probably to the chemical sense for this specific behaviour. This also requires further study.

The hydrogen sulfide tolerance is a phenomenon independent of darkness. High hydrogen sulfide contents and chemoautotrophically based cave ecosystems are known from the Movile Cave in Romania (Sarbu et al. 1995), Capo Palinuro and Verzino caves in Italy (Southward et al. 1996, Latella et al. 1999a,b), Lower Kane Cave, Cesspool Cave and Sulfur Cave in the United States (Egemeier 1981). But their inhabitants have not been studied for a possible hydrogen tolerance. In general the cave molly is an excellent model to study neutral evolution, as well as natural and sexual selection in the laboratory and field.

Acknowledgements

Thanks to the Deutsche Forschungsgemeinschaft (Pa 148) for financial support over many years and the Mexican Government for the permits (242.-2190 276/36, 210696-213-03) which allowed the study of the cave Molly in laboratory and field conditions. My Mexican colleague Hector Espinoza Perez from the Biological Institute of the UNAM was kind to support my activities and those of my colleagues for many years. M. Hänel prepared the drawings. I. Schlupp and two anonymous reviewers read the MS and made valuable suggestions.

References cited

Bechler, D. 1983. The evolution of agonistic behaviour in amblyopsid fishes. Behav. Ecol. Sociobiol. 12: 35–42.

Bisazza, A. & A. Pilastro. 1997. Small male mating advantage and reversed size dimorphism in poeciliid fishes. J. Fish Biol. 50: 397–406.

Egemeier, S.J. 1981. Cavern development by thermal waters with a possible bearing on ore deposition. Nat. Speleol. Soc. Bull. 43: 31–51.

Ercolini, A., R. Berti & A. Cianfanelli. 1981 . Aggressive behaviour in *Uegitglanis zammaronoi* Gianferrari (Clariidae, Siluriformes) an anophtalmic phreatic fish from Somalia. Monit. Zool. Ital. 5: 39–56.

Gordon, M.S. & D.E. Rosen. A cavernicolous form of the poeciliid fish *Poecilia sphenops* from Tabasco, Mexico. Copeia 1962: 360–368.

Hausberg, C. 1995. Das Aggressionsverhalten von *Astyanax fasciatus* (Characidae, Teleostei): Zur Ontogenie, Genetik und Evolution der epigäischen und hypogäischen Form. Doctoral Dissertation, University of Hamburg, Hamburg. 137 pp.

Hose, L. & J.A. Pisarowicz. 1999. Cueva de Villa Luz, Tabasco, Mexico: reconnaissance study of an active sulfur spring cave and ecosystem. J. Cave Karst Stud. 61: 13–21.

Hüppop, K. 1987. Food finding ability in cave fish (*Astyanax fasciatus*). Int. J. Speleol. 16: 59–66.

Körner, K.E. 1999. Zur sexuellen Selektion höhlenlebender Atlantikkärpflinge (*Poecilia mexicana* Steindachner 1863). Verlag Dissertation, Berlin. 104 pp.

Langecker, G., H. Wilkens & J. Parzefall. 1996. Studies on the trophic structure of an energy-rich Mexican cave (Cueva del Azufre) containing sulfurous water. Mém. Biospéol. 23: 121–127.

Latella, L., M. Cobolli & M. Rampini. 1999a. La fauna delle grotte nei gessi dell'Alto Crotonese (Calabria) (Cave fauna of Alto Crotonese). Thalassia Salentina 23: 103–113.

Latella L., C. Di Russo, L. De Pasquale, L. Dell'anna, G. Nardi & M. Rampini. 1999b. Preliminary investigations on a new sulphurous cave in central Italy. Mém. Biospéol. 53: 131–135.

Partidge, B.L. & T.J. Pitcher. 1980. The sensory basis for fish schools: relative roles of lateral line and vision. J. Comp. Physiol. A 162: 543–550.

Parzefall, J. 1969. Zur vergleichenden Ethologie verschiedener *Molliensesia*-arten einschließlich einer Höhlenform von *Mollienesia sphenops*. Behaviour 33: 1–37.

Parzefall, J. 1970. Morphologische Untersuchungen an einer Höhlenform von *Mollienesia sphenops* (Pisces Poeciliidae). Z. Morph. Tiere 68: 323–342.

Parzefall, J. 1973. Attraction and sexual cycle of poeciliids. pp. 177–183. *In*: J.H. Schröder (ed.) Genetics and Mutagenesis of Fish, Springer Verlag, Berlin.

Parzefall, J. 1974. Rückbildung aggressiver Verhaltensweisen bei einer Höhlenform von *Poecilia sphenops* (Pisces, Poeciliidae). Z. Tierpsychol. 35: 66–84.

Parzefall, J. 1979. Zur Genetik und biologischen Bedeutung des Aggressionsverhaltens von *Poecilia sphenops* (Pisces, Poeciliidae). Untersuchungen an Bastarden ober- und unterirdisch lebender Populationen. Z. Tierpsychol. 50: 399–422.

Parzefall, J. 1982. Changement of behaviour during the evolution of cave animals. Mém. Biospéol. 8: 55–62.

Parzefall, J. 1993. Schooling behaviour in population-hybrids of *Astyanax fasciatus* and *Poecilia mexicana* (Pisces, Characidae and Poeciliidae). pp. 297–303. *In*: J.H. Schröder & J. Bauer (ed.) Trends in Ichthyology: An International Perspective, GSF-Forschungszentrum für Umwelt und Gesundheit in association with Blackwell Scientific, Oxford.

Parzefall, J. 2000. Ecological role of aggression in the dark. pp. 221–228. *In*: H. Wilkens, D. Culver & B. Humphreys (ed.) Ecosystems of the World – Subterranean Biota, Elsevier, Amsterdam.

Parzefall, J., U. Gagelmann & M. Schartl. 1997. Aggressive behaviour and optomotor response in different populations of *P. mexicana* (Pisces, Poecilidae). Mém. Biospéol. 24: 63–69.

Peters, N. & G. Peters. 1968. Zur genetischen Interpretation morphologischer Gesetzmäßigkeiten der degenerativen Evolution. Z. Morph. Tiere 62: 211–244.

Peters, N., G. Peters, J. Parzefall & H. Wilkens. 1973. Über degenerative und konstruktive Merkmale bei einer phylogenetisch jungen Höhlenform von *Poecilia sphenops* (Pisces, Poeciliidae). Int. Revue ges. Hydrobiol. 58: 17–436.

Sarbu, S.M., T.C. Kane & B.K. Kinkle. 1995. A chemoautotropically based cave ecosystem. Science 272: 1953–1955.

Schemmel, C. 1967. Vergleichende Untersuchungen an Hautsinnesorganen ober- und unterirdisch lebender *Astyanax*-Formen. Z. Morph. Tiere 61: 255–316.

Schemmel, C. 1980. Studies on the genetics of feeding behaviour in the cave fish *Astyanax mexicanus* f. *anoptichthys*. Z. Tierpsychol. 53: 9–22.

Southward, A.J., M.C. Kennicut II, J. Alcala-Herrera, M. Abbiati, L. Airoldi, F. Cinelli, C.N. Bianchi Morri & E.C. Soutward. 1996. On the biology of submarine caves with sulfur springs: appraisal of 13C/12C ratios as a guide to trophic relations. J. Mar. Biol. Assoc. 76: 265–285.

Taborsky, M. 1994. Sneakers, satellites and helpers: parasitic and cooperative behavior in fish reproduction. pp. 1–100. *In*: P.J.P. Slater, J.S. Rosenblatt, C.T. Snowdon & M. Milinski (ed.) Advances in Study of Behavior 23, Academic Press, San Diego.

Trajano, E. 1991. The agonisitc behaviour of *Pimelodella kronei*, a troglobitic catfish from southeastern Brazil (Siluriformes, Pimelodidae). Behav. Processes 23: 113–124.

Walters, L.H. & V.W. Walters. 1965. Laboratory observations on a cavernicolous poeciliid from Tabasco, Mexico. Copeia 1965: 214–233.

Wilkens, H. 1986. The tempo of regressive evolution: studies of the eye reduction in stygobiont fishes and decapod crustaceans of the Gulf Coast and West Atlantic region. Stygologica 2: 130–143.

Wilkens, H. 1987. Genetic analysis of evolutionary processes. Int. J. Spel. 16: 33–57.

Wilkens, H. 1988. Evolution and genetics of epigean and cave *Astyanax fasciatus* (Characidae, Pisces). Evol. Biol. 23: 271–367.

Zeiske, E. 1968. Prädispositionen bei *Mollienesia sphenops* (Pisces, Poeciliidae) für einen Übergang zun Leben in subterrranen Gewässern. Z. vergl. Physiol. 58: 190–222.

Part 4. Special cases

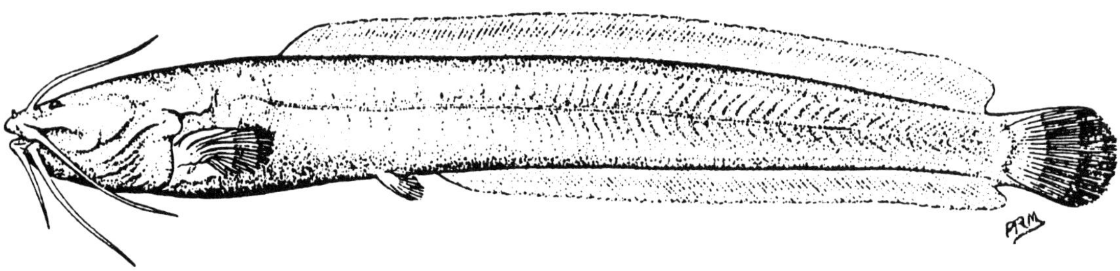
PRM

Environmental Biology of Fishes **62**: 279–296, 2001.
© 2001 *Kluwer Academic Publishers. Printed in the Netherlands.*

Troglomorphic sculpins of the *Cottus carolinae* species group in Perry County, Missouri: distribution, external morphology, and conservation status

Brooks M. Burr[a], Ginny L. Adams[a], Jean K. Krejca[b], Regina J. Paul[a] & Melvin L. Warren, Jr.[c]
[a]*Department of Zoology and Center for Systematic Biology, Southern Illinois University, Carbondale, IL 62901-6501, U.S.A. (e-mail: burr@zoology.siu.edu)*
[b]*Section of Integrative Biology, University of Texas, Austin, TX 78712, U.S.A.*
[c]*Center for Bottomland Hardwoods Research, U.S. Forest Service, Southern Research Station, Oxford, MS 38655, U.S.A.*

Received 3 April 2000 Accepted 15 January 2001

Key words: Cottidae, grotto sculpins, cave distribution, density estimate, cave adaptation

Synopsis

The existence of cavernicolous sculpin (here allocated to *Cottus carolinae*, banded sculpin, and referred to as grotto sculpin), in the karst regions of Perry County, Missouri, first came to our attention in 1991. Examination of 35 caves in Missouri, 96 in Illinois, 17 in Tennessee, two in Indiana, and 11 in Arkansas revealed that banded sculpin are common in cave habitats; however, grotto sculpin are limited to two karst areas of Perry County, Missouri, where they are known from only six cave systems. These caves and their streams are extensive and apparently provide a unique habitat compared to other karst systems; this may be a critical factor in the present restricted distribution of the grotto sculpin. Grotto sculpin occupy pools and riffles of cave streams, and occur over a variety of substrates, from sediment to breakdown. Density estimates in Mystery and Running Bull caves were 0.29 and 0.63 individuals m^{-2}, respectively. Grotto sculpin have small eyes (1–6% SL vs. 6–10% SL in epigean samples), significantly reduced pigmentation (including nearly complete loss of dorsal saddles), a reduction in pelvic fin ray number (from $4+4$ elements to often $4+3$, or $3+3$), and enlarged cephalic lateralis pores (e.g., mandibular pores of cavernicolous samples are 2–3 times those of epigean stream samples). Multivariate analyses of body shape revealed statistically significant separation of epigean and hypogean samples, with eye size highly variable, but smallest in the Running Bull Cave population. We interpret these results as representative of losses associated with long-term cave habitation. Caves of Perry County provide ample habitat for grotto sculpin, but because the caves are located downgradient of the city of Perryville and an intensively farmed landscape, point and non-point source pollution threaten their continued existence. Escape of farm-pond fishes through the extensive sinkhole network in Perry County has increased potential predation pressure on grotto sculpin by channel catfish, *Ictalurus punctatus*, and other species normally excluded from cave environments.

Introduction

> '*A certain predisposition in habit and structure must be present to enable a species to dispense with light and to live in caves.*'
>
> Carl H. Eigenmann (1909, p. 9)

The karst terrain and aquifers of North America provide habitats for some of the most spectacular examples of hypogean fishes found anywhere in the world including, an endemic family of cavefishes (Amblyopsidae) and troglobitic ictalurid catfishes from deep artesian waters of Texas and caves of Mexico (Page & Burr 1991, Walsh & Gilbert 1995). New hypogean species continue to be discovered and described (e.g., Walsh & Gilbert 1995) from the North American continent.

Fish species that bridge existence between epigean and hypogean habitats are relatively few in the New World. These include the most thoroughly investigated case of any epigean-hypogean species to date, the Mexican tetra, *Astyanax fasciatus* (Characidae) of Atlantic Slope karst in Mexico (Mitchell et al. 1977). Less studied is the cavernicolous population of Mexican molly, *Poecilia mexicana* (Gordon & Rosen 1962) from Tabasco, Mexico. Sculpins (Cottidae: *Cottus*), benthic cool-water fishes, are known to frequent springs, cave outflows, and the twilight and aphotic zones of caves in the karst topography of eastern United States (Mohr 1950, Cooper 1978, Williams & Howell 1979, Poly & Boucher 1996). These quasi cave-dwelling sculpin do not exhibit physical characteristics reported for typical stygobites (i.e., loss of pigmentation, reduced eyes, and enlarged lateral line receptors). Even an albinistic sculpin, *Cottus carolinae*, described from a single individual in a West Virginia cave, had normal eye size albeit unusual morphology (Williams & Howell 1979).

Periodic study of the fish fauna in Perry County caves and surface streams over the past 10 years has revealed unique populations of sculpin (allocated here to *Cottus carolinae*) comprised of reproducing, subterranean individuals exhibiting phenotypic variability in structural and physiological features normally restricted to fish populations inhabiting either epigean or hypogean environments.

Here we document the geographic distribution of the cave-inhabiting sculpins, provide population density data for two caves located 6 km apart, and describe the external morphology of the cave-inhabiting sculpins and their epigean relatives. We also document local environmental threats to the groundwater in Perry County. The cave populations may not represent a species different from *C. carolinae* but our observations establish their uniqueness and vulnerability to extinction from various groundwater pollution problems. To facilitate discussion we use the common name grotto sculpin for the unique cave-inhabiting sculpins.

Materials and methods

Study area

Perry County, Missouri, is located in the karst region of the Salem Plateau in the northeast Ozarks (N 37° 35′ to 37° 55′, W 90° 05′ to 89° 45′). Perry County contains several large caves and one of the largest concentrations of caves in North America (about 650 identified caves, Elliott 2000), which are generally found in limestones and dolomites of Ordovician age. Cinque Hommes Creek, which drains into the nearby Mississippi River, is the resurgence stream for many caves in the area (Figure 1). Some caves in Perry County contain grotto sculpin that have access to a source pool of epigean sculpin. Hydrological information is insufficient to infer whether sculpin move between epigean and cave populations. Mean temperatures of caves and their associated streams in the region are about 13°C, and they contain permanent, slow-flowing streams over rubble and silt substrates.

Distribution

We undertook cave exploration in five states to determine the geographic range of the grotto sculpin. Our primary exploration focused on nine karst regions within Perry County, Missouri (House 1976), where grotto sculpin were known to occur (Figure 2). We used standard caving equipment (Rea 1987), specialized gear (e.g., wetsuits, ropes, and vertical gear), and vertical techniques (Padgett & Smith 1987) as conditions dictated.

We searched for sculpin in cave streams by walking slowly upstream in shallow water, catching fish by hand or small dip nets, and temporarily holding them in jars and buckets. We used a portable aerator, thermometer, ruler, and camera to aid in documentation of habitat and fish. We collected surface sculpin using standard seining techniques (Jenkins & Burkhead 1994).

Population density estimates and habitat

We estimated sculpin density in Mystery and Running Bull caves in October 1991 and 1993 when low water levels afforded high water clarity. We measured the lengths and widths of shallows and riffles with a Keson meter tape and took 10 random depths with a meter stick within each macrohabitat. We recorded current speed (nearest 1 m sec^{-1}) by floating a neutrally buoyant object over a measured distance. At fifteen sites in each of both Mystery and Running Bull caves we surveyed a total of 180 and 101 m of stream, respectively. Within a given macrohabitat, we turned over all rocks and counted sculpin. We qualitatively noted eye size (small vs. large [comparison shown in Figures 4, 5]) and size class (juvenile [<40 mm TL] or adult

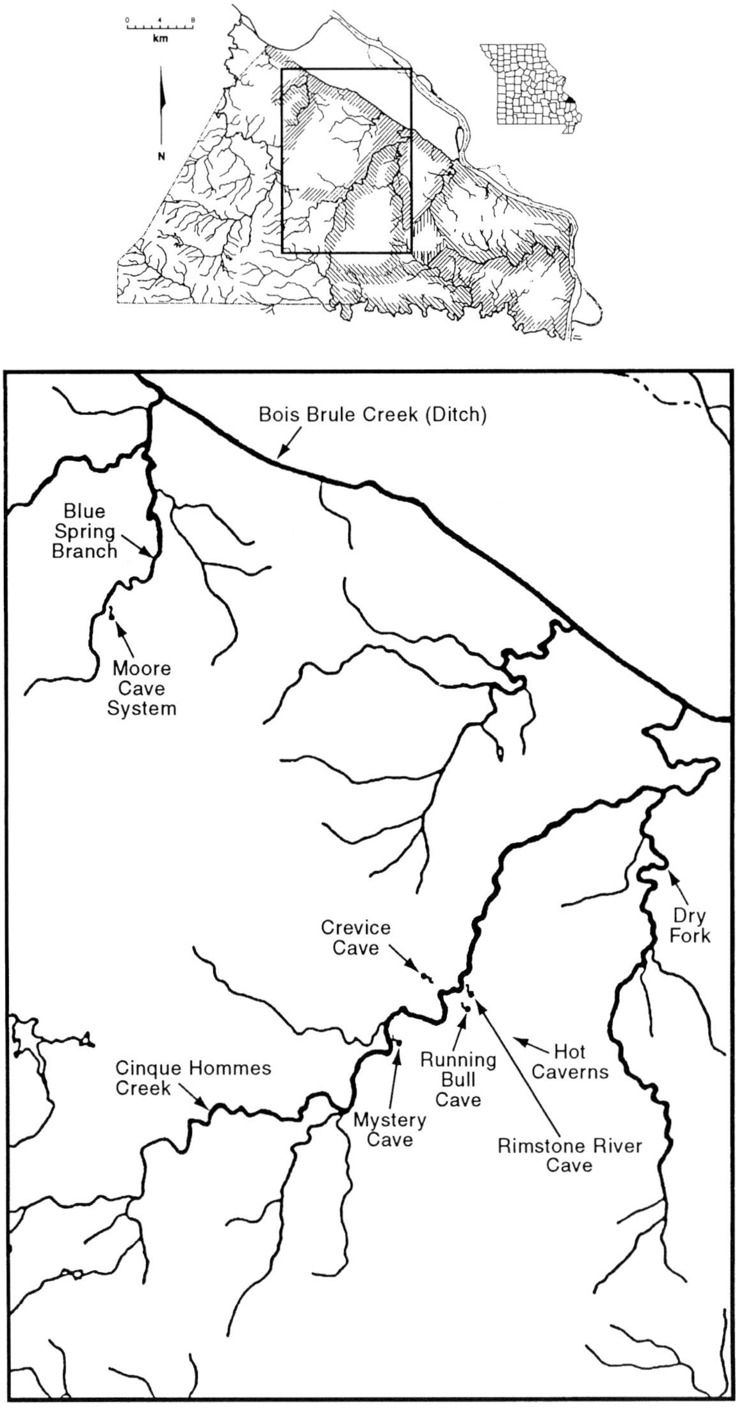

Figure 1. Drainage relations of central Perry County, Missouri.

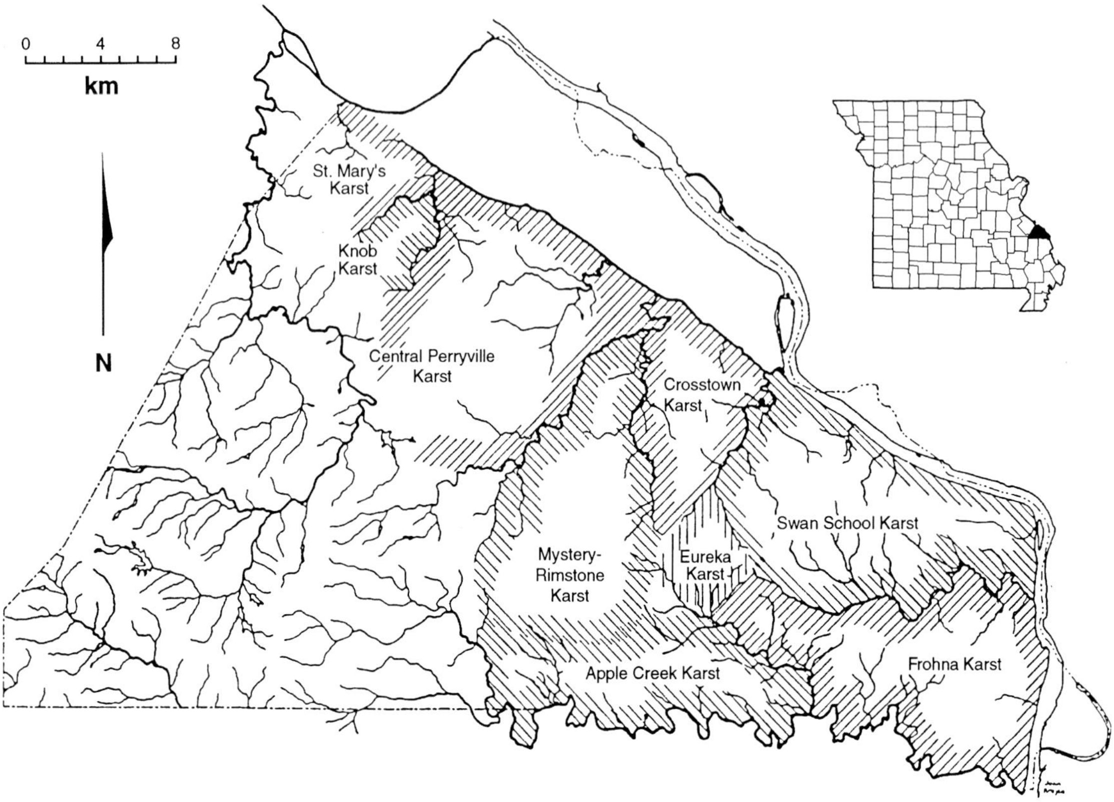

Figure 2. Karst regions of Perry County, Missouri, as defined by House (1976).

[>45 mm TL]) for each specimen. We based extrapolation of total numbers of grotto sculpin on conservative estimates of available stream passage within each cave.

External morphology

We made seven measurements (eye length, eye depth, head depth in line with jaw tips, head depth in line with edge of preopercle spine, length of longest preopercle spine, length of pelvic fin, interorbital length) on the left side of 84 specimens using dial calipers (nearest 0.01 mm) following methods in Hubbs & Lagler (1974). An additional 24 angular distances were taken from one homologous landmark to another (Figure 3) following truss geometric protocol (Strauss & Bookstein 1982). We subjected the distance values to multivariate analyses (i.e., principal components, canonical variates, discriminant function) to elucidate phenotypic differences among populations. We then subjected each of the highest loading components from the principal component analyses to bivariate regression analyses with standard length serving as the independent variable. We conducted univariate and multivariate analyses on the Southern Illinois University at Carbondale (SIUC) mainframe computer using SAS (1985) and a Macintosh personal computer using programs in Statistica (Statsoft 1994).

Comparisons of nine meristic features between hypogean and epigean samples were nonsignificant. We retained one feature, number of pelvic fin rays, for further analysis. We counted all rays from both the right and left pelvic fins. We examined size of cephalic lateralis pores using a Wild binocular microscope fitted with an eyepiece micrometer. We measured the longest axis for length and width for each of pores 3, 4, and 6 on the mandibular canal, using the chin pore as pore number 1. We followed general terminology and illustrations of head pore arrangements from Robins & Miller (1957). We vouchered all specimens used in the analyses (Table 1) in the Fluid Vertebrate Collection of SIUC, where they are stored either in 70% ethanol or as frozen carcasses.

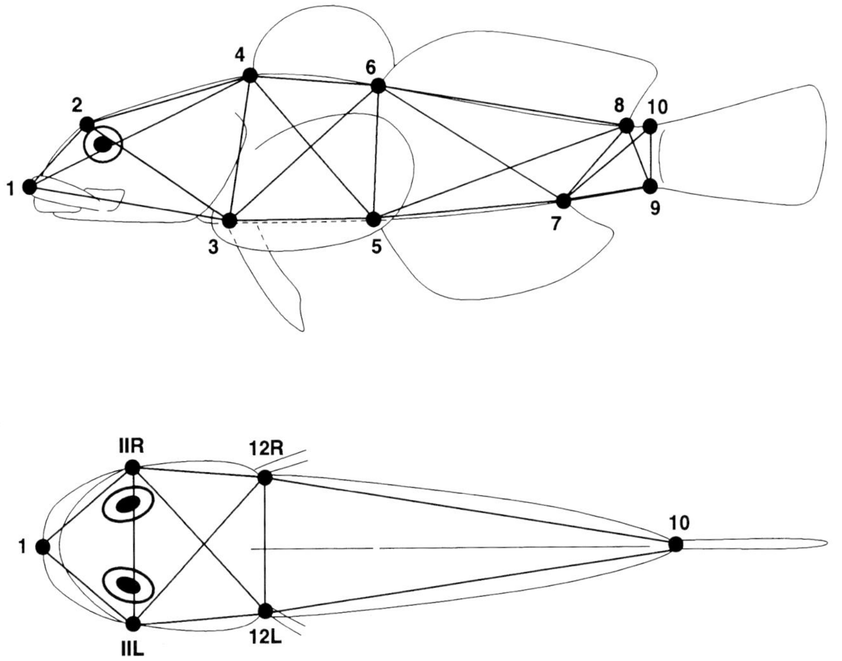

Figure 3. Truss geometric design for a sculpin (modified from Strauss & Bookstein 1982).

Table 1. General location, macro-environment, sample size (N), and SIUC catalog number of specimens of *C. carolinae* used for counts and measurements. All specimens are from southeast Missouri; all resurgence stream and hypogean samples are from Perry County.

Location	Macro-environment	N	Catalog number
Black River, Reynolds Co.	Epigean stream	4	SIUC 18702
Big Piney River, Texas Co.	Epigean stream	3	SIUC 3216
Big Creek, Wayne Co.	Epigean stream	5	SIUC 3252
Big Creek, Wayne Co.	Epigean stream	4	SIUC 3158
Cinque Hommes Cr.	Resurgence stream	20	SIUC 19279
Mystery Cave	Hypogean stream	20	SIUC 18923
Rimstone River Cave	Hypogean stream	23	SIUC 19839
Running Bull Cave	Hypogean stream	7	SIUC 21260
Running Bull Cave	Hypogean stream	12	SIUC 37892
Crevice Cave	Hypogean stream	2	SIUC 21261
Hot Caverns	Hypogean stream	4	SIUC 21258
Tom Moore Cave	Hypogean stream	9	SIUC 20916

Results

Distribution

Grotto sculpin are restricted to two karst areas of Perry County: Central Perryville Karst and Mystery-Rimstone Karst (Figure 2). Only six of 27 cave streams sampled in six karst regions contained grotto sculpin. In the Central Perryville Karst, viable populations occur in the Moore Cave system and Crevice Cave (Figure 1). In the Mystery-Rimstone Karst, Hot Caverns, Mystery Cave, Rimstone River Cave, and Running Bull Cave (Figure 1) all harbor relatively large populations of grotto sculpin. Grotto sculpin also may occur in a few other caves that are hydrologically connected to these known sites but were not accessible for searches.

Results of specific distribution data are divided between the two known karst areas of occurrence.

In the Central Perryville Karst, the Moore Cave system contains 31.18 km of mapped passage; we have observed grotto sculpin in the main cave stream from its source (known as Tom Moore) to the Siphon area downstream (Berome Moore), a distance of 3.45 km. Crevice Cave, with 45.47 km of mapped passage, is comprised of four major cave streams; grotto sculpin occur in the major southeast trending Eternity Passage. In the Mystery-Rimstone Karst, the main trunk of Rimstone River Cave contains 8 km of underground stream, referred to as the Nile, and sculpin were collected over a distance of 604 m beginning at the Flaming River entrance. Hot Caverns, a closely associated cave, contains a tributary of the Nile River, referred to as the Labyrinth Stream. Sculpin were seen or collected in this section of the Nile and Labyrinth streams. Running Bull Cave has 374 m of mapped passage; sculpin were found from between the first crawlway downstream of the entrance for a distance of about 252 m. Mystery Cave has 27.35 km of surveyed passage. Grotto sculpin were found in Main Stream for about 5 km downstream of the gated sinkhole entrance.

Of 12 surface stream sites sampled, we found epigean sculpin in Blue Spring Branch, Cinque Hommes Creek, and Dry Fork. Blue Spring is the principal resurgence stream for caves and sculpin in the Moore Cave system. Cinque Hommes Creek is the principal resurgence stream for caves in the Mystery-Rimstone Karst and Crevice Cave in the Central Perryville Karst that harbor grotto sculpin (Figure 1).

Our field searches of 153 additional caves verified that sculpin frequent caves, but cavernicolous features were lacking in every case except populations discussed above from Perry County, Missouri. Field searches by us and colleagues revealed sculpin identified as *C. carolinae* in 25 caves in seven states (Table 2), but sculpin were not found in 11 Arkansas caves, 91 Illinois caves and karst springs, one Indiana cave, 28 Missouri caves, and 15 Tennessee caves and karst springs.

Population density estimates and habitat

We found 60 grotto sculpin in Mystery Cave (15 sites, 180 m of stream) at a mean density of 0.29 individuals m^{-2} (SE = 0.087). Of these, 24 had small eyes and 36 had larger eyes but not as large as those of epigean samples taken outside of Perry County (Figure 4, Table 3). Sampled sites (mean reach of 12 m) averaged about 1 sculpin per 3 m. We found no correlation between eye size and pool length, pool width, pool area, and volume of site ($p > 0.05$). Total number of sculpin and depth of water were correlated positively ($r = 0.39$, $p < 0.05$). Assuming that at least 8 km of stream is available, and that sculpin are distributed uniformly with no barriers to dispersal, we estimate 2667 sculpin occur in Mystery Cave.

In Running Bull Cave (15 sites, 101 m of stream), mean density of grotto sculpin was 0.63 individuals m^{-2} (SE = 0.191). We counted 51 sculpin, 43 of which had small eyes and eight had larger eyes. Sampled sites (mean reach of 6.8 m) averaged about 1 sculpin per 2 m. We observed six other sculpin in areas where measurements were not made. We estimate at least 150 sculpin occur in the 300 m of stream available in Running Bull Cave.

Grotto sculpin frequent sustained, modest flow of pools and riffles with little reliance on physical cover or formation of large aggregations. The main streams in caves harboring grotto sculpin averaged 1.1–3.3 m in width and between 4 and 33 cm depth during autumn low-water levels. Sculpin occupy pools and riffles; mean current speed of riffles was 0.05 m sec^{-1}. They occur over a variety of substrates including silt, gravel, cobble, breakdown, and bedrock, most frequently over a rocky bottom, and are just as likely to be in open water as hidden under rocks. Most adults occur singly but occasionally as many as three individuals are found together, only a few centimeters apart.

Perry County has hundreds of sinkholes and numerous farm ponds that may unexpectedly drain through a sinkhole transferring the pond water, flora, and fauna into a cave. We have found epigean farm-pond fishes in all of the caves where cavernicolous sculpin occur including an unsuitably large population of channel catfish, *Ictalurus punctatus*, in Mystery Cave. Other fishes found with sculpin in Perry County caves include common carp *Cyprinus carpio*, fathead minnow *Pimephales promelas*, yellow bullhead *Ameiurus natalis*, green sunfish *Lepomis cyanellus*, and bluegill *Lepomis macrochirus*. All of these species are potential predators on eggs and young of sculpin. The two catfish species are the most significant potential predators of adult sculpin.

Identification and taxonomy

We recognize four phenotypes of sculpin in southeastern Missouri, two epigean stream phenotypes and the

Table 2. Caves where sculpin (*Cottus carolinae*) have been found, but where specimens do not exhibit phenotypic features of cave adaptation (i.e., small eyes, depigmentation, reduced meristics, enlarged cephalic canal pores).

State Cave, County	Date	Source	Voucher
Illinois			
Fogelpole Cave, Monroe Co.	1 Jul 92	J. Krejca, S. Taylor	SIUC 21305
Crystal Cave, Hardin Co.	7 Aug 91	J. Krejca	None
Indiana			
Seiberts Well, Crawford Co.		H. H. Hobbs III, J. Lewis	None
Batey's Cave, Green Co.		H. H. Hobbs III	None
Binkley's Cave, Harrison Co.		H. H. Hobbs III	None
Connerley's Cave, Lawrence Co.		H. H. Hobbs III	None
Donnehue's Cave, Lawrence Co.		H. H. Hobbs III	None
Popcorn Spring Cave, Lawrence Co.		H. H. Hobbs III	None
Sullivan's Cave, Lawrence Co.	29 Oct 92	H. H. Hobbs III, J. Krejca	None
Boone Cave, Owen Co.		H. H. Hobbs III	None
3 small caves, Washington Co.		H. H. Hobbs III	None
Fredericksburg Cave, Washington Co.		J. Lewis	None
Kentucky			
Cool Spring Cave, Barren Co.		H. H. Hobbs III	None
Mammoth Cave, Echo & Mystic R. Edmonson Co.		J. Lewis, M. Sutton	None
Missouri			
Saranac Spring Cave, Crawford Co.	25 Sep 92	J. Krejca, S. Taylor	None
Blowing Spring Cave, Laclede Co.	15 Oct 93	J. Krejca, S. Taylor	SIUC 23158
Lewis Cave, Ripley Co.	28 Jan 67	R. A. Brandon	SIUC 18948
Cropper Cave, Oregon Co.		M. Sutton, J. Gardner	None
Powder Mill Creek Cave Shannon Co.		M. Sutton, S. Ervine	None
Tennessee			
Yates Cave, Robertson Co.	24 Apr 94	J. Krejca, S. Taylor	None
Virginia			
Butler Cave, Bath Co.		H. H. Hobbs III	None
Tawneys Cave, Giles Co.		H. H. Hobbs III	None
West Virginia			
Buckeye Creek Cave Greenbriar Co.	several dates	H. H. Hobbs III	None

grotto sculpin represented by two hypogean phenotypes. We distinguish phenotypes based on relative eye size, degree of pigmentation, relative cephalic canal pore size, and pelvic fin-ray number. One epigean stream phenotype (Figures 5a,b) ranges outside of Perry County and the Cinque Hommes Creek system and is typical of the midlands race of *C. carolinae* as described by Robins (1954). The other epigean stream phenotype occurs in Cinque Hommes Creek, the major stream of resurgence for several caves, and is distinguished by reduced pelvic fin-ray number (52% with 3 + 3 or 3 + 4 rays), moderate eye size (4–6% of SL), moderate cephalic canal pore size (Table 4), and dark pigmentation similar to midlands race sculpin. The two hypogean phenotypes co-occur in deep cave habitats of Perry County, Missouri. One hypogean phenotype is depigmented with reduced pelvic fin-ray numbers (47–75% with 3 + 3 or 3 + 4 rays) and enlarged cephalic canal pores, but eye size (5–6% of SL) is nearly equivalent to the epigean phenotypes (Figures 4b,f). The second hypogean phenotype is also depigmented (Figures 5c,d), has enlarged cephalic canal pores and reduced pelvic fin-ray numbers, but eye size is tiny (1–4% of SL) relative to other phenotypes (Figures 4a,e). We allocate these four phenotypes to the banded sculpin, *Cottus carolinae*, using the key characteristics and descriptive features outlined by Pflieger (1997) for Missouri sculpin and Page & Burr (1991) for North American sculpins. The two features shared by all four populations include: (1) usually a complete lateral line ending near the base of the caudal fin, and (2) dorsal fins usually not connected.

External morphology

Trenchant features characterize individuals of grotto sculpin when compared to the wide-ranging midlands race of *Cottus carolinae carolinae* described in detail by Robins (1954). Further, sculpin from Cinque Hommes Creek (a resurgence stream), although a surface population, also have a unique set of morphological traits (see above) compared to other surface populations outside Perry County. Here we describe in detail the external features of grotto and resurgence stream sculpin as a basis for comparison in the Discussion to the midlands race in surface streams.

Size. Mean standard length (SL) of grotto sculpin was 59.98 mm (N = 8, SD = 20.53) for Tom Moore Cave; 68.18 mm (N = 4, SD = 13.57) for Hot Caverns; 70.13 mm (N = 20, SD = 11.33) for Mystery Cave; 69.83 mm (N = 21, SD = 13.51) for Rimstone River Cave; and 68.18 mm (N = 15, SD = 9.50) for Cinque Hommes Creek. The largest grotto sculpin was 104 mm SL from Mystery Cave. Specimens used for comparison from outside Perry County averaged 60.23 mm (N = 16, SD = 17.24). We detected no significant difference (p = 0.12) in mean SL among any of the populations studied.

Pigmentation and skin prickling. Grotto sculpin have dorsal, caudal, and pectoral fins and distal edge of anal fin lightly spotted with melanophores, forming weak bands to slightly tessellated; pelvic fin unpatterned; margin of first dorsal fin sometimes pale yellow, not edged with orange. Ground color light tan to bleached tan; throat, belly, and lower caudal peduncle unpigmented. Three saddles on back reduced to vestigial spotted blotches, not extending downward and forward on side. Lower sides without reticulations. Two dark bars on cheek reduced in size and intensity of pigmentation. Small dark blotch near middle of pectoral fin base absent. Chin only slightly mottled to unmottled. We did not stain skin prickles or examine their shape, but grotto sculpin appear to be nearly devoid of prickles. Individuals from the resurgence stream (Cinque Hommes Creek) have pigmentation similar to that described for epigean stream populations (Robins 1954).

Lateral line system. Lateral line usually complete to caudal base (three specimens with lateral line ending under posterior edge of second dorsal fin). Lateral line pores in the resurgence stream population (Cinque Hommes Creek) 32–36, modally 35; preoperculomandibular pores 10–12, modally 11; canals never fused anteriorly into a median chin pore; infraorbital pores 8–9, modally 9. Lateral line pores in cavernicolous individuals (Mystery Cave) 32–36, modally 33; preoperculomandibular pores 10–11, modally 11; canals often fused anteriorly into a median chin pore (e.g., 10 of 20 specimens from Rimstone River Cave); postmaxillary (or postmandibular) pore occurs at pore 5, missing in only 3 of 32 specimens examined; infraorbital pores 7–10, modally 9. Canal pore sizes large, especially those on the head (i.e., mandibular pores 3, 4, and 6) (Table 4).

Morphometry. Body deepest under first dorsal fin, tapering sharply in either direction; venter flat; caudal peduncle slender (N = 68), comprising 6.4% (±0.47% SE) to 7.6% (±0.77% SE) of SL. Eye length variable, ranging from eyes reduced to pin-holes to near eye length of epigean fish, comprising 1.2–6.5% of SL. Greatly reduced eye size is frequent (e.g., 24 of 60 individuals, Mystery Cave); however, the ratio of large-eyed and reduced-eyed individuals was not significantly different from 1 : 1 (adjusted G-test, G = 2.40, p = 0.5). Preopercular spines present, blunt and reduced in length. Dorsal fins slightly connected or entirely separate. Caudal fin truncate or subtruncate (sensu Jenkins & Burkhead 1994).

All cavernicolous sculpin and the samples from Cinque Hommes Creek have smaller eyes than those from epigean stream samples regardless of length of

Figure 4. a and b: Dorsal view of tiny-eyed and larger-eyed grotto sculpin co-occurring in Mystery Cave, Perry County, Missouri, 1 Oct 1991. c: Close-up view of tiny-eyed individual from Mystery Cave, same data. Photos by J.J. Swayne. d: Frontal view of tiny-eyed individual from Mystery Cave showing enlarged mandibular pores. Photo by C.A. Lee. e and f: Lateral view of tiny-eyed and larger-eyed grotto sculpin co-occurring in Mystery Cave, Perry County, Missouri, 1 Oct 1991. Photos by B.M. Burr.

Table 3. Numbers of grotto sculpin in Mystery and Running Bull Caves, Perry County, Missouri, on 22 Oct 1991 and 11 Oct 1993, respectively. Fifteen sites representing a total of 179.5 and 101.2 m of stream in Mystery and Running Bull Caves, respectively, were characterized and sampled.

	Length (m) of pool/run	Width (m) of pool/run	Depth (cm) of pool/run	Number of sculpins	
				large eye	small eye
Mystery Cave					
Mean	12.0	1.9	15.2	2.4	1.6
Range	3.5–22.8	1.3–3.3	6.5–25.1	0–5	0–4
Total				36	24
				60	
Running Bull Cave					
Mean	6.8	1.3	11.3	0.5	2.9
Range	2.4–15.5	0.5–2.5	4.0–21.8	0–3	0–7
Total				8	43
				51	

Figure 5. a: Lateral view of *C. carolinae* from Blowing Springs Cave (78 mm SL), Laclede County, Missouri, 15 Oct 1993 showing coloration and features of epigean stream sculpin. b: Dorsal view of Blowing Springs Cave specimen showing dark saddles across back and normal eye size. c: Lateral view of grotto sculpin (73 mm SL) showing tiny eyes and reduction in pigmentation, Mystery Cave, Perry County, Missouri, 1 Oct 1991. d: Dorsal view of Mystery Cave grotto sculpin showing lack of saddles and shrunken eye sockets. The Blowing Springs Cave individual was found past the twilight zone of the cave. Photos by B.M. Burr.

Table 4. Mean size of cephalic lateralis canal pores in *C. carolinae* complex as determined by straight line measurements of greatest length (L) and width (W) of pore. Pores measured are those from the mandibular canal, numbers 3, 4, and 6, using the chin pore as pore number 1 (N = 8, each population). Units are micrometers rounded to the nearest whole number.

Location	L, pore 3 (± SE)	W, pore 3 (± SE)	L, pore 4 (± SE)	W, pore 4 (± SE)	L, pore 6 (± SE)	W, pore 6 (± SE)
Mystery Cave	819 (94)	442 (96)	640 (71)	390 (64)	515 (34)	299 (49)
Rimstone R. Cave	975 (70)	485 (33)	825 (78)	450 (51)	540 (59)	408 (47)
Cinque Hommes Cr.	600 (28)	360 (25)	538 (68)	305 (26)	415 (39)	268 (22)
Epigean streams outside Perry Co.	298 (42)	205 (29)	260 (56)	185 (32)	240 (24)	168 (12)

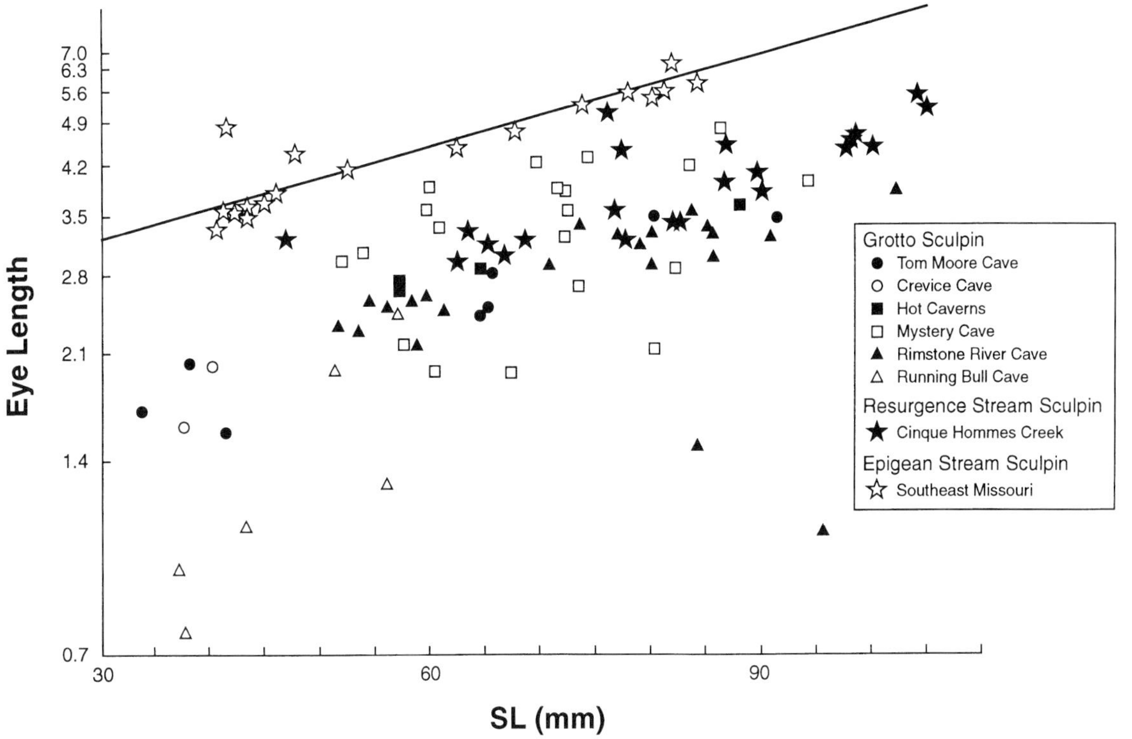

Figure 6. Relationship between eye length and standard length in six populations of grotto sculpin, one resurgence stream population, and epigean stream samples from southeast Missouri, outside Perry County. The regression line is fitted to individuals representing typical epigean populations of *Cottus carolinae* from southeast Missouri.

sculpin (Figure 6). In a principal component analysis, eye length loaded highest on PC axis 2 and separated populations in Running Bull Cave (smallest eyes) from several caves (intermediate in eye lengths) and epigean surface stream samples (largest eyes). Mystery Cave contained two distinct phenotypes, one with small eyes and one with eye sizes approaching those on the surface (Figure 7). Discriminant function analysis revealed all populations, except Rimstone River Cave and Hot Caverns which share drainage systems, were significantly different ($p < 0.05$, squared Mahalanobis distances, Table 5). Only three of 84 individuals were misclassified based on a discriminant function classification matrix. Two of the three individuals were from Hot Caverns and were misclassified with Rimstone River Cave individuals. Canonical variates analysis separated samples in Mystery Cave from those in other caves and epigean samples from southeast Missouri, on the basis of head size and shape and body width (Figure 8, Table 6).

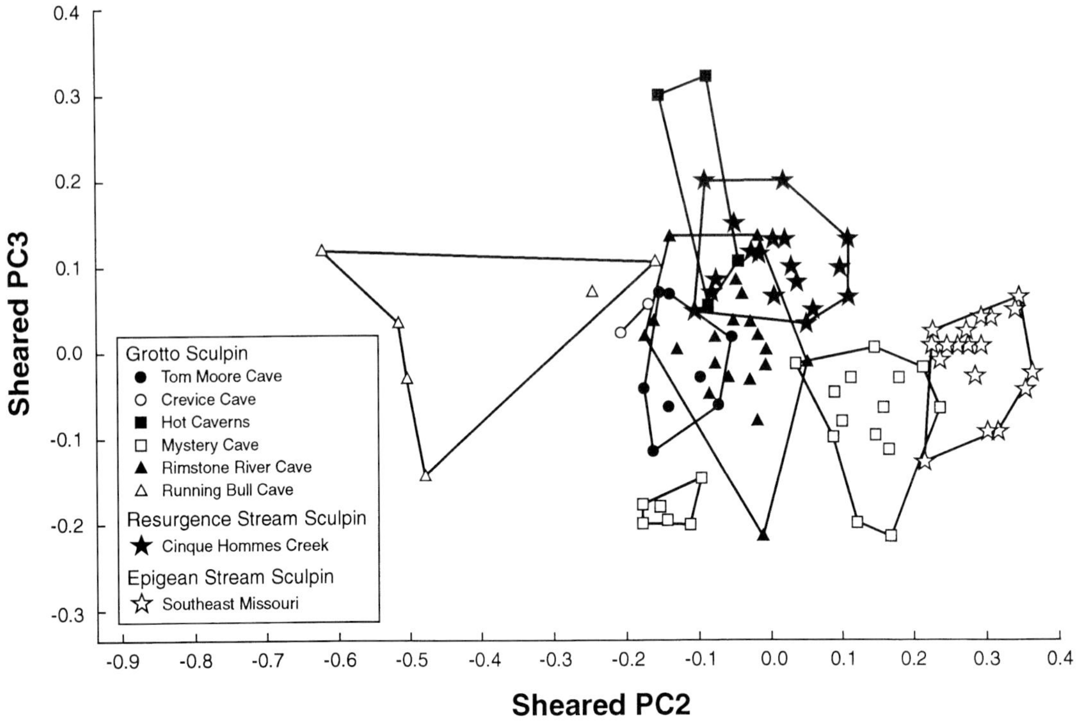

Figure 7. Scores of sculpin samples from six cave streams, a resurgence stream, and epigean surface streams on sheared principal component axes 2 and 3 for 31 morphometric variables. Note that grotto sculpin from Mystery Cave exhibit two distinct phenotypes based on relative eye size.

Table 5. Matrix of squared Mahalanobis distances between group centroids of a discriminant function analysis based on 24 morphometric variables. Six populations representing the four phenotypes of the *C. carolinae* complex in southeastern Missouri were examined.

Location	Tom Moore	Hot Caverns	Mystery Cave	Rimstone River Cave	Cinque Hommes Cr.
Tom Moore Cave					
Hot Caverns	33.75				
Mystery Cave	33.90	26.35			
Rimstone River Cave	27.36	11.29	20.22		
Cinque Hommes Cr.	31.93	14.46	28.17	8.75	
Southeast Missouri	41.95	22.22	9.65	21.01	35.68

Meristic features. The resurgence stream population (Cinque Hommes Creek) had 7 (25%) or 8 (75%) dorsal fin spines, 17 (18%) or 18 (82%) dorsal fin rays; 12 (8%), 13 (83%), or 14 (8%) anal fin rays; pectoral rays were 14 + 15 (1), 15 + 15 (1), 16 + 16 (8), and 16 + 17 (1); branched caudal fin rays were consistently 9. Grotto sculpin from Mystery Cave had 7 (50%), 8 (45%), or 9 (5%) dorsal fin spines, 16 (15%), 17 (65%), or 18 (20%) dorsal fin rays; 12 (25%), 13 (60%), or 14 (15%) anal fin rays; pectoral rays were 14 + 15 (1), 15 + 15 (4), 15 + 16 (3), 16 + 16 (9), 16 + 17 (2), and 17 + 17 (1); branched caudal fin rays were 8 (1), 9 (18), or 10 (1). A reduction in pelvic fin ray number from 4 + 4 to 3 + 4 or 3 + 3 occurs in

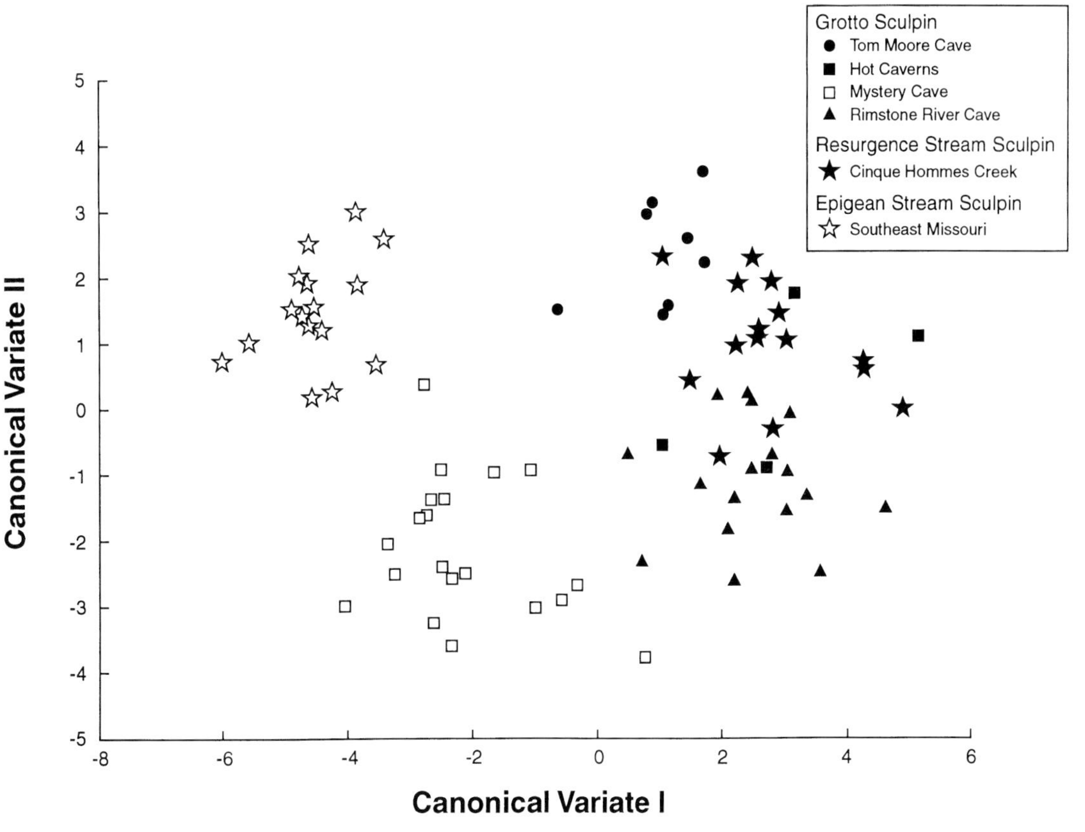

Figure 8. Scores of sculpin samples from four cave streams, a resurgence stream, and epigean surface streams on canonical variate axes I and II for 24 morphometric variables.

about 50% of grotto sculpin and the resurgence stream population (Table 7).

Discussion

Distribution

Grotto sculpin are limited in distribution to six major caves or cave systems in central Perry County, Missouri. A similar karst topography in terms of presumed age and structure occurs just east of Perry County across the Mississippi River in Monroe County, Illinois (Bretz & Harris 1961). Some caves in Monroe County have enough permanent water to support populations of grotto sculpin, but we have found only one adult (120 mm SL) sculpin, in Fogelpole Cave, Monroe County, moderately depigmented, but otherwise with epigean characteristics.

We have never captured grotto sculpin in surface streams and suspect dispersal to epigean waters is low to non-existent. Habitat for the grotto sculpin also may be limited. Perry County has the four longest caves in the state, each exceeding 16 km in length, and containing sizable stream passages (Unklesbay & Vineyard 1992). Grotto sculpin have been found in all four of these major caves. These caves, formed beneath a sinkhole plain that provides substantial organic input, may be the only habitats that provide enough food and sustained flow to support sculpin populations. Inventories of the invertebrate cave fauna in this area indicated an abundance of isopods, amphipods, flatworms, and snails (Peck & Lewis 1978).

Population density estimates and habitat

Our population density estimates for grotto sculpin (0.3–0.6 individuals m^{-2}) are comparable to those for a surface population of *C. carolinae* in Little River, eastern Tennessee (0.4–0.9 sculpin m^{-2}) (Greenberg &

Table 6. Standardized canonical variate coefficients for 24 morphometric variables taken from six populations representing the four phenotypes of the *C. carolinae* complex in southeastern Missouri.

Variable	Canonical variate I	Canonical variate II	Canocical variate III
1. Lip to Coronal Pore (1–2)	0.84892	0.46649	−0.05148
2. Lip to Pelvic Fin Ins. (1–3)	−0.69874	0.04699	0.64716
3. Lip to 1st Dorsal Fin Spine (1–4)	0.79944	2.05690	−0.35249
4. Coronal Pore to D1 Spine (2–4)	0.07899	−0.16379	−0.46668
5. Coronal Pore to Pelvic Fin Ins. (2–3)	2.23308	−0.54269	−0.64243
6. D1 Spine to Pelvic Fin Insertion (4–3)	−0.51647	−1.42064	−1.66432
7. D1 Spine to Base Anal Spine (4–5)	0.08928	2.15818	2.97467
8. Length of D1 Fin (4–6)	0.41621	−0.32719	0.71057
9. Base length of D2 fin (6–8)	−0.21571	−2.69564	−2.57967
10. D2A to Front of Anal Fin (6–5)	1.45704	1.61546	−0.46662
11. Front of D2 Fin to Front Anal Fin (6–3)	−1.85940	−2.18929	1.55320
12. Front of D2 Fin to Back of Anal Fin (6–7)	−0.16915	2.11797	0.33814
13. Least Depth Caudal Peduncle (10–9)	−0.08950	−1.58638	−0.91326
14. Lower Edge Caudal Peduncle (7–9)	−0.06081	−0.46937	−0.24279
15. Posterior Anal Fin to D2P (7–8)	−0.03114	−0.48110	0.27565
16. Base length of anal fin (5–7)	−0.52984	0.07512	0.16596
17. Anterior Anal to D2P (5–8)	1.36360	0.05764	−0.30940
18. P2 to Anterior Anal (3–5)	−0.12309	0.44399	−0.15687
19. Head Width at Jaw	0.72743	0.91085	−0.63998
20. Body Width at Pectoral Fin	−0.12748	−0.86602	0.91807
21. Eye Length	−0.85656	1.75497	−0.30613
22. Eye Depth	−0.33876	−0.86974	−0.16707
23. Head Depth at Jaw	−0.89976	−0.01912	0.57359
24. Head Depth at Pelvic Spine	−1.41893	−0.18526	0.35479
Eigenvalue	9.08743	2.58380	2.36978

Table 7. Frequency of pelvic fin ray number (left and right fins) in sculpins of the *C. carolinae* species group, including those of the midlands race (from Robins 1954), and Perry County, Missouri, epigean and hypogean populations.

Population	Pelvic ray number				
	3 + 3	3 + 4	4 + 4	Total	% with reduced rays
C. carolinae (epigean streams) Midlands Race	—	2	1459	1461	0.1
C. carolinae (resurgence stream) Cinque Hommes Creek	9	9	16	34	52
Grotto sculpin, *C. carolinae* (hypogean streams)					
Tom Moore Cave	6	8	6	20	75
Hot Caverns	1	1	2	4	50
Mystery Cave	10	5	15	30	50
Rimstone River Cave	6	10	15	31	52
Running Bull Cave	3	6	10	19	47

Holtzman 1987). These population densities are much greater than typical densities of amblyopsid cavefishes (0.005–0.15 fish m^{-2}; Helfman et al. 1997), suggesting caves in Perry County are richer in nutrients than caves occupied by amblyopsids. We speculate, however, that grotto sculpin populations are relatively unstable because catastrophic events (e.g., food shortage, drought) could result in short-term depletion or local extirpation.

Identification and taxonomy

Prior to more recent (e.g., Mitchell et al. 1977) studies of populations of fishes that bridge the gap between the epigean-hypogean environment, it was common practice to emphasize the phenotypic distinctiveness of cave fishes, many of which were placed in new genera. For example, Hubbs & Innes (1936) described a new genus and species, *Anoptichthys jordani*, for the cave-inhabiting population of the Mexican tetra, now referred to as *Astyanax fasciatus*. In that case, the parental epigean stock and hypogean populations introgressed at some locations (Avise & Selander 1972, Romero 1983). Because reductions in pigmentation, eye size, and other features are commonplace among stygobites, including the grotto sculpin studied here, most modern workers apply the zoological name of the epigean relative to its presumed cave descendants. Unfortunately, without a unique taxonomic name, an unusual biological phenomenon (i.e., cave adaptation or phenotypic plasticity) is buried under a single name, and as a consequence, only specialists are aware of the distinctive populations in caves. Moreover, when cave populations are treated nomenclaturally the same as epigean relatives, these unique animals may not be considered for actions needed to conserve them.

External morphology

Sculpin in Perry County represent three fundamental phenotypes: (1) epigean resurgence stream sculpin with essentially normal surface pigmentation, moderate eye size, moderate cephalic pore size, and a reduced pelvic fin-ray count; and (2) two hypogean sculpin phenotypes differing in eye size but sharing reductions in pigmentation and pelvic fin ray number, and enlarged cephalic canal pores. Sculpin outside of Perry County are typical epigean examples of the midlands race assigned to *Cottus carolinae* (Robins 1954). In contrast to the grotto sculpin described here, midlands race individuals have all fins spotted with dark brown, forming strong tessellations. The margin of the first dorsal fin is edged with orange-red, and the ground color is mottled dusky brown or rusty brown. Three dark brown and conspicuous saddles cross the back extending downward, usually well below the lateral line. The lower sides are reticulated, and there are usually two dark bars on the cheek and a small dark blotch near the middle of the pectoral fin base. The chin is usually darkly mottled. Hypogean sculpin from Mystery Cave retained their depigmented appearance even after several weeks in the laboratory in a 375-l aquarium with a 12L : 12D photoperiod.

Similar to Perry County sculpin, midlands race sculpin usually have a complete lateral line, with from 28 to 37 pores (N = 572), modally 33. Robins (1954) reported 11 preoperculomandibular pores and 9 infraorbital pores for midlands race sculpin, the modal numbers for Perry County sculpin, respectively. Cephalic lateralis pore size is larger in hypogean and resurgence stream sculpin when compared to sculpin outside Perry County (Table 4). In fact, mandibular pores of grotto sculpin are readily visible with the unaided eye and, on the average, are one to three times the size of those from epigean stream samples outside Perry County.

General body shape of midlands race sculpin is similar to sculpin in Perry County. Caudal peduncle depth (N = 22) was 8.2% (±0.61% SE) of SL in midlands race individuals (Robins 1954) and is not significantly ($p = 0.7, F_{5,77} = 3.20$) different from any Perry County population (ANCOVA, all assumptions met). Mean eye length in midlands race sculpin was 7.7% (±0.30% SE) of SL (range 6–10%), significantly ($p < 0.001, F_{5,77} = 29.98$) higher than the 2–4 % of SL found in grotto sculpin; additionally epigean sculpin from southeastern Missouri outside of Perry County were significantly ($p < 0.01$) different in eye length (7.7% [±0.19% SE] of SL, range 6–9%) from all Perry County populations (ANCOVA, all assumptions met). Preopercular spines appear sharper in epigean samples compared to those in caves, but we are unable to convincingly quantify this character. We initially observed a significant gap between the dorsal fins in samples from Perry County caves, but this feature is highly variable in midlands race fish and is not consistently different when samples are compared. We found no quantifiable difference in caudal fin shape among samples from caves, the resurgence stream, or the midlands race.

Multivariate analyses of shape quantitatively demonstrated differences we observed by eye. That is, head shape (affected by eye size and presumed differences in brain and optic lobe morphology) and eye size were often significantly discriminating features. Measures of body width were also discriminators, but those differences may be related to condition of cave fishes (i.e., subjected to limited food resources) relative to epigean populations. One meristic feature, pelvic fin-ray number, distinguishes Perry County samples from those of the midlands race. In about 50% of Perry County sculpin, the pelvic fin ray count is 3+4 or 3+3 (both sides) compared to a count of 4+4 in 1459 individuals of epigean midlands race *C. carolinae* (Table 7; Robins 1954 in part) and 60 specimens from Buckeye Creek Cave, West Virginia (Williams & Howell 1979). Reduction or loss of pelvic fins is common in cavernicolous fishes (Woods & Inger 1957, Cooper & Kuehne 1974).

The occurrence of *C. carolinae* in subterranean waters is well known (Poly & Boucher 1996), and we have documented such occupation in about 25 caves from several states with known karst environments (Table 2). None of these sculpin show evidence of cave adaptation and none are known to be permanent cave residents; it might, however, be expected because some cave-inhabiting sculpin occur several kilometers from a known surface stream source. An unusual albino specimen reported from Buckeye Creek Cave, Greenbriar County, West Virginia (Williams & Howell 1979), is an exception warranting further searches and study. We emphasize that only a single specimen is known and attempts to document a 'population' by numerous independent investigators were unsuccessful. The albino sculpin (67 mm SL) has a frenum, enlarged cephalic lateralis pores (not quantified), fused post-mandibular pores, and only 6 dorsal spines and 3 + 3 pelvic rays. None of the grotto sculpin from Perry County have a frenum or 6 dorsal spines (N = 40), but they do have enlarged cephalic pores. Fusion of median chin pores into a single large pore occurs locally in high frequency (10 of 20 specimens in Rimstone River Cave), complete loss of the postmandibular pore occurs in low frequency (3 of 32 specimens). Thus, three common features of cave adaptation are shared between grotto sculpin and the single albino specimen, namely reduced pigmentation, reduced pelvic fin-ray number, and enlarged cephalic canal pores. Clearly, the discovery of a permanent residential population of cave-adapted sculpin in Buckeye Creek Cave would be highly significant for further comparisons to grotto sculpin.

Grotto sculpin were somewhat smaller (means ranged from 60 to 70 mm SL) than typical adult sizes (75–110 mm SL) in Virginia populations of epigean *C. carolinae* (Jenkins & Burkhead 1994). The largest known Virginia specimen is 144 mm SL (185 mm total length). Our largest cave-inhabiting specimen was 104 mm SL; the single albino sculpin from Buckeye Creek Cave was 67 mm SL (Williams & Howell 1979). We are unaware of primary literature addressing the significance of adult size in cave-adapted fishes, but we predict a strong positive correlation between adult size and those caves that are richest in nutrient volume and varied food resources.

Conservation status

Prior to our work, cave adaptation was unknown in cottid fishes (Groombridge 1992). Some 90 species of fishes in 14 families are subterranean residents (Helfman et al. 1997), and show evidence of cave adaptation (i.e., trends toward eyelessness, lack of pigment, enlarged lateral line receptors, and low metabolic rates). Troglomorphic populations of sculpin may be jeopardized from a conservation standpoint because of their limited distribution (total range less than 260 km^2) and the impact of human activity on the surface. Caves are located downgradient of the city of Perryville and dye trace studies of water movement in the Central Perryville Karst indicated that urban runoff from Perryville and vicinity enters the cave systems. Ammonia, nitrite + nitrate, chloride, and potassium, primarily resulting from cultivation on the surface, were detected within caves at levels high enough to be detrimental to aquatic life (Vandike 1985). For example, we recently observed a mass mortality of grotto sculpin in Running Bull Cave, presumably resulting from point source pollution similar to that described above.

Aquifers in karst terrains are exceptionally sensitive and have higher potential for contamination than other aquifers (Helling 1986, Field 1989). Sinkholes, a prominent feature of karst terrain, allow sediments and chemicals to be directly injected into the groundwater without the filtration associated with non-karst regions (Mitchem et al. 1988, Field 1989). No less than one-half of sinkholes in Perry County contain anthropomorphic refuse, ranging from household cleansers and sewage to used pesticide and herbicide containers. In these places contaminants can amass in the soil and epikarst zone during dry periods to be recharged into the cave water later (Williams 1985, Hobbs & Smart[1], Smart & Friederich 1987, Field 1989).

Spear (1995), a representative of the U.S. Fish and Wildlife Service, considered that distinctness and significance would be the primary qualities addressed in defining a unique population segment of any species before ultimately protecting it under the U.S. Endangered Species Act. Grotto sculpin clearly meet several of the criteria (sensu Spear 1995) for distinctness and significance, as judged from the unique morphology and the unusual ecological settings. Because grotto sculpin hold the unique position as the only representatives of cottids showing cave adaptation we argue for their recognition, if not as a distinct taxonomic species, then at least as a distinct population segment or a biologically isolated unit. We recommend that the U.S. Fish and Wildlife Service

[1] Hobbs, S. L. & P. L. Smart. 1986. Characterization of carbonate aquifers: a conceptual base. International Congress of Speleology (Barcelona, Spain), Communications 9: 43–46.

consider the grotto sculpin for listing under the U.S. Endangered Species Act.

Cooperative agreements with local landowners to maintain access to caves and to reduce contamination and sedimentation of the cave streams are needed. Communication with and assistance from organized speleological clubs and societies should be maintained because of their concern and knowledge of karst systems and groundwater. Educating the citizens of Perry County would increase understanding and appreciation for how karst systems function and might ultimately reduce contamination of groundwater. Finally, all fish predators (e.g., channel catfish) introduced into Perry County caves either intentionally or accidentally should be systematically captured and removed from the cave environment.

Future directions

We have studies underway comparing grotto sculpin to epigean relatives in aspects of physiology, brain morphology, life history traits, and reproductive biology. Additionally, we are continuing analyses of body shape, the lateral line system, pigmentation, and skin prickling. We will use frozen and ethanol-stored samples for an examination of population genetics, assessment of speciation models, and relationships of caves and karst history. Additional population density estimates are needed and a long-term monitoring program established and supported by the State of Missouri. We are searching for undiscovered populations of grotto sculpin in Perry County and in the similar karst topography of Monroe County, Illinois.

Acknowledgements

We are immensely grateful to S.R. Adams, C. Lee, M.L. Mugele, P.L. Moss, J.J. Swayne, S.J. Taylor, and R. Young for countless hours of assistance in difficult subterranean environments and their photographic expertise under less than ideal conditions. The SIUC Little Egypt Grotto granted permission to conduct field work in Mystery Cave and assisted with gaining landowner permission to explore other caves. The Missouri Speleological Survey (MSS Inc., c/o Missouri Geological Survey, P.O. Box 250, Rolla, Missouri) provided cave locations which are property of the MSS. We are most appreciative to our home institutions for providing space, secretarial assistance, curatorial support, and adequate time to pursue the captivating study of fishes. Aspects of this research were supported by grants to GLA from the Raney Fund and the Explorers Club.

References cited

Avise, J.C. & R.K. Selander. 1972. Evolutionary genetics of cave-dwelling fishes of the genus *Astyanax*. Evolution 26: 1–9.

Bretz, J.H. & S.E. Harris, Jr. 1961. Caves of Illinois. Ill. State Geol. Surv. Rep. Invest. (215): 1–87.

Cooper, J.E. 1978. American cave fishes and salamanders. Nat. Speleol. Soc. Bull. 40: 89.

Cooper, J.E. & R.A. Kuehne. 1974. *Speoplatyrhinus poulsoni*, a new genus and species of subterranean fish from Alabama. Copeia 1974: 486–493.

Eigenmann, C.H. 1909. Cave vertebrates of America. Carnegie Institution of Washington, Washington, D.C. 241 pp.

Elliott, W.R. 2000. Below Missouri karst. Miss. Conserv. 61: 4–7.

Field, M.S. 1989. The vulnerability of karst aquifers to chemical contamination. pp. 130–142. *In:* J.E. Moore, A.A. Zaporozec, S.C. Scallany & T.C. Varney (ed.) Proceedings of the International Conference on Recent Advances in Ground-Water Hydrology, American Institute of Hydrology, Minneapolis.

Gordon, M.S. & D.E. Rosen. 1962. A cavernicolous form of the poeciliid fish *Poecilia sphenops* from Tabasco, Mexico. Copeia 1962: 360–368.

Greenberg, L.A. & D.A. Holtzman. 1987. Microhabitat utilization, feeding periodicity, home range and population size of the banded sculpin, *Cottus carolinae*. Copeia 1987: 19–25.

Groombridge, B. (ed.) 1992. Global biodiversity, first edition. Chapman & Hall, London. 585 pp.

Helfman, G.S., B.B. Collette & D.E. Facey. 1997. The diversity of fishes. Blackwell Science, Inc., Malden. 528 pp.

Helling, C.S. 1986. Agricultural pesticides and ground water quality. pp. 161–175. *In:* Agricultural Impacts on Ground Water (Omaha, Nebraska), Proc. Nat. Water Well Assoc., Dublin, Ohio.

Hubbs, C.L. & W.T. Innes. 1936. The first known blind fish of the family Characidae: a new genus from Mexico. Occas. Pap. Mus. Zool. Univ. Mich. (342): 1–7.

Hubbs, C.L. & K.F. Lagler. 1974. Fishes of the Great Lakes region. University of Michigan Press, Ann Arbor. 213 pp.

House, S. 1976. Karst and cave distribution in Perry County. Southeast. Caver 2: 13–14.

Jenkins, R.E. & N.M. Burkhead. 1994. Freshwater fishes of Virginia. American Fisheries Society, Bethesda. 1079 pp.

Libra, R.D., G.R. Hallberg, B.E. Hoyer & L.G. Johnson. 1986. Agricultural impacts on ground water quality: the Big Spring basin study, Iowa. pp. 253–273. *In:* Agricultural Impacts on Ground Water (Omaha, Nebraska), Proc. Nat. Water Well Assoc., Dublin, Ohio.

Mitchell, R.W., W.H. Russell & W.R. Elliott. 1977. Mexican eyeless characin fishes, genus *Astyanax*: environment, distribution, and evolution. Spec. Publ. Mus. Texas Tech. Univ. (12): 1–89.

Mitchem, P.S., G.R. Hallberg, B.E. Hoyer & R.D. Libra. 1988. Ground-water contamination and land management in the karst area of northeastern Iowa. pp. 442–458. *In:* D.M. Neilsen & A.I. Johnson (ed.) Ground-Water Contamination: Field Methods, American Society for Testing and Materials Special Technical Publication 963, ASTM, Philadelphia.

Mohr, C.E. 1950. Ozark cave life. Nat. Speleol. Soc. Bull. 12: 3–11.

Padgett, A. & B. Smith. 1987. On rope. North American vertical rope techniques. National Speleological Society, Huntsville. 341 pp.

Page, L.M. & B.M. Burr. 1991. Field guide to freshwater fishes, North America north of Mexico. Houghton Mifflin Co., Boston. 432 pp.

Peck, S.B. & J.J. Lewis. 1978. Zoogeography and evolution of the subterranean invertebrate fauna of Illinois and southeastern Missouri. Nat. Speleol. Soc. Bull. 40: 39–63.

Pflieger, W.L. 1997. The fishes of Missouri, revised edition. Missouri Department of Conservation, Jefferson City. 372 pp.

Poly, W.J. & C.E. Boucher. 1996. Nontroglobitic fishes in caves: their abnormalities, ecological classification and importance. Amer. Midl. Nat. 136: 187–198.

Rea, T. (ed.). 1987. Caving basics, revised edition. National Speleological Society, Huntsville. 128 pp.

Robins, C.R. 1954. A taxonomic revision of the *Cottus bairdi* and *Cottus carolinae* species groups in eastern North America (Pisces, Cottidae). Doctoral Dissertation, Cornell University, Ithaca. 248 pp.

Robins, C.R. & R.R. Miller. 1957. Classification, variation, and distribution of the sculpins, genus *Cottus*, inhabiting Pacific slope waters in California and southern Oregon, with a key to the species. Calif. Fish Game 43: 213–233.

Romero, A. 1983. Introgressive hybridization in the *Astyanax fasciatus* (Pisces: Characidae) population at La Cueva Chica. Nat. Speleol. Soc. Bull. 45: 81–85.

SAS Institute, Inc. 1985. SAS user's guide: statistics. Statistical Analysis Systems Institute, Inc., Cary. 584 pp.

Smart, P.L. & H. Friederich. 1987. Water movement and storage in the unsaturated zone of a Carboniferous limestone aquifer, Mendip Hills, England. pp. 59–87. *In:* Environmental Problems in Karst Terranes and Their Solutions (Bowling Green, Kentucky), Proceedings, National Water Well Association, Dublin, Ohio.

Smart, P.L. & S.L. Hobbs. 1986. Characterization of carbonate aquifers: a conceptual base. pp. 1–14. *In:* Environmental Problems in Karst Terranes and Their Solutions (Bowling Green, Kentucky), Proceedings, National Water Well Association, Dublin, Ohio.

Spear, M. 1995. Considerations in defining the concept of a distinct population segment of any species of vertebrate fish or wildlife. pp. 423–424. *In:* J.L. Nielsen (ed.) Evolution and the Aquatic Ecosystem: Defining Unique Units in Population Conservation, American Fisheries Society Symposium 17, Bethesda.

Statsoft. 1994. Statistica for the Macintosh, volumes 1–3. Statsoft, Inc., Tulsa. 1062 pp.

Strauss, R.E. & F.L. Bookstein. 1982. The truss: body form reconstructions in morphometrics. Syst. Zool. 31: 113–135.

Unklesbay, A.G. & J.D. Vineyard. 1992. Missouri geology: three billion years of volcanoes, seas, sediments, and erosion. University of Missouri Press, Columbia. 189 pp.

Vandike, J.E. 1985. Movement of shallow groundwater in the Perryville Karst area, southeastern Missouri. Water Res. Rept. (40): 1–56.

Walsh, S.J. & C.R. Gilbert. 1995. New species of troglobitic catfish of the genus *Prietella* (Siluriformes: Ictaluridae) from northeastern México. Copeia 1995: 850–861.

Williams, P.W. 1985. Subcutaneous hydrology and the development of doline and cockpit karst. Z. Geomorph. 29: 463–482.

Williams, J.D. & W.M. Howell. 1979. An albino sculpin from a cave in the New River drainage of West Virginia (Pisces: Cottidae). Brimleyana 1: 141–146.

Woods, L.P. & R.F. Inger. 1957. The cave, spring, and swamp fishes of the family Amblyopsidae of central and eastern United States. Amer. Midl. Nat. 58: 232–256.

Environmental Biology of Fishes **62**: 297–313, 2001.
© 2001 *Kluwer Academic Publishers. Printed in the Netherlands.*

Milyeringa veritas (Eleotridae), a remarkably versatile cave fish from the arid tropics of northwestern Australia

William F. Humphreys
Terrestrial Invertebrate Zoology, Western Australian Museum, Francis Street, Perth, Western Australia 6000, Australia (e-mail: bill.humphreys@museum.wa.gov.au)

Received 13 April 2000   Accepted 10 January 2001

Key words: stable isotope, allozymes, physico-chemical environment, hydrogen sulphide, anoxia, food, anchialine

Synopsis

The blind cave gudgeon *Milyeringa veritas* is restricted to groundwaters of Cape Range and Barrow Island, northwestern Australia. It occurs in freshwater caves and in seawater in anchialine systems. It is associated with the only other stygobitic cave vertebrate in Australia, the blind cave eel, *Ophisternon candidum*, the world's longest cave fish, and a diverse stygofauna comprising lineages with 'tethyan' tracks and widely disjunct distributions, often from North Atlantic caves. The cave gudgeon inhabits a karst wetland developed in Miocene limestones in an arid area. There is an almost complete lack of information on the basic biology of this cave fish, despite it being listed as threatened under the Western Australian Wildlife Conservation Act. Allozyme frequencies and distributions indicate significant population sub-structuring on the Cape Range peninsula such that the populations are essentially isolated genetically suggesting that more than one biological species is present. Further, they suggest that the vicariant events may have been associated with a series of eustatic low sealevels. Analysis of intestinal contents indicates that they are opportunistic feeders, preying on stygofauna and accidentals trapped in the water, at least at the sites sampled which were open to the surface, a conclusion supported by the results of stable isotope ratio analysis. The gudgeons are found in freshwater caves and throughout deep anchialine systems in which they occur in vertically stratified water columns in which there is a polymodal distribution of water chemistries (temperature, pH, salinity, dissolved oxygen, redox, dissolved inorganic nitrogen series, hydrogen sulphide).

Introduction

The arid tropical northwest of Australia is notable for the diversity of the subterranean fauna that inhabits the karstic Cape Range peninsula and Barrow Island. It includes a terrestrial fauna with wet forest affinities (Humphreys 1993a,b, 2000a,b), and an aquatic fauna with both limnic and anchialine components. The latter includes many disjunct Tethyan lineages (Humphreys 1993a, 2000b, Jaume & Humphreys 2001, Jaume et al. 2001) whose congeners or sister taxa inhabit caves on either side of the North Atlantic, in the Caribbean region and the Canary Islands (see below).

Included in this subterranean karst wetland (ANCA 1996, Humphreys 2000b) are the only known stygobiontic cave vertebrates in Australia. They are the blind cave gudgeon, *Milyeringa veritas* Whitley, 1945 (Perciformes: Eleotridae) and the blind cave eel, *Ophisternon candidum* (Mees, 1962) (Synbranchiformes: Synbranchidae) (Figure 1) which occur sympatrically over much and possibly all of their ranges. The latter species is considered by Romero & Vanselow (2000b) and it will not be discussed further here. Globally, there are few occurrences of sympatric species of cave fishes (Thinès & Proudlove 1986). Proudlove[1] (personal communication) records 13 instances of two species and three instances of three species occurring in sympatry. Kuhajda & Mayden

[1] Proudlove, G.S. 1997. A synopsis of the hypogean fishes of the world. *In:* Proceedings of the 12th International Congress of Speleology, La Chaaux de Fonds, Switzerland, August 1997, 3: 351–354.

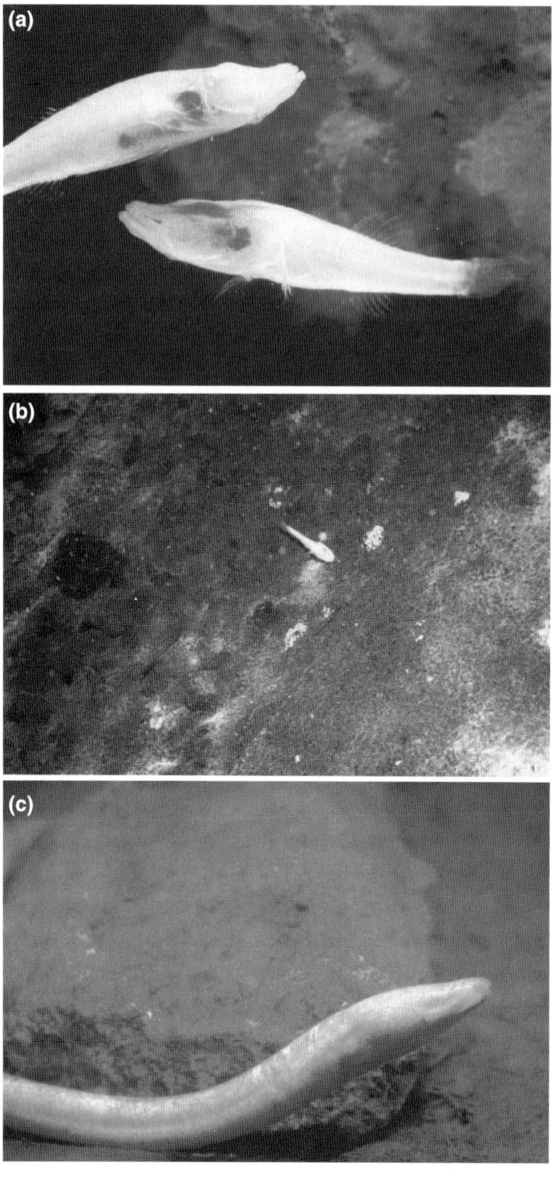

Figure 1. a – *Milyeringa veritas* from Kubura Well (photo: Douglas Elford, Western Australian Museum); b – *M. veritas* foraging over various sulphur bacteria in Bundera Sinkhole, an anchialine cave (photo: Stefan Eberhard, Western Australian Museum); c – *Ophisternon candidum* from Kubura Well (photo: Douglas Elford, Western Australian Museum).

(2001) report the first known occurrence of sympatry between North American cave fish species.

While some cave fishes, particularly the Mexican characid *Astyanax fasciatus*, and the American Amblyopsidae, have been extensively researched (e.g. Wilkens 1988, Bergstrom 1997, review in Weber et al. 1998b), little is known of those inhabiting the Cape Range peninsula. This is partly because even the general affinities of *M. veritas* are unknown, precluding the interesting questions that can be posed about the evolution of those cave fishes that are part of lineages with extant epigean sister taxa. The identification of the phyletic affinities of *M. veritas* would be the most effective information to stimulate research on the Australian cave fishes. *Milyeringa veritas* is the only stygobitic species of eleotrid known in the world (Romero & Vanselow 2000a) and occupies a wide range of physico-chemical environments (Humphreys 1999b). While no research has focused on these Australian cave fishes, a variety of data has been collected in association with other projects and a synthesis of this information is presented here to stimulate and facilitate research on this versatile species.

History and affinities

Milyeringa veritas was described from subterranean waters of the Cape Range peninsula in the monotypic family Milyeringidae (Gobioidea) (Whitley 1945). Mees (1962) recognized the eleotrid affinities of *Milyeringa*, a family widespread in tropical and subtropical shallow marine to fresh waters mainly in the Indo-Pacific region (Nelson 1984). The monotypic genus *Milyeringa* is endemic to the Cape Range peninsula and Barrow Island but its phylogenetic relationships within the Eleotridae have not been established. *Milyeringa veritas*'s closest affinity is possibly with the eleotrid genus *Butis*, a mangrove dwelling genus of brackish and fresh waters of the Indo-Australian Archipelago (i.e. Malay Peninsula to northern Australia including Indonesia, Philippines and most of Melanesia) (G.R. Allen personal communication). Mangroves are still present on the Cape Rage peninsula although they were formerly more widespread (Morse 1993).

On consideration of the geological and tectonic history of the area Humphreys & Adams (1991) suggested that a southern extension of the range of *M. veritas* was unlikely owing to the salinity of the groundwater but that 'it may be worth searching wider afield for this fauna ...' to the northeast (see also Humphreys 1993a, b). In 1996, elements of the tethyan crustacean fauna were found as predicted where the Precambrian plateau (the Western Shield,

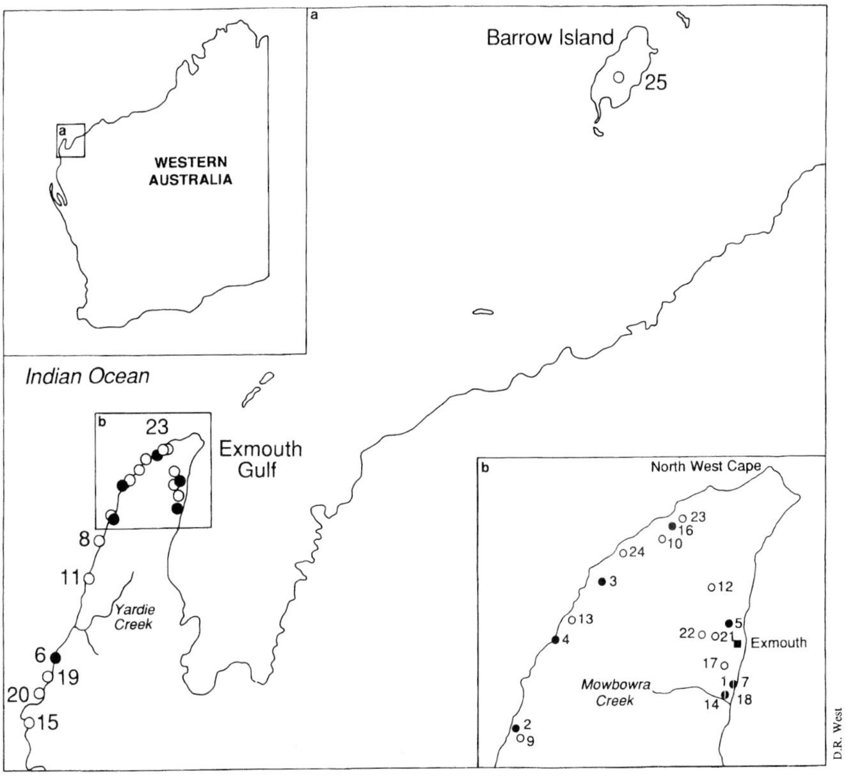

Figure 2. The distribution of blind cave fishes in Australia. Filled circles denote *Ophisternon candidum*. *Milyeringa veritas* is present at all locations denoted by open or filled circles except location 4 (Tantabiddi Well). Top of the map is north.

Beard 1998) approaches the shore line to the south of Karratha, and in 1999 *M. veritas* itself was collected from deep groundwater on Barrow Island (Humphreys 1999c).

Barrow Island lies on the shallow North West Shelf (Figure 2) which, until the early Holocene, ca. 8000 years ago, would have been connected by land to the Cape Range peninsula owing to the generally lower sea level over much the last million years (Chappell & Thom 1977). Hence, for much of that time there would have been a broad karstic plain, encompassing the North West Shelf from the tip of the peninsula to the Fortescue River and including Barrow Island, that would have allowed widespread dispersal of the subterranean fauna (Humphreys 1993b). Indeed, both terrestrial (Harvey & Humphreys 1995) and aquatic (Adams & Humphreys 1993, Humphreys 2000a) conspecific subterranean fauna occur on Barrow Island and the northeast coast of the Cape Range peninsula (Humphreys 1993b).

Regional context

Distribution

The known distribution of *M. veritas* is shown in Figure 2 and details of the sites and the history of their discovery are given in Humphreys (1999c). The species' range encompasses the coastal plain from Exmouth to Javis Well (location 15), a distance along the coast of about 136 km, and one site on Barrow Island which lies 160 km to the northeast of North West Cape across the shallow North West Shelf.

There has been extensive sampling for stygofauna along the coast of both the mainland and the Cape Range peninsula, as well as from Barrow Island[2] (Pesce et al. 1996, figure 29, W.F. Humphreys unpublished

[2] Humphreys, W.F. 1994. The subterranean fauna of the Cape Range coastal plain, northwestern Australia. Report to the Australian Heritage Commission and the Western Australian Heritage Committee. 202 pp.

data) and the species appears to be endemic to the Cape Range and Barrow Island karst formed on anticlines of Tertiary limestones (Wyrwoll et al. 1993, McNamara & Kendrick 1994). However, *M. veritas* has rarely, and *O. candidum* has never been collected using methods not involving direct sighting prior to capture, hence the mainland sampling, which was conducted in bores and pastoral wells using haul nets and traps without visual contact, could have missed them. It is notable that other elements of the Cape Range and Barrow Island stygofauna do occur on the mainland in the calcrete and alluvial fans near the mouths of the Robe and Fortescue rivers (Humphreys 1999a). One tethyan element, the thermosbaenacean *Halosbanea,* has even been recorded from an altitude of 300 m in the Robe Valley, the approximate high sea level stand in the Late Eocene on the Yilgarn (G.W. Kendrick personal communication 1999).

Geology and geomorphology

The Cape Range peninsula comprises an anticline of marine limestones of Late Eocene–Miocene age rising to an altitude of 314 m. The peninsula is fringed by a coastal plain interrupted on the west coast by a series of erosion terraces cut into the Tertiary limestones during sea level stadia and overlain in places by Quaternary carbonates forming a karst pavement except where they are overlain by Pleistocene and Holocene dunes (Wyrwoll et al. 1993). The western shore is bordered by the Ningaloo Reef, the longest fringing reef in Australia, and considered an integral part of the Cape Range karst province[3]. Further information on the geology and hydrogeology may be sought in Allen (1993).

Climate and vegetation

The area is in the arid tropics with high evaporation (mean 3219 mm) and receives its rainfall (median 280 mm) as a result of several meteorological processes (Beard 1975, Gentilli 1972, 1979). The predominantly episodic heavy rainfall may occur through much of the year but with very low predictability (Humphreys et al. 1989) and it is responsible for the principal recharge of the freshwater aquifers through the surface karst which affects the elevation of the groundwater surface for prolonged periods. Temperatures are high (mean 27°C) and the mean monthly maximum temperature exceeds 35°C for four months of the year.

On the coastal plain the vegetation is naturally dominated by tussock grasses with eucalypt shrubs and *Ficus* associated with caves but which occurs more widely on the karst. Further details on the flora and vegetation associations of the range can be found elsewhere[4] (Beard 1975, Keighery & Gibson 1993).

Regional groundwater

The general hydrogeological model for oceanic islands has been applied to the groundwater of Cape Range peninsula (Allen 1993). Following the Ghyben–Herzberg principle (Ford & Williams 1989), a freshwater lens overlies saltwater so that locally a wedge of saltwater intrudes under the freshwater contained in the limestone, and between which occurs a zone of mixing that broadens towards the coast. In the Exmouth area, on the northeastern part of the Cape Range peninsula, the freshwater–saltwater transition, at about 5 km from the coast, is exceptionally far inland. In the Exmouth area the inland limit to the salt-water interface appears to be controlled by the presence of solution cavities and channels below the water table (Martin 1990). Such aquifers are vulnerable to saltwater upconing as a result of pumping water from the thin freshwater lens, and to the inland migration of seawater.

Habitat

General

Milyeringa veritas has been recorded exclusively from inland karst systems from within 150 m of the coast and up to 4.3 km inland. They are found in freshwater caves as well as in anchialine ecosystems; the latter comprise fresh to brackish groundwaters overlying seawater that has no surface connection to the sea but which is affected by marine tides (Sket 1981, 1986, 1996,Stock

[3] Hamilton-Smith, E., K. Kiernan & A. Spate. 1996. Karst management considerations for the Cape Range karst province, Western Australia. W.A. Department of Environmental Protection, Perth. 80 pp.

[4] Payne, A.L., P.J. Curry & G.F. Spencer. 1987. An inventory and condition survey of rangelands in the Carnarvon Basin, Western Australia. Western Australian Department of Agriculture Technical Bulletin No. 73. 478 pp.

et al. 1986). Much of the habitat of *M. veritas* is known to be affected by marine tides (Humphreys et al. 1999, figure 1) but the tidal influence is much more marked on the west coast of the peninsula (W.F. Humphreys & R.D. Brooks unpublished).

Milyeringa veritas may be found swimming in sunlight at the surface in small karst windows in the anchialine system and at depths of up to 33 m in caves containing seawater (Humphreys 1999b). It has been sampled from groundwaters through bore holes in which the water table is up to 47 m below the ground and remote from known openings to the surface.

Physico-chemical environment

Humphreys & Adams (1991) and Humphreys[2] reported the occurrence of *M. veritas* from a range of water chemistries (Table 1) and subsequently from more saline waters (Yager & Humphreys 1996) with complex physico-chemical stratification (Humphreys 1999b, Humphreys et al. 1999). *M. veritas* populations are known from freshwater caves (0.3 g l^{-1} TDS: Poore & Humphreys 1992) but also inhabit the anchialine system with salinities up to 34 g l^{-1} TDS (Yager & Humphreys 1996, Humphreys 1999b) – they are reported to be susceptible in captivity to fungus infections (*Saprolegnia*) in freshwater (Young 1986). The fish occur in hyperoxic surface water to suboxic (see Sket 1996) waters below the pycnocline (Humphreys 1999b) and they have been seen foraging immediately above sediments (Figure 1) covered by black and white sulphur bacteria.

At one location, Bundera Sinkhole, *M. veritas* inhabits an anchialine cave with marked physico-chemical and biological stratification (Figure 3). The fish occur below multiple layers of hydrogen sulphide and redox and oxygen minima associated variously with sulphur bacteria and a cascade of dissolved inorganic nitrogen series (Humphreys 1999b). While *M. veritas* inhabits waters with a remarkable range of physico-chemical characteristics (Yager & Humphreys 1996), it is not known how rapidly they move between these different water masses: the same individuals could move

Table 1. Water analysis for the sites inhabited by *Milyeringa veritas* (n = 20). The data for individual sites are contained in Humphreys[2]. All units are in mg l^{-1} unless otherwise stated.

Parameter	Mean	SE	Min	Max	CV
pH	7.45	0.06	7.02	8.08	3
Turbidity NTU	3.83	1.00	0.6	16	116
Colour TCU	4.7	1.7	0	34	162
Conductivity mS m^{-1}	1295	307	35	4700	106
TFS	8168	1930	250	26 500	106
TFS-CO_2	8013	1927	210	26 500	107
Fe-unfiltered	0.20	0.08	0	1.6	185
Mn-unfiltered	0.04	0.02	0	0.45	256
Al	0.12	0.03	0.02	0.6	119
Na	2352	58 948	7780	112	
K	94.0	22.7	2	310	108
Ca	161	22.3	15	390	62
Mg	288	68.7	3.4	940	107
Hardness as $CaCO_3$	1580	336	50	4840	95
Alkalinity	4.5	0.34	1.6	8.2	34
Cl	4321	1088	54	15 050	113
SO_4	579	145	7	1950	112
Si – as SiO_2	13.8	0.9	2.8	23	29
Filterable organic carbon	19.6	4.1	1	74	93
Total phosphorous	0.05	0.01	0.009	0.185	91
Free reactive phosphorous	0.026	0.007	0.004	0.13	121
Total Kjeldahl nitrogen	0.36	0.09	0.04	1.7	113
Nitrite as nitrogen	0.02	0.009	0	0.19	202
Nitrate as nitrogen	3.2	1.54	0.02	31.5	215
Ammonia as nitrogen	0.06	0.02	0	0.36	135

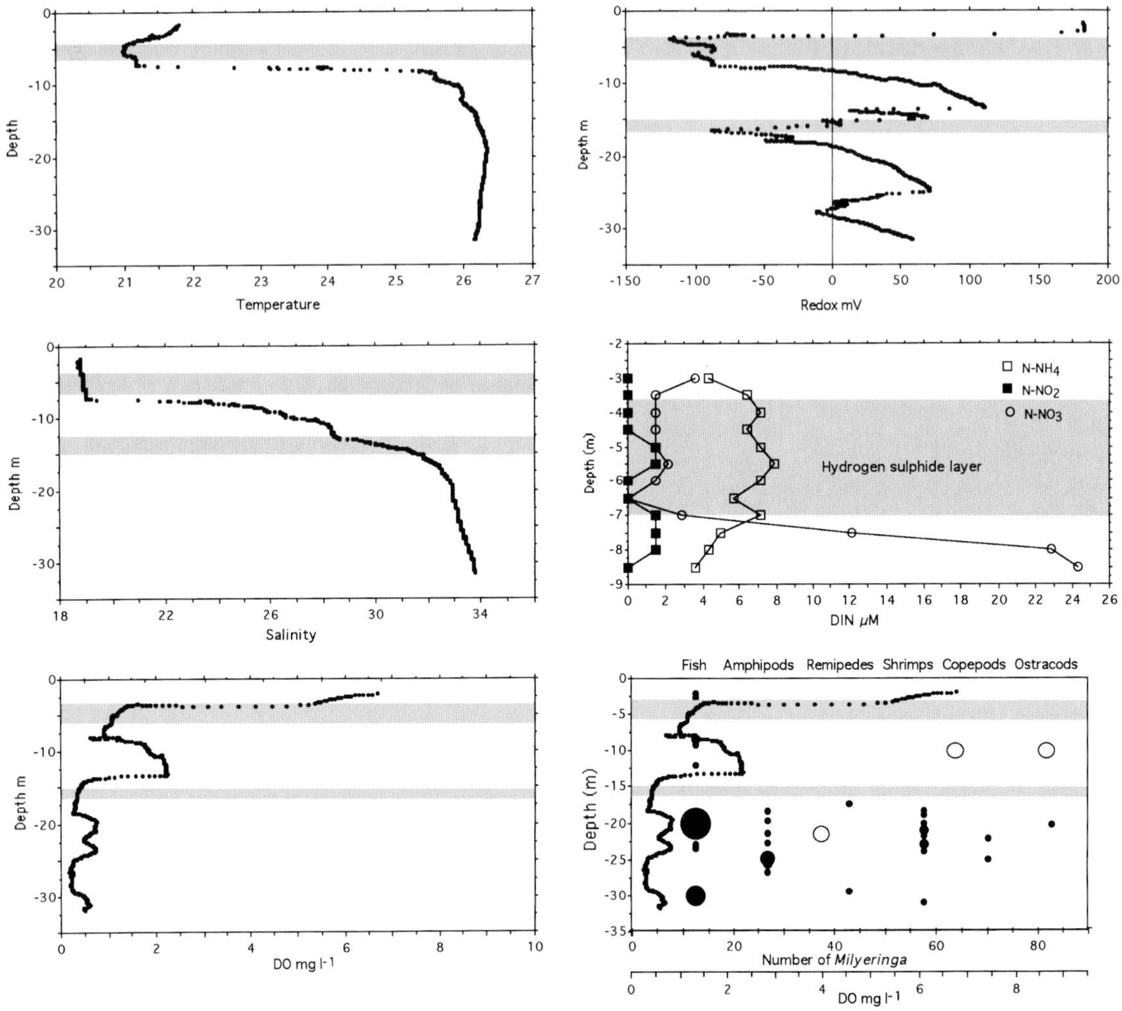

Figure 3. Vertical profile with depth in Bundera Sinkhole in 1997 where the grey bands denote the zones of elevated hydrogen sulphide concentration (see Humphreys 1999b: Figure 2f): (a) temperature (°C), (b) salinity (g l^{-1} TDS), (c) dissolved oxygen (mg l^{-1} DO), (d) oxidation reduction (redox) potential (mV), (e) dissolved inorganic nitrogen species (DIN μM) $N\text{-}NH_4$, $N\text{-}NO_2$, $N\text{-}NO_3$, (f) vertical distribution of sightings of the fauna in Bundera Sinkhole reported by divers [(a) the closed circles denote the vertical distribution of the taxon named at the head of the column, (b) the open circles denote the number of blind gudgeons, *Milyeringa veritas*, counted at one occasion at each of three depths (second bottom scale), (c) the curve denotes the DO concentration (bottom scale)]. These data are illustrative only as no regular sampling has been conducted through time. Further details in Humphreys (1999b) and Humphreys et al. (1999).

regularly or rarely between the water types, there may be life interval movements or different stocks may inhabit the different water masses.

Associated fauna

Milyeringa veritas is part of a rich stygofaunal community that contains a number of ecogeographic relicts and phyletic relics with tethyan affinities[2] (Humphreys 1993a, 2000a, Danielopol et al. 2000). These include the only member of the class Remipedia (Yager & Humphreys 1996) and the order Thermosbaenacea (Poore & Humphreys 1992) known from the southern hemisphere which are congeneric with species occurring in caves on either side of the North Atlantic.

In the general groundwater of the Cape Range peninsula both epigean and stygal taxa cohabit with *M. veritas*. The stygal species *Halosbaena tulki* Poore &

Humphreys (Thermosbaenacea), *Haptolana pholeta* Bruce & Humphreys (Isopoda: Cirolanidae) and *Nedsia douglasi* Barnard & Williams occur in freshwater. *Stygiocaris lancifera* Holthuis and *S. stylifera* Holthuis (Malacostraca: Atyidae) are found in freshwater and anchialine systems. *Grandidierella* sp. nov., a non-stygal species, occurs in open anchialine pools in which *M. veritas* may occasionally forage. A number of the sites with *M. veritas* are also the type localities for other cave restricted species, some of which are prey of the fish, such as Thermosbaenacea and atyid shrimps (Humphreys & Feinberg 1995).

On Barrow Island *M. veritas* is known from a single bore that also contains calanoid, cyclopoid and harpacticoid copepods including *Inermipes humphreysi* Lee & Huys (Harpacticoida: Ameiridae), *Halosbaena tulki* (Thermosbaenacea), the amphipods *Nedsia sculptilis* Bradbury & Williams (Melitidae) and *Bogidomma australis* Bradbury & Williams (Bogidiellidae), *Haptolana pholeta* (Cirolanidae) and *Stygiocaris stylifera* (Atyidae).

The vertical stratification of the faunal associations is best documented from Bundera Sinkhole because it is accessible by divers to a depth of 33 m and *M. veritas* occurs through the full depth of the water column (Figure 3f). Bundera Sinkhole is the only deep anchialine system known in Australia and the only known continental anchialine system in the southern hemisphere (see Iliffe 2000, figure 1) as well as the only known site for the crustacean class Remipedia in the southern hemisphere (Yager & Humphreys 1996). It is the most complex habitat known to be inhabited by *M. veritas*. Several taxa are known only from below the thermohalocline in the cave, namely *Lasionectes exleyi* Yager & Humphreys (Remipedia), *Danielopolina kornickeri* Danielopol et al. 1999 (Ostracoda: Halocyprida: Thaumatocyprididae), *Liagoceradocus branchialis* Bradbury & Williams (Crustacea: Hadziidae), *Stygiocaris stylifera*, and the copepods *Bunderia misophaga* Jaume & Humphreys, 2001 (Calanoida: Epacteriscidae), *Stygocyclopia australis* Jaume et al., 2001 (Calanoida: Pseudocyclopiidae) and *Speleophria bunderae* Jaume et al., 2001 (Misophrioida: Speleophriidae).

Above the thermohalocline in the photic zone this anchialine system also contains a number of non-stygal plants and animals (the algae *Rhizoclonium tortuosum* (Dillwyn) (Chlorophyta: Cladophoraceae) and *Lamprothamnium papulosum* J. Groves (Charophyta: Characeae), and the invertebrates *Prionospio* sp. nov. (Spionidae: Polychaeta), *Halicyclops longifurcatus* Pesce et al. (Copepoda: Cyclopidae), ostracods, *Iravadia* sp. (Mollusca: Iravadiidae, Slack-Smith 1993), water striders (Gerridae: Hemiptera) and *Kiefferulus intertinctus* Skuse (Chironomidae, Diptera). A fuller description of the epigean fauna associated with these karst wetlands is given elsewhere (Humphreys 2000a,b).

Although no systematic sampling has been conducted, it is clear that both the distribution and abundance of the fauna, including *M. veritas*, differs between years (Humphreys 1999b). While *M. veritas* has been recorded throughout the water column, the vertical distribution differs markedly between years. In 1994, there was a large concentration of the fish at the extreme depth of the cave, while they were entirely lacking from this area in 1997 when there were concentrations just below the pycnocline (ca. 10 m) and at ca. 20 m, between the H_2S layers (Figure 3f). This change in location of the principle predator (Humphreys & Feinberg 1995), and of *D. kornickeri*, suggests that the ecosystem is variable through time.

Like *M. veritas*, other fishes with reduced or vestigial eyes, which are sometimes found in total darkness, are known to inhabit caves exhibiting a density interface (thermohalocline: sensu Humphreys 1999b). Species of the genus *Lucifuga* variously occur both above (Cohen & McCosker 1998) and below (Díaz Perez 1987b) the density interface at depths of up to 40 m (Díaz Perez 1987b), in freshwater (Díaz Perez 1987a) or saltwater (34 mg^{-1}: Palmer 1985). They may also occur with relictual crustaceans (Yager 1981, Palmer 1985, Díaz Perez 1987a) such as Remipedia, Thermosbaenacea and *Thaumatocypris* (Ostracoda: Yager 1981, Díaz Perez 1987a).

Biology

Appearance and behaviour

Milyeringa veritas is eyeless, depigmented and translucent, and *in vivo* the brain is visible through its braincase, the scales are reduced on the body and absent from the head, all indicative of a high degree of troglomorphy. The general morphology of *M. veritas* is convergent with other cave dwelling fish such as *Amblyopsis rosae*, the Ozark cave fish of Missouri (Bergstrom 1997). Although visible eyes are overtly lacking it is not known whether the eyes of *M. veritas* are entirely regressed as is suggested by the behavioural observation that the fish in caves show no rapid response to sudden bright light; however, they sometimes swim in sunlight in anchialine pools. The degree of eye reduction in *M. veritas* does suggest that the lineage has been troglobitic for a long time since the degree of eye reduction may be considered indicative of the time since troglobite evolution (Wilkens et al. 1989). The latter correlated the degree of eye reduction in *Lucifuga* and *Ogilbia* with the phylogenetic age of the lineages in the New World tropics as deduced from geological and eustatic information.

Milyeringa veritas often hangs more or less motionless in the water or perches on rock ledges for long periods. When the water is disturbed the fish may swim directly to investigate the disturbance, gradually distance itself from the area, or swim to the bottom and move behind obstacles, especially to move beneath ledges. Despite the reported sluggish nature of these fish (Knott 1993), they may exhibit strong escape behaviour from a net, are strong swimmers for short distances, even leaping from the water and, if stranded, can flip their bodies into the air. Young (1986) reported that, in captivity, they displayed 'aggression' with the 'typical gill flaring and body waggle of a gudgeon' and that when familiar with their surroundings could be quite evasive.

In shallow conduits of the anchialine system that drain at low tide, *M. veritas* may be seen closely following the advancing tidal front, apparently foraging over the newly submerged substrate, an observation consistent with the apparent opportunistic intake of food revealed by analysis of the intestinal contents. Although *M. veritas* occur in particular parts of caves, often in some numbers (see below), they do not overtly group. They move widely though the water column within a given cave but it is not known whether individuals cross the pycnocline.

Numbers

Several *M. veritas* are commonly seen in caves, but there are some sites where more than 100 may be seen (Humphreys 1999b). Formal determination of the number of *M. veritas* at any location is limited to Kubura Well, actually an open cave that is connected to an entirely flooded chamber extending to a depth of 11 m. The minimum size of the population of *M. veritas* in Kubura Well was estimated by dilution sampling to be 48 (95% CI 45 and 52) and 43 (95% CI 38 and 50) fish in June and July 1993, respectively (Humphreys[2]).

Food and stable isotope ratio analysis

Young (1986) reported that, in captivity, *M. veritas* ate mosquito larvae, live brine shrimps, and that eventually they would take frozen brine shrimps and would take prey up to one-third of their body length. He observed that *M. veritas* often missed live prey on the surface film but rarely missed them when live prey were in the water column; they took live prey by a sudden gulp while nearly stationary and with the fins flared. When offered food they frequently ignored it and allowed tubificid worms to move over them without overt response even though they eventually ate them. Allen (1989) considered *M. veritas* to feed opportunistically upon detritus, algae and whatever organisms, including insects, that accidentally fell into the water. Examination of the intestinal contents of existing collections showed that the diet of *M. veritas* included terrestrial (79%) and aquatic species and that amongst the aquatic species both stygobitic (10%) and epigean species were ingested (Humphreys & Feinberg 1995). The predominant prey of terrestrial origin consumed by *M. veritas* differed between the east and west coast of the peninsula, reflecting the land use. Exotic cockroaches (*Periplaneta*) and native isopods (*Buddelundia*) predominated in the diet of fish, respectively, inhabiting the settled east coast and the largely unpopulated west coast (Humphreys & Feinberg 1995). Hence, the fish appear to feed opportunistically on terrestrial invertebrates accidentally introduced into the aquatic system (Humphreys & Feinberg 1995). Recent observations (1999) of *Buddelundia* foraging underwater in Kubura Well means that their presence in water is not necessarily accidental. *M. veritas* also includes in its diet aquatic larvae of aerial species (chironomid and caddis larvae) as well as specialized members of the stygofauna (*Stygiocaris* sp.) (Humphreys & Feinberg 1995). The bigger west coast fish were associated with bigger food items (isopods). However, as terrestrial Crustacea generally have an energy density only about 70% that of terrestrial insects (from Cummins & Wuycheck 1971), then this apparent advantage in food size may not confer an energetic advantage.

The gut contents examined by Humphreys & Feinberg (1995) were predominantly from large individuals all of which had been sampled from wells or open caves where there was direct open access between the ground surface and the groundwater and hence where epigean prey species would be present. Throughout most (>99.99%) of the inferred distribution of this obligate groundwater community (stygofauna) there are no known openings to the surface. Extensive sampling down boreholes has shown that stygofauna, including *M. veritas*, is present in such areas, from where there is a total absence of epigean species (Humphreys & Feinberg 1995). Hence, in general, the fish are probably dependent on the stygofauna for food rather than on epigean species accidentally in the water.

The fish share their habitat with a diverse stygofauna, many of which were not represented in their gut contents (Humphreys & Feinberg 1995), probably owing to the specimens having been sampled at sites open to the surface where stygofauna would have been scarce and where prey of epigean origin was abundant. The smallest taxa, such as copepods and ostracods, may lie outside the size range of prey items, which for *M. veritas* is known to include the length range of 2.8–14 mm, and for *O. candidum* 3 mm (*Halosbaena*) to an estimated 8 mm (Odonata). Melitid amphipods are abundant and sympatric with *M. veritas* but they were not identified from the gut contents despite being within the known size range of prey (Humphreys & Feinberg 1995).

The prey identified is consistent with the behaviour of the fish. *M. veritas* moves widely through the water column, often hovering in mid- to surface waters where prey with hygrophobic integuments, such as cockroaches, would be encountered when they fell into the water. In contrast, *O. candidum* inhabits the surface of, and burrows into, the flocculent faecal ooze characteristic of crustacean-rich stygal habitats, and all the prey items identified were bottom dwellers or, in the case of the isopod, would sink to the bottom when it fell into the water.

To further examine the food of *M. veritas* stable isotope ratio analysis was conducted on elements of the stygofauna and surrounding habitats (methods in Humphreys 1999b). The locations of the ∂C13 and ∂N15 values and the trajectory of the ∂C13 and ∂N15 coordinates between tropic levels provides information on the trophic level of the species as well as the origin of the food source for each trophic level (detailed in Humphreys 1999b).

The stable isotope signature (Figure 4) provides strong evidence that *M. veritas* is a predator and does not rely to any extent on detritus or algae as a food source as suggested by Allen (1989). Furthermore, the $\partial^{13}C$ ratios suggest that they depend on the energy derived from plants using the C4 photosynthetic pathway, a mixture of C3 and C4 plants, and from chemoautotrophic energy fixation by sulphur bacteria via stygal shrimps (Atyidae) scraping the biofilm in the anoxic/oxic ecotone in the anchialine sinkhole at Bundera (details in Humphreys 1999b).

Size, growth, breeding and condition

Milyeringa veritas ranged from 5 to 57 mm standard length (SL) and those collected from the east coast were bigger in every respect from those on the west coast but the meristic variables do not differ between areas (Humphreys 1994, figure 13). For example, the mean SL of fish from the west coast was 32.8 mm (7.9 = SD; 125 = sample size) compared with a mean elsewhere of 24.6 mm (10.4, 23, $F_{s1,146} = 18.67$, $p < 0.001$).

Milyeringa veritas isolated in a cave pool for 42 days grew in length at a rate of 0.6 mm per month (95% limits 0.2–1.0 mm) but during this time lost weight indicating that their conditions may not have been optimal.[2] These sparse data on growth rate are consistent with the apparent cohort size indicated from length frequency distributions and suggest that the larger fish are 3+ years old (Figure 5).

The only data available on seasonal change in condition in *M. veritas* are derived from museum specimens. The condition index differs significantly through the year being higher in August and September than in other months (Table 2). There are inherent problems with the use of such material, including preservation artifacts and poor distribution in the samples of the size-classes and seasons. However, the results are of interest as small individuals (less than 18 mm SL, cf. maximum 57 mm SL) have been observed in December–March in several years (K. Cameron personal communication) suggesting that there may be a distinct breeding season. The style of breeding in *M. veritas* is unknown. The smallest individuals caught are 4.5 mm SL.

Annual, even seasonal breeding cycles may be unexpected in tropical cave animals. What could be the cue to seasonal breeding in *M. veritas* suggested by the 'cohorts' in Figure 5 and the apparent periodic appearance of small individuals? The growth data and size class frequency analyses were derived from *M. veritas* sampled from caves and open pastoral wells and so they are not dependent on the stygal ecosystem as they gain most of their food directly from animals inhabiting terrestrial habitats (Humphreys & Feinberg 1995). Such *M. veritas* populations, closely tied to surface openings, may be expected to respond differently to those populations inhabiting groundwaters more remote from surface openings and dependent

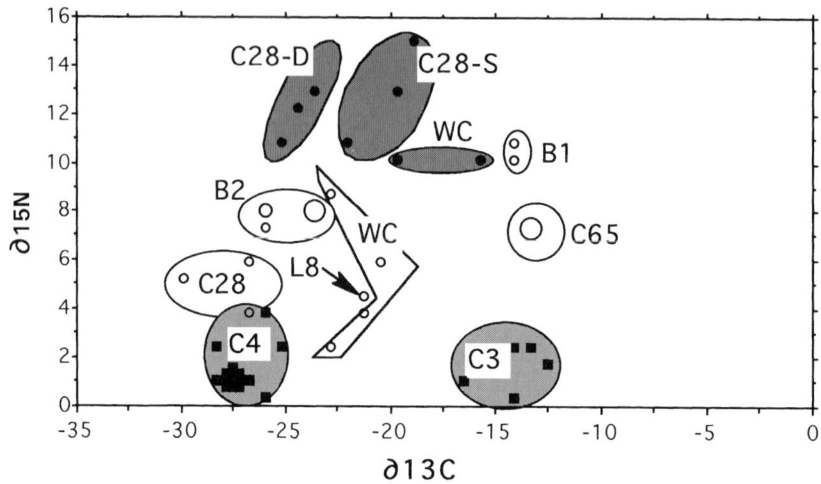

Figure 4. Stable isotope analyses (∂13C and ∂15N) of *Milyeringa veritas* (dark shading: C28-D and C28-S = deep and shallow samples, respectively, from Bundera Sinkhole, west coast of the Cape Range peninsula (Humphreys 1999b); WC = east coast of the Cape Range peninsula and some elements of the associated system. Light shading are plants with C3 and C4 photosynthetic pathways. Unshaded = *Stygiocaris stylifera* from various sites (C28 = Bundera Sinkhole; WC = east coast of the Cape Range peninsula; B2, L8 and C65 = anode protection wells on Barrow Island; B1 = anchialine cave on Barrow Island). The location of the *M. veritas* data appears to reflect the widely varying stable isotope signature of the largest prey, the atyid shrimp *Stygiocaris*. It has been argued from these and other data that in places the atyids depend on chemoautotrophic energy fixed by sulphur bacteria associated with the oxic/anoxic interface (Humphreys 1999b). Data for *Stygiocaris* and *M. veritas* from location WC are from Davies, P.M. 1996. Stable isotope analyses of the food-webs associated with stygofauna of Exmouth borefield. A report to the Water Corporation Appendix 1, 13 pp. *In:* Supplementary Investigations of the Effects of Public Water Supply Abstraction on the Stygofauna and Aquifer of the Cape Range, Water Corporation, Perth.

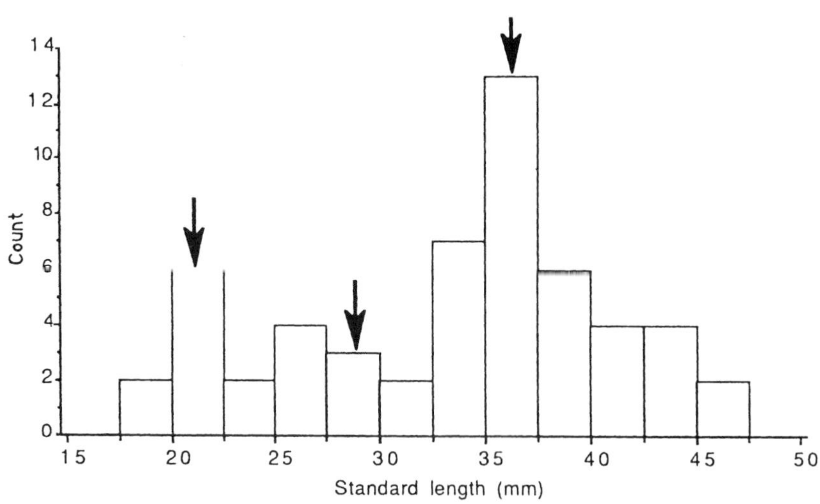

Figure 5. The pooled size class distribution of *Milyeringa veritas* from Kubura Well in 1993. Arrows denote the projected movement of the lower 'cohort' over successive years as estimated from the measured growth rate (see text) in Kubura Well.

Table 2. Condition index (standard length (mm)/ $3\sqrt{}$ weight (g)) for *Milyeringa veritas* determined from museum specimens.

Month	Mean	SE	*N*
May	38.2a	0.28	46
June	37.0a	0.36	19
July	38.3a	0.46	31
August	39.8b	0.49	24
September	39.8b	0.81	18
October	38.1	0.60	5

The means differ significantly ($F_{5,137} = 4.373$, $p = 0.001$; October samples excluded owing to small sample size). Rows with common letters are not significantly different by the Fisher PLSD test. The samples were fixed in formaldehyde and represent combined samples from several years of collecting in the 1960s.

on stygal prey species. Cave and well populations of *M. veritas* (and other stygofauna, such as *Stygiocaris* spp.) may be expected to respond closely to changes in food resources rather than to variation in the stygal community itself. The episodic heavy rainfall associated with summer cyclones may serve to wash copious food into the caves, or else the humid caves may serve as a refuge for terrestrial invertebrates during the very dry and hot (frequently $> 40^\circ$C) summer period. Clearly, detailed work on the fish is required.

Morphometrics

The morphometric development of cave fish has largely been neglected other than the general observations that they often have reduced body size and larger head and fins (Poulson 1964) and that this may be a 'neotenic' trend (Weber et al. 1998a). However, no thorough examination of the heterochrony (McNamara 1990) of cave fish has been undertaken with respect of their lineages or considered generally as a stygomorphy (cf. troglomorphy: Christiansen 1962, Humphreys 2000c).

A synopsis of the morphological trends in *M. veritas* is presented here (Figure 6, Table 3) for comparison with other species of cave fish. The general trends apparent in these data are that as the fish grow they become more bulky, the tail becomes shorter, the body becomes deeper and narrower and the buccal region shorter. How this development pattern relates to habits is unknown.

Genetics

Seven populations of *M. veritas*, from throughout its distribution on the Cape Range peninsula, were subjected to a comprehensive allozyme analysis at 43 loci (Adams & Humphreys 1993), a synopsis of their findings is presented here. Of the 43 loci scored, 28% of loci were polymorphic (P) and the average level of heterozygosity per population (H) was 0.028 ± 0.013 ($n = 7$). Compared with marine fish, generally, *M. veritas* exhibits low H but high P: for 169 populations of marine fish, after removal of studies involving few individuals (<20), Planes (1998) found $H = 0.058 \pm 0.037$ and $P = 21.37 \pm 14.09\%$. Planes (1998) found no effect of taxonomic placement in marine fish on the magnitude of H or P. Low values of H have sometimes been associated with small populations (Soulé 1976, Nevo et al. 1984), stable environments (Levins 1969, Van Valen 1965) or habitat specialization (Nevo 1978) associated with *M. veritas*. However, as no comparable allozyme data are available for other eleotrids, such comparisons are premature.

The average Nei's D for all populations was 0.04, and an average of 2.3%, with a maximum of 7% of loci had fixed differences. The mean F_{ST} between populations was 0.457 (95% CI = 0.211–0.599) while the mean FIS between populations was 0.208 (95% CI = 0.071–0.300). Whilst some fixed differences were encountered, the overall levels of genetic differentiation between populations and the geographic pattern of allele distribution were both consistent with the presence of a single biological species (Figure 7). However, analysis of the F-statistics demonstrates the presence of significant population substructuring in this species. The average gene flow per generation (*Nm*) can be determined from the conditional average frequency, F_{ST} (Slatkin 1981). In an island population model *Nm* represents the average number of individuals exchanged between populations in each generation and populations do not diverge unless $Nm < 1$ (Wright 1931). Generally, if $F_{ST} > 0.2$ ($Nm < 1$) considerable divergence will occur between populations, and with values of $F_{ST} > 0.33$ ($Nm < 0.5$) populations will be essentially unconnected genetically (Trexler 1988, Slatkin & Barton 1989). Hence, the population of *M. veritas* is not genetically uniform and there is evidence of restricted gene flow on the peninsula (Humphreys & Adams 1991, Adams & Humphreys 1993), as is also the case in the sympatric shrimp *Stygiocaris stylifera*. As some populations of

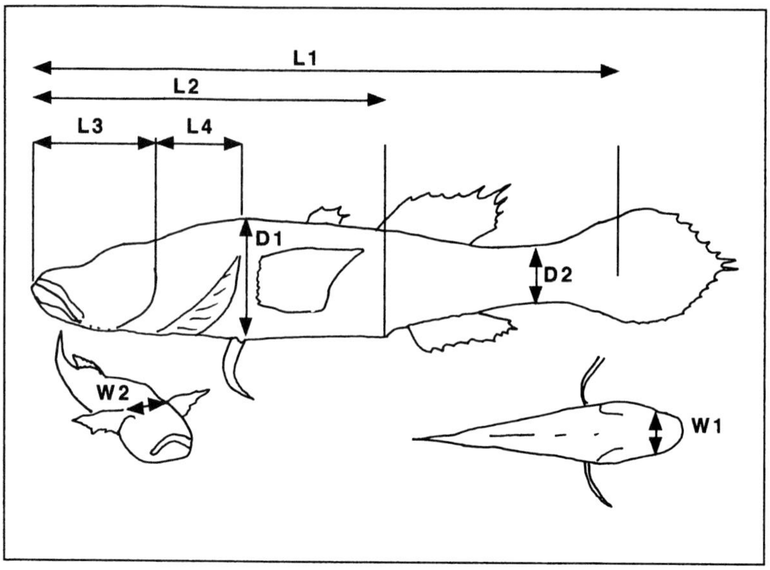

Figure 6. Morphometric measurements on *Milyeringa veritas* (see Table 3) from specimens in the Western Australian Museum.

Table 3. Morphometric relationships: regressions of linear measurements and weight $^{-3}$ of *Milyeringa veritas* ($N = 148$) summarized in the matrix of slope (upper right) and r^2 (lower left) of regressions of log X (vertical) on log Y (horizontal) of variables L1 = standard length (size range examined was 5–57 mm); L2 = snout vent length; L3 = length 3 (see Figure 6); L4 = length 4; W1 = width 1; W2 = width 2; D1 = depth 1; D2 = depth 2. The slope is shown only when the 95% confidence intervals for the slope do not include a slope of 1.0.

	Wt	L1	L2	L3	L4	D1	D2	W1	W2	L1–L2
Wt	—	0.95	—	0.90	—	—	1.04	—	0.96	0.93
L1	0.98	—	—	0.94	1.06	1.05	1.08	—	—	—
L2	0.97	0.98	—	0.91	—	—	—	—	—	0.92
L3	0.96	0.97	0.95	—	1.09	1.10	1.13	—	1.05	—
L4	0.97	0.95	0.95	0.91	—	0.94	—	0.91	0.91	0.87
D1	0.98	0.93	0.92	0.92	0.89	—	—	0.94	0.91	0.87
D2	0.98	0.94	0.93	0.93	0.90	0.92	—	0.90	0.89	0.85
W1	0.94	0.87	0.87	0.86	0.84	0.89	0.88	—	0.90	0.85
W2	0.96	0.96	0.95	0.94	0.94	0.93	0.94	0.90	—	0.93
L1–L2	0.90	0.94	0.85	0.89	0.85	0.86	0.86	0.79	0.93	—

M. veritas are genetically isolated from others, this may suggest that more than one biological species is present.

The genetic relationships amongst populations are displayed visually in Figure 7. Two major clusters are evident and these reflect a general 'western' versus 'eastern' geographic split. Populations C-27 and C-361 from the east coast of Cape Range peninsula are invariant at all loci examined, and display 5–7% fixed differences when compared to all but one of the populations on the west coast. The exception is the geographically intermediate well C-273 at the northern end of the peninsula, which is genetically intermediate between the two groupings. This within-species geographic trend broadly mimics the distribution of the two species of shrimp, which overlap in C-24, C-25, and C-273 (Humphreys & Adams 1991, Adams & Humphreys 1993). The gudgeons are the only group examined in which the F-statistics suggest there may not be random mating occurring within subpopulations

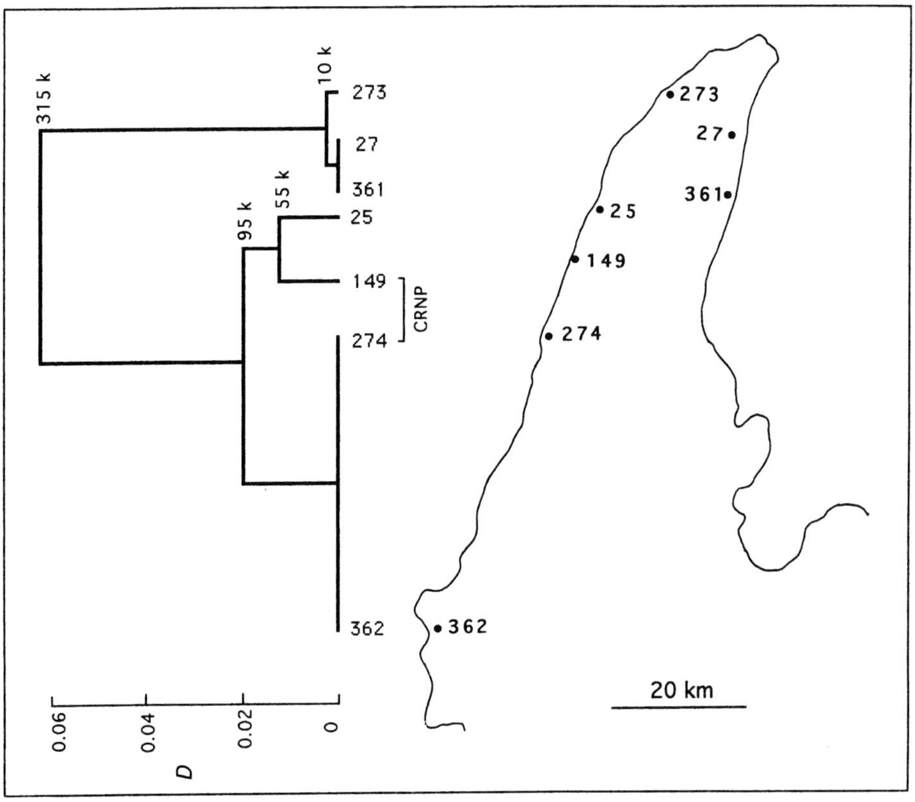

Figure 7. Dendrogram depicting the genetic relationship (Nei's *D*) amongst the seven populations of *Milyeringa veritas* (data from Adams & Humphreys 1993) plotted alongside a map depicting the position of the populations sampled. The numbers on the dendrogram are the molecular clock ages (thousand years) derived from Nei's relationship (Nei 1987) as discussed in the text. The numbers alongside the dendrogram and on the map denote the site numbers as recorded in the karst index for Cape Range. CRNP denotes the two populations sampled that are included within Cape Range National Park. Top of the map is north.

but it is not possible to explore the significance of this result any further without additional sampling.

Many cave organisms appear to conform to the molecular clock (Sbordoni et al. 1990) proposed by Nei (1987) in which one unit of Nei's *D* is equivalent to 5 million years of divergence between populations. Using this relationship, the estimated divergence times between the *M. veritas* populations (Figure 7) are concordant with major periods of low sea level, up to 130 m below that now present. Three of the last four extreme sea level minima occurred at about 95 000, 55 000 and 20 000 years ago (Wyrwoll et al. 1993), the estimated times of the three most recent divergences between the *M. veritas* populations. There was another major sea level minimum at ca. 315 000 years ago (Hays et al. 1976), coinciding with the estimated time of the initial divergence between the *M. veritas* populations. These data are consistent with the hypothesis that the regional genetic structure between *M. veritas* populations was determined by vicariance associated with eustatic events.

Conservation status

The remote location of these cave fish and their subterranean habitat restricts observation in life and hence very little is known about the biology of either sympatric species. Nonetheless, *M. veritas* and *O. candidum* have, respectively, been classified as rare and recommended for total protection, and vulnerable (Michaelis 1985). Both fish species (as well as some stygal crustacean species) have been listed as endangered under Schedule One of the Wildlife

Protection Act of Western Australia owing to their small geographic distribution, low populations and vulnerability of their habitat. Bundera Sinkhole and Camerons Cave, sites inhabited by *M. veritas*, are classified as endangered communities by the Western Australian wildlife authorities. The conservation status of these species is discussed by Romero & Vanselow (2000a,b).

While there has been a significant extension of the range of *M. veritas* to Barrow Island (Humphreys 1999c), the inclusive known range of the two species on the Cape Range peninsula has not increased since 1991 (Humphreys & Adams 1991). The number of sites from which *M. veritas* has been recorded has increased steadily over time, from two sites in 1962 to 25 sites in 1999 (Humphreys 1999c). These additional sites have markedly increased the known range of habitats and water types inhabited by the species. However, 19 of these sites are insecure or have been lost – six of these access sites have been lost in the intervening period due to infilling, drying or siltation of the sites, and a further 10 sites are close to planned developments or are within an urban area, and three are within a military bombing range reserve. Only six sites are within Cape Range National Park where infrastructure developments are also planned. *M. veritas* is certainly more widespread in the karst than indicated by the point samples. Despite this, the loss of access sites is crucially important as they provide the only windows into the karst system to enable the biological study and monitoring of cave fish populations and their associated ecosystems.

Owing to the genetic diversity and restricted gene flow in *M. veritas* on the peninsula (Humphreys & Adams 1991, Adams & Humphreys 1993), the presence of *M. veritas* within Cape Range National Park does not in itself provide an adequate representation of the genetic diversity of the species (Figure 7), nor does it adequately represent their habitat diversity.

The stygal community on the Cape Range peninsula is unusual in having two species of sympatric cave fish within a rich community of stygal Crustacea, many belonging to relict groups. As the fish and crustaceans are part of the same food web (Figure 4), the broad conservation implications are clear–the stygofauna is a functional community, changes to part of which may affect the remainder. Such relatively simple ecosystems are potentially vulnerable to disruption by 'top-down' effects (e.g. removal of *M. veritas* as the top predator), and *M. veritas* susceptible to 'bottom-up' effects (e.g. changes to biofilms by nutrients), factors, amongst others, discussed by Boulton (2000) for subsurface systems.

Prospects

Although information on *M. veritas* is sparse, and of variable quality, it is clear that this species inhabits waters with an extraordinary physico-chemical and biological variety. There are clear interruptions to the gene flow within the Cape Range peninsula that may reflect earlier eustatic events.

Few sites have been examined in respect of the allozyme variation in *M. veritas* and a more complete sampling may provide a more complex picture of the variation between local populations, and of the vicariance hypothesized to be associated with eustatic events. In addition, the non-random mating apparent from the allozyme data suggests that detailed population genetics using molecular markers would provide insights into the population processes of the cave fish.

Individual long term marking is needed to provide the basic population data required intrinsically and for possible management needs, such as growth rate, dispersal, dispersion, population size and structure, and breeding. This would also facilitate examination of the physico-chemical environment variation for a given population, which, with their physiological tolerance or ability to conform, are at present unknown.

The degree of ocular regression is unknown, as is the degree and nature of the stygomorphies. Finally, the affinities of the species within the Eleotridae are unknown, the resolution of which would stimulate research on the Australian cave fishes. It is now known that the populations of *M. veritas* are of sufficient size to sustain a carefully planned major research effort. Despite the general remoteness of the area, the populations occur at sufficiently accessible locations to facilitate sampling.

Acknowledgements

I acknowledge with pleasure the sustained technical support of Julianne Waldock, the voluntary assistance of Caroline Lawrence who measured the fish, and the cooperation over many years of Darren Brooks, Ken Cameron, Brian Vine and Keith Hallett. Graham Proudlove provided unpublished material and useful references. Part of the work derived from funding by the National Estate Grants Scheme, Australian National

Parks and Wildlife Service, Water Authority of Western Australia, the Water Corporation, West Australian Petroleum Pty. Ltd. and Ampolex Ltd.

References cited

Adams, M. & W.F. Humphreys. 1993. Patterns of genetic diversity within selected subterranean fauna of the Cape Range peninsula, Western Australia: systematic and biogeographic implications. Rec. West. Austr. Mus. (Suppl.) 45: 145–164.

Allen, A.D. 1993. Outline of the geology and hydrogeology of Cape Range, Carnarvon Basin, Western Australia. Rec. West. Austr. Mus. (Suppl.) 45: 25–38.

Allen, G.R. 1989. Freshwater fishes of Australia. T.F.H. Publications Inc., Neptune City. 240 pp.

ANCA. 1996. A directory of important wetlands in Australia, 2nd edn. Australian Nature Conservation Agency, Canberra. 478 pp.

Beard, J.S. 1975. Pilbara: vegetation survey of Western Australia 1: 1 000 000 vegetation series. Explanatory notes to sheet 5, the vegetation of the Pilbara area. University of Western Australia Press, Perth. 120 pp.

Beard, J.S. 1998. Position and development history of the central watershed of the Western Shield, Western Australia. J. Roy. Soc. West. Austr. 81: 157–164.

Bergstrom, D.E. 1997. The phylogeny and historical biogeography of Missouri's *Amblyopsis rosae* (Ozark cavefish) and *Typhlichthys subterraneus* (southern cavefish). M.Sc. Thesis, University of Missouri–Columbia, Columbia. 62 pp.

Boulton, A.J. 2000. The subsurface macrofauna. pp. 337–361. *In:* J. Jones & P. Mulholland (eds), Streams and Ground Waters, Academic Press, New York.

Chappell, J. & B.G. Thom. 1977. Sea levels and coasts. pp. 275–291. *In:* J. Allen, J. Golson & R. Jones (ed.) Sunda and Sakul: Prehistoric Studies in Southeast Asia, Melanesia & Australia, Academic Press, London.

Christiansen, K. 1962. Proposition for the classification of cave animals. Spelunca 2: 76–78.

Cohen, D.M. & J.E. McCosker. 1998. A new species of bythitid fish, genus *Lucifuga*, from the Galápagos Islands. Bull. Mar. Sci. 63: 179–187.

Cummins, K.W. & J.C. Wuycheck. 1971. Caloric equivalents for investigations in ecological energetics. Mitt. internat. Verein. Limnol. 18: 1–158.

Danielopol, D.L., A. Baltanás & W.F. Humphreys. 2000. *Danielopolina kornickeri* sp. n. (Ostracoda: Thaumatocypridoidea) from a western Australian anchialine cave – morphology and evolution. Zool. Scripta 29: 1–16.

Díaz Perez, P.A., A.M. Lima & E.G. Machado. 1987a. Morphologia externa de ejemplares machos de *Lucifuga simile* Nalbant, 1981 (Ophidiiformes, Bythitidae). Revista Biol. (Havana) 1: 77–84.

Díaz Perez, P.A., E. Nieto Misas & G. Abio Virsida. 1987b. Peces ciegos del genero *Lucifuga* (Ophidiiformes: Bythitidae) en dos casimbas cubanas. Rev. Invest. Mar. 8: 41–47.

Ford, D.C. & P.W. Williams. 1989. Karst geomorphology and hydrology. Unwin Hyman, London. 601 pp.

Gentilli, J. 1972. Australian climatic patterns. Nelson, Melbourne. 285 pp.

Gentilli, J. 1979. Epitropical westerly jet advective storms. Q. Geogr. J. 5: 1–20.

Harvey, M.S. & W.F. Humphreys. 1995. Notes on the genus *Draculoides* Harvey (Schizomida: Hubbardiidae), with the description of a new troglobitic species. Rec. West. Austr. Mus. (Suppl.) 52: 183–189.

Hays, J.D., J. Imbrie & N.J. Shackleton. 1976. Variations in the Earth's orbit: pacemaker of the ice ages. Science 194: 1121–1132.

Humphreys, W.F. 1993a. Stygofauna in semi-arid tropical Western Australia: a Tethyan connection? Mém. Biospéol. 20: 111–116.

Humphreys, W.F. 1993b. The significance of the subterranean fauna in biogeographical reconstruction: examples from Cape Range peninsula, Western Australia. Rec. West. Austr. Mus. (Suppl.) 45: 165–192.

Humphreys, W.F. 1999a. Relict stygofaunas living in sea salt, karst and calcrete habitats in arid northwestern Australia contain many ancient lineages. pp. 219–227. *In:* W. Ponder & D. Lunney (ed.) The Other 99%, The Conservation and Biodiversity of Invertebrates, Transactions of the Royal Zoological Society of New South Wales, Mosman.

Humphreys, W.F. 1999b. Physico-chemical profile and energy fixation in Bundera Sinkhole, an anchialine remiped habitat in north-western Australia. J. Roy. Soc. West. Austr. 82: 89–98.

Humphreys, W.F. 1999c. The distribution of the Australian cave fishes. Rec. West. Austr. Mus. 19: 469–472.

Humphreys, W.F. 2000a. The hypogean fauna of the Cape Range peninsula and Barrow Island, north-west Australia. pp. 581–601. *In:* H. Wilkens, D.C. Culver & W.F. Humphreys (ed.) Ecosystems of the World, Vol. 30, Subterranean Ecosystems, Elsevier, Amsterdam.

Humphreys, W.F. 2000b. Karst wetlands biodiversity and continuity through major climatic change – an example from arid tropical Western Australia. pp. 227–258. *In:* B. Gopal, W.J. Junk & J.A. Davis (ed.) Biodiversity in Wetlands: Assessment, Function and Conservation, Vol. 1, Backhuys Publishers, Leiden.

Humphreys, W.F. 2000c. Background and glossary. pp. 3–14. *In:* H. Wilkens, D.C. Culver & W.F. Humphreys (ed.) Ecosystems of the World, Vol. 30, Subterranean Ecosystems, Elsevier, Amsterdam.

Humphreys, W.F. & M. Adams. 1991. The subterranean aquatic fauna of the North West Cape peninsula, Western Australia. Rec. West. Austr. Mus. 15: 383–411.

Humphreys, W.F., M. Adams & B. Vine. 1989. The biology of *Schizomus vinei* (Chelicerata: Schizomida) in the caves of Cape Range, Western Australia. J. Zool. Lond. 217: 177–201.

Humphreys, W.F. & M.N. Feinberg. 1995. Food of the blind cave fishes of northwestern Australia. Rec. West. Austr. Mus. 17: 29–33.

Humphreys, W.F., A. Poole, S.M. Eberhard & D. Warren. 1999. Effects of research diving on the physico-chemical profile of Bundera Sinkhole, an anchialine remiped habitat at Cape Range, Western Australia. J. Roy. Soc. West. Austr. 82: 99–108.

Iliffe, T.M. 2000. Anchialine cave ecology. pp 59–76. *In:* H. Wilkens, D.C. Culver & W.F. Humphreys (ed.) Ecosystems of the World, Vol. 30, Subterranean Ecosystems, Elsevier, Amsterdam.

Jaume, D., G.A. Boxshall & W.F. Humphreys. 2001. New stygobiont copepods (Calanoida, Misophrioida) from Bundera Sinkhole, an anchialine cenote on north-western Australia. Zool. J. Linn. Soc., Lond. (in press).

Jaume, D. & W.F. Humphreys. 2001. A new genus of epacteriscid calanoid copepod from an anchialine sinkhole in northwestern Australia. J. Crust. Biol. 21: 157–169.

Keighery, G. & N. Gibson. 1993. Biogeography and composition of the flora of the Cape Range peninsula, Western Australia. Rec. West. Austr. Mus (Suppl.) 45: 51–85.

Knott, B. 1993. Stygofauna from Cape Range peninsula, Western Australia: tethyan relicts. Rec. West. Austr. Mus. (Suppl.) 45: 109–127.

Kuhajda, B.R. & R.L. Mayden. 2001. Status of the federally endangered Alabama cavefish, *Speoplatyrhinus poulsoni* (Amblyopsidae), in Key Cave and the surrounding caves, Alabama. Env. Biol. Fish. 62: 215–222 (this volume).

Levins, R. 1969. Evolution in changing environments. Princeton University Press, Princeton. 120 pp.

Martin, M.W. 1990. Exmouth town water supply investigation report and recommendations for future work. Hydrogeology Report No. 1990/36, Western Australian Geological Survey, Perth. 11 pp.

McNamara, K.J. (ed.), 1990. Evolutionary Trends. Belhaven Press, London. 368 pp.

McNamara, K.J. & G.W. Kendrick. 1994. Cenozoic molluscs and echinoids of Barrow Island, Western Australia. Rec. West. Austr. Mus. (Suppl.) 51: 1–50.

Mees, G.F. 1962. The subterranean fauna of Yardie Creek station, North West Cape, Western Australia. J. Roy. Soc. West. Austr. 45: 24–32.

Michaelis, F.B. 1985. Threatened fish. A report on the threatened fish of inland water of Australia. Australian National Parks and Wildlife Service, Report Series (3): 1–45.

Morse, K. 1993. Who can see the sea? Prehistoric aboriginal occupation of the Cape Range peninsula. Rec. West. Austr. Mus. (Suppl.) 45: 227–242.

Nei, M. 1987. Molecular Evolutionary Genetics. Columbia University Press, New York. 512 pp.

Nelson, J.S. 1984. Fishes of the World, 2nd edn. Wiley-Interscience, New York. 532 pp.

Nevo. E. 1978. Genetic variation in natural populations: patterns and theory. Theor. Pop. Biol. 13: 121–177.

Nevo, E., A. Beiles & R. Ben-Sholmo. 1984. The evolutionary significance of genetic diversity: ecological, demographic, and life history correlates. pp. 13–213. *In:* G.S. Mani (ed.) Lecture Notes in Biomathematics, Vol. 53, Springer-Verlag, Berlin.

Palmer, J. 1985. The blue holes of the Bahamas. Jonathon Cape, London. 185 pp.

Pesce, G.L., P. De Laurentiis & W.F. Humphreys. 1996. Copepods from ground waters of Western Australia, Part I, The genera *Metacyclops, Mesocyclops, Microcyclops* and *Apocyclops* (Crustacea Copepoda: Cyclopidae). Rec. West. Austr. Mus. 18: 67–76.

Planes, S. 1998. Genetic diversity and dispersal capabilities in marine fish. Evol. Biol. 30: 253–298.

Poore, G.C.B. & W.F. Humphreys. 1992. First record of Thermosbaenacea (Crustacea) from the Southern Hemisphere: a new species from a cave in tropical Western Australia. Invert. Taxon. 6: 719–725.

Poulson, T.L. 1964. Animals in aquatic environments: animals in caves. pp. 749–771. *In:* D.B. Dull (ed.) Handbook of Physiology, American Physiological Society, Washington.

Romero, A. & P.B.S. Vanselow. 2000a. Threatened fishes of the world: *Milyeringa veritas* Whitley, 1945 (Eleotridae). Env. Biol. Fish. 57: 36.

Romero, A. & P.B.S. Vanselow. 2000b. Threatened fishes of the world: *Ophisternon candidum* (Mees) (Synbranchidae). Env. Biol. Fish. 58: 214.

Sbordoni, V., A. Caccone, G. Allegrucci & D. Cesaroni. 1990. Molecular island biogeography. Atti Convegn. Lincei 85: 55–83.

Sket, B. 1981. Fauna of anchialine (coastal) cave waters, its origin and importance. pp. 646–647. *In:* B.F. Beck (ed.) Proceedings of the Eighth International Congress of Speleology, Bowling Green, Kentucky, Vol. 6, National Speleological Society, Huntsville.

Sket, B. 1986. Ecology of the mixohaline hypogean fauna along the Yugoslav coast. Stygologia 2: 317–338.

Sket, B. 1996. The ecology of anchihaline caves. Trends Ecol. Evol. 11: 221–255.

Slack-Smith, S.M. 1993. The non-marine molluscs of the Cape Range peninsula, Western Australia. Rec. West. Austr. Mus. (Suppl.) 45: 87–107.

Slatkin, M. 1981. Estimating level of gene flow in natural populations. Genetics 99: 323–335.

Slatkin, M. & N.H. Barton. 1989. A comparison of three indirect methods for estimating average levels of gene flow. Evolution 43: 1349–1368.

Soulé, M. 1976. Allozyme variation: its determinants in space and time. pp. 60–77. *In:* F.J. Ayala (ed.) Molecular Evolution, Sinauer Associates, Sunderland.

Stock, J.H., T.M. Iliffe & D. Williams. 1986. The concept 'anchialine' reconsidered. Stygologia 2: 90–92.

Thinès, G. & G. Proudlove. 1986. Pisces. pp. 709–733. *In:* L. Botosaneanu (ed.) Stygofauna Mundi, A Faunistic, Distributional, and Ecological Synthesis of the World Fauna Inhabiting Subterranean Waters, E.J. Brill/Dr. W. Backhuys, Leiden.

Trexler, J.C. 1988. Hierarchical organization of genetic variation in the sailfish molly, *Poecilia latipinna* (Pisces; Poeciliidae). Evolution 42. 1006–1017.

Van Valen, L. 1965. Morphological variation and width of ecological niche. Amer. Nat. 94: 377–390.

Weber, A., G.S. Proudlove & T.T. Nalbant. 1998a. Pisces (Teleostei). A. Morphology, systematic diversity, distribution, and ecology of stygobitic fishes. pp. 1179–1190. *In:* C. Juberthie & V. Decu (ed.) Encyclopaedia Biospeologica II, Société de Biospéologie, Moulis & Academie Roumaine, Bucarest.

Weber, A., G. S. Proudlove, J. Parzefall, H. Wilkens & T.T. Nalbant. 1998b. Pisces (Teleostei). pp. 1177–1213. *In:* C. Juberthie & V. Decu (ed.) Encyclopaedia Biospeologica II, Société de Biospéologie, Moulis & Academie Roumaine, Bucarest.

Whitley, G.P. 1945. New sharks and fishes from Western Australia, Part 2. Austr. Zool. 11: 35–37.

Wilkens, H. 1988. Evolution and genetics of epigean and cave *Astyanax fasciatus* (Characidae, Piscesae, Pisces): support for the neutral mutation theory. Evol. Biol. 23: 271–367.

Wilkens, H., U. Strecker & J. Yager. 1989. Eye reduction and phylogenetic age in ophidiform cave fish. Z. zool. Syst. Evolut. -forsch. 27: 126–134.

Wright. S. 1931. Evolution in Mendelian populations. Genetics 16: 97–159.

Wyrwoll, K.H., G.W. Kendrick & J.A. Long. 1993. The geomorphology and Late Cenozoic geological evolution of the Cape Range–Exmouth Gulf region. Rec. West. Austr. Mus. (Suppl.) 45: 1–23.

Yager, J. 1981. Remipedia, a new class of Crustacea from a marine cave in the Bahamas. J. Crust. Biol. 1: 328–333.

Yager, J. & W.F. Humphreys. 1996. *Lasionectes exleyi* sp. nov., the first remipede crustacean recorded from Australia and the Indian Ocean, with a key to the world species. Invert. Taxon. 10: 171–187.

Young, M. 1986. Bringing the blind gudgeon into captivity. Fish. Sahul 4: 145–148.

Environmental Biology of Fishes **62:** 315–337, 2001.
© 2001 *Kluwer Academic Publishers. Printed in the Netherlands.*

Mexican blindcats genus *Prietella* (Siluriformes: Ictaluridae): an overview of recent explorations

Dean A. Hendrickson[a], Jean K. Krejca[b] & Juan Manuel Rodríguez Martinez[c]
[a]*Texas Memorial Museum, Texas Natural History Collections, University of Texas, PRC 176/R4000, 10100 Burnet Road, Austin, TX 78758-4445, U.S.A. (e-mail: deanhend@mail.utexas.edu)*
[b]*Section of Integrative Biology, PAT 140/C0930, University of Texas, Austin, TX 78712, U.S.A.*
[c]*González Ortega 333 Ote., Complejo Habitacional La Finca, Edificio No. 4, Departamento 9, Monterrey, NL 64000, México*

Received 5 July 2000 Accepted 12 December 2000

Key words: cave fish, stygobite, fish conservation, distribution, behavior, habitat

Synopsis

The ictalurid genus *Prietella* was described from a single locality in northern México (Coahuila) in 1954, and until very recently went largely unstudied. Cave explorers have recently uncovered new localities and a second species much farther to the south (México: Tamaulipas). Our team visited over 50 sites, including all of the previously known sites possible, and explored many new sites, expanding the known range of *Prietella* and describing their habitat. We identified geological units and mapped caves, identified associated troglobitic invertebrates, estimated population sizes and measured water chemistry parameters. We also comment on laboratory diet, parasites, sensory biology, behavior (such as jaw locking and periods of inactivity), reproduction and systematics based on preliminary genetic data. *Prietella phreatophila* is listed as endangered, and due to the recent discovery of many more sites (formerly documented from three localities, now known from twelve sites, though some are hydrologically connected) we recommend threatened status, with careful attention to growing threats such as over pumping and contamination of the aquifer it lives in. Should these patterns continue unchecked, re-listing this species as endangered may be called for. *Prietella lundbergi* was also described from one site but is now known from two, though it is quite rare at both (only five specimens have ever been seen). *P. lundbergi* was described after the most recent revision of the Mexican endangered species list and should probably be considered as endangered.

Introduction

Twenty-three of the 85 known hypogean fishes are catfishes. Thus 0.96% of siluriforms are hypogean, a proportion surpassed at the ordinal level only by cypriniformes with 1.28% hypogean (based on Nelson 1994, Axel et al. 1999, and G. Proudlove personal communication). Among the eight catfish families with hypogean species, the exclusively North American Ictaluridae have an especially high proportion of hypogean representatives, with 3 (37.5%) of its eight genera and four (8%) of its 50 or so species being exclusively hypogean (Lundberg 1992). The four stygobitic ictalurids, *Prietella lundbergi*, *P. phreatophila*, *Satan eurystomus*, and *Trogloglanis pattersoni*, all live in North America's largest karst region (Carillo-Bravo 1971, Smith & Veni 1994, Fish 1977, Walsh 2000, Culver et al. 2000), the Cretaceous limestones surrounding the northwestern Gulf of México.

The distribution of *Prietella* now is known to extend from the very northern part of the state of Coahuila, México, very near the international boundary, to southernmost Tamaulipas, 750 km away to the southeast. In the 600 km between *P. lundbergi* and the closest populations of *P. phreatophila* at Múzquiz, Coahuila, are substantial mountain ranges, making hydrologic connections between the ranges of the two species, even ancient ones, seemingly quite improbable, and posing interesting questions about the biogeography and evolutionary history of the genus.

Across this great area, blindcat localities lie alongside and within the edges of the Sierra Madre Oriental, known for its deep, active, and extensive karst systems (e.g. Russell & Raines 1967, Lesser & Lesser 1988). Though much of the surface is desert, caves harboring *P. lundbergi* are surrounded by mesic, semi-tropical vegetation. Throughout the region, considerable portions of many drainages are subterranean, forming sinkholes, pits, cave passages, and springs that resurge along the bases of the mountains, and the entire region, including the karst of adjacent Texas, is well known for its rich subterranean fauna (Culver et al. 2000).

None of the hypogean ictalurids are well studied, but our recent explorations and laboratory observations provide new insights into ecology, biology and biogeography of the genus *Prietella*, and we are optimistic that our studies might shed light on regional hydrologic relationships among aquifers, something which is very little understood in this region. We review previous studies and collections of this genus, and relate our own explorations and collections, which have added nine new sites to the formerly documented three for *P. phreatophila* and one more to the range of *P. lundbergi*, previously known from only a single specimen. We also provide a general discussion of our laboratory observations and conclude with a discussion of the conservation status of the two species in this genus.

Previous collections

Prietella phreatophila was described by Carranza in 1954, based on collections made in the same year at El Potrero de Doña Mariana, near Múzquiz, Coahuila. His specimens all came from this single spring cave locality, though he reported that locals advised that the species occurred elsewhere in the immediate area. The type series consisted of sixty-six specimens, of which Carranza (1954) indicated that eighteen were deposited as paratypes in major collections, three each to IPN, BMNH, MNHN, USNM, UMMZ, and Stanford (now CAS) (Museum symbolic codes follow Leviton et al. 1985 and Leviton & Gibbs 1988). Forty-six specimens, including the holotype, were indicated to have been kept in his personal collection, and two were given to co-collector, C. Bolivar. Our inquiries verified that all paratypes except those at BMNH and MNHN are extant, and it would appear that specimens perhaps intended for the two European collections were later given instead to Tulane [TU 10810, three specimens and TU (OC) 178, one skeleton] (Bart & Taylor 1993) and Universidad Autónoma de Nuevo León (UANL 1014, three specimens). Carranza later deposited what he referred to as the remaining eleven specimens (paratypes) in his possession, plus the holotype, at the Instituto Biológico of the Universidad Nacional Autónoma de México (IBUNAM-P 97, six paratypes; IBUNAM-P 8833 holotype) (H. Espinosa personal communication, IBUNAM, 1995) and UMMZ (223109, five paratypes).

Subsequent collections of specimens from the type locality were made in 1961 (TU 43872, 43879, 84464), 1969 (UMMZ 187684), 1984 (UANL uncatalogued), and 1989 (UANL 9009 (from 'near Múzquiz'). In 1984, local residents' contention that the species also could be found at El Socavón, 5.15 km west of the plaza in Melchor Múzquiz, was confirmed by the capture of a single specimen (TCWC 5144.01) (Amemiya et al. 1986).

The distribution of the species was extended further north with the 1986 collection of a single specimen (TNHC 12100) from 48 km NW of Múzquiz, and a 1992 collection of nine specimens (IBUNAM 7557) from Sótano de Amezcua, about 180 km N of Múzquiz and 50 km W of Ciudad Acuña, Coahuila, just across the Río Grande from Del Río, Texas. It was these specimens, delivered live to the first author, which initiated our studies. A specimen at UANL documenting an eastern extension of the range, to near the town of Morelos, Coahuila, about 80 km NE of Múzquiz, was apparently lost (S. Contreras-Balderas personal communication).

As the known distribution of *P. phreatophila* was being extended to the north, a second species of the genus was being analyzed following the 1989 collection of a single specimen far to the south by a recreational cave diver, Tom Morris. Based on this, the only specimen available, Walsh and Gilbert (1995) described *P. lundbergi* from the small spring cave at San Rafael de los Castro, 15 km WNW of Ciudad Mante, Tamaulipas.

The fact that *P. lundbergi* had not been reported before was somewhat surprising. The area had been very thoroughly explored for cave fishes for years by the many biologists and cavers studying what has become probably the most studied cave fish in the world, *Astyanax fasciatus*, in the plethora of caves in Tamaulipas (e.g. Mitchell & Elliott 1977). Caves with populations of blind *Astyanax* are located within 20 km to both north and south of San Rafael de los Castro (Mitchell & Elliott 1977, Borowsky 1996). Additionally, cave divers seeking SCUBA depth records have spent probably hundreds of man hours in a large, hydrologically and morphologically very similar spring, Nacimiento del Río Mante, only 8 km south of San Rafael de los Castro, and thoroughly explored other promising cave dives in the area (Exley 1979), without ever reporting blindcats.

Recent explorations

Since 1993, working with a variable, but always highly trained team of cave explorers and cave divers, we have searched for *Prietella* and associated stygobitic invertebrates throughout the known, and what we believe to be the potential, range of the genus in Tamaulipas, Nuevo León and Coahuila, México, as well as in the adjacent south Texas border area, United States. All localities explored by our team are listed in Table 1.

The first author first collected *Prietella phreatophila* at Sótano de Amezcua in 1993, about one year after the species was discovered there by cavers from Austin, Texas, who brought live specimens to him. About the same time, news of discovery of *P. lundbergi* reached us, and the first author took cave divers to the type locality for this species in April 1994. Two divers spent approximately two hours in the cave, penetrating an estimated 213 m horizontally and 40 m maximum depth, collecting invertebrates, but observing no cave fishes. On the same expedition, divers also explored the previously little-explored large submerged caves of San Felipe springs in Del Rio, Texas (locally rumored to have blind fishes and not far from Sótano de Amezcua) and the Nacimiento del Rio Sabinas, not too far north of the known distribution of *P. lundbergi*, failing to find cave fishes at both sites. This expedition met with misfortune in the form of a nearly disastrous vehicle accident, forcing us to abandon additional collection plans extending throughout much of Coahuila.

In 1996, the third author, a geohydrologist previously well acquainted with regional aquifers and springs, having conducted water supply and well surveys throughout the region for many years, spent a total of about two weeks on several trips with the first author, searching for potential blindcat habitats in Coahuila and Nuevo León. In the same year, the first author made several additional trips with the same objectives, either alone or with small crews. All trips focused on mapped springs, caves and sinks and surrounding areas and relied heavily on interviews with local residents. Several potential new localities requiring vertical descents and SCUBA were located during these surveys, and landowner permissions were obtained in preparation for a major expedition in 1997, which included participation of crews specifically trained in both vertical cave exploration and cave diving. Since then, the authors and collaborating cave divers and cavers have continued to make occasional, shorter (3–5 day) exploration of selected areas of southern Texas, northern Coahuila, Nuevo León and Tamaulipas. Results of all of these expeditions are discussed, site by site, below.

Methods

Methods used to explore and sample subterranean habitats in the range of *Prietella* are often extremely technical and require special training in vertical cave exploration techniques and cave diving (Padgett & Smith 1987, Prosser & Grey 1992). All collecting was by hand using a small plastic vial during a dive, or by dipnets deployed from banks. Water chemistry data were taken by a Hydrolab Data Sonde 3 Multiprobe Logger. Cave mapping was done using standard techniques (Dasher 1994).

In order to facilitate a broad range of research, specimens were taken alive to the laboratory, frozen in the field in liquid nitrogen, or preserved in 10% buffered formalin or 95% ethanol. All fish specimens collected by us are catalogued and housed in the Texas Natural History Collection (TNHC) of the Texas Memorial Museum at the University of Texas at Austin,[1] though part of our collections will be transferred in the near future to Mexican collections.

Unless otherwise noted, each site was visited by our team only a single time. Three sites are considered here as proven *P. phreatophila* localities though museum specimens do not exist. We do this on the basis of extremely convincing accounts of presence of blindcats as related to us by local residents who have perfectly described the species. Additionally, their reports of occasional sightings of blindcats after floods are congruent with our expectations that they occasionally are pushed out of their deeper, more inaccessible habitat during extreme spates.

[1] (http://www.utexas.edu/depts/tnhc/)

Table 1. List of localities with word descriptions. The numbers correspond to Figure 1, and the site names correspond to descriptions in text. Italicized sites are documented blindcat sites (asterisks indicate no museum specimens are available, as they have only been reported from these sites – see text in Methods section for explanation). Abbreviations: Cd. = Ciudad; Co. = County; Coah. = Coahuila; Mex. = México; Tamps. = Tamaulipas; Tx. = Texas. When descriptions of localities are within 15 km of a town, the distance is given from the center of the town.

1.	Dandridge Spring and Richter Cave	40 km SW Sonora, Sutton Co., Tx.
2.	Devil's Sinkhole	10 km NE Rocksprings, Edwards Co., Tx.
3.	Big Tree Cave and Emerald Sink	5 km NNE Langtry, Val Verde Co., Tx.
4.	Emilio's Cavern	Comstock, Val Verde Co., Tx.
5.	Goodenough Springs	38 km NW of Del Rio, Val Verde Co., Tx.
6.	San Felipe Springs	Del Rio, Val Verde Co., Tx.
7.	*Sótano de Amezcua* and La Rajada	40 km W of Ciudad Acuña, Coah., Mex.
8.	*Noria de San Pedro**	32 km WSW of Santa Eulalia, Coah., Mex.
9.	La Vinata Well	18 km SW of Santa Eulalia, Coah., Mex.
10.	Cueva de Rancho Las Pilas	90 km SW of Ciudad Acuña, Coah., Mex.
11.	El Abra and Tinaja Azul	75 km W of Piedras Negras, Coah., Mex.
12.	*Ojo del Yermo*	60 km N of Melchor Múzquiz, Coah., Mex.
13.	Poza San Miguel	35 km NW of Allende, Coah., Mex.
14.	*El Consuelo**	16 km NW of Allende, Coah., Mex.
15.	*La Tembladora**	5 km SW of Allende, Coah., Mex.
16.	Nacimiento Kikapoo and Falcón well	33 km NW of Melchor Múzquiz, Coah., Mex.
17.	*Cueva de Juana*	22 km W of Melchor Múzquiz, Coah., Mex.
18.	*El Socavón area: El Socavón, above El Socavón 1 and 2, El Cedral*	5 km SW of Melchor Múzquiz, Coah., Mex.
19.	*El Potrero area: 1 and 2*, Tiro Palmito, and vertical mine shaft near El Potrero	8 km SE of Melchor Múzquiz, Coah., Mex.
20.	Cueva Cabrito	9 km WSW of Estación Obayos, Coah., Mex.
21.	Cueva La Zumbadora	4 km NW of La Madrid, Coah., Mex.
22.	Cueva El Venado	El Venado, 22 km SE of Cuatro Ciénegas, Coah., Mex.
23.	Mojarral East and West, La Campana	15 km SSE of Cuatro Ciénegas, Coah., Mex.
24.	Gruta de Carrizal	Candela, Coah., Mex.
25.	Cueva de El Tule and Cueva de la Espantosa	Lampazos, Nuevo Leon, Mex.
26.	El Ebanito	15 km SW of Linares, Nuevo León, Mex.
27.	Purificacion area Caves	30 km NW of Cd. Victoria, Nuevo León and Tamps., Mex.
28.	El Sótano	3 km ENE of El Carrizo, Tamps., Mex.
29.	Ojo Encantado	30 km W of Cd. Victoria, Tamps., Mex.
30.	Manantial La Penita and Cueva del Manantial La Penita	5 km W of Cd. Victoria, Tamps., Mex.
31.	Guayalejo Spring	Juamave, Tamps., Mex.
32.	*Cueva del Nacimiento del Río Frío* and Nacimiento del Río Frío	32 km NW of Cd. Mante, Tamps, Mex.
33.	Springs near San Rafael de los Castro	San Rafael de los Castro, Tamps., Mex.
34.	*Manantial de San Rafael de los Castro* and Cueva del Manantial de San Rafael de los Castro	11.5 km W of Cd. Mante, Tamps., Mex.

Descriptions of *Prietella* habitats

Locations of wells, springs, caves, and mines we have explored are mapped in Figure 1 and listed in Table 1. Access varies from extremely remote and requiring pack animals and/or long hikes on foot over rugged terrain, and often technical vertical ascent and descent techniques as well as cave diving, to others to which one can easily drive within meters and simply walk in. Site descriptions begin with passage configuration and geological information, then continue with a habitat and biota section that describes exactly where at each site blindcats were found, including microhabitat and biotic associations. Water chemistry parameters are given when available.

Sites listed in Table 1 are only those visited by our team. Many neighboring sites harboring aquatic cave organisms, including habitats of the best-studied

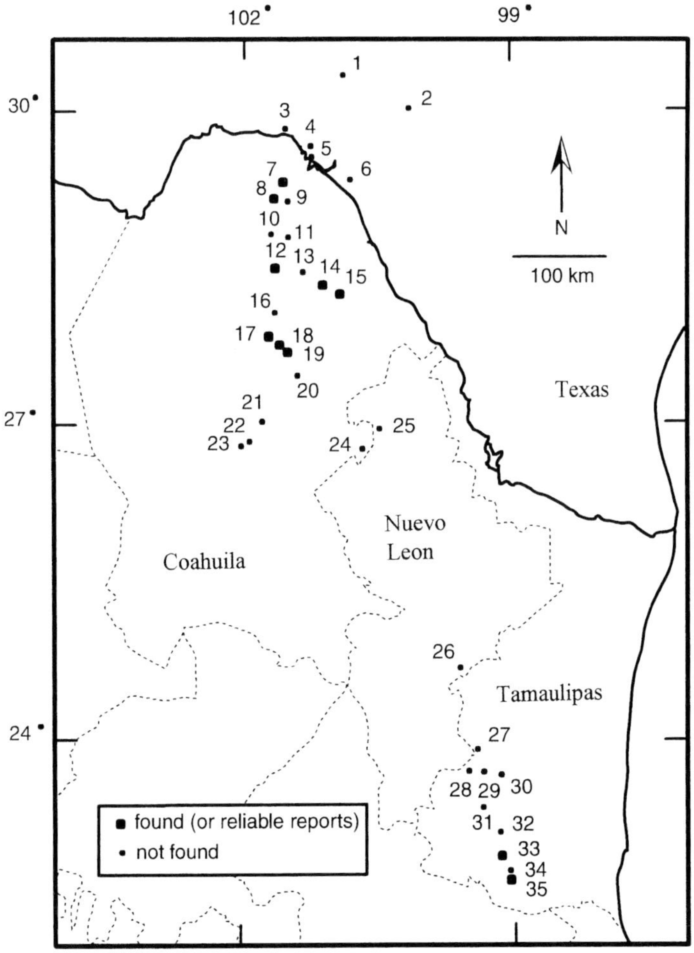

Figure 1. Map of south Texas and Northeastern México showing *Prietella* localities, also summarized in Table 1.

cave fish, *Astyanax fasciatus* (Mitchell et al. 1977), have been thoroughly collected by others, but *Prietella* was not reported. It is likely that most of these other aquatic cave sites are not blindcat sites, since presumably a biologist collecting a stygobitic crustacean or *Astyanax* would also report blindcats if seen or taken. We have not included any of these sites as 'negatives', however, since in our experience for at least some sites, specimens were not at the surface and collections required cave diving. Few caves in the range of *Prietella* have been explored by divers.

While the configuration and genesis of each site is variable, habitats of the blindcats share many characteristics. Most specimens were taken well into the dark zone, in habitats that likely have relatively little short-term fluctuation in environmental parameters. For the most part blindcats are found in still pools; they have never been seen in shallow running water, but are sometimes in high-flow systems such as El Socavón. Often they are associated with a silt substrate, but this may be simply a correlate of preference for still water. All documented blindcat localities are at relatively low elevations as compared to surrounding, and often adjacent mountain ranges. Careful observations made in the Sistema Purificación area, and in higher altitude areas of Coahuila (e.g. Cueva la Zumbadora) failed to find them. Water chemistry ranges we measured in waters occupied by blindcats sites are: temperature $= 21 - 31.5$°C, pH $= 7.45 - 7.9$, dissolved oxygen $= 0.75 - 5.4\,mg\,l^{-1}$, specific conductivity $= 0.425 - 0.75\,mS\,cm^{-1}$, redox potential $= 335 - 362\,mV$.

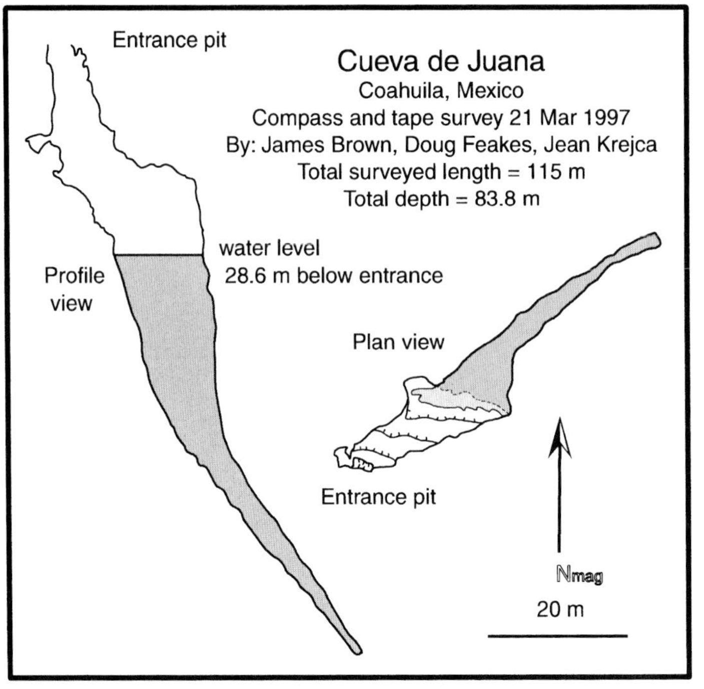

Figure 2. Map of Cueva de Juana.

Prietella phreatophila localities

Included in this list are all sites from which specimens have been collected and deposited in museum collections, as well as three sites (marked by asterisks) where we consider there to be extremely convincing reports of presence of the species.

Cueva de Juana

This cave is formed in lower Cretaceous limestone of the Aurora formation. The entrance is a 3 m diameter vertical shaft that enlarges and continues nearly vertically to the water surface, which was 30 m below the entrance during our visit. The shaft continues underwater 2–12 m in diameter at a nearly vertical angle to a water depth of 55 m (85 m below the entrance). The underwater passage consists of a complicated maze of connecting vertical shafts where great care should be taken in navigation (Figure 2). The walls and ledges are bedrock, with some silt. No flow was detected and ceiling debris was quickly dislodged by divers' bubbles. Blindcats were found in the shallower areas (20 m or less), possibly due to the fact that visibility was quickly reduced at depth. Also because most of the food input is probably from entrance debris, this shallower area may be where the fish forage. Water chemistry was essentially homogeneous from the water surface to a depth of 45 m as follows: temperature = 21°C, pH = 7.81, dissolved oxygen = 5.4 mg l^{-1} (the highest recorded for any blindcat localities), specific conductivity = 0.425 mS cm^{-1} (the lowest recorded for any blindcat localities), redox potential = 335 mV.

El Cedral

This site is located in the barite and zinc mining area of El Potrero 1 and 2 (below), in the lower Cretaceous limestone and barite of the Cupido formation (= Sligo formation in Texas). Connections among these sites (El Cedral and El Potrero 1 and 2) are not documented, but possible given their relative proximity. The entrance to this site is a 3.5 m diameter concrete-walled vertical shaft located in a building that has been used as a chlorination tank for water that is currently used in a coal-washing plant in nearby Palau. A ladder descends approximately 20 m to the water surface. Water depth in the tank was 6 m during our visit, with a horizontal conduit at 5 m in depth and 1–2 m in diameter through which water flows into the tank. The horizontal conduit can be followed for 100 m, at the end of which it intersects a natural cave passage. All of the blindcats from this site were taken from the side passage that has walls of concrete, metal, and excavated rock. No flow was noticeable. In March of 1997 it was reported to not have been used for chlorination for about one year.

**El Consuelo*

This is a hand-dug, vertical well round in cross section and approximately 10 m deep and 1 m in diameter. At the time of our visit, the water level was about 6 m below ground level, and accesible submerged passage continued another 3 m or so. The well's walls are shored up with stones, and a few meters below the stone work an inaccessible natural cavity, about 0.3 m in diameter, can be seen continuing downward, with no noticeable flow. The natural cavity had bedrock walls and ledges with a layer of silt. Some organic debris had fallen in from the well entrance. The owner reports blindcats used to be captured in well buckets, but since a pump was installed about 30 years ago, no specimens have been seen. On one occasion in the 1950s lime was added to the well in an attempt to sterilize it after a goat fell in and drowned. Large numbers of dead blindcats were seen shortly afterward. Our dive here yielded only the collection of a single, apparently troglobitic snail and a toad.

El Potrero area

This is a region of barite and zinc mines in lower Cretaceous limestone and barite of the Cupido formation (= Sligo formation in Texas). What we call El Potrero 1 and El Potrero 2 are within 100 m of one another. Hydrologic connections seem likely, but differences in water chemistry indicate perhaps otherwise, at least during low-flow periods.

El Potrero 1: This is the type locality for *P. phreatophila* (Carranza 1954). The entrance has been modified since Carranza's visit in 1954, being now covered by large steel doors adjacent to a steel shed and windmill. The first 5 m of vertical shaft are walled in concrete (3 m in diameter, matching Carranza's description) and ladders descend to water, approximately 7 m from the ground surface at the time of the team's visit. Earlier inspections by the first author found water level to vary as much as 3–4 m above, to several meters below, this level. Carranza described the water level 2.5 m below the surface in 1954. Several electrical pumps are now installed below water level in the modified area, being moved regularly up and down as water levels fluctuate. The pumps typically reside in a natural crevice that continues downward from the modified area.

Carranza (1954) based his description of the species on 66 specimens taken from here using traps baited with decomposing frog legs. We did not find any fishes in two dives during the team's visit, but the first author had previously left traps with the local caretaker, who collected two specimens of *P. phreatophila* by baiting traps with commercial pet foods and locally caught insects. Local residents report seeing blindcats after floods, sometimes even outside of the cave following extremely high discharges. At no time, however, did we see blindcats so abundant that we could ever collect anywhere near 66 specimens in two days, as Carranza apparently did, nor do local residents indicate that they are ever so abundant.

The substrate is bedrock and some breakdown, covered by fine silt. Water flow has been unnoticeable during all of our visits, and we did not detect any hint of the relatively intense cross-current described by Carranza as being 2 m from the 'pozo'. As mentioned above, however, the system occasionally has large discharges to the surface, and clearly at such times, considerable currents. Visibility was initially clear, but quickly silted with divers in the narrow passage. A constriction at 8 m in water depth was passed to an ultimate depth of 14 m. The passage was seen to continue downwards from here, but visibility was too poor to continue the descent.

Carranza (1954) reported temperatures ranging here from 27–31.0°C, pH = 7, and total hardness = 366 ppm $CaCO_3$. Our single sample taken during the team's visit from a shallow point was temperature = 31.12°C, pH = 7.73, dissolved oxygen = 1.58 mg l^{-1}, specific conductivity = 0.72 mS cm^{-1}, redox potential = 345.8 mV (Botosaneanu et al. 1998). A temperature data logger recording at two hour intervals from 20 March to 15 June 1997 recorded a range of 27.25–31.5°C (the highest we recorded at any blindcat site).

Water level now seems to be consistently lower than described by Carranza, but historic data on water levels other than that from Carranza's description are not available. Carranza mentioned that the site was pumped during times of water scarcity to supply the town of Nueva Rosita. Water from here is now used on a more consistent, but not continual, basis to supply a large coal washing plant in Palau. Aquatic organisms collected here and at El Potrero 2 include snails and the troglobitic cirolanid isopod, *Speocirolana thermydronis* (Cole & Minckley 1966), an interesting range extension from the only previously known localities in the Cuatro Ciénegas basin (Botosaneanu et al. 1998).

Local residents report periodic massive surface discharges from this cave during heavy rains and continuing afterward for a day or two. At times, specimens of *P. phreatophila* are said to be observed in surface pools downstream in the normally dry arroyo that receives the cave's discharge. We can substantiate the flooding claim since subsequent to the 15 June 1997 retrieval of data from the temperature data logger mentioned above, a large discharge from El Potrero 2 (below) removed the device from that cave. It had been tethered with heavy nylon cord to a stone weighing 2–3 kg, indicating considerable hydraulic scour of that system. Locals reported that both Potrero 1 and 2 flowed similarly during that event.

El Potrero 2: This may be the site that Carranza refers to as being 30 m distant from, and 1.5°C warmer than, the main pozo (which we believe to be our El Potrero 1), on the two days of his visit. While we did not find another pozo as near as 30 m to El Potrero 1, this modified natural cave is approximately 100 m from that pozo, on the far bank of the arroyo at the base of a small cliff. The entrance has been altered by shoring it up with concrete and covering it with a hinged steel lift door that is typically padlocked. A date in the concrete indicates the construction was done in 1952, two years prior to Carranza's visit. Once inside, a ladder leads down about 4 m to the floor of a 5 m-diameter room, filled with about 1 m of water at the time of our team visit. A 4 m crawlway just above the water surface led to another room where the passage continued underwater. This narrow passage was explored to a depth and penetration of approximately 10 m, but was seen to continue.

The passage is 1 m wide by 3 m long, with bedrock walls and ledges. Isolated small accumulations of silt and surface debris, such as twigs and sticks, were noted at intervals. In addition to specimens of blindcats taken during the team's visit, the same invertebrates taken in El Potrero 1 were collected here. Invertebrates were taken near or on the sticks, and blindcats were noted to be in cooler water, also near these piles of organic debris. Two water masses differing in water quality were noted in the cave, a surface layer (temperature = 29.15°C, pH = 7.92, the highest recorded for any blindcat locality, dissolved oxygen = 3.48 mg l^{-1}, specific conductivity = 0.72 mS cm^{-1}, redox potential = 344.18 mV extending to 3.6 m depth) and a transition layer (temperature = 28.25°C, pH = 7.79, dissolved oxygen = 1.76 mg l^{-1}; conductance and redox = upper layer) from 3.7 to at least 5.4 m or more (Botosaneanu et al. 1998). The difference in water chemistry parameters between Potrero 1 and 2 indicate at least very limited or perhaps distant connections between the two water bodies, at least during periods of normal water levels.

El Socavón area

Though Carranza (1954) failed to collect blindcats at any site other than El Potrero, he visited this site and mentioned that local residents reported the presence of blindcats here. We explored three sites in close proximity: El Socavón, and two apparently un-named local sites that we named 'above El Socavón 1' and 'above El Socavón 2', all of them in lower Cretaceous limestone of the Aurora formation. The first two are clearly hydrologically linked, as indicated by historic accounts, and corroborated by our explorations. According to an unpublished report by Rodriguez Guerra on file in the Múzquiz municipal water supply office, in 1888–1890 a long, nearly horizontal passage (El Socavón) was excavated in an attempt to de-water a flooded lead and silver mine, Mina de San Juan. The workers intersected a large water-filled chamber before reaching the mine, thus flooding the Socavón with 2000 l s^{-1} stopping the project. We named what we believe to be the point at which the Socavón hit the natural passage, 'above El Socavón 1.' The site can now be accessed from the surface. Water at above El Socavón 1 is now permanently diverted through the Socavón tunnel to surface at what is now known locally as 'El Socavón'. Though the connection between El Socavón and above El Socavón 1 has never been proven (as no one lives to tell exactly where, from the mine side, the breakthrough occurred, and no one has traversed the entire, now-submerged passage), our underwater survey data, a previous survey deposited in the municipal water supply office in the plaza in Múzquiz, combined with the striking similarity of water chemistry data (below) strongly corroborate such a connection.

El Socavón: This is the downstream outlet of water from above El Socavón 1, and is now the municipal water supply for the city of Melchor Múzquiz. The large pool just beyond the entrance to the flooded tunnel is covered by a concrete, steel-roofed building (ca. 10 m × 10 m × 3 m high) from which water is diverted. The submerged tunnel is about 2 m in diameter and continues upstream for about 130 m to a breakdown constriction which has been passed by divers using streamlined equipment to counter the high velocity flow (recorded at 200 l s^{-1} in a 1981 unpublished report by Rodriguez Guerra on file in Múzquiz municipal water supply office).

The floor of the passage is rock and breakdown. Blindcats have been taken from the building and city's diversion structure in the past (TCWC 5144-1) and we collected them from the submerged passage upstream of the building and downstream of the breakdown constriction. Water chemistry data (temperature = 26.75°C, pH = 7.77, dissolved oxygen = 4.51 mg l^{-1}, specific conductivity = 0.64 mS cm^{-1}, redox potential = 350.63 mV) are taken from a single point, as they are expected to be homogenous in this high flow system. A native surface catfish, *Ictalurus lupus*, was taken far into the passage, syntopic with *Prietella*.

Above El Socavón 1: This is apparently the natural spring or cavity that was accidentally diverted to El Socavón in 1890. Two entrances are within 5 m of each other. One is shored up with stonework, and a vertical drop of about 10 m leads directly into a pool of water. The second leads to a short climb down and vertical drop to the same pool, which, at the time of our visit, filled a 4 m diameter passage, penetrating 13 m to a depth of 9 m. In the wall of this passage, about 1 m from the water surface, the drain hole that originally flooded El Socavón was obvious. Through this small drain hole, which is now partially braced with timbers, a diver can look downstream to see that the water emerges on the other side as a waterfall into air-filled passage. Locals report that at times of extremely high discharge, water still flows from this spring, indicating that discharge of the system occasionally exceeds the hydraulic capacity of the drain hole.

The floor of the passage consists of organic debris fallen from the entrance covering cobbles from which a moderate flow emanates. Blindcats were collected from an adjacent silt bank. Water chemistry is essentially homogenous, as expected given the large discharge: temperature = 26.78°C, pH = 7.77, dissolved oxygen = 2.77 mg l^{-1}, specific conductivity = 0.64 mS cm^{-1}, redox potential = 348.69 mV. Differences in dissolved oxygen between here and El Socavón are likely due to the aeration provided by the intervening drain hole.

Above El Socavón 2: This is a flooded mine several hundred meters from El Socavón 1, and it is unknown how it is hydrologically linked to the other El Socavón area sites. It may be part of the old 'Mina de San Juan' mentioned by Rodriguez Guerra (unpublished data). One of our divers experienced a serious passage collapse while diving. We advise that future explorers exercise considerable caution in this site.

Silt-bottomed tunnel with abundant, scattered timbers fallen from the roof. Our brief exploration produced both a single blindcat specimen and troglobitic amphipods.

**La Tembladora*

Two hand-dug, vertical wells, 1–2 m in diameter, and less than 10 m deep have walls of un-mortared rockwork descending to intersect natural bedrock joints filled with water which has been about 1 m deep at the time of our visits. The joints, less than 0.3 m wide, and too small to be explored, can be seen continuing horizontally. A large spring is a few kilometers distant. The natural joints have bedrock walls with silt-covered floors and ledges. Flow was unnoticeable. Though no specimens have been collected here, the well owners provided reliable reports of blindcats being captured incidentally in their well buckets, and report that at one time they held two specimens alive for us for a week or more before releasing them.

Ojo del Yermo

The entrance to this cave is described as a 45 m vertical shaft that intersects a horizontal stream channel. At one end of the channel is a debris constriction followed by a small sump. At the other end of the stream channel is a lake with large diameter submerged passage explored to 46 m depth and seen to continue. Water level is reported to be highly variable. At least once water was partway up the shaft when a diver attempting to dive to the intersecting horizontal passage encountered swift, life-threatening currents. Descriptions indicate large diameter passage that in some places has considerable flood debris, and water that is extremely clear and deep. Explorers report abundant blindcats in the lake and its submerged passage. The smaller sump beyond the debris constriction reportedly had no blindcats. A single specimen of *P. phreatophila* (TNHC 12100) was collected here in 1986 by divers who provided the site description we relate above, but we have not been allowed access by the landowner.

**Noria de San Pedro*

This well, built in 1918, was originally a narrow joint through which water would sometimes flow to the surface. The joint was manually enlarged and shored up with rocks in a square cross section to a depth of approximately 25 m from the surface. Now it is about 1.5 m in diameter and can be easily entered with a rope. At the bottom is a silt ledge from which a natural water-filled crack 2 m long and 0.5 m wide can be seen. Dives were attempted on two different occasions, once getting to a depth of 10 m, but both were aborted due to the narrow passage configuration and heavy siltation problems. According to locals, water sometimes flows out of the top of the well for weeks. What can be seen from the silt ledge at the bottom of the well is a vertically oriented water-filled crevice, 0.5 m wide by 2 m long, extending as far as we could see (using masks, fins, and dive lights) through very clear water. After an initial constriction just underneath the water surface, the passage widens to more than 1 m and appears to continue to enlarge. Walls are bedrock with ledges covered with silt. Locals report that blindcats and other surface fish have shown up in the well bucket during or after rain events that raise the level of the water in the well, or even make it flow out over the surface. Though blindcats have never been collected here, the combination of convincing accounts from locals and proximity to Sótano de Amezcua lead us to include this as a known site. A single cirolanid isopod was taken from here. Water chemistry data are: temperature = 23.99°C, pH = 7.14 and dissolved oxygen = 7.01 mg l^{-1}.

Sótano de Amezcua

The entrance to this cave is in the bottom of a 20 m deep sinkhole in the Segovia Limestone. A 70 m vertical shaft leads into a large chamber with a rocky bottom stream crossing alongside. The stream passage can be explored both up and downstream. Upstream the air-filled passage terminates quickly in a sump pool ('catfish parlour'), while the downstream passage can be explored approximately 200 m through a crawlway (= BBB Lake) that intersects canyon passage leading to the downstream sump. Both sumps have been dove, but only the upstream one successfully traversed by divers, who found it to continue at shallow depths (5–10 m) to connect after about 100 m to air-filled canyon passage which again sumps (thanksgiving sump) and re-emerges in an air-filled chamber before sumping a third time (Figure 3). Blindcats were discovered here in 1992, and many visits have been made since, both by members of our team and others. This, plus two years of continual temperature data, extensive SCUBA explorations and thorough cave survey, as well as an ongoing capture-mark-recapture study of the blindcats, make this now the most-studied *Prietella* locality.

Blindcats in Sótano de Amezcua have been taken from silt-bottomed pools while aquatic crustaceans have been collected in riffles of the downstream section of cave, particularly those that lack fish and have bits of organic debris such as twigs and sticks. The aquatic fauna includes at least two families of isopods, Cirolanidae: *Cirolanides texensis* (identified by Botosaneanu, Iliffe personal communication), Asselidae: *Lirceolus* (Reddell personal communication), and amphipods (Botosaneanu et al. 1998).

Upstream: The first sump pool encountered in the upstream direction from the entrance, catfish parlour, is the most accessible and apparently permanent blindcat habitat. The pool is 2 m × 4 m × 0.5 m deep, and silt floored. Blindcats have been observed here on every visit. Typically flow is unnoticeable in the pool, and once the silt bottom is disturbed, it can take days for the water to clear. It is expected that blindcats here feed on the sometimes very abundant cave crickets that fall into the pool from the ceiling and walls. Water temperature in this pool, measured every two hours between March 1997 and August 1999, ranged from 21.47–24.69°C. Continuing upstream from this pool is a 100 m section of submerged passage that is largely silt-floored, except for some constrictions that have a gravel floor. Throughout this pooled passage blindcats have been seen at depths of 0.01–5 m. The upstream, air-filled canyon passage is also silt bottomed, and when explored by members of our team on two occasions had very shallow trickling water. No blindcats have been seen in this reach but a small pool just below the second ('thanksgiving') sump provided our most upstream observations of blindcats. During the single exploration of the underwater passage beyond 'thanksgiving sump', no blindcats were found. All passage above the sump at 'catfish parlour' appears to have comparatively very low invertebrate

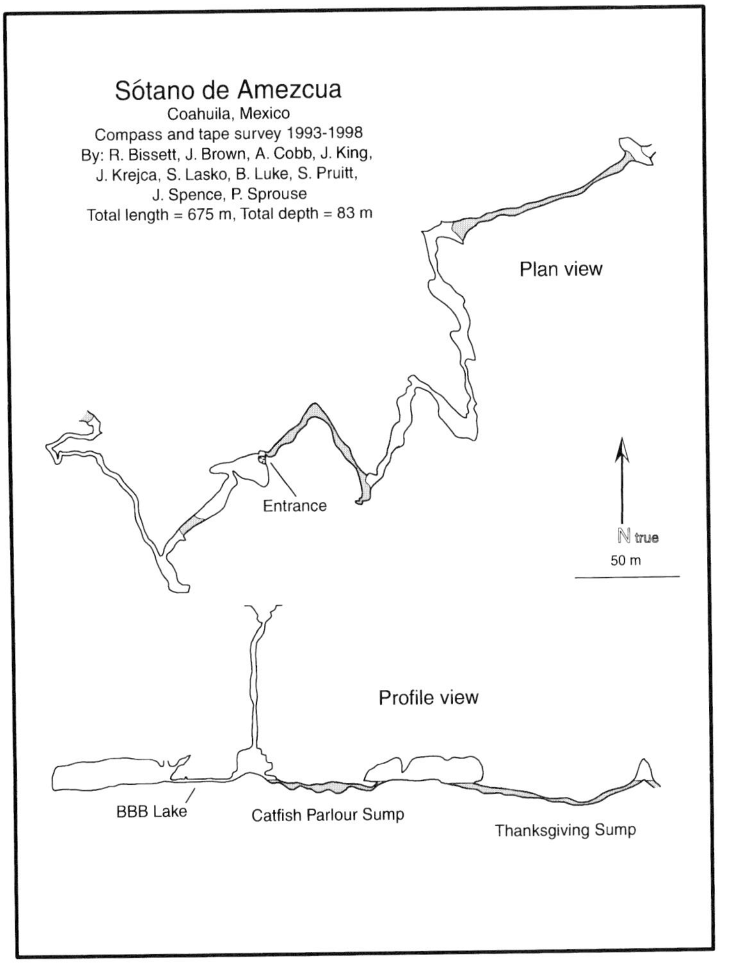

Figure 3. Map of Sótano de Amezcua.

population densities, and thus probably very limited food for blindcats.

Downstream: Between the entrance chamber and the downstream sump, the stream is typically very shallow (0.01–0.1 m) with riffles. Substrates include calcite, gravel and cobbles of various sizes. Sometimes flows drop into the gravel such that BBB Lake is entirely dry, while other times this low-ceiling section of passage is nearly or entirely submerged. A single blindcat was seen twice in this section, in a small pool in the upper canyon passage. It is in the canyon passage that most of the aquatic invertebrates were collected. Our explorations end at the downstream sump pool, which is silt and gravel bottomed, with a surface of about 5×2.5 and 2 m deep. Blindcats have been seen here on every visit, often swimming under calcite rafts floating on the surface of the heavily mineralized water, though at times these rafts are absent.

Prietella lundbergi localities

Manantial de San Rafael de los Castro

This cave spring is formed in middle Cretaceous dolomitic limestone of the El Abra formation. The entrance is a 2 m by 5 m opening bisected by a natural bridge. A 4 m drop leads to water, where a pump has been recently installed. The submerged entrance cave area is wide and maze-like, including some pits dropping to 16 m below water surface and at least two air-filled rooms. Beyond this area, a narrow slot can be followed to 45 m below water surface, where the passage levels then rises to 23 m, before finally plunging to 90 m and continuing beyond our explorations. The site has been well explored, yet no blindcat specimens other than the holotype have been collected or seen. Our explorations began in April 1994, when two divers spent two hours in the cave, penetrating to 40 m maximum depth, collecting invertebrates, but observing no cave fishes. During our 1997 explorations, four divers spent many additional hours in this cave during two days of diving, and finally, the second author and another diver dove the site again in early January 2000, reaching 90 m depth.

The passage diameter varies from 1–10 m, and substrates are gravel, breakdown, bedrock and silt. Though this is the type locality for the species, six dives subsequent to collection of the type specimen failed to yield additional specimens or even sightings. Troglobitic invertebrates are extremely abundant here, including two stygobitic isopods, a cirolanid, *Sphaerolana interstitialis* (Botosaneanu et al. 1998), and a potentially new species of asellid isopod, *Mexistenasellus* (J. Reddell personal communication), as well as troglobitic amphipods and a troglobitic shrimp, *Troglomexicanus huastecae*, for which this is the type and only known locality (Villalobos et al. 1999). Juvenile shrimp are usually on silt, while adults are seen in the water column. Isopods are typically on rocks and amphipods in the water column. Since this is a flowing spring (though discharge is often quite small), we found little variation in water quality with temperature = 27.3°C, pH = 7.47 (the lowest for any blindcat site), dissolved oxygen = 0.9 mg l^{-1} (the lowest for any blindcat site), specific conductivity = 0.76 mS cm^{-1} (the highest for any blindcat site), redox potential = 335 mV (the lowest for any blindcat site) (Botosaneanu et al. 1998) except for notably lower temperature in a section of the cave to the left of the entrance. The lower temperatures may indicate that this may be a dead end passage where cooler water from the entrance settles. The temperature, pH, dissolved oxygen, and redox potential observed were the lowest, while specific conductance here was the highest, of all *Prietella* localities we sampled. In surface waters below the spring we observed (our permit did not allow collections of surface organisms) fishes previously recorded from the area (e.g. Darnell 1962). Divers reported individuals of an unidentified bass and a large surface catfish extending far into the dark zone of the cave.

Cueva del Nacimiento del Río Frío

First described by Russell & Raines (1967) the entrance to this cave is an inconspicuous 1 m diameter hole in a dry arroyo very close to the Nacimiento del Río Frío upstream from Poza Azul. Quickly the passage diameter opens up to 4+ m diameter, continuing about 50 m to a 40 m vertical drop into water. The passage continues underwater to a depth of 60 m, then immediately turns upward to return to the water surface below a large air bell (Figure 4). The submerged passage is 3–10 m in diameter and blindcats were found only at depths ranging from 50 to 60 m. Other aquatic fauna includes the troglobitic shrimp *Troglomexicanus tamaulipasensis*, for which this is the type and only known locality (Villalobos et al. 1999), the parasitic (on *Speocirolana bolivari*) branchiobdellid worm, *Cambarincola acudentatus*, as identified by Stuart Gelder (Tom Iliffe personal communication), troglobitic cirolanid isopods *Speocirolana bolivari*, and *S. pelaezi,* mysids, shrimp (Atyidae), and large snails (Botosaneanu et al. 1998). At the time of our visit, the cave contained two distinctly different, but homogeneous water masses separated by a sharp boundary layer. The upper layer extends from the surface to 39.8 m, while the lower layer begins at 41 m and continues to at least 45 m. Overall organisms were more abundant in the lower layer. Water chemistry data for the shallow layer and deep layers, respectively are: temperature = 21.52°C and 21.63°C, pH = 7.76 and 7.72, dissolved oxygen = 3.4 and 4.4 mg l^{-1}, specific conductivity = 0.43 and 0.52 mS cm^{-1}, redox potential = 361 and 363 mV. Redox potential here was the highest recorded for any blindcat site (Botosaneanu et al. 1998). Due to substantial differences in water quality, especially temperature and specific conductance, lack of obvious outflow from the cave, and failure of divers to detect currents, we conclude that cave water represents an isolated, perched sump, unrelated to water emerging from springs 200–1500 m from the cave entrance (Villalobos et al. 1999).

Sites examined where no *Prietella* were found

Sites listed here are all those visited by our team which appeared to have habitat which might harbor blindcats, or showed promise of providing access to such habitats, but at which our explorations failed to find specimens. We do not mention here other sites visited which appeared unlikely to have, or have had in the

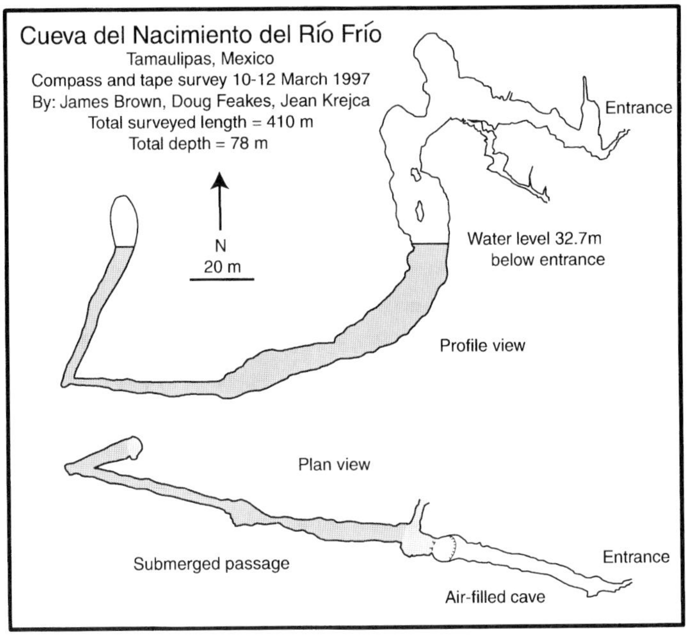

Figure 4. Map of Cueva del Nacimiento del Río Frío.

past, suitable habitat. Invertebrates collected at many of these sites will be reported upon elsewhere. This list is provided so that readers might fully understand the extent of our explorations and for the benefit of future researchers. We fully expect that blindcats may one day be found at least in some of these.

Cueva de la Espantosa

This is a 1–2 m diameter, silt floored spring cave carrying a small stream. It can be followed for about 30 m until reaching a sump. Though no blindcats have been seen, this site warrants a dive to check the submerged passage beyond the sump. This is possibly a new southernmost locality for the troglobitic cirolanid isopod, *Cirolanides texensis* (Iliffe & Botosaneanu personal communication).

Cueva del Nacimiento San Rafael de los Castro

In the sierra immediately above Nacimiento San Rafael de los Castro, within a couple hundred meters of the spring, a natural bridge spans a 15 m vertical drop to the bottom of a large open-air pit floored with an impressive layer of small mammal, reptile, and bird bones. Cave passage can be followed for approximately 100 m to a breakdown terminus. No suitable fish habitat was seen.

Cueva el Venado

A cave with a 1.5 m entrance from which water flows. About 5 m into the cave the passage ends and the water can be seen issuing from an un-enterable hole near the ceiling.

Cuatro Ciénegas area

Rumors among locals of blindcats in some of the springs of the Cuatro Ciénegas basin, and stygobitic organisms, including four species of isopods, two amphipods and four snails known from cave-like microhabitats of this basin (Reddell 1981, Cole 1984, Hershler 1984), make it seem likely to harbor blindcats. Our dives at the only sites with accessible passages, however, yielded no troglobites. Mojarral East (Minckley 1969) is a body-sized spring outlet in about 7 m of water that can be entered only about 3 m before becoming too constrained for divers. Mojarral West (Minckley 1969, now locally known as Poza Azul) has a 1.5 m diameter siphon in about 7 m of pooled water which a diver descended 5–10 m, but extremely strong currents make this site too dangerous to enter farther. The most promising site in the basin that we have discovered so far is locally known as La Campana. A pool of water lies about 9 m below a cenote-like, belled-out vertical entrance. Submerged 3–4 m diameter passage extends in two directions a total length of 130 m, terminating in both directions where the silt floor rises to meet the ceiling. No stygobitic organisms were taken from the conduit passages explored at Mojarral East, West, and La Campana.

Cueva Cabrito

This short cave has a 4 m diameter passage followed by a vertical drop of 5 m and ends in a large guano deposit partially covered by highly enriched, deoxygenated and stagnant water at the time of our visit. No suitable blindcat habitat was seen.

Cueva de El Tule

This otherwise dry cave contains a sump with underwater passage that goes beyond two air bells. It was explored to 16 m depth and 130 m penetration and still continues. Isopods were collected.

Cueva de Rancho Las Pilas

The entrance to this cave is located in the wall of a steep valley and leads into about 100 m of 1–3 m diameter passage at the end of which a dome harboring a bat colony was noted. A 5 m drop leads from this room to a 0.5–1 m diameter sandy crawlway that can be followed to a room and a crawlway with flowing water. A troglobitic ricinuleid was found in this wet crawlway. Approximately 100 m of passage was explored.

Cueva La Zumbadora

This cave in the lower Cretaceous limestone of the Aurora formation has been altered with a concrete dam and aqueduct in an attempt to bring water from the cave to the town of La Madrid. A short distance into the cave a sizeable stream is encountered, dropping downstream into a tightly enclosed waterfall, but possible to explore upstream about 100 m to a siphon. Divers explored 150 m more of submerged passage, with a maximum depth of 12 m. Passage diameter is typically 2 m, with bedrock walls and floor. This cave is the type and only known locality for the stygobitic cirolanid isopod *Speocirolana zumbadora*. It also harbors *Sphaerolana affinis*, and a snail (Botosaneanu et al. 1998).

Dandridge Spring

This high flow spring on the Devil's River has about 30 m of 1 m (or less) diameter passage. Troglobites collected include one isopod and a harvestman. Ostracods and some hydrobiid snails were also taken.

El Abra

This cave is a 0.5 m diameter hole in the bottom of a bedrock creek bottom that leads down 3 m to a 1 m diameter horizontal tube carrying a small stream. The stream passage was followed at least 50 m and continues.

El Ebanito

A steep passage, 2–5 m in diameter, descends to a boulder choke some 40 m vertically below the entrance. Water level is highly variable, with the boulder choke at times 5 m above water, while on another visit it was at 37 m of depth. Cirolanid isopods and shrimp have been collected here.

El Sótano

This 10+ m diameter cave can be followed approximately 100 m before two drops, totaling about 40 m, are reached. At the bottom of this vertical section is a sump pool. No blindcats were seen, but this site warrants a dive to check the deeper waters (Proyecto Espeleológico Purificación identification number 118, Peter Sprouse personal communication).

Emilio's Cavern

This small blind pit was explored to a depth of 9 m. No water or suitable habitat was found.

Falcón Well

This 2–4 m diameter well below a windmill was easily entered on rope to water at a depth of about 10 m from the surface. The pool at the bottom was stagnant, water visibility was poor, and there appeared to be no enterable passage or flow.

Goodenough Spring

The third largest spring in Texas (Brune 1981) now sits inundated at 20–30 m depth below the surface of Lake Amistad, a reservoir on the Río Grande. Divers have explored the cave passage to a depth of 65 m, and report strong flow, warm water, good visibility, and organisms such as epigean catfish (R. D. Milhollin personal communication).

Gruta de Carrizal

This is a spring flowing out of dry cave with two sumps inside – one warm and the other cold. The warm water sump and downstream area harbors epigean catfish (*I. lupus*) as well as amphipods. A series of four sumps were explored to a termination in breakdown. Exploration of submerged passage reached 14 m in depth, and 130 m penetration.

Guayalejo Spring

This cave spring is on the south bank of the Rio Guayalejo where it flows through a mountain pass. A cenote-like entrance with a 100 m long submerged cave passage (16 m depth) leads to air filled chambers and possible dry caving leads. A troglobitic cirolanid isopod, likely an undescribed species of *Speocirolana* (Botosaneanu personal communication) was collected.

La Rajada

Also known as El Indio or Los Indios, this is a naturally enlarged joint in a cliff descending about 10 m to a small pool of water 1.5 m deep with a deep mud bottom. A windmill above the entrance extracts water from the pool. No cave passage or troglobitic organisms could be found.

La Vinata Well

This is a hand-dug well fitted with a diesel-powered pump that sits in a little house adjacent to the well. The pump is no longer operational and excessive grease from it coated the walls and surface of the water at the bottom. A ladder leads down about 10 m to water no more than 2 m deep. There is evidence, including reports from locals, that indicate that this well flows during heavy rains.

Manantial La Penita

This municipal water source for Ciudad Victoria consists primarily of springs that are pumped into the city. We visited first a small cave in the limestone wall of the creek bed. The entrance and passage has been modified with steps and concrete platforms connecting to a 1–2 m diameter passage that drops 4–5 m to water with sumps in both upstream and downstream directions. The site was dove for a short distance, but no aquatic fauna was found. The second site explored here was a large concrete box buried in the stream cobbles at the edge of the stream bed just across from the cave. A small steel door allows entry to a 4–5 m drop to a small creek flowing over stream cobbles and gravel through a small room. Troglobitic cirolanid isopods were collected.

Nacimiento del Río Frío

This series of springs forms the Río Frío, with the uppermost springs being about 200 m from the entrance to Cueva del Nacimiento del Río Frío (above), with many more outlets extending downstream to Poza Azul, though none were found which could be entered for exploration. Continuing downstream as far as La Florida are more spring outlets. Water chemistry clearly indicates that these springs are not connected to the water in Cueva del Nacimiento del Río Frío.

Nacimiento del Río Sabinas

First explored by Exley (1979) and others, our 1994 expedition dove this site again to 40 m depth, but failed to collect or observe stygobites.

Nacimiento Kikapoo

This heavily vegetated spring pool forming the headwaters of the Río Sabinas had no accessible passage, but significant discharge indicates substantial subterranean aquatic habitat.

Ojo Encantado

This is a resurgence cave at the bottom of a deep canyon. Just inside the 10 m diameter entrance is a long, deep pool where we collected cirolanid isopods preliminarily identified as *Speocirolana pelaezi* and *S. bolivari* (Botosaneanu personal communication).

Water level fluctuates extremely here, such that sometimes water comes out of the entrance as a spring and the cave can not be entered without diving, while other times the water is about 10 m below the entrance. Passage continues upstream for at least 100 m.

Poza San Miguel

This is a large spring pool diveable to about 14 m where a narrow bedding plane constriction prevents further exploration. We observed several epigean fishes in the pool, including *Gambusia* sp., *Lepomis* sp.

Purificación area caves

Many caves have been explored in this high altitude karst plateau in the Sierra Madre Oriental west of Ciudad Victoria, including the country's largest, and some very deep systems such as Sistema Purificación, Tecolote, Río Corona, Sima Chupacable, and Calenturas. Cavers and biologists have explored and collected these sites for many years and have found the aquatic fauna limited to troglobitic cirolanid isopods *Speocirolana bolivari* and *S. endeca,* as identified by Botosaneanu (T. Iliffe personal communication). The combination of little food input and high elevation make these caves perhaps less likely than others to harbor blindcats.

Richter Cave

This is a 4 m deep blind pit located up the hill from Dandridge Spring. No water is present. Fauna included a harvestman (*Lieobenum*), a spider (*Cicurina varians*), a Texas cliff chirping frog (*Syrrhophus marnockii*), and small mammal bones.

San Felipe springs

The east spring (= San Felipe Spring #3 in Brune 1981), with several large pump intake structures extracting water for the city of Del Rio, was dove to a boulder constriction at a depth and penetration of 10 m. The west spring (= San Felipe Spring #2 in Brune 1981), rumored to harbor blindcats, was also dove to about 10 m, to a small restriction where the flow velocity prevented further penetration. The other springs mentioned by Brune (1981), including #1 and #4–10 were field checked and all except for #10 were un-enterable. Permission to dive San Felipe spring #10 could not be obtained.

Springs near El Nacimiento de San Rafael de los Castro

Following rumors of other springs and caves, we spoke with locals who described all of the caves on the N side of the Cañon La Servilleta as dry and formerly mined for guano. The canyon was partially checked for springs, and none were found. A small dammed spring head at La Presa was checked but appears to be inaccessible.

Tinaja Azul

This small spring is pumped by a windmill. The entrance has been shored up with concrete to 1–2 m diameter. Water depth is about 1 m, with a silt and cobble floor. Submerged passage continues horizontally from the floor of the spring as a low, inaccessible crack along a bedding plane. Locals reported past sightings of blindcats here, but we saw only epigean *Astyanax*.

Tiro el Palmito

This mine shaft is in the lower Cretaceous limestone of the Cupido formation (= Sligo Formation in Texas). A railroad track drops 50 m vertically at a 45 degree angle through the 1–3 m diameter tunnel. The pool at the bottom was explored by a diver for only a short distance because visibility was very quickly reduced by ceiling debris dislodged by the diver's bubbles.

Vertical mine shaft near El Portrero

This 1.5 m diameter mine shaft begins horizontally then intersects a vertical shaft about 25 m deep. The shaft intersects water, but was cluttered at various levels by precariously perched, fallen timbers, and thus judged too dangerous to enter.

Abundance

In general, vertebrate troglobites in temperate environments are rare relative to epigean organisms. The desert landscape that overlays most of the range of *Prietella* probably contributes little energy to the system in the way of organic debris, and such input may be largely restricted to rare flood events. It is unusual to see more than six blindcats on any given visit to one of these sites, and it is easy to see that higher abundances are near cave entrances that serve as energy input points. The most recorded specimens collected from any site are from Carranza's (1954) collection of 66 from El Potrero 1. Never have we seen abundance approaching this, though on our first visit to Sótano de Amezcua, we collected 25 specimens from the two sumps and intervening small pools. Preliminary capture-mark-recapture data analyzed using the Fisher-Ford estimation and subsequently taken during nine sampling periods over 13 months in the upstream and downstream sumps of Sótano de Amezcua indicate a fairly small population with 9–48 fish in each sump. At Ojo del Yermo, however, divers reported on the order of many tens of blindcats. It is almost certainly the case that sites we have been able to access represent a miniscule proportion of the total habitat available to these organisms, and the potential of an energy source in the deep aquifer, such as the one that may exist below San Antonio in Texas (Longley 1981), may sustain reasonably large populations.

Laboratory observations

Transport and laboratory conditions

Live specimens of *P. phreatophila* have been maintained in a variety of aquaria and wading pools in the first author's laboratory nearly continually since 1992. Except for occasional isolation of individuals, all have been kept in tanks housing 2–10 individuals. All specimens have been maintained in dechlorinated tap water in tanks with under-gravel and/or external filters, and a diversity of large, highly porous limestone rocks and laboratory flasks of assorted sizes and shapes intended to provide cavities such as might be used for reproduction. Water temperature has typically been maintained at laboratory temperature, a relatively constant 21–22°C, but aquarium heaters have been employed to hold selected specimens at constant and varying temperatures in the range of 21–27°C.

Transport to the laboratory from the field has evolved with the first specimens collected by cavers being carried in ice chests to more recent collections of the first author with transport for up to four days in plastic bags filled with pure oxygen and water from the habitats in which the specimens were captured, together with Stress coat™ and small pieces of Polyfilter™ to remove wastes. These last measures seem to reduce problems and reduce mortality, but in general, even under the best of conditions, specimens seem to suffer fin abrasions and other transport-related damage. Hemorrhaging between fin rays is commonly seen following transport. All recent collections have attempted to minimize transport time and distance, with consequential marked reductions in mortality.

Diet in captivity

Wild-caught specimens readily adapt to, and seem to do well on, laboratory diets consisting of prepared flake foods (Tetramin™), supplemented by occasional frozen blood worms, mosquito larvae and adult brine shrimp.

Parasites and diseases

One specimen was brought to the laboratory from the wild with what appeared to be a nematode encysted in the epaxial musculature. The specimen survived many months, but died within a day of rupture of the cyst. The parasite could not be found in the aquarium, which contained other blindcat specimens which may have consumed it, and which appeared to have attacked the site of the rupture, increasing trauma to the specimen. On several occasions captive individuals have developed bacterial infections with initial symptoms of slight ulceration in fin membranes and/or fin erosion, progressing to liver discoloration (visibly yellow as seen through the ventral body wall), septacemia and external ulceration. Such infections proved difficult to treat with a variety of antibiotics tested, but if treatment began early enough some individuals responded to long term 120 mg l^{-1} baths of Chloramphenicol. Fin erosion and associated hemorrhaging between fin rays (above), also responds to Chloramphenicol baths.

Light sensitivity

P. phreatophila specimens we have observed appear incapable of perceiving light through any remnant eyes they might have (we have not done dissections or histologic sections for detailed examination of eye morphology). In total darkness, they show no obvious reaction to the turning on and off of a 500 W bulb held within 5–10 cm of their head. We are currently conducting rigorous light sensitivity experiments with *P. phreatophila*, but most specimens kept to date have been held in normal laboratory and office spaces lit for at least 8–12 hours daily with no obvious adverse impacts. Recently, however, four individuals have been maintained in total darkness for six months.

Sensory biology

P. phreatophila obviously has acute auditory sensitivity, reacting immediately to taps on aquarium glass and other laboratory noises. Other observers of specimens in Sótano de Amezcua have reported individuals feeding on crickets thrown onto the surface of the water by swimming around the prey in a path with continually decreasing radius until it is encountered and consumed; a behavior likely mediated by lateral line function.

Chemosensory functions appear to be well developed in *P. phreatophila*, as attested to by the rapidity with which animated searching behavior begins after non-moving foods (frozen blood worms, brine shrimp, mosquito larvae, or flake foods) are placed in their tanks. It is obvious that odors or tastes given off by these foods are very rapidly perceived. Attempts to quantify olfactory sensitivity to a variety of amino acids and steroids in the laboratory of Norman Stacey (University of British Columbia) detected very low responses to these substances, but results may have been affected by a variety of technical difficulties including the need to perform surgery to install electrodes, and prior treatments of infections with antibiotics (Norman Stacey personal communication).

Figure 5. Photographs showing varying conditions of *Prietella*.

Starvation tolerance

Individuals with strikingly low, but unquantified, condition factor have been collected (Figure 5). All *P. lundbergi* collected were in this condition, but not as extremely low as were some *P. phreatophila* collected together with individuals in 'normal' condition. Individuals in the laboratory have been intentionally not fed for as long as 44 months, during which time condition factor obviously decreased, but not to levels equal that of some specimens obtained from the wild, which were far more emaciated. When food was made available, these individuals generally quickly resumed feeding and condition factor increased.

Jaw locking

On several occasions, aggressive encounters involving extended periods of two individuals biting one another's jaws have been observed, (Figure 6). All such encounters have been observed following movements of individuals to new tanks that result in new combinations of individuals coming into contact with one another. These encounters take the form of two individuals biting at one another and then locking jaws. The two specimens will then often remain in mouth to mouth contact for many hours, during which they occasionally swim and writhe vigorously, but briefly, pulling the other fish with them. These periods of activity are interspersed with extended periods during which both specimens rest motionless but maintain their grip on the other. Opercular movements are often exaggerated and accelerated while specimens maintain these jaw locks, indicating obvious respiratory stress. Both individuals may suffer abrasions to the mouth, barbels, nares and orbit regions produced by the teeth of the other individual. Specimens involved in jaw locks are not always of equal sizes. In some cases heads of smaller individuals have been seen to be nearly completely inside the mouth of the larger specimen during these encounters. Jaw locking incidents typically last < 30 min, but at least one jaw lock was maintained continuously for > 12 but <18 h, with only brief separations of 2–3 s, during which jaw lock position was altered slightly. In all cases, specimens involved in jaw locking appeared to be males, but sex was not subsequently verified by dissection. We assume these

Figure 6. Photographs showing jaw-locking behavior in *Prietella phreatophila*.

encounters to be related to establishment of social dominance hierarchies. Once two individuals have jaw locked, subsequent jaw-lockings between them have not been observed, indicating that perhaps long-term dominance relationships are established by jaw locking shortly after individuals first come into contact, and subsequently never again tested.

Periods of inactivity

Specimens, especially larger individuals and those under some form of stress (e.g. infection, lack of food, temperature), are often observed resting motionless on substrates or simply drifting with light currents in their tanks. During these periods opercular pumping rates sometimes drop to low levels (10 or fewer per minute) and specimens often lay on their sides or in inverted position (Figure 7). A light finger tap on the tank glass quickly 'awakens' them and they begin normal swimming. Such behavior likely reflects lack of predators in their habitats, as such behavior would seem to leave them extremely vulnerable to predation.

Figure 7. Photograph showing periods of inactivity where *Prietella phreatophila* lay on their side or inverted.

Failure to induce reproduction in captivity

Typically 4–6 specimens of unverified, but presumably mixed, sexes have been maintained together in the laboratory in hopes that they might reproduce. Assorted substrates such as locally obtained limestone with multiple cavities and conduits of various sizes, as well as assorted sizes of glass laboratory flasks, have been provided should this species be a cavity spawner. Little attention has typically been paid to cavities, however, and reproduction has not yet been observed.

Water changes

Since caves harboring *P. phreatophila* have signs of periodic flooding interspersed with periods of stable or slowly dropping water levels, we hypothesized that floods, which might introduce new organic material from the surface into the cave habitats renewing food resources, may trigger reproduction. In an attempt to mimic floods in the laboratory, we allowed tank water to evaporate for extended periods followed by rapid refilling with de-ionized water. Our several repetitions of such cycles produced no response.

Flow-through well water (Dallas Aquarium)

Our laboratory relies on dechlorinated or deionized municipal water and flow-through tank water supply is not possible. Four specimens were housed for about one year, however, in flowing well-water at the Dallas Aquarium where specimens were fed live invertebrates resident in the aquarium water supply. Behavior remained similar to what had been observed in our laboratory, and no signs of reproduction were noted (D. Schlesser & C. Yancey personal communication).

Hormone injections

Failing to induce reproduction with environmental changes, the first author injected three adult specimens (12–15 g) with Pimozide (500 $\mu g\,kg^{-1}$) and LHRHa (100 $\mu g\,kg^{-1}$) (Prentice & Thomas 1987). The total volume of the intramuscular injections was about 0.07 ml. Specimens seemed unaffected by injections, feeding voraciously within 24 hours. With no obvious response to initial injections, fish were again injected with the same dosage at 72 and 144 hours after the first injections. Again, no response was observed.

Presumptive spawning behavior

As this manuscript was being drafted, one pair of 3 specimens being housed together in a 208 l tank began associating with one another much more closely than is typically observed. Normally, individuals which encounter one another display a fright response and flee in different directions, however, these two specimens began to be seen often in close proximity or touching one another. The presumed male had developed pronounced swelling of the dorsal head region, especially of the tissues covering the orbits. What appeared to be large, yellowish or cream-colored eggs in the posterior, dorsal area of the body cavity could be barely seen through the body walls of the presumed female, which lacked any signs of swelling of head tissues. These two individuals would frequently together enter a 100 ml beaker laying on its side on the tank bottom. The male would flex the posterior half of his body into a u-shaped arch and quiver as the female vigorously nuzzled, nipped and pushed at his anal region with her mouth, often pushing the male against the bottom of the beaker or, if oriented in the opposite direction, pushing him out of the beaker. This behavior would last 18–23 seconds, and was repeated at 10–30 minute intervals throughout approximately eight days with no signs of actual spawning or emission of gametes by either individual. The same behavior was occasionally observed in a nearby, larger beaker, but the one, smaller beaker (one of two equal beakers in the tank) was much preferred. Upon completion of this sequence the two individuals would separate and typically swim only 10–20 cm and rest motionless (as described above) on the tank bottom or in supportive structures provided by limestone.

Throughout this period both individuals fed voraciously. After four days of frequent observations of this behavior, the authors had to leave the laboratory for several days, during which time the behavior ceased. Personnel providing fish care did not notice eggs or any other signs of spawning, but time-lapse video of this period has yet to be analyzed. On return of the authors, the belly of the female was noted to be somewhat less distended than previously, and presumed eggs were not so obvious through the body wall as they were before. Searches for eggs and/or larvae failed to find either, and neither individual showed any signs of mouth brooding.

Evolutionary history, biogeography and conservation status

Position within Ictaluridae

Lundberg (1982, 1992) and previous workers working with morphological data considered *Prietella* to be sister to the genus *Noturus*, to which it bears striking morphological similarities. Preliminary results of phylogenetic analyses of mtDNA sequence data (Garcia de León et al. unpublished data), however, do not tend to support this relationship. We are currently in the process of expanding our molecular data set in attempts to increase phylogenetic resolution.

Intrageneric relationships

Preliminary analyses of mtDNA sequences indicate large genetic distances between *P. phreatophila* and *P. lundbergi*, but relatively little differentiation among populations of *P. phreatophila* indicate probable high levels of interconnections among aquifers from throughout its range.

Conservation status

P. phreatophila was listed as an endangered species in México by the federal government in 1994 (Secretaría de Desarrollo Social 1994), and was listed as endangered by Williams et al. (1989). *P. lundbergi* was described subsequent to the last revisions of the Mexican endangered species list.

P. phreatophila should probably be considered threatened. Its distribution is much more extensive than previously realized, and it can be found at many localities, indicating that it is probably less endangered than previously thought. Populations almost surely extend throughout the extensive aquifer it inhabits, occupying many habitats inaccessible to humans or still undiscovered. The total numbers of individuals in this species is probably very large. The arid region it inhabits, however, is facing rapid human population growth, with resultant increasing groundwater exploitation for municipal, agriculture and industrial uses (Contreras-Balderas & Lozano-Vilano 1994). Our interviews with many local residents clearly indicate that declines in water tables over the past few decades are well known throughout the area. Increasing industrialization and recent proposals to use desert areas as waste repositories also threaten to contaminate aquifers occupied by this species.

P. lundbergi appears to have a very restricted distribution. Habitat at one of the two localities from which it is known is very restricted and totally isolated from surrounding waters. Only one specimen has been taken at the type locality despite considerable effort. Unless it occupies deep aquifers to the east of the localities from which it is now known, or alongside the sierra, it is definitely rare. Its presence there may have gone undetected due to inaccessibility of its habitats and the fact that wells, such as those in which *P. phreatophila* is sometimes found in Coahuila, are not common in much more mesic southern Tamaulipas. Based on what we know about the limited distribution, we recommend that this fish be considered endangered.

Future directions

We continue to search for and explore potential new *Prietella* habitats as possible, and to study *P. phreatophila* in the laboratory. We are conducting phylogenetic studies of Ictaluridae, with focus on the position of *Prietella* within the family (Garcia de León et al. unpublished data), and relationships among its populations using microsatellite DNA techniques. Osteological studies of *P. lundbergi* have been hampered by lack of specimens, however, we recently obtained high-resolution CAT scans of the holotype (low resolution versions may be viewed via internet links[2]). Behavioral observations continue in the laboratory, and we continue attempting to induce reproduction in captivity. The second author's research focuses on phylogeography of blindcats and associated invertebrates (based on molecular genetic techniques) over relatively small geographic areas, and application of such data in hydrology as indicators of inter-aquifer connections. Population studies using capture-mark-recapture techniques and habitat monitoring in Sótano de Amezcua continue.

[2] From http://uts.cc.utexas.edu/~deanhend/.

Acknowledgements

We thank the following expedition participants for their field and technical assistance: Celene Denev Acuña Leal, Melonie Asplaugh, Don Broussard, James Brown, Allan B. Cobb, Adam Cohen, Salvador Contreras-Balderas, Brett Dodson, Héctor Espinosa Pérez, Doug Feakes, Tania Yssel Flores Martinez, Francisco Garcia de Leon, Marcus Gary, Steve Gerrard, Andy Gluesenkamp, Aldo Antonio Guevara Carrizales, Dean, Garrett and Jacob Hendrickson, Tom Iliffe, Jean and Christine Krejca, Mike Lang, Mike Loeffler, Carol Malcolm, Annmarie Mikelski, R.D. Milhollin, Collin Morris, Tom Morris, John Moseley, Rae Nadler-Olenick, Walt Olenick, Mary Poteet, Jorge Homero Rodríguez Castro, Azael Salazar, Yara Sanchéz Johnson, Greg Sedbrook, Lara Storm, David Sweetin, Steve Taylor, Bruce Tipps, Bill Tucker, Anastasio Urbina, Fernando Vanage Eligio, Mike and Cindy Warton. Peter Thomas provided the hormones and advise for the reproduction experiments. Major funding for expeditions and laboratory work was granted by the U.S.–México Foundation for Science to DAH (project 54-E-94), and the Smithsonian Institution (for 1994 expedition to Robert Hershler and DAH). Homero Amezcua, José Falcón, Junta Administradora de Agua Potable y Alcantarillado de Múzquiz, Coahuila (Marco Antonio Cardozo M. and José H. Múzquiz Quedea), Hugo Martinez, Salvador Martinez Garza, Guillermo Osuna Saenz, Javier Valdez, and many other friendly landowners in México and Texas provided logistical support and access to their properties. Other general support was provided by Texas Memorial Museum at the University of Texas at Austin (financial and general support), Illinois Natural History Survey (financial support, equipment), Vertebrate Paleontology Laboratory at the University of Texas at Austin (loan of equipment trailer), Robert Hershler (co-PI for 1994 expedition), Hydrolab Corporation (DataSonde datalogger loan), Eastman Kodak Corporation (digital camera loan), University of Texas Institute for Latin American Studies Tinker Fellowship to JKK, University of Texas Zoology Department fellowship to JKK, Phi Kappa Phi fellowship to JKK, Cave

Research Foundation Karst Research Grant to JKK, NightRider (dive lights), Suzanna Walaszek (cartography), James Reddell (invertebrate identifications), David Riskind (many spring and cave leads and landowner contacts), Solveig Turpin (cave leads and landowner contacts), Peter Sprouse (cave leads and landowner contacts), the National Speleological Society, Proyecto Espeleológico Purificación, Texas Speleological Survey. Collections of specimens were authorized under permits A00 700 (2).-06083, A00 700 (2).-06085, A00.700 (2) 00072, A00 700 (2).-00362, DAN-03004, DAN 02748, D000 750.-1576, 150896-213-03, and 041099-213-03 granted by SEDESOL, SEMARNAP and Secretaria de Relaciones Exteriores. Angelica Narvaez and Francisco García de León helped obtain permits. Additional photos and information may be viewed on the web.[3]

References cited

Amemiya, C.T., S.W. Kelsch, F.S. Hendricks & J.R. Gold. 1986. The karyotype of the Mexican blindcat, *Prietella phreatophila* Carranza (Ictaluridae). Copeia 1986: 1024–1028.

Axel, W., G. Proudlove, J. Parzefall, H. Wilkens & T.T. Nalbant. 1999. Pisces (Teleostei). pp. 1177–1213. *In*: C. Juberthie & V. Decu (ed.) Encyclopaedia Biospeologica, Societe de Biospeologia, Moulis-Bucarest.

Bart, H.L. Jr. & M.S. Taylor. 1993. Type specimens of fishes in the Tulane University Museum of Natural History. Tulane Stud. Zool. Bot. 29: 29–72.

Borowsky, R. 1996. The Sierra de El Abra of northeastern México: blind fish in the world's largest cave system. Trop. Fish Hobb. 46: 178–188.

Botosaneanu, L., T.M. Iliffe & D.A. Hendrickson. 1998. On a collection of stygobitic cirolanids (Isopoda: Cirolanidae) from northern México, with description of a new species. Biologie 68: 123–134.

Brune, G. 1981. Springs of Texas. Branch-Smith, Fort Worth. 566 pp.

Carillo-Bravo, J. 1971. La plataforma Valles-San Luis Potosí. Bol. Asoc. Mex. Geól. Petr. 23: 1–113.

Carranza, J. 1954. Descripción del primer bagre anoftalmo y depigmentado encontrado en aguas Mexicanas. Ciéncia (México) 14: 129–136.

Cole, G.A. 1984. Crustacea from the Bolsón of Cuatro Ciénegas, Coahuila, México. J. Ariz-Nev. Acad. Sci. 19: 3–12.

Contreras-Balderas, S. & M.L. Lozano-Vilano. 1994. Water, endangered fishes, and development perspectives in arid lands of México. Conserv. Biol. 8: 379–387.

Culver, D.C., L.L. Master, M.C. Christman & H.H. Hobbs, III. 2000. Obligate cave fauna of the 48 contiguous United States. Conserv. Biol. 14 : 386–401.

[3] http://uts.cc.utexas.edu/~deanhend/ and http://www.inhs.uiuc.edu/~sjtaylor/cave/México/México.html.

Darnell, R.M. 1962. Fishes of the Rio Tamesí and related coastal lagoons in east-central México. Publ. Inst. Mar. Sci. Univ. Texas (8): 299–365.

Dasher, G. 1994. On station: a complete handbook for surveying and mapping caves. National Speleological Society, Huntsville. 242 pp.

Exley, S. 1979. ¡Nacimientos! Diving the big springs of the Sierra Madre. Ass. Mex. Cave Stud. Bull. 10: 23–31.

Fish, J.E. 1977. Karst hydrogeology and geomorphology of the Sierra de El Abra and the Valles-San Luis Potosi region, México. Ph.D. Dissertation, McMaster University, Hamilton. 469 pp.

Hershler, R. 1984. The hydrobiid snails (Gastropoda: Rissoacea) of the Cuatro Ciénegas basin: systematic relationships and ecology of a unique fauna. J. Ariz.-Nev. Acad. Sci. 19: 61–76.

Lesser, J.M. & G. Lesser. 1988. Region 9, Sierra Madre Oriental. pp. 89–92. *In*: W. Back, J.S. Rosenshein & P.R. Seaber (ed.) Hydrogeology (volume O-2), Geological Society of America, Boulder.

Leviton, A.E., R.H.J. Gibbs, E. Heal & C.E. Dawson. 1985. Standards in herpetology and ichthyology: part 1. Standard symbolic codes and institutional resource collections in herpetology and ichthyology. Copeia 1985: 802–832.

Leviton, A.E. & R.H.J. Gibbs. 1988. Standards in herpetology and ichthyology. Standard symbolic codes for institutional resource collections in herpetology and ichthyology. Supplement No. 1: additions and corrections. Copeia 1988: 280–282.

Longley, G. 1981. The Edwards Aquifer: Earth's most diverse groundwater ecosystem? Int. J. Speleol. 11: 123–128.

Lundberg, J.G. 1982. The comparative anatomy of the toothless blindcat, *Trogloglanis pattersoni* Eigenmann, with a phylogenetic analysis of the ictalurid catfishes. Misc. Publ. Mus. Zool. Univ. Mich. (163): 1–83.

Lundberg, J.G. 1992. The phylogeny of ictalurid catfishes: a synthesis of recent work. pp. 392–420. *In*: R. L. Mayden (ed.) Systematics, Historical Ecology, and North American Freshwater Fishes, Stanford University Press, Stanford.

Minckley, W.L. 1969. Environments of the Bolsón of Cuatro Ciénegas, Coahuila, México. Science Series, University of Texas, El Paso, Texas (2): 1–65.

Mitchell, R.W., W.H. Russell & W.R. Elliott. 1977. Mexican eyeless characin fishes of the genus *Astyanax*: environment, distribution and evolution. Mus. Tex. Tech Univ. Sp. Publ. (12): 1–89.

Nelson, J.E. 1994. Fishes of the world. John Wiley and Sons, New York. 600 pp.

Padgett, A. & B. Smith. 1987. On rope: North American vertical rope techniques. National Speleological Society, Huntsville. 341 pp.

Prosser, J. & H.V. Grey. 1992. NSS cave diving manual. The Cave Diving Section of the National Speleological Society, Branford. 377 pp.

Prentice, J.A. & P. Thomas. 1987. Successful spawning of orangemouth corvina following injection with des-Gly10, [d-Ala6]-Luteinizing hormone-releasing hormone (1–9) Ethylamide and Pimozide. Prog. Fish-Cult. 49: 66–69.

Reddell, J.R. 1981. A review of the cavernicole fauna of México, Guatemala, and Belize. Bull. Tex. Mem. Mus. (27): 1–327.

Reddell, J.R. 1994. The cave fauna of Texas, with special reference to the western Edwards Plateau. pp. 31–49. *In*: W.R. Elliott & G. Veni (ed.) The Caves and Karst of Texas, National Speleological Society, Huntsville. 342 pp.

Russell, W.H. & T.W. Raines. 1967. Caves of the Inter-American highway, Nuevo Laredo, Tamaulipas to Tamazunchale, San Luis Potosi. Ass. Mex. Cave Stud. Bull. (1): 1–126.

Secretaría de Desarrollo Social. 1994. Norma Oficial Mexicana NOM-059-ECOL-1994, que determina las especies y subespecies de flora y fauna silvestres y acuáticas en peligro de extinción, amenazadas, raras y sujetas a protección especial, y que establece especificaciones para su protección. Diario Oficial De La Federación, México 488(10): 2–60.

Smith, A.R. & G. Veni, 1994. Karst regions of Texas. pp. 7-12. *In*: W.R. Elliott & G. Veni (ed.) The Caves and Karst of Texas, National Speleological Society, Huntsville.

Villalobos, J.L., F. Alvarez & T.M. Iliffe. 1999. New species of troglobitic shrimps from México, with the description of *Troglomexicanus*, new genus (Decapoda: Palaemonidae). J. Crust. Biol. 19: 111–112.

Walsh, S.J. 2000. Conservation of aquatic karst biotas: shedding light on troubled waters. pp. 106–108. *In*: R.A. Abell, D.M. Olson, E. Dinerstein, P.T. Hurley, J.T. Diggs, W. Eichbaum, S. Walters, W. Wetengel, T. Allnutt, C.J. Loucks & P. Hedao (ed.) Freshwater Ecoregions of North America: A Conservation Assessment, Island Press, Washington, D.C.

Williams, J.E., J. Johnson, D.A. Hendrickson, S. Contreras-Balderas, J.D. Williams, M. Navarro-Mendoza, D.E. McAllister & J.E. Deacon. 1989. Fishes of North America endangered, threatened or of special concern: 1989. Fisheries 14: 2–20.

Environmental Biology of Fishes **62:** 339–344, 2001.
© 2001 *Kluwer Academic Publishers. Printed in the Netherlands.*

A new blind cave fish population of genus *Astyanax*: geography, morphology and behavior

Luis Espinasa[a], Patricia Rivas-Manzano[b] & Héctor Espinosa Pérez[c]
[a]*CEAMISH, Universidad Autónoma del Estado de Morelos, Avenida Universidad #1001, Colonia Chamilpa, Cuernavaca. Morelos 62210, Mexico (e-mail: espinasl@buzon.uaem.mx)*
[b]*Facultad de Ciencias,* [c]*Instituto de Biología, Universidad Nacional Autónoma de México, Mexico*

Received 20 July 2000 Accepted 8 December 2000

Key words: A. fasciatus, A. mexicanus, A. aeneus, troglobite, troglomorphism, vision, eye development, aggression

Synopsis

A new population of blind, cave dwelling tetra fish of the genus *Astyanax* was discovered in Granadas Cave, in the Balsas drainage, southern México. All blind Mexican tetras previously described are from Tampico and San Luis Potosí, northern México. The discovery of a new blind morph thus represents an independent colonization and convergent adaptation to the cave environment by this fish. Individuals of this population display variability of their troglomorphic features. Some individuals presented asymmetrical degeneration of the eyes, where one was normal, but the other somewhat reduced in size and complexity. Loss of pigmentation and eye reduction, although sometimes correlated, were not always linked; reduced eyes were found on pigmented fish and unpigmented fish often possessed normal eyes. Some individuals had reduced lens size or an absence of lens altogether. Retina is highly modified with photoreceptors sometimes absent. Eye reduction was correlated with a diminished size of the optic lobes and an increase of the prosencephalon. Modifications of the skull involve closing in of the circumorbital series of bones. Certain aspects of behavior are also modified.

Introduction

Until now, all troglomorphic (blind, depigmented) Mexican tetras previously reported have been from Tampico and San Luis Potosí, northern México (Mitchell et al. 1977). Although other cave populations of the genus *Astyanax* have been reported for the Yucatán peninsula (Hubbs 1938), Costa Rica (Romero 1984), and Brazil (Trajano 1987), individuals of those populations did not show the troglomorphic features that characterize so many hypogean fishes.

There is still some uncertainty about the specific name for the troglomorphic *Astyanax* from northern México. Originally, the first three cave populations were assigned three different specific names each within a new genus, *Anoptichthys*. As Mendelian and molecular studies provided evidence for high levels of genetic similarity between the hypogean and epigean populations of these fish, typological thinking was replaced by a populational one and the cave populations were assigned to the species *Astyanax fasciatus* (Cuvier 1819) (cf. Romero 2001).

Yet, this issue has yet to be fully resolved because there is also some uncertainty about the specific status of *Astyanax* in northern México. Eigenmann (1917) described five species and subspecies for that country. Some believe that *Astyanax* from northern México is *A. mexicanus*, while fish from south of Punta del Moro, Veracruz are *A. aeneus* (e.g., Obregon-Barbosa et al. 1994). The status of the *Astyanax* from the river Balsas drainage to which the new cave population belongs is not totally clear either. According to Contreras-Valderas (1988) they are either *A. aeneus* or a separate species altogether. Others think that *Astyanax* has so much variability, that separation at the species level is very difficult (Bussing 1985).

Wilkens (1988) concluded that as long as the research on the taxonomy of the different *Astyanax* populations in México and South America has not been conclusive, the oldest name *A. fasciatus* (Cuvier 1819) must be maintained. For this paper, we will refer to the new cave population and the epigean fish of the Balsas drainage as *Astyanax* sp. in order to avoid further confusion until the genus *Astyanax* for Mesoamerica is more thoroughly reviewed.

The new population of *Astyanax* we are reporting here, displays in variable degrees troglomorphic features characteristic of the ones found among the cave populations in northern México. This new cavernicole morph is interesting as it may provide evidence for both, independent colonization of and convergent adaptation to the hypogean environment by *Astyanax* in México.

This population was found in the 'Gruta de las Granadas' cave in Guerrero, southern México (Espinasa 1986). This is a cave resurgence of approximately 180 m from the entrance to the final sump, from which water flows year round. Water springs from a spacious sump at the rear of a lake room where the current is almost negligible. From here the stream meanders gently out on a gravel floor until reaching the entrance. A pile of breakdown rocks at the entrance of the cave serves as a partial barrier, which impedes surface fish from entering the cave except in the rainy season when the river flows above the breakdown.

Astyanax sp. in the stream emerging from this cave is abundant. Hundreds of individuals are seen in the pools of this surface stream. They are all fully eyed and pigmented as epigean *Astyanax* elsewhere in other surface rivers. Approximately 300 m downstream from the cave entrance there are two waterfalls of 32 m and 20 m in height, respectively. These waterfalls effectively isolate this section of surface stream and the cave populations from downstream fishes. The stream eventually joins the Amacuzác River in the Balsas drainage.

The Granadas *Astyanax* consist of a population of fully eyed and pigmented fish mixed with troglomorphic fish with a gradation of different levels of eye development and pigmentation. The eyed fish appear to come from the surface and is the most common morph in the cave.

In order to characterize the cave population, some aspects of fish morphology and behavior were studied and compared with those of blind tetras from northern México. The latter are not only typically troglomorphic but also display reduced scotophilia, schooling behavior, and aggressive behavior when compared to epigean populations (Wilkens 1988). Troglomorphic blind tetras from northern México also have a specific feeding behavior, they search for food while swimming at a 55 degree angle from the bottom. Conversely, the epigean morph can only feed while maintaining a steep angle of about 80 degrees (Wilkens 1988).

Materials and methods

Troglomorphic fish were collected from the Gruta de las Granadas cave, northeast of Taxco, Guerrero, México, 99°33′7″W, 18°34′30′N. The epigean fish were collected from the surface stream that comes out of this cave, which is a tributary to the Río Amacuzác, in the Balsas drainage. Some collected individuals were preserved in either 95% ethanol or 10% formaldehyde, while others were kept alive in the laboratory.

Eye development and levels of pigmentation were studied in the field or in the laboratory using a dissecting microscope. For the morphology of the brain, the top of the skull was opened under a dissection microscope to expose the different cranial lobes. Measurements were also done under the microscope. For the histology of the eye, six adult epigean and six adult hypogean fish with high degree of pigment and eye degeneration (troglomorphic) were anesthetized and sacrificed. The cerebra of these specimens, together with the eyes, were fixed in 10% formaldehyde, embedded in paraffin and sagitally sectioned in 5 mm slices, which in turn were stained with hematoxilin–eosin. For the osteological description of the skull, fish were cleared and stained following the procedure described in Taylor & Van Dyke (1985), a technique for staining bone and cartilage. Drawings were done using a camera lucida.

The responses to light for both epigean and cave morphs were studied by constructing an opaque observation tank of 18 cm by 130 cm with a black cover to create light and dark halves of equal size. Two 100 watt light bulbs were placed 70 cm above the tank to illuminate one side only. Single fish was placed in the tank and, after a 10 min to adjust to the new environment, the amount of time they spent in the illuminated side out of a 20 min period was recorded. Fourteen troglomorphic and fourteen epigean individuals were tested. A two-tailed t-test was performed to establish whether there was significant difference in time spent in either side of the tank.

For studying schooling behavior a grid of eight sectors of equal area was laid in a tank of 60 by 90 cm and the tank then filled with water to a height of 20 cm. Two 100 watt light bulbs were placed 2 m above the center of the tank for illumination. Six epigean and six troglomorphic fish were placed in the tank at the same time and at 20 sec intervals for 10 min, the number of fish in the sector having the most fish was recorded. We conducted four replicates for each of the six-fish groups. The expected distribution of fish in the absence of schooling was determined by a Monte Carlo simulation (hypothetical mean of maximum number of fish in any sector = 2.21) and was used to determine which populations were significantly different from the expected 'no schooling' mean.

We measured the angle at which fish search and pick up food from the bottom by videotaping six epigean and six troglomorphic fish in a five gallon tank with food scattered on the bottom. They were videotaped for 10 min and the film was later analyzed. Feeding behavior was recorded and the angle of their bodies to the tank floor measured. The data were analyzed using a two tailed t-test.

We measured aggressive behavior (a fish swimming vigorously towards another and trying to bite it) by placing six epigean and six troglomorphic fish in a five gallon tank and the number of ramming attempts was recorded over periods of 10 min. Twelve replicates substituting different individuals were done for the troglomorphic fish and ten replicates for the epigean fish. The data were analyzed using a two tailed t-test.

Results

Eye development and pigmentation

The levels of eye development were classified according to the following categories (Figure 1): (a) eyeless: eye sockets overgrown with tissue to level of cheek; (b) sunken eye: eye socket appearing as a small pit; eye size reduced somewhat; (c) reduced pupil: iris tissue enlarged to almost occlude the pupil; eye size may or may not be reduced; (d) eyed: eyes normal or nearly normal, but always convex.

Since 1983 we have collected fish from Granadas at least once every year, with 30 to 100 fish collected each time. Close to 90% of all individuals collected belong to the eyed category. Fish with sunken eyes and fish with reduced pupils are more or less equally common and together they constitute ca. 10% of the population. Eyeless fish are very rare, with only two individuals collected each on a separate occasion.

We observed asymmetrical degeneration of the eyes, such that one eye was normal while the other was highly degenerated (Figure 1). We also observed different levels of pigmentation among the Granadas cave fish, although none showed a total loss of pigmentation as is seen in some blind tetras of northern México. Reduction in eye size/structure and pigmentation, although often correlated, are not necessarily linked. Highly reduced eyes can be found in fully pigmented fish, and vice versa.

Eye histology

The eye in the epigean fish is not different than that of a typical diurnal teleostean. The sclera is thin, consisting of connective tissue, without bone or cartilage. The anterior sclera forms the cornea. The cornea itself is thick. The coroid consists of connective tissue, with rather fibrous argentea, blood vessels, nerve fibers and a little falciform process that extends into the vitreous body. The retina shows the normal, ten well-developed layers. Cones are in high density. The pigmentary epithelium cells have the characteristic cylindrical form.

In the troglomorphic fish eyes, the following modifications can be found: lens reduced or even absent; reduction in the size of the pupillary opening; thickening of the cornea; modification in size of the chambers; flattening of the pigmentary epithelium; reduction of size, organization, number of layers and even absence of the retina. Modifications in different structures may be independent from one another, such as eyes in which the retina is reduced, but the lens remains of proper size and morphology.

Brain morphology

Although there is variability within populations, troglomorphic fish generally have smaller mesencephalon (optic lobes) and tectum opticum, but larger prosencephalon than their epigean counterparts (Figure 2).

Skull

The circumorbital bones are modified in those individuals that show eye reduction although the degree at

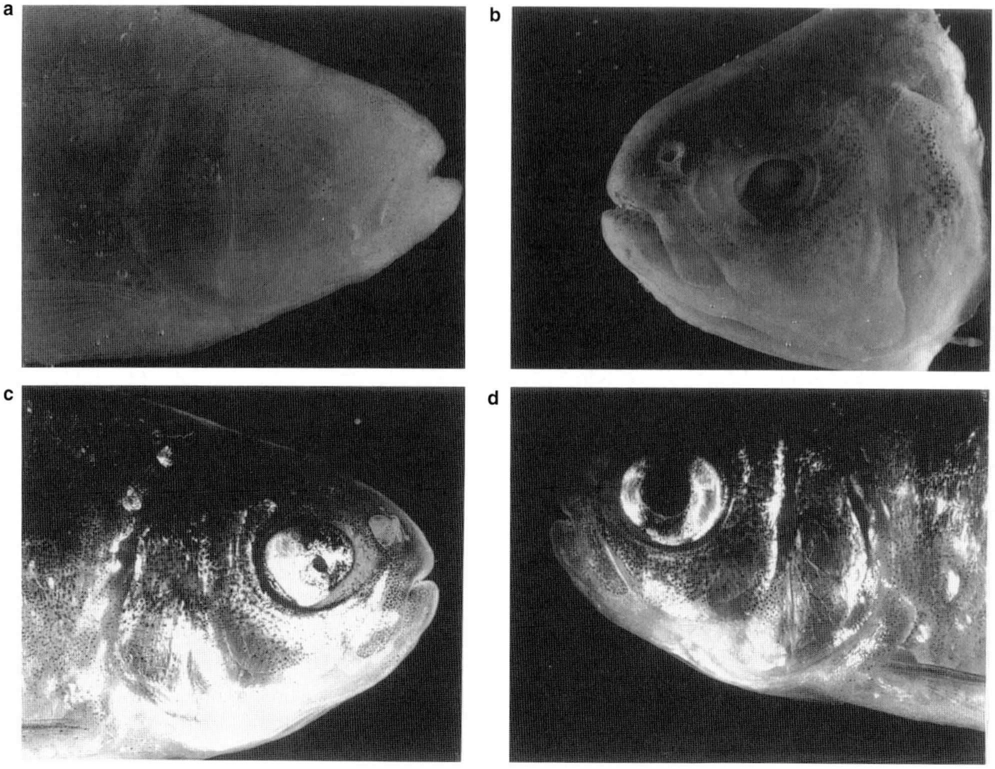

Figure 1. Levels of eye development in the Granadas Cave population: a – eyeless, b – sunken eye, c – reduced pupil, d – eyed. Pictures c and d are right and left sides of the same fish to show asymmetrical degeneration.

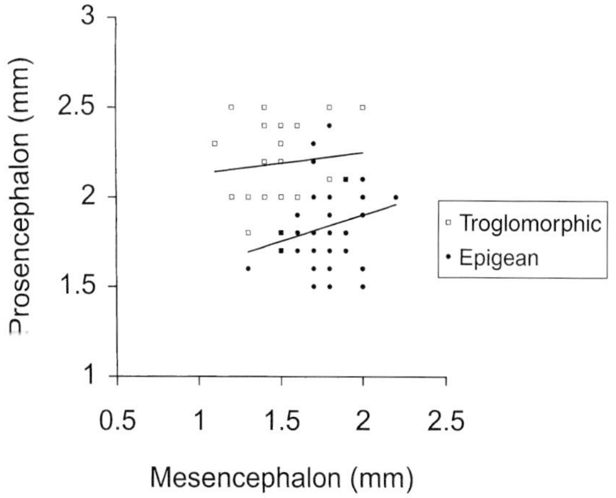

Figure 2. Length of optic lobes and prosencephalon in troglomorphic and epigean fish.

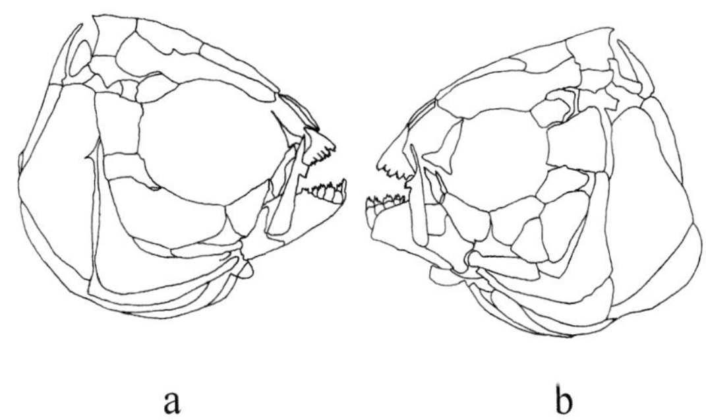

Figure 3. Skull of (a) epigean fish and (b) troglomorphic fish.

which that occyrs is highly variable (Figure 3). Even within individuals we found asymmetry in the number of fragments of particular circumorbital bones from one side to the other.

Responses to light

Epigean *Astyanax* sp. showed scotophilia ($p < 0.001$), spending an average of less than 20% of the time in the illuminated side of the tank while troglomorphic individuals showed no preference for light or darkness, spending close to 50% of the time in the illuminated side ($p > 0.01$).

Schooling behavior

The epigean fish aggregate in groups much more than the troglomorphic fish do. The troglomorphic fish do not aggregate significantly more than the Monte Carlo expected mean of 2.21 fish in the sector having the most fish. The epigean fish aggregate significantly more ($p < 0.001$) than the troglomorphic one.

Feeding behavior

Fifteen feeding events were recorded for the epigean fish and 50 for the troglomorphic. The two morphs search for food at significantly different angles ($p < 0.001$) with respect to the horizontal. Troglomorphic fish use a smaller angle (mean = 50.9 degree from the horizontal, SD = 11.6) and epigean fish use a larger angle (mean = 72 degree from the horizontal, SD = 14.8).

Aggressive behavior

The troglomorphic fish display reduced aggressive behavior (mean = 18.9, SD = 16.4 SD attacks in 10 min) relative to the epigean morph (mean = 140.4, SD = 99.9 attacks in 10 min) which is highly significant ($p < 0.001$).

Discussion

For both morphology and behavior, the troglomorphic fish of Granadas are intermediate between surface fish and the fully blind fish of northern México. The new population of cave dwelling *Astyanax* sp. is similar in many ways to the hybrid swarms formed by surface and cavernicole fish from the lower pools of La Cueva Chica (Breder 1942, Romero 1983), one of the classic cave localities of blind tetras. The eye morphology of this cave population also shows different levels of degeneration such as eyeless, sunken eyed, and eyed. The level of eye development referred to in this paper as 'reduced pupil', although present in Cueva Chica, is not as abundant as in the Granadas cave population.

A greater size reduction in individual eye structures is more prevalent in populations of northern blind cave fish (Wilkens 1988) than in the Granadas fish. Furthermore, our histological analysis corroborates previous reports (Fack & Wilkens 1988) that within the *Astyanax* and other vertebrate eyes, there are two subunits of differentiation. On one hand, the lens plays a dominant role and influences the degree of development of the pupillary opening, cornea, and vitreous body. On the other hand, the retina is closely tied to the development of the pigmentary epithelium, the lens muscle, and probably the vitreous body.

The brain morphology of Granadas fish is similar to other cave fish. Reports in which the mesencephalon (optic lobes) is reduced while the prosencephalon is enlarged are found for the blind tetras of northern México (Schmatolla 1972) and Texas blind catfishes (Langecker & Longley 1993) among others.

Results of all the behavioral experiments for *Astyanax* sp. are similar to those previously reported for northern *Astyanax* (Wilkens 1988). We utilized cave and surface fish from northern México as controls in our experiments and basically obtained the same results as previously reported (unpublished data).

In our studies, troglomorphic fish did not display scotophilia. This is not consistent with similar studies carried out on other troglomorphic populations in northern México (e.g., Romero 1985), in which scotophilia is very weak. This contradictory results may be due to the use of different statistical techniques and/or experimental approaches. Romero (op. cit.) reported variability in the degree of scotophilia among different cave populations. Thus, more studies are needed to ascertain the actual degree of scotophilia for the Granadas troglomorphic fish.

The troglomorphic fish are less likely to form schools. This observation is consistent with previous observations on the cave populations of northern México (Parzefall 1992). Both in the northern caves and in the Granadas cave, fish search for and pick up food while swimming at a low angle from the bottom (55 degree for northern cave populations and 51 degree for Granadas population). In contrast, eyed forms feed on the bottom while maintaining a steep angle (80 degree and 72 degree for the northern and Granadas epigean populations). It has been proposed that this low feeding angle in northern cave fish give them the advantage of being able to efficiently apply to the floor the extended taste bud area on the ventral side of the head (Wilkens 1988).

As for some northern cave populations, aggressive behavior is highly reduced in the troglomorphic Granadas fish. The reasons for a reduction of aggressive behavior in the cave morphs have not yet been elucidated, although it is possible that the epigean aggressive behavior is released almost exclusively by vision, thus it is no longer influenced by stabilizing selection and becomes genetically reduced

(Wilkens 1988). Yet, studies conducted by us in other Mexican caves (unpublished data) have suggested that blind fish display aggressive behavior. Thus, chemicals cannot be ruled out as triggering signals for this type of behavior.

We believe that the troglomorphic features in the Granadas fish are the result of genetical changes, not epigenetics. This conclusion is based on the following observations: (1) individuals of epigean *Astyanax* bred and kept in darkness do not experience a reduction in size of the eyeball, and only a reduction in the thickness of the retina has been reported (Wilkens 1988); (2) the troglomorphic fish of Gruta de las Granadas have reduced eyeballs (Figure 1) and the retina not only has fewer layers, but may also be entirely absent; (3) dramatic morphological changes in the skull and brain are found in the Granadas population; (4) the angle at which troglomorphic and epigean *Astyanax* fish search and pick up food is genetically based (Schemmel 1967, 1980).

Acknowledgements

Adriana Barona and Gerardo Obispo performed the brain measurements, Clemente Eduardo Granados and Jason Kimball did the behavioral counts, and Leticia Parra and Rosario Hortíz did the histological preparations. Monika Baker read the manuscript and made valuable suggestions. We also thank Oscar Dorado for general support.

References cited

Breder, C.M. 1942. Descriptive ecology of La Cueva Chica, with special reference to the blind fish *Anoptichthys*. Zoologica 27: 7–15.

Bussing, W.A. 1985. Patterns of distribution of the Central American ichthyofauna. pp. 453–473, *In:* F.G. Stehle & S.D. Webb (ed.) Topics in Geobiology, Vol. 4, Plenum Press, New York.

Contreras-Balderas, S. & M. de L. Lozano-Vilano. 1988. Problemas nomenclaturales de las formas mexicanas del género *Astyanax* (Pisces: Characidae). Zoología Informa (38): 1–13.

Eigenmann, C.H. 1917. The American characidae. Mem. Mus. Comp. Zool. 43: 1–558.

Espinasa, L. 1986. Exploraciones bioespeleológicas en Guerrero. Tepeyollotli. Gaceta SMES 1: 3–4.

Fack, H. & H. Wilkens. 1988. Eye reduction in hybrids and a naturally variable cave form of *Astyanax fasciatus* (Characidae, Pisces). Mém. Bioespéol. 15: 1–20.

Hubbs, C. 1938. Fishes from the caves of Yucatan. Carnegie Inst. Wash. Publ. 491: 261–295.

Langecker, T.G. & G. Longley. 1993. Morphological adaptations of the Texas blind catfishes *Trogloglanis pattersoni* and *Satan eurystomus* (Siluriformes: Ictaluridae) to their underground environment Copeia 1993: 976–986.

Mitchell, R.W., W.H. Russell & W.R. Elliot. 1977. Mexican eyeless characin fishes, genus *Astyanax*: environment, distribution, and evolution. Spec. Publ. Mus. Texas Tech. Univ. 12: 1–89.

Obregon-Barboza, H., S. Contreras-Balderas & M. Lozano-Villano. 1994. The fishes of northern and central Veracruz, México. Hydrobiologia 286: 79–95.

Parzefall, J. 1992. Behavioral aspects in animals living in caves. pp. 327–376. *In:* A.I. Camacho (ed.) The Natural History of Biospeleology, Museo Nacional de Ciencias Naturales, Madrid.

Romero, A. 1983. Introgressive hybridization in a population of *Astyanax fasciatus* (Pisces: Characidae) at La Cueva Chica. Nat. Speleol. Soc. Bull. 45: 81–85.

Romero, A. 1984. Behavior in an 'intermediate' population of the subterranean-dwelling characid *Astyanax fasciatus*. Env. Biol. Fish. 10: 203–210.

Romero, A. 1985. Ontogenetic change in phototactic responses of surface and cave populations of *Astyanax fasciatus* (Pisces: Characidae). Copeia. 1985: 1004–1011.

Romero, A. 2001. It's a wonderful hypogean life: a guide to the troglomorphic fishes of the world. Env. Biol. Fish. 62: 13–41 (this volume).

Schemmel, C. 1967. Vergleichende Untersuchungen an des autsinnesorganen ober-und unterirdisch lebender *Astyanax*-Formen. Z. Morphol. Tiere 61: 255–316.

Schemmel, C. 1980. Studies on the genetics of feeding behavior in the cave fish *Astyanax mexicanus* f. *Anoptichthys*. An example of apparent monofactorial inheritance by polygenes. Z. Tierpsychol. 53: 9–22.

Schmatolla, E. 1972. Dependence of tectal neuron differentiation on optic innervation in teleost fish. J. Embryol. Exp. Morphol. 27: 555–576.

Taylor, W.R. & G.C. Van Dyke. 1985. Revised procedures for staining and clearing small fishes and other vertebrates for bone and cartilage studies. Cybium 9: 107–109.

Trajano, E. 1987. Fauna cavernicola brasileira: composiçâo e caracterizaçâo preliminar. Rev. Bras. Zoologia 3: 533–561.

Wilkens, H. 1988. Evolution and genetics of epigean and cave *Astyanax mexicanus* (Characidae, Pisces). Support for the neutral mutation theory. Evol. Biol. 23: 271–367.

Environmental Biology of Fishes **62:** 345–364, 2001.
© 2001 *Kluwer Academic Publishers. Printed in the Netherlands.*

Adaptations of cave fishes with some comparisons to deep-sea fishes

Thomas L. Poulson[a]
Biological Sciences, University of Illinois-Chicago, Chicago, IL 60607-7060, U.S.A.
[a]*Current address: 318 Marlberry Circle, Jupiter, FL 33458-2850, U.S.A. (e-mail: tomandliz@adelphia.net)*

Received 21 August 2000 Accepted 22 January 2001

Key words: amblyopsid fishes, Mexican cave characins, rigor, variability, predictability, food supply, lack of light, progenesis, preadaptation, punctuated vs. gradual evolutionary change, convergence, life history patterns, egg size and number, mesopelagic, bathypelagic, bathybenthic, seamount

Synopsis

I provide my retrospective and prospective views on adaptations of cave fishes. I emphasize the history of my insights into cave adaptation from 45 years of research using surface, cave-spring, and cave species of amblyopsid fishes. My approach has been to use natural experiments and to always consider multiple hypotheses. To clarify evolutionary adaptations, I show the importance of a broad comparative approach which includes studies of morphology, metabolic physiology, foraging behavior, life history, and ecology. And I show that the most important agents of selection, of darkness and attendant low food supply, are best understood in the context of rigor, variability, and predictability. I also present my insights from what I consider the most insightful contributions on deep-sea fishes. The contributions are those of Marshall in studies of morphology in relation to energy economy of pelagic and benthic species, Childress in studies of physiological and biochemical adaptations with depth for pelagic species, and Koslow in studies on population biology and life history of bathybenthic and benthic sea-mount species. Compared to caves, I suggest that the extremes of metabolic and life history adaptations of deep-sea fish are explained by a longer evolutionary history and a much greater habitat range, food supply, and predation risk. Finally, I take a retrospective view of what we have learned about cave fishes. I discuss possible evolutionary mechanisms that can explain the trends with increasing cave adaptation in amblyopsid fishes, especially progenesis and the pleiotropic effects of the stress resistance syndrome. Finally, based on insights from deep-sea fishes, and emerging new techniques, I suggest what cave fish biologists should do in the future.

Introduction

Let me make it clear that this is not meant to be a review of all work on adaptations of either cave fishes or deep-sea fishes. Rather it will be a presentation of retrospective and prospective ideas from my own work. I will emphasize why my broad, long-term, and comparative studies of amblyopsid fishes have given me insights over 45 years of study and suggest how my approach should shape studies of other cave fishes, especially the Mexican cave characins. I do not include the important topic of 'regressive evolution' of eyes and pigment which has been extensively reviewed by many authors (e.g., Culver 1985, Culver et al. 1995). I will consider only those students of deep-sea fishes from whom I have gained important insight and perspective about cave adaptation.

Only a few workers have made extensive and insightful contributions to our understanding of metabolic, foraging, life history, and morphological adaptations of cave fish or deep-sea fish. In the 1950s and 1960 N.B. Marshall studied so-called progressive evolution of deep-sea fishes and I studied progressive evolution of amblyopsid cave fishes (Marshall 1984). In the 1970s Jim Childress and his many colleagues contributed a monumental amount of careful work on the growth and metabolic adaptations of a wide variety of pelagic deep-sea organisms (Childress et al. 1980, 1995). And in the 1990s, Tony Koslow and his colleagues added important perspectives on benthic deep-sea fishes, especially those that aggregate around sea mounts (Koslow 1996, 1997). I will show how the synthetic overviews by these authors of their own and each other's past work, have influenced my thinking about adaptations to caves among amblyopsid fishes (Poulson 1985 and this contribution).

A retrospective of influences on my research

Formative years through publication of my dissertation (1950s–early 1960s)

I got interested in cave fishes while an undergraduate at Cornell University but I had always been involved in natural history. I collected and raised butterflies and moths starting at age three. I watched, photographed, and banded birds starting at age 12. And first heard about caves from Charles Mohr while a trail guide at the Audubon Camp in Connecticut in 1950. At that time I was still most interested in birds and only knew of his work in caves on bats (in 1966 Charles and I co-authored 'Life of the cave' in McGraw Hill's 'Living world of nature' series which sold over 125 000 copies). In 1952 I won second prize in the New York State Science Fair for a project based on inducing and testing alternative hypotheses (in chemistry!) and went to study birds at Cornell. By the time I graduated I knew that I would do a comparative study of cave fishes at the University of Michigan for my Ph.D. What changed my mind? In 1952 I went on my first cave trip with the Cornell 'Grotto' of the National Speleological Society to a cave in New York State, squirmed into a tight hole and got really excited when I found troglobitic amphipods in a rimstone pool. A second cave connection came when I helped Don Griffith and his students to band and study bats (I taught free-flying big brown bats to eat a mealworm-cheese glop in my room and got kicked out of the dorm). At the same time I became impressed with the natural history knowledge of ichthyologist Ed Raney and went on a spring break cave collecting trip with his students to the Carolinas. Kicking through a weed bed at nightfall netted us some funny looking little swampfish, named *Chologaster cornuta*, that the graduate students told me were not of interest for Systematic Ichthyology since it belonged to an obscure and small family, the Amblyopsidae, in which most of the species occur in caves. Bingo, another cave connection! The graduate students preserved their fish but I brought some swampfish back alive and watched them day and night in an aquarium on my desk. I learned immeasurably more about their feeding behavior and morphology than I found when I later went to the library and found Eigenmann's monograph on cave vertebrates. Ed Raney was annoyed that I had abandoned his assigned systematic research project counting fin rays and scales of silversides. I told him it was dull compared to swampfish behavior but he would not let me off the hook on his project. I wrote up what I thought was an excellent paper on the swampfish and he gave me the only C grade I received in science. Undaunted and more determined, I went to the American Society of Ichthyologists and Herpetologists meeting with a paper on the swampfish and there proposed to Bob Miller that I come to the University of Michigan at Ann Arbor to do my Ph.D. with him on a comparative study of cave adaptation in the Amblyopsidae.

At this juncture, influential mentors enlarged my focus again, from Bob's interest in how behavior and morphology related to systematics of the amblyopsids to a broader approach that included how morphology, behavior, metabolic physiology and life history related to cave adaptation. Howard Schneiderman's Comparative Physiology course at Cornell had enthused me with its rigor and emphasis on testing hypotheses and then Bill Dawson's Environmental Physiology course at Michigan showed me that I could combine natural history with physiology. I approached Bill and asked him to sponsor my project but he was reticent. Fish were not his expertise and I would be his first student. He promised that he would accept me if I did well in the three day, 18-hour, all-subject diagnostic essay exam for entering students. Due to my self-designed broad curriculum at Cornell with 160 h (only four years), I did exceedingly well and Bill reluctantly agreed to be my dissertation advisor. We decided that I should include ecological approaches to metabolic physiology, a la Fred Fry at Toronto, since I had a favorite hypothesis that food limitation in caves was the strongest agent of selection. At this point the ecology faculty at Michigan made me aware of competition and life history theory and I proposed to my committee that I include this perspective. My committee demurred but I stood firm and, in retrospect, I was able to avoid the negative tradeoff of academic rigor but shallow depth and so set the stage for the next 35 years of my contributions to understanding adaptation to caves by fish.

Before my dissertation on amblyopsid fishes, studies of cave fishes had been fragmentary. The pioneering studies of the biology of cave fishes were done from the 1940s to 1960s but they were narrow in scope (reviewed by Poulson 1964). In the 1940s the most detailed studies were on Mexican cave characins by Charles Breder, Priscilla Rasquin, and their colleagues. These studies mainly concerned the cave species and their surface ancestor with overwhelming emphasis on the effects of regression of eyes and pigment and only a little work on chemosensory organs. Only some behavior and physiology has been added recently by Horst Wilkens, Jake Parzefall, and Katherine Huppop (in Culver 1985). In the 1960s I followed up the fragmentary studies on amblyopsid fish feeding and sense organs by students of Eigenmann (e.g., Payne in Eigenmann 1909) and on general morphology by Eigenmann (1909) with broad studies of morphology, sense organs, behavior, metabolic physiology, and life history[1] (Poulson 1961, 1963).

I compared the one surface species, *Chologaster cornuta*, one cave-spring species *Chologaster agassizi*, and three cave species as a natural evolutionary experiment to infer important agents of selection in temperate zone caves. I reasoned that the increasing reduction of size and eye structures from *Typhlichthys subterraneus*, to *Amblyopsis spelaea*, to *Amblyopsis rosae* (Eigenmann 1909) reflected increasing evolutionary time in caves (Poulson 1963, 1985). From this evolutionary series I inferred evolutionary patterns of adaptation in obligate cave fishes. From Eigenmann's drawings and my observations of museum specimens, I was convinced from the start that increasing relative head size and attendant elaboration of the free lateral line organs and their brain centers was of central importance in enabling the fish to find what I thought would be scarce living prey in caves. After all, caves have no light and so there is no photosynthetic primary production. I used a combination of plankton nets and timed areal searches for isopods and amphipods under rocks to document the similar and low food supply in amblyopsid caves (Poulson 1961). Only recently have I reviewed the data in depth on food limitation in temperate and tropical caves and lava tubes (Poulson & Lavoie 2000).

[1] Poulson, T.L. 1969. Population size, density, and regulation in cave fishes. Proc. IV Int. Cong. Speleol. Ljubljana 4–5: 189–192.

The rise and fall of the constant cave paradigm with insights from the deep sea

In addition to the mystique of poorly known environments with fish of odd morphology, I saw simplistic scientific parallels between fish in caves and the bathypelagic realm of the deep-sea (Poulson 1961). Both were essentially without light and so should be food-limited. Both had cool to cold constant temperatures and constant water chemistries. And both cave fishes and bathypelagic fishes had evolutionarily convergent elaborations of sense organs that allow them to find scarce food in the dark. Externally the fishes differed only in color and eyes. With bioluminescence in the deep-sea, fish were brown or black and retained small eyes. And without bioluminescence cave fishes were 'white' and blind with extreme reductions of pigment cells and eyes.

Along with scientists studying the deep-sea in the 1960s, I made the point that relative constancy of the abiotic environment made the deep sea and caves attractive 'natural laboratories' for studying ecological and evolutionary patterns and principles. I gained perspective on caves with my review that emphasized physiological, especially metabolic, adaptations to the cave environment (Poulson 1964). And I became excited by interdisciplinary work with colleagues in The Cave Research Foundation that led to an invited lead article for Science (Poulson & White 1969) in which the importance of environmental stability was one working hypothesis. At the same time Howard Sanders and his colleagues were finding fantastic densities and species diversities of sediment in-fauna in the deep sea and championing their time-stability hypothesis as an explanation for evolution of species diversity with parallels from the tropics (e.g. Sanders & Hessler 1969). Stability was also a theme of a symposium on Cave and Deep-sea Comparisons that I organized in 1970 for the American Association for the Advancement of Science (Poulson 1971).

In the late 1960s I became convinced that the 'constant' cave paradigm was wrong. I tested predictions from the hypothesis using physiological and ecological approaches. I falsified my hypothesis that there would be increased sensitivity to temperature change with increasing evolutionary time in caves among amblyopsid fish. I showed that, from 5°C to 25°C, standard and active metabolic rate varied very little and that there was no difference from cave-spring to cave species or among the cave species (Poulson unpublished). In retrospect I realize that the fish had to be physiologically adapted to the rare winter and summer floods that change water temperature and bring in detrital food. Then, in a baiting study in the field, my first student and I showed that rigor and predictability of flooding and drying had significant negative influence on species diversity of terrestrial cave organisms (Poulson & Culver 1969). We were initially surprised at these results because the environmental gradients in the cave were subtle and the rigors were low. But in retrospect we concluded that we had missed the idea of appropriate spatial and temporal scales. With little variability, subtle abiotic gradients become especially important in terrestrial cave environments and rare 'disturbances' like floods become important in aquatic cave environments. Much of my subsequent work in caves has borne out this revised view (reviewed in Poulson 1992 and Poulson & Lavoie 2000) and has clarified differences between caves and the deep sea.

Multiple hypotheses and long-term studies

From the 1970s, my insights about adaptations of cave fishes and cave ecology have grown with my continued study in the Mammoth Cave region (Poulson 1992, Poulson & Lavoie 2000) and with my studies of old-growth forests and of dunes succession. The efficacy of my use of multiple hypotheses and long-term studies has been confirmed again and again. My major take-home message for colleagues and students is that this approach ensures serendipity with occasional 'eurekas'.

The use of multiple hypotheses is central to all my research. Every time I go to the field I try to distinguish among multiple hypotheses. Not only does this avoid the trap of a favorite hypothesis but it also forces me to ask what critical observational or manipulative data are needed to falsify predictions from each hypothesis. And I use new observations to generate revised or new hypotheses.

I champion the use of natural experiments to test hypotheses. I use natural experiments as an efficient way to falsify most hypotheses and indicate which manipulative experiments are needed. On an evolutionary scale the natural experiments are different species of cave amblyopsids. On an ecological scale the natural experiments are replicates of the same habitat within a cave, different streams in the same cave, and different caves. I also use 'trajectory experiments' where I follow the same population of cave fish over time. Trajectory experiments are critical for organisms that live much longer than the time constants involved with renewal of research grants and the granting of tenure for academic scientists. And long-term observations have shown me that unpredictably rare 'good' and 'bad' times are critical for long-lived species, such as amblyopsid cave fishes.

Let me exemplify all these points with several instances drawn from 40 years of observation of amblyopsid fishes.

In the 1960s I started periodic censuses of body length and location for each individual of different cave species in streams. The streams were shallow enough to see all individuals with carbide lamps, not only fish but also crayfish, and where all likely microhabitats were 'accessible' for census. In ca. 100 caves only five met these conditions. Younger, more vigorous, and less hypothermia-prone colleagues have continued these censuses. The continued pattern of missing size classes allowed me to demonstrate that successful reproduction is infrequent and based on rare renewal of detritus by '50–100 year floods'.

In 40 years of periodic censuses of cave fish populations, I have noted rare good and rare bad years, with + and − tradeoffs. For *Amblyopsis spelaea*, in the Twin-Bronson-Donaldson cave streams in Spring Mill State Park Indiana, little or no reproduction occurred in 13 different census years when there were little or no floods. Then in one year of high floods with possible injury to fish (−) there was a massive input of detritus (+). In this year my colleague Bill Pearson observed a high proportion of adult females carrying eggs or young in their gill chambers (+). In four Mammoth Cave streams, where *Typhlichthys subterraneus* is found, there was renewal of particulate organic matter in 2 of 40 years (+). In one of these years the associated floods completely changed the proportions of silt and cobble bottoms and eliminated many of the best deep pool habitats along a kilometer of one stream in one year (−). But along another stream habitats were unchanged and the subsequent appearance of small fish and crayfish indicated successful reproduction (+). With longevities of 20 to 30 years for cave fishes, such statistically '100-year' floods are essentially unpredictable for individuals but have important population consequences, within and among species. For example, in a Kentucky cave there was successful recruitment after an episode of increased reproduction and decreased density-dependent cannibalism. I had used this cave as a source for live specimens for laboratory studies and so had reduced the numbers of adults and greatly decreased the metabolic demand of the population with the prediction that there would be more reproduction.

An unanticipated case of unexpected insight came from long-term censuses of several streams in Mammoth Cave along an upstream to downstream gradient where the smaller *Typhlichthys subterraneus* is replaced by the larger *Amblyopsis spelaea*. The gradient (Poulson 1992) is of decrease in stream velocity, substrate particle size, and food supply and increasing depth and water volume. At the end of a 30-hour trip we visited a new stream as an afterthought despite fatigue and incipient hypothermia, and I was surprised to observe a hiatus of no fish at the upstream to downstream transition. The 'eureka' idea was that ongoing competition with character displacement could be an alternative to a hypothesis that the upstream to downstream species replacement was a matching of species size and water depth. If character displacement is the correct hypothesis then I predict a reversal of the *Amblyopsis* > *Typhlichthys* metabolic rate seen among other caves (Poulson 1963, 1992). The data have yet to be gathered.

A final example of a 'eureka' comes from studies on growth rate allowed by improved micro-tagging techniques developed for small cave fish by both Bill Pearson and Jim Johnson and their students. Recapture and growth rate results for *Amblyopsis rosae*, with moderately high average growth rates and projected longevities of ca. 10 years, were not consistent with my hypothesis that the species is longer-lived. Since most adults had projected growth of <1 mm per year, my projected average longevities were closer to 20 years. My 'eureka' insight came from a combination of rare very high growth rates (as much as 20 mm a year in one 35 mm fish!), from capture-mark-recapture data, and my data on the rarity of large prey items, from gut contents of nearly 200 museum specimens. All the specimens had copepods in their guts, four had amphipods, but only one had a large salamander. My exciting hypothesis is that the low metabolic rate of *Amblyopsis rosae* enables high growth efficiency and so rare high growth rates in response to rare large meals. This is consistent with what Pearson and his students have found in the past five years (personal communication) for *Amblyopsis spelaea* and *Typhlichthys subterraneus*. They calculated that almost all adults had growth rates of <1 mm per year with only one or two of ca. 100 in the 2–6 mm per year range.

To me, bathypelagic fishes are similar to cave fishes in that they occasionally have high growth rates allowed by high efficiencies of growth. Bathypelagic sit-and-wait predators often attract prey with bioluminescent lures, have huge mouths with stabbing teeth, and distensible stomachs that allow them to swallow prey as large or larger than themselves. Their astounding growth efficiencies allow very high growth rates (Marshall 1960, Koslow 1997). I see the data in figures 1 and 2 of Childress et al. (1980) as consistent with my hypothesis of occasional large meals. The error bars in their graphs show much greater variability of growth rates of bathypelagic fishes than mesopelagic fishes with normal morphologies. An extension of my occasional large meal hypothesis for bathypelagic fishes is that adult females with a very large meal turn it all into reproduction followed by death, i.e. semelparity.

Adaptations of deep-sea fishes and cave fishes

We now know the nature of morphological, physiological, and life history specializations of deep-sea fishes in detail and I compare these to cave fishes in Tables 1, 2 and 3. The deep-sea habitats of interest are pelagic and benthic. Within the pelagic, from the epipelagic to the mesopelagic to bathypelagic habitats, food supply declines and swimming ability declines due to simplified body organization (Marshall 1960). These trends are associated with increased metabolic economy (Childress 1995). The benthic habitats are bathybenthic and seamounts and have both higher and lower food supplies and higher current velocities than the bathypelagic with attendant differences in swimming ability, morphology, metabolic physiology and life history (Koslow 1996, 1997).

By considering pelagic and benthic deep-sea habitats and caves, I am able to narrow down possible agents of selection that might, within any one habitat, account for the morphological, physiological, and life history patterns of the fishes in these environments. I will only deal with those agents that are demonstrably important because Childress (op. cit.) has superbly analyzed and rejected the importance of several abiotic factors that

Table 1. Energetic and life history patterns for deep-sea and cave fishes. For deep sea pelagic fish: epi = epipelagic, meso = mesopelagic, bathy = bathypelagic. For deep-sea benthic-benthopelagic species = benthic = dispersed species and seamount = aggregated species. The first column shows the traits and ranges of values while the second column give the order from high to low for the different deep-sea habitats and for caves. See the text for references from Koslow and Childress for the deep-sea fish data and Poulson for cave fish. The deep-sea data on reproductive patterns come from Mead et al. (1964).

Traits	Ranks from high to low
Food supply/productivity (adult habitat)	epi > mount = meso ≫ cave >= bentho > bathy
Metabolic rate (scaled to 5 g and 5C, 0.07–0.003)	epi > mount = meso ≫ cave = bentho = bathy
Caloric energy density of bodies wet wt. (1.3–1.7 > 0.9–1.2 ≫ 0.4–0.6)	mount >= epi > bentho = meso = cave ≫ bathy
Predation pressure including cannibalism	epi > meso >= mount > bathy > bentho> cave
Competition (intra- > inter-specific)	mount > epi >= meso > bentho > bathy >= cave
Current velocity	mount > epi > meso >= cave >= bentho > bathy
Efficiency of growth (0.28–0.68 > 0.08–0.15 > 0.02–0.05)	bathy > cave = bentho > meso > epi > mount
Reproductive pattern (few, large eggs–many, small eggs)	mount >= cave >= bentho > epi >= meso >= bathy
Longevity in years (~100 > 20–30 > 5–10 > 1–5)	mount > bentho >= cave ≫ epi >= meso >= bathy

Table 2. Tradeoffs (+ and −).

Large body size	
+ resistance to starvation	+/− high lifetime energy needs
+ eat large prey = sudden growth advantage	+/− high maintenance costs
+ higher maximum fecundity	− less chance of any reproduction
+ higher maximum sprint swimming	+/− more power needed for same speed
Schooling	
+ decreased per capita chance of predation (statistical dilution effect), confuses predator with large numbers, detect predator sooner	− attract predators
+ find food patches easier	− strong intraspecific competition
+ energetic advantage swimming (drafting)	− energetic cost of continual swimming
Long life, low r, and few large eggs (opposite tradeoffs for short life, high r with small & many eggs) [K = carrying capacity]	
+ slow population decline in bad times	− poor 'tracking' of K in good times
+ large & few eggs better early survival	− low spread of death risk among young
+ high spread of risk of reproductive failure among years	

might explain the observed patterns of metabolic rate with depth in the ocean. I concur for caves. Childress rejects incredible increase in pressure as important to deep-sea organisms and I believe caves are so shallow that pressure is not even conceivably a factor. And Childress (op. cit.) rejects the oxygen minimum layer at mid depths as important and I found that most caves with fishes have oxygen concentrations near saturation (Poulson 1961, 1964).

The remaining abiotic and biotic agents of selection are inextricably related. Without light, productivity will be limited and food supply low but predation by visual predators will be absent. This can lead to poor swimming ability except when current acts as

Table 3. Tradeoffs of studying cave fish and deep-sea fish to understand adaptation.

Category	Deep-sea fish	Cave fish
Annual growth (otoliths) → easy study of growth efficiency	+ yes (clear annual signal of surface productivity)	– no
Census + mark recapture → population size & 'density' + growth rate	– no	+ yes
Study of fish in situ	+/– no	+ yes
Realistic studies in the lab (issues of scale)	– no (schooling, ontogenetic migration, pressure, & huge areas/volumes)	+ yes
Variety habitats = evolutionary experiments	+ huge differences (e.g. among pelagic & among benthic)	+/– subtle differences among caves
Taxonomic control (phyletic constraints)	+/– often different families in different habitats (since Cretaceous)	+ cases with same species & family different caves (only since Pliocene)

a countervailing agent of selection either because of high current velocities or need to follow odor plumes up-current to find food.

Morphology, food supply, metabolic rate, predation risk, and swimming ability

Childress (1977) provided a critical and insightful analysis of the technical methodologies of studying metabolic rates of deep-sea organisms. He showed how methods should be matched to the questions asked. His work should still be required reading for anyone who studies either deep-sea fishes or cave fishes.

Two extremes of morphology, body composition, and food supply

Marshall (1960) showed that bathypelagic fishes never develop a robust adult morphology after they, as larvae, leave the food-rich epipelagic and metamorphose in the depths. Instead they retain poorly developed muscle and skeletons with gelatinous tissues in the subdermal spaces and, at best, are weak swimmers. They live in an environment that has virtually no current and without a swim bladder their high water content allows near neutral density. In addition to a tiny and fat-filled swim bladder, they also have extreme reductions in other organs. They have small brains, gills with few and very short filaments, kidneys with sometimes only single nephrons, scales that are tiny and embedded or absent, and tiny eyes. Luminescent photophore systems are not as numerous or elaborate as in mesopelagic species that migrate to the epipelagic at night to feed. But bathypelagic species often have elaborate appendages with photophores at the tip presumably to lure prey. Many species with lures have huge jaws with stabbing teeth and a very distensible stomach; together these allow the capture and ingestion of prey almost as large as the predator. Their black color makes them invisible to prey or predators that are bioluminescent. With lower food supplies, bathypelagic fish also have more elaborated extra-optic sensory systems than mesopelagic fishes, especially the lateral line and chemoreceptors (both distance reception by olfactory rosettes and body contact chemoreception).

In parallel to bathypelagic species, other deep-sea species and cave species have elaborated sense organs but in contrast have robust body builds with well-developed muscles and good to excellent swimming ability (Childress 1995, Koslow 1997, Poulson 1971). Benthic deep-sea and cave fishes also have few reductions of organ systems. Where it has been measured, they have high energy densities of 1–1.8 kcal g^{-1} wet weight associated with high protein contents, low water content, and variable fat contents whereas the bathypelagic species have low energy densities of 0.4–0.6 kcal g^{-1} with low protein and high water contents. All seamount-associated deep-sea fishes cope with variable and strong currents, have the highest energy densities, and are the best at both sustained and sprint swimming. All of the cave species have a robust body build with well-developed muscles which is, for amblyopsids at least, adaptive to the strong currents that occur during floods.

Metabolic rates and metabolic scope

Except for amblyopsid and *Astyanax* cave fishes that have been studied live in the laboratory, and weak-swimming bathypelagic fishes that are easily coaxed into a respirometer by biologists in deep-sea submersibles, most measures of metabolic rate are indirect because they are based on muscle enzyme activities. Citrate synthetase is part of the aerobic pathways and its activity gives an index of aerobic power or scope. Aerobic scope is the difference between standard and sustained active metabolic rates. It is the energy available, beyond maintenance costs, for foraging, growth, and reproduction and thus ecologically is a very important variable. Lactic dehydrogenase is part of the anaerobic pathway and its activity gives an index of anaerobic power or scope. Anaerobic scope is the energy available for short-term sprint or burst swimming which are both important in the capture of prey or the avoidance of predators. Note that it is the absolute energy available between routine metabolic rate, that includes maintaining position in a current, and maximum metabolic rate that is important; not the ratio of maximum to minimum energy.

Major differences in maintenance cost make it unclear which habitat has species with the greatest metabolic scope. Bathypelagic species are at the low extreme of swimming ability and muscle development with virtually no maintenance cost and they have a very low routine metabolic rate because they do not face currents and they have sit-and-wait foraging. With such low costs, virtually all extra metabolic capacity goes to growth and reproduction and this is reflected in extremely high efficiency of growth. But the metabolic scope is low because the maximum metabolic rate is low. Seamount species are at the high extreme of swimming ability and muscle development with high maintenance cost and they have a high routine metabolic rate because they face fast currents and they have pursuit foraging. So, despite high active metabolic rates, there is little extra energy available for growth and reproduction; this is reflected in extremely low efficiency of growth. Fishes from caves and from bathybenthic habitats are intermediate in efficiency of growth because they have intermediate routine metabolic costs and intermediate active metabolic rates. I have summarized information on metabolic rates in Table 1 (corrected for differences in body size and temperature). Epipelagic fish have, by far, the highest metabolic rates. Mesopelagic vertical migrators and seamount species, cave and bathybenthic species and bathypelagic species have sequentially lower metabolic rates.

Is low food supply a primary or secondary agent of selection in the bathypelagic?

Despite mostly indirect evidence everyone agrees that food supply declines greatly with depth in the pelagic zone and that somewhat higher food supplies are present in benthic habitats and caves. However, the quantitative documentation of food supply in the deep-sea is still more indirect and more difficult to document than in caves (Poulson & Lavoie 2000).

The indirect evidence for Marshall's inferred difference in food supply in the bathypelagic and bathybenthic was impressive from the start. Overall the density of in-sediment fauna decreases dramatically from 1000s inshore to <100s in the abyssal plains, an incredibly greater biomass and species diversity exists in the benthic sediments than in the bathypelagic zone which is just above the bottom and extends many kilometers up to the mesopelagic (Sanders & Hessler 1969). The frequency of seeing fish from submersibles or taking pictures from remote cameras is dramatically greater in the mesopelagic and bathybenthic than in the broad depth range of the bathypelagic in between (Hessler personal communication).

Childress (1995) agrees with most others that food limitation is the most intuitive hypothesis to explain the evolutionary trends in metabolic economy seen in bathypelagic fishes but he argues that it is not the primary agent of selection for loss of robust body build. For example, metabolic rate of bathypelagic species does not seem to correlate with surface productivity. Childress further argues that swimming ability will be evolutionarily lost since organisms with image-forming eyes cannot see predators approaching at a distance and so there is no selection for escape by burst swimming. I agree but suggest that this favorite hypothesis is not mutually exclusive of an energy economy hypothesis (admittedly my favorite hypothesis but one that I cannot falsify; Poulson & Lavoie 2000) and that some of his arguments for the unimportance of food limitation are weak. Thus the asymptote of metabolic rate decline at ca. 1000 m may reflect a limit of reduction even though food supply continues to decline with depth. It is hard to imagine that there could be further reductions in organs and muscle than that in bathypelagic fish at 1000 m (Marshall 1960). And I suggest that the lack of decline of metabolic rate of medusae and chaetognaths, which Childress (op. cit.) ascribes to the absence of image-forming eyes, could also be explained by the already very low metabolic rate of their equally gelatinous surface species relatives (as low or lower rates than those of any bathypelagic fish, Childress 1995, figure 2). Finally, not considered by Childress are deep-sea habitats that are in constant darkness but that have significant currents; fishes in this habitat are good swimmers and have normal body builds (e.g. Marshall 1960). However to give due credit, Childress (1980) predicted that if a high-current niche existed, it would have deep-sea fish with excellent swimming ability; a high current niche around seamounts was subsequently found and studied by Koslow (1996).

Bioluminescence as a side issue

Bioluminescence is important in the oceans but not in freshwater caves. No freshwater organisms are bioluminescent and all cave fish show convergence of extreme reduction of eyes and pigment. Unlike cave fishes that have vestigial eyes, many deep-sea fish permanently live below the maximum depth of light penetration from the surface and still retain moderately developed eyes. The reason is bioluminescence. Bioluminescence is important in the deep sea for species recognition, luring prey, and detection of prey and predators. Bathypelagic predators are black because this color absorbs light from the bioluminescence of potential prey and predators and so makes them difficult to see.

The influence of current velocities on swimming ability

Current velocity is an important agent of selection for some deep-sea fishes and cave fishes. At one extreme, species that aggregate around seamounts face variable and locally strong currents that concentrate food around seamounts by advection (Koslow 1996). Not surprisingly they have well-developed musculature and are strong swimmers. At the other extreme are bathypelagic fishes that are very weak swimmers and are barely able to stop from sinking even with near neutral buoyancy. They have very poorly developed muscles and a gelatinous consistency but these are not a problem in the essentially zero current velocity environment of the bathypelagic. In contrast, cave environments periodically have strong currents. Amblyopsid cave fishes have large heads with a hypertrophied lateral line system and so they are not streamlined. This means that they cannot swim well against fast currents despite well-developed musculature. Their normal musculature allows them to forage in areas of slow current and behaviorally avoid fast currents. I have watched cave fish (*Amblyopsis spelaea*) seek quiet areas under ledges and in back eddies when stream velocities are high during floods and Pearson (personal communication) has seen *Typhlichthys* hide when faced with only subtle increases in stream velocity.

Another niche that requires good swimming ability, not reviewed by either Koslow or Childress, is of opportunistic feeders on very large dead food items. These include giant amphipods (Hessler et al. 1972), hagfish, and macrourid fish that have been photographed by cameras suspended over large bait cans. Their food bonanzas are rare in time and space and so these scavengers should need to swim long distances up-current to follow odor plumes to the food location (e.g. Priede et al. 1991).

Metabolic costs in relation to rates and efficiencies of growth and to longevity

Some of the most exciting recent insights have been the incredible differences in demonstrated rates of growth and longevities and inferred efficiencies of growth in deep-sea fishes and cave fishes. These differences are summarized in Table 1 and I have alluded to data that suggest high growth efficiencies for cave and bathypelagic fishes in the earlier section on my insights from long-term studies and multiple hypotheses.

Both in seamount species and bathybenthic fishes, Koslow (1996, 1997) has reviewed the evidence pertaining to metabolic costs and growth rates and how these are related to longevity and food supply. Growth and longevity differ between the extremes of seamount benthic, and bathypelagic species with bathybenthic and cave fishes intermediate (Table 1).

Fishes in all of the benthic habitats have well-developed musculature and the seamount species are especially good at sustained and sprint swimming. Unlike the very short-lived species (1–<10 years) with adults in the bathypelagic, almost all of the benthic species spend their entire lives in the deep sea and have much longer lives (bathybenthic and benthopelagic of several decades) to very much longer lives for seamount species (>100 years). Cave fishes, where extensive data are available on the Amblyopsidae, are close to the bathybenthic species in morphology, physiology, and longevity.

An advantage for study of deep-sea fish growth and longevity is that annual growth rings on scales and otoliths clearly reflect annual surface productivity. A disadvantage is that data on overall size–age distributions are lacking for all species, even for those species commercially harvested (net size precludes catching small individuals). In many caves we can do nearly complete visible censuses of all fish sizes. However unlike the deep-sea we cannot get unambiguous ages from scales or otoliths. Even in the absence of unambiguous annual signals, direct censuses and mark recapture studies for cave amblyopsids over long times (ca. 5 years to several decades) have shown tradeoffs (Table 2) of growth rate, metabolic rate, population density, body size, longevity, and number and size of eggs. Such complete data have not been gathered for any deep-sea species.

The diversity of life history patterns in caves and in the deep sea

Various tradeoffs complicate interpretations of the ecological and evolutionary bases for patterns of life history. These tradeoffs involve body size, schooling, and life history mode (Table 2). The tradeoffs are associated with very different costs and modes of foraging and body builds that I have already reviewed in the context of very different costs of existence and efficiencies of growth. Bathypelagic fish are at one extreme with seamount deep-sea species at the other extreme. Cave fishes and bathybenthic fishes are intermediate.

Additional complications stem from the diversity of models that might explain patterns of life history. Model extremes involve deterministic and stochastic processes. The stochastic models better fit a metaphor of gambling for existence than a metaphor of competitive struggling for existence (deterministic) which is the basis for the classic dichotomy of r-selected vs. K-selected life histories. Dave Mertz and I convened a symposium on gambling vs. struggling at the American Association for the Advancement of Science in 1975. Speakers criticized the r- selected vs. K-selected dichotomy of life histories, based on the logistic equation, and the r- vs. K-selected dichotomy has not stood the test of time. Instead many stochastic models can explain differences in life history. The best of these models consider the amount and variability of early mortality compared to adult mortality and the frequency of successful reproduction (e.g. Murphy 1968, Stearns 1992). Given these controversies I have opted to consider the consequences of a non-controversial high-r vs. low-r life history dichotomy where r can be determined from age specific schedules of survivorship and reproduction.

Survivorship and reproductive patterns dictate low-r and high-r life histories

The overall life history pattern, on a high-r to low-r continuum [r = per capita rate of population growth $= \ln Ro\, T^{-1}$, where $T = (\Sigma lxmxx)\ (\Sigma\ lxmx)^{-1}$], depends on age specific survivorship (lx) across all ages and age specific fecundity of adult females (mx). Age specific mortality rate (qx) is the slope of the survivorship curve. Any good ecology text will explain the details (e.g. Krebs 1994, Begon et al. 1996). My perspective on + and − tradeoffs for high-r and low-r life history extremes is shown in Table 2.

For high-r fish with longevities generally < to five years the mortality rate starts high at perhaps 0.80 per year for year one and only diminishes to about 0.40 at an early age at first reproduction. The youngest adult females have a high reproductive output and a high reproductive value (Vx). Reproductive value only increases a little with age because reproductive lifespan is short and most extra energy goes to reproduction with little toward growth. Often many small eggs are spawned at one time with high early mortality (−) and reproductive spreading of risk among many small offsprings (+). The extreme of this pattern is one-time or big bang reproduction where the female dies after spawning (semelparity). Thus reproductive risk is not spread over years ( ). This high-r life history extreme is shown by bathypelagic fishes which live only 5–7 years.

Although direct data are few, low-r fish are long-lived, have a late age of first reproduction, and should as adults have lower mortality rates. As females continue to grow, the mortality rate should continue to decline. These older and larger fish have higher fecundities and continued high reproductive values. Reproductive lifespan is long with extra energy going to both growth and reproduction. Often few large eggs are spawned with only moderate early mortality (+) but little spreading of reproductive risk among the few offspring at each spawning (−). Instead, spreading of risk occurs over a number of years (+) of repeated reproduction (iteroparity). A corollary of this spreading of risk among years is occasional strong year classes. This low-r life history extreme is shown by amblyopsid cave fishes and many bathybenthic species that live several decades. Seamount species live much longer but have intermediate sizes and numbers of eggs (Table 1 and Koslow 1997).

Ecological and conservation implications of high-r and low-r life histories

Life history and r are population attributes. So differences in life history and r have population level consequences (population as the unit of evolution) and the basis for their evolution is selection on individuals (the most likely unit of selection). I suggest that rigor, variability, and predictability of abiotic and biotic factors are always agents of selection (Poulson & Culver 1969). And r is much more sensitive to differences in age at first reproduction than to differences in fecundity (mx), mortality rate (qx), or longevity (e.g. recent ecology textbooks like Krebs 1994, Begon et al. 1996).

With a low-r life history the positive tradeoff is that the rate of population decline and chances for local extinction are low over a long series of bad years while the negative is that such species cannot respond quickly by reproduction in a series of good reproductive years (Mertz 1971). The tradeoffs are the opposite for species with high r. To me, a corollary is that low-r species cannot quickly track changes in carrying capacity (K in the logistic equation). This leads to what I label for students a 'fizz and fizzle' population growth cycle. The opposite is a 'boom and bust' cycle in high-r species. A high r allows these species to closely track carrying capacity but if there are several years in a row with no reproduction they can become locally extinct, because females have a short reproductive lifespan.

Conservation consequences are that long-lived adults of low-r species are much more susceptible, than short-lived adults of high-r species, to anthropogenic mortality. Toxins bioaccumulate with age and biomagnify greatly in species at the top of the food chain such as amblyopsid cave fishes. Overfishing is likely with nets that select for the largest and oldest fish and this greatly impacts very long-lived fishes such as seamount species. These kinds of anthropogenic mortality factors for older and larger fish are additive because natural mortality rates are already low. And the older and larger females have the higher reproductive values because they have higher fecundity. In contrast for high-r species the larger and older females, that are fished by size selective nets, often have no higher fecundity than smaller adults and they are likely to die soon anyway because of high natural mortality rates. Thus any anthropogenic mortality is complementary rather than additive.

Difficulties of evaluating predation risk and competition as agents of selection

Vulnerability to predation is clearly important. I have already mentioned Childress' (1995) hypothesis that ability to swim fast confers no advantage in an environment with no current and no visible predators. According to Childress (op. cit.), in the absence of light relaxation of selection by visual predators led to loss of swimming ability in bathypelagic species whose ancestors had image-forming eyes. I suspect that, even without predators that hunt by vision, predation is probably an important agent of selection since the bathypelagic species have larvae that live in the epipelagic and small maximum size as adults. This means that they are always vulnerable to predation. This helps explain why many bathypelagic species are short-lived and semelparous (Table 1). Cave fish maximum sizes are no bigger than bathypelagic species but cave species are always the top predator and so the only risk of predation is by cannibalism on smaller individuals.[1] This is consistent with their long lives and repeated reproduction (iteroparity), at least of amblyopsids. Seamount species are not the top predators but live even longer, perhaps because their extremely well-developed schooling and sprint swimming make them relatively invulnerable to predation.

Both intraspecific competition, especially in schooling species, and interspecific competition may add to other causes of energy limitation. I am talking about resource competition which is indirect via a shared food resource. At the high competition extreme both intra- and inter-specific competition may contribute to the great longevity and extremely slow growth of seamount species. Many seamount species occur in immense schools and Koslow and I agree that this leads to intense intra-specific competition since reproductive effort increases when commercial fishing reduces size of schools (Koslow 1996, 1997). In addition strong inter-specific competition occurs since many species take advantage of local food concentrations around seamounts. At the low competition extreme are adult bathypelagic species. Though their larvae and other species in the epipelagic zone are abundant, food is also extremely abundant and may not be limiting. However intra- and inter-specific competition are unlikely to affect adult bathypelagic fishes because they occur at great depths where food is very scarce and densities of all organisms are very low.

Intermediate levels of competition are present for bathybenthic species and cave species, based on low density of fishes in their habitats. However we have data on the impacts of intra-specific competition only for cave fishes. Among cave fishes the Mexican cave characins occur in a wide range of densities associated with a wide range of food supplies but we only know that schooling is absent in cave populations. Parzefall (1985) relates this loss of schooling to absence of light. But I suggest that it more likely relates to absence of predators because seamount species whose ~800 m minimum depth of occurrence is below the maximum depth of penetration of light have predators and occur in immense schools. Not even surface relatives of the cave amblyopsids school so that the absence of that behavior in cave amblyopsids cannot explain their density dependent competition.

My data[1] from one cave over many years shows that density dependent population regulation in *Amblyopsis spelaea* was due to intraspecific resource competition based on energy limitation. Over 15 years of multiple censuses in one cave, fish numbers varied the most (2.1 fold), biomass change was intermediate (1.3 fold), and overall metabolic demand of the population changed the least (1.05 fold). In another cave, when overall metabolic demand was lowest spawning was more frequent and density dependent cannibalism of young of the year was rare. Young of the year are only 10–15 mm long and show a well-developed freezing reaction which makes them difficult for adults to find even with well-developed lateral line sense organs.

Both cannibalism and aggressive behavior are examples of direct competition (interference competition), both of which can harm the loser and are usually intraspecific. In *Amblyopsis spelaea* cannibalism of very small fish occurred only at high population densities[1] and in very high density populations of *Chologaster agassizi* cannibalism was frequent at all times (Hill 1966). Aggression is a less extreme form of interference than cannibalism. Bechler (1983) has shown that among species of amblyopsid fish aggressive behavior decreased with increasing cave adaptation. Within a species the degree of aggression increased with difference in size between a resident fish and another fish introduced into the resident's aquarium. Among cave species, not only were particular kinds of aggressive acts less intense but also the kinds of acts were fewer and each was less frequent along the series of increased cave adaptation. Bechler's (op. cit.) interpretation for amblyopsids was the same as Parzefall's for Mexican cave characins (Parzefall 1985), i.e. that neutral loss mutations accumulate over evolutionary time in caves.

Parzefall (op. cit.) suggested that aggressive acts that require vision are the first to be lost as neutral mutations accumulate. I suggest additional interpretations. One is that aggressive acts are not adaptive when food is abundant or where there are no defendable microhabitats. The spring-cave fish *Chologaster agassizi* shows intense territorial behavior and more kinds of aggressive acts when areas in their aquarium are defendable; they actively defend the few hiding places under rocks that are also hiding places for amphipod prey (Poulson unpublished). However in this spring-cavefish's natural habitat we might not expect strong competition. Spring-cavefish densities and food densities are often high and the habitat also has unlimited numbers of rocks. In caves where amblyopsids occur, there is sometimes little habitat structure that would allow territorial defense but more often there seem to be unlimited numbers of hiding and foraging places under rocks or ledges.

There is evidence for competition in seamount deep-sea species that live to ca. 100 years and have extremely inefficient growth (Table 1 about 0.03, similar to epipelagic species). Their inefficient growth is due to very high costs of swimming and great intraspecific and probably also inter-specific competition (Koslow 1996, 1997). Though food is locally concentrated around seamounts, it is quite variable over short spatial scales and the cost of foraging must be high since currents are locally strong where food is likely to be concentrated by advection. And the increased proportion of the population with mature ova after fishing exploitation shows that intra-specific competition in huge schools is severe when there is no fishing. Finally the fast retreat of schools when underwater cameras are lowered is consistent with the hypothesis that sprint swimming is needed to effectively mitigate strong predation pressures. The effectiveness of escape from predation helps explain the extreme longevities. In addition, high metabolic costs and competition enforce very low growth rates which are usually associated with extreme longevities. Associated with long life is a reproductive pattern which includes repeated attempts at reproduction (iteroparity) with relatively few, large eggs at each reproductive event.

Based on the literature (Mead et al. 1964) and my data (Poulson 1964) most benthic deep-sea and cave fishes have few, large eggs and longevities of several decades (Table 1). They have repeated reproduction (iteroparity), because they face predictably difficult energetic problems at all life history intervals and because the adults are relatively invulnerable to predation. Unlike the bathypelagic deep-sea fish species, both benthic deep-sea fishes and cave fishes face low food supplies throughout their lives and they either outgrow vulnerability to predation or are protected from predation. Seamount fishes are protected by schooling and sprint swimming and adult cave fish become the top predators in their habitats as they grow larger. Finally, unlike bathypelagic fish, deep-sea benthic and cave fishes need to retain good sustained and/or sprint swimming abilities to either escape predators and/or cope with strong currents.

Childress et al. (1980) had a difficult time relating the differences in life histories among pelagic fish species to life history theory and problems of competition and predation. To me the problem is that their analysis pertains only to subtle differences among pelagic species, which are all very short-lived compared to benthic deep-sea and cave species (Table 1). Epipelagic species live ca. 10 years and have repeated reproduction (iteroparous), mesopelagic species live ca. 1–5 years and are either annual or have repeated reproduction, and bathypelagic species live ca. 5–10 years and reproduce only in their last year of life (semelparous). All of these species have larvae that live in the food-rich epipelagic (+) but face many predators so juvenile survival should be low and unpredictably variable (−). Also maximum adult size is small and so they should always have high risk of predation and cannibalism. These traits should select for short life and high numbers of small eggs and most of the pelagic species show this life history. The subtle differences are explained, I believe, by the differences in energetics and growth efficiency discussed above and by differences in maximum body size.

The tradeoffs of small and large body size

Increasing body size provides positive and negative tradeoffs (Table 2). The + and − tradeoffs are in reproduction, locomotion, predation, and overall energetics. The positive reproductive tradeoff of large size is higher fecundity which is balanced by the negative tradeoff of great chance of reproductive failure because of the high costs of high fecundity. The positive locomotory tradeoff of large size is better ability to swim, forage, and escape predators in currents which is balanced by the negative tradeoff of increased power needed for sprint swimming. However the positive predation and energetic tradeoffs of the ability to eat large prey items, resistance to starvation, and low vulnerability to being prey do not seem to be lessened by negative tradeoffs.

I disagree with Childress that higher lifetime energetic costs associated with larger body size are important because they do not affect energetic short-term tradeoffs and so at best reflect 'soft' selection. They do however have population consequences as I have reviewed for two species of cave fish in the genus *Amblyopsis* that have similarly high population densities[1] (Poulson 1985). In brief, the smaller species, *Amblyopsis rosae* (maximum total length ca. 60 mm), has a lower metabolic rate but also a higher proportion of the adult females have mature ova and probably reproduce more regularly as inferred from evenly distributed size classes. *Amblyopsis spelaea* (maximum total length ca. 140 mm) may need the larger body size because it experiences stronger floods but may pay the cost of larger size by the smaller proportion of adult females with mature ova and more infrequent reproduction as inferred from quite uneven distribution of size classes.

Perspectives on cave fishes

I have gained insights about cave fishes from comparisons with deep-sea fish because of the similarities and differences in agents of selection and the availability of new data and techniques since I reviewed the subject in 1971. There are also tradeoffs for what can and cannot be studied both in cave fishes and in deep sea fishes (Table 3). Most of the tradeoffs relate to direct vs. indirect methods of study.

Mark recapture, growth efficiency, and longevity

Our inability to reliably age cave fish from scales or otoliths (see above) are more than balanced by being able to census whole populations and to follow growth of individuals with mark-recapture methods. In the deep-sea the annual signal of food supply even at abyssal depths is reflected in clear annual rings of growth of fish otoliths (Childress et al. 1980, Koslow 1996). This, coupled with large sample sizes in a number of species, has enabled scientists to gather much more reliable data on growth rates, reproductive patterns, and longevity than for cave fishes. In caves, despite the absence of clear annual differences in detrital food input, there is seasonal reproduction in many aquatic species.[2] Detrital renewal in the cave occurs most often when snowmelt causes flooding above ground but this can occur in late fall, in winter, or in spring. The associated rise in water level triggers breeding below ground and synchronizes circannual rhythms of reproductive readiness (Jegla & Poulson 1971). The circannual rhythms allow organisms to be prepared to reproduce whenever detrital renewal occurs. But the amounts of these detrital renewals are not very predictable in time or space (Poulson & Lavoie 2000) and so it is not surprising that growth rings on scales are subtle and may not be formed every year (Poulson 1963, 1985). Compared to difficulties of early short-term studies with tattoo and elastomer tags (Norbert Welch personal communication) the recent success at miniaturizing tags has now allowed several long-term mark recapture studies (personal communication from Bill Pearson and students of Jim Johnson). These recent studies have given direct data on growth rates and from these, and my stomach content data, I can now make more informed inferences about efficiencies of growth and longevity in amblyopsid cave fishes.

It is now clear to me that amblyopsid cave fish live several decades and this is much longer than I originally estimated from scale marks (Poulson 1961, 1963). The typical pattern for adult fish is <1 to 2 mm growth per year with only slightly higher rates for juvenile fish. *Amblyopsis rosae* has the smallest maximum size (60 mm total length) and for its population that has the fastest individual growth in gray bat caves I still estimate longevities of ca. 20 years. *Amblyopsis spelaea*, the largest species (ca. 135 mm maximum length) probably lives over 30–40 years in caves with more usual food supplies. As I outlined in an earlier section on life history, these great longevities have important implications for conservation.

[2] Poulson, T.L. & P.M. Smith. 1969. The basis for seasonal growth and reproduction in aquatic cave organisms. Proc. IV Int. Cong. Speleol., Ljubljana 4–5: 197–201.

I have already reviewed my most exciting inference about growth in all amblyopsid cave fishes that, despite usual slow growth, growth can occasionally be fast and is always very efficient. My own data on stomach contents of museum specimens and food density in the field is that food is scarce and most of the prey, in frequency and volume, are copepods. Medium sized prey, isopods and amphipods, are unusual in stomachs and large prey, such as crayfish or salamanders, are exceedingly rare. Coupling these data with new growth rate data on individuals in the field, I infer efficient growth. Growth must be very efficient because fish grow very quickly (10–20 mm in one year or less) on those occasions when, I suggest, they get lucky and catch a large prey item like a salamander, crayfish, or another fish. This variability and unpredictability of growth is another example of why I now eschew the constant cave paradigm (Poulson 1991, Poulson & Lavoie 2000) that I championed early in my career (Poulson & White 1969).

Differences in temporal and spatial scale for caves and the deep sea

Temporal and spatial scales are greater in the deep-sea than in caves and this leads me to expect greater specialization of fishes to different habitats in the deep-sea. For 90 million years the deep-sea has been oxic and, except for vents, deep sea temperatures have been cold for 50 million years (Childress 1995). In contrast limestone caves in the temperate zone are thought to be 1–3 million years old at maximum. Thus available evolutionary time is longer in the deep sea. And there is greater spatial and temporal stability in the deep sea because of the vast volume and great spatial extent. Childress (1995) points out that 79% of the volume occupied by life on earth lies at ocean depths greater than 1 km! In combination with a huge range of environmental conditions in the deep sea, huge volumes and areas increase the chances of evolutionary specialization to particular environments. But the chances for local specialization are decreased by the absence of absolute barriers to dispersal and gene exchange among habitats. In contrast, even for caves in the same geographic region there are near absolute barriers to dispersal and so great chances for local specialization undiluted by gene flow. All of this gives perspective of the degree of evolutionary specialization that we might expect for cave fishes.

It is now clear that, because of vast differences in spatial scale, different deep-sea habitats are much more constant and predictable, in space and time, than cave habitats or cave streams (Table 3). I suggest that a useful framework is the rigor, variability, and predictability of both abiotic and biotic agents of selection (Poulson & Culver 1969).

Though aquatic cave habitats are much less variable in temperature, current velocity, and food supply than surface streams in the same region,[2] variability and its predictability are important for cave fish at a variety of temporal and spatial scales. Two important aspects of predictability are constancy and contingency (Colwell 1976). Caves are constant with respect to the rigor of the low food quality of detrital wash-in and permanent darkness. And caves are contingent with respect to the winter- to spring-time when detrital input occurs by infiltration and wash-in (Poulson & Lavoie 2000). Differences among caves over many years and among cave streams at one time are natural experiments of variability and predictability which have given me important insights about the biology of cave fishes, especially in the Mammoth Cave region (Poulson 1992).

Over time, variability in caves is important at a scale of a year to many decades. Both good and bad years are unpredictable at a scale of a decade. I have already mentioned my unpredictably lucky large meal hypothesis for individual fish at the scale of years to decades. When they are small and immature a moderate meal leads to dramatic growth (e.g. from a size of 30 mm to a size of 50 mm in one year). And, by extension, when they are mature a large meal would lead to greatly enhanced fecundity or make the difference between not reproducing at all in one year and reproducing.

A multidecadal scale of variability is of the amount and timing of renewal of detrital food into caves. This is usually associated with winter or spring snowmelt coupled with rain but the timing is not predictable within or among years. This unpredictability is, as I have previously suggested (Jegla & Poulson 1970), the adaptive basis for synchronization of circannual rhythms of reproductive cycles that allow cave crayfish and cave fish to mate and be ready to lay eggs whenever detrital input occurs. This may not happen very often in the lifespan of a cave fish and is associated with a (+) tradeoff of food renewal and a (−) tradeoff of potential injury from floods. Flooding effects vary spatially at scales from a few meters to 1000s of meters and both within and among streams. I illustrate this with data from 40 years of study in the Mammoth Cave region. I contrast streams that are hydrologically interconnected, though sometimes remotely, but are differentially susceptible to flooding.

At one extreme are more open caves of the sinkhole plain that have multiple input points from the surface and regularly flood. In these caves there are big differences for fish between adjacent rocky riffles and silty pools. Riffles are often too shallow to be used by fish and so provide a refuge for amphipod and isopod prey that occur at high densities under rocks where the current brings fine detrital food to them. During floods some prey are washed into deeper pools where they may hide under rocks or ledges but are accessible to foraging fish. The ledges along with undercut banks and larger rocks also provide refuges for fish during floods.

At the other extreme are more closed caves that have few input points because of the protective sandstone cap rock of the Mammoth Cave Plateau. Unlike the short repeated riffle and pool sequence of sinkhole plain cave streams, the streams in caves under the plateau have major differences in habitat and food along a long spatial scale from the headwater streamlets out of vertical shafts to base-level rivers near spring resurgences on the Green River. Along this gradient habitats have a lot of local heterogeneity in depth and substrates but a general decline in current velocity, substrate heterogeneity and particle sizes, detrital food supply, and densities of live prey eaten by amblyopsid cave fishes. These changes are associated with a decline in fish and crayfish numbers and a change from a community of amphipod-isopod-crayfish-*Typhlichthys* cave fish (maximum fish length 55 mm and hypothesized higher metabolic rate) to a community of copepod-shrimp-*Amblyopsis* cave fish (maximum fish length 140 mm and a hypothesized lower metabolic rate). On a still larger scale the Plateau streams differ greatly in detrital food input, substrate heterogeneity, and both density and population sizes of cave fishes. And at the largest scale among caves there are even greater differences mainly because detrital input and yearly floods are much more predictable and intense in sinkhole plain caves than in plateau caves.

At a regional scale, different species of *Amblyopsis* cave species are found in structurally different caves that may have constrained their evolution. *A. spelaea* occurs in a karst region of the interior low plateau where limestone beds are thick and caves are often large with deep streams. Its maximum size is twice that of *A. rosae* which occurs in a karst region of the Ozark uplands were limestone beds are thin and caves are usually small with shallow streams. An alternative hypothesis is that body size constrains the habitats that can be occupied.

Consequences of differences in scale for adaptive radiation and local adaptation

Deep-sea fishes show incredibly more adaptive radiation and specialization than cave fishes. There are at least two reasons for this difference. First we judge that adaptation to caves by fishes has proceeded only from late Pliocene whereas adaptation to the deep-sea by fishes has proceeded from at least the Cretaceous. Often whole deep-sea families are in only one habitat so it is hard to make evolutionary inferences. But in caves even the same species can show differing degrees of adaptation in different caves. Second, the deep-sea habitats are much more different from each other than cave habitats are from each other. This has resulted in some striking differences in adaptations. But the similarity in cave habitats has led to some striking convergences. The overall corollary is that it is easier to infer evolutionary patterns in cave fishes than in deep-sea fishes.

One of the advantages of cave fish studies is that there are cases where closely related taxa occur in different caves with different food supplies. Among deep-sea fish, species are widespread and the only comparative studies of closely related species that I know of are of different *Cyclothone* spp. in the food rich mesopelagic and the food-poor bathypelagic. In caves we have different populations of the same species, *Astyanax fasciatus*, that on the basis of eye and pigment degeneration, have been isolated in caves for different periods of evolutionary time (e.g. Wilkens 1985, 1988). The great advantage of this system is that the populations are not reproductively isolated and so hybridization studies can give insight into genetic mechanisms associated with evolutionary adaptation to caves. Unfortunately most of the many elegant studies have concentrated on eye and pigment degeneration (Wilkens 1985, 1988) and have not included parallel studies of adaptations to cave life such as increased density of taste buds and metabolic economies; both of these increase with evolutionary time (e.g. Huppop 1985). And the studies have not considered the effects of very different food supplies in different caves or multiple independent isolation in caves from the same common ancestor as in the elegant studies of a cave amphipod by Culver et al. (1995).

Even though no hybridization studies are possible, we know most about cave adaptation among four species of cave fish in the family Amblyopsidae, that have, on the basis of relative eye and pigment degeneration, been isolated in caves for increasing evolutionary time (Poulson 1963, 1985). This is because I have documented differences in the morphology, physiology, behavior, and life history of all six species in the family.

Typhlichthys is the most likely amblyopsid taxon in which to find adaptations to local conditions. As presently recognized it is the only taxon with a wide geographic range, high morphological variation among caves, and large differences in food supply among caves (Poulson 1961). I suggest that, as has been documented for a cave amphipod (Culver et al. 1995), we will find that *Typhlichthys* represents many independent isolations in caves with subsequent convergent evolution of morphology and physiology. In Missouri *Typhlichthys* populations face extreme differences in food supply. One habitat is a huge karst window with a very high food supply. These fish grow only to ~45 mm but much more quickly than nearby cave populations. In the nearby big spring caves the food supply is much lower and cave fish grow slowly to ~80 mm. East of the Mississippi River the ecology of *Typhlichthys* caves are also quite different. In one large Alabama lake system at the water table there is rare food renewal and, based on the extremely slow growth and 100 year longevity of one of the crayfish species, food supply is very low. The fish grow to ca. 70 mm and probably also to an old age. This contrasts to a maximum fish size of 55 mm and maximum crayfish longevity of 10–15 years in cave streams with frequent renewal of detrital food and dense populations of isopod prey. These streams are under the sinkhole plain in the Mammoth Cave region. In adjacent cave streams under the sandstone caprock of the Mammoth Cave plateau, I suggest that *Typhlichthys* (maximum size 55 mm) shows character displacement due to competition with *Amblyopsis* (maximum total length 140 mm). Briefly (see Poulson 1992 for details), *Typhlichthys* in upstream areas is replaced by *Amblyopsis* in downstream areas, with occasional zones without either species. Along this gradient of upstream rocky riffle and shallow pool habitat to downstream deep river with silt-sand habitat there is a great decrease in kind and density of food supply from isopods and amphipods to copepods.

Evolutionary mechanisms of adaptation

One area of strong overlap between studies of deep-sea fishes and cave fishes is in suggestions that evolutionary adaptation may be rapid (also called punctuated) rather than gradual evolutionary adaptation. Early on there were inferences about progenesis as a mechanism of rapid change.

I have documented that cave adaptation within the Amblyopsidae shows a trend of progenesis with body proportions and morphology of adults in one cave adapted species reflecting the morphology of larvae in less cave adapted species (Poulson 1963). The result is a trend toward relatively larger heads with more elaboration of the superficial lateral line system which is used both for detection of moving prey and avoidance of obstacles. On this basis I predicted that any amblyopsid that was more cave adapted would show a continuation of this progenetic trend. I also predicted that such a species would occur in geographically peripheral karst areas with no other amblyopsids. Subsequently colleagues discovered just such a species and named it in my honor as *Speoplatyrhinus poulsoni* (Cooper & Kuehne 1974). To add to my pride, about 1997 my namesake became #81 in a Burger King series of cards for kids on endangered species.

An even more extreme progenesis has been suggested by Marshall (1984) as the evolutionary basis for the extreme reductions in organ systems and locomotory ability of bathypelagic fish (Marshall 1960). Unaware of my conclusion that progenesis was important in cave adaptation in amblyopsids (Poulson 1963), Marshall later recognized these reductions as progenetic evolution that result in fish that are sexually mature even though they still retain larval characters (Marshall 1984).

An exciting parallel to this hypothesis, of fast evolution of energy economy by progenesis, is the fast evolution of a 'stress resistant syndrome' in plants by the pleiotropic effect of relatively few quantitative genes that have much larger than average effects (Chapin et al. 1990). Chapin et al. (op. cit.) provide strong evidence that a variety of energy economies, associated with long lives and slow growth, are selected in plant habitats where the high rigors of drought, salinity, low soil nutrient availability, and soil toxicity act as stressors. To me, there is no reason why not to expect parallels among cave fishes with limited food availability as the main agent of selection.

In addition I suggest a positive feedback for evolution of increasing metabolic economies among amblyopsid cave species. Once the initial advantage of energy economy is achieved evolutionarily, a positive feedback would explain further increases in metabolic economy with increasing evolutionary time in those caves that have relatively similar low food supplies. According to my hypothesis the differences in body size and metabolic rate of the two *Amblyopsis* species is due to a positive feedback of increasing energy economy. The result is more frequent reproduction associated with longer evolutionary adaptation to caves by *A. rosae* with its smaller adult body size and lower metabolic rate. Given the same absolute low food supply and based on relative changes in food limitation, the positive feedback could proceed as follows. An initial selection for lower metabolic rate would allow increased population density (+) but also increased intra-specific competition (−). Increased competition would result in further reductions of metabolic rate (+) and then decreases in body size could select for further energy economies (+), etc. A strong test of my positive feedback hypothesis would be to see a similar trend within a species with increasing evolutionary time in caves, as for example in Mexican caves with *Astyanax fasciatus*. For these Mexican cave characins, food supply varies incredibly from sinking streams, that only occasionally carry detritus into caves, to bat caves, where energy input is regular and large, to sulphur based chemoautotrophic systems, where there is an even higher and more constant availability of food. The effects of this food heterogeneity have not been studied by researchers but it is an advantage of caves not to be found in any one habitat in the deep-sea.

'Preadaptation' to caves

Most obligate cave fish show evidence that their putative ancestors were 'preadapted' to life in constant darkness. 'Preadaptation' means that the ancestors were nocturnal and had sensory systems that made it possible to find food in the dark. Thus many species of cave fish in families or genera, especially of catfishes, have surface species with small eyes, nocturnal behavior, and well-developed non-visual sense organs. Several families of 'catfish' and in loaches exhibit tactile and contact chemosensory organs on lips and barbels as well as distance chemoreceptors in the olfactory rosette. In amblyopsid fish free head lateral line organs are the most important sensory system, both for locating live prey and avoiding obstacles. And catfish, loaches and amblyopsids have facultative cave species that live in springs and cave entrances.

Exceptions that prove the rule are cases where a single species with large-eyed ancestors have been isolated in caves. The best known cases are in Mexico with *Astyanax fasciatus* and *Poecilia mexicana*. The known ancestor of both species was a schooling fish with large eyes that relied only a little on non-visual information in foraging. And in both cases evolutionary reduction of eyes and pigment has been rapid because the cave populations are not reproductively isolated from the surface species. And in both cases food in caves may not have been limiting. *Astyanax* may have been catastrophically isolated in caves as surface streams contacted thick-bedded limestones and were 'captured' by cave drainages. In a sense *Astyanax* was 'preadapted' since surface fish immediately change foraging behavior to rely on taste organs in the dark (Parzefall 1985). In many caves these detritivore feeders had high food supplies available as either bat guano or washed-in organic matter; thus selection for metabolic economies was not consistent.

Prospectus for the future

I conclude with suggestions on directions research on cave fishes should take in the next decade. I use the differences in scale in the deep-sea and caves as one theme. As another theme I consider where we should borrow techniques from students of deep-sea fish. And as another theme I suggest how my approach to study of amblyopsid fish should be adopted for study of other cave fishes, especially the Mexican characins. A final theme is unanswered questions – a challenge to students.

Based on lesser temporal and spatial scale in caves, we should explore my prediction that there is local specialization in caves greater than in the deep sea. As I suggest in the accompanying paper in this volume, we should be able to use molecular techniques to document independent invasion of caves by the same 'species'. And we should do parallel studies of metabolic physiology, foraging behavior, and morphology to see if there is local adaptation. Above I suggested lines of evidence that support local adaptation within species of amblyopsid cave fishes. In Mexican cave characins we should be able to use the hybridization techniques already used by the Hamburg group for eyes, pigment, schooling, and feeding behavior to look at the genetic basis for local adaptation for these traits and extend the analysis to metabolic and sensory physiology. As a corollary we should study the ecological conditions that I hypothesize will lead to local adaptation. In particular we should document food supply, substrate, current, and other co-occurring species.

Researchers interested in other cave fishes should apply the broad approach and multiple hypotheses that I have so successfully used for amblyopsids. The minimum data should be size distributions and population densities since this gives important clues about life history patterns. In caves that are easily accessible with large fish populations, we should use new tagging techniques to get data on growth rates and longevity. Simple studies of metabolic rate are also useful. And we should use existing specimens to study morphology of the brain and sensory systems in conjunction with gut contents and measures of food supply in the field. These approaches will be most useful with the Mexican cave characins because we already know so much about aspects of the 'regressive' evolution of cave populations that have been isolated for different periods of evolutionary time.

Finally we should get together and 'brain-storm' about new questions, techniques, and approaches. A pet proposal of mine is to mount an expedition of colleagues with different expertise to understand the biology of my namesake *Speoplatyrhinus poulsoni* soon, before it is no longer endangered but extinct.

Acknowledgements

Too often we forget or are unaware of hypotheses advanced and data presented by our predecessors or contemporaries; in a sense little or 'nothing is new under the sun'. In this context I acknowledge colleagues who have had major influences on the ideas I present about cave fishes in this review. In addition to Carl Eigenmann's monograph of blind vertebrates at the turn of the century, which included lots of data on amblyopsid fishes, N.B. Marshall's 1960 monograph on deep sea fish excited me by the parallels I saw in amblyopsid cave fishes; I am really disappointed that I never even corresponded with him to exchange views and express my appreciation for his seminal work. I did, early on, correspond with Jim Childress about our mutual interest in possible metabolic economies in deep-sea and cave fishes but I here want to express my gratitude for his very careful and extensive physiological and biochemical studies of pelagic deep-sea organisms. Also his elaboration of his favorite hypothesis for reduced metabolic rates and locomotory abilities forced me to consider my favorite hypothesis more carefully. Recently I have corresponded with Tony Koslow about his incredible insights on the relationship of the biology of seamount and other benthic deep sea fishes to the biology of pelagic deep sea fishes. I also thank my colleagues whose work on other cave fishes has allowed me to hone my ideas about amblyopsid cave fish species. I especially thank Horst Wilkens and Jake Parzefall, and their students, who have so effectively used modern techniques and concepts to build on the pioneering work of Charles Breder and Pricilla Rasquin on Mexican cave characins. Finally, I thank the many colleagues who have helped me in the field over the past 40 years with amblyopsid and extended my field studies to further levels of sophistication. Among the latter, Jim Johnson's students and especially Bill Pearson and his students have used much improved capture-mark-recapture techniques and frequent censuses of fish in particular caves to allow me to suggest new insights. I extend my heartfelt thanks to you one and all!

References cited

Bechler, D.L. 1984. The evolution of agonistic behavior in amblyopsid fishes. Behav. Ecol. Soc. 12: 35–42.

Begon, M., J.L. Harper & C.R. Townsend 1996. Ecology: individuals, populations, and communities. Blackwell Science, Oxford. 1068 pp.

Chapin, F.S. III, A. Kellar & F. Pugnaire. 1993. Evolution of suites of traits in response to environmental stress. Amer. Nat. 142 (Suppl): 61–84.

Childress, J.J. 1977. Physiological approaches to the biology of midwater organisms. pp. 301–324. *In*: N.R. Andersen & B.J. Zahuranec (ed.) Oceanic Sound Scattering Prediction, Plenum Press, New York.

Childress, J.J. 1995. Are there physiological and biochemical adaptations of metabolism in deep-sea animals? Trends Ecol. Evol. 10: 30–36.

Childress, J.J., S.M. Taylor, G.M. Coillet & M.H. Price. 1980. Patterns of growth, energy utilization and reproduction in some meso- and bathy-pelagic fishes off southern California. Mar. Biol. 61: 27–40.

Colwell, R.K. 1974. Predictability, constancy, and contingency of periodic phenomena. Ecology 55: 1148–1153.

Cooper, J.E. & R.A. Kuehne. 1974. *Speoplatyrhinus poulsoni*, a new genus and species of subterranean fish from Alabama. Copeia 1974: 486–493.

Culver, D.C. 1985. Introduction to special issue on regressive evolution. Bull. Nat. Speleol. Soc. 47: 70–147.

Culver, D.C., T.C. Kane & D.F. Fong. 1995. Adaptation and natural selection in caves: the evolution of *Gammarus minus*. Harvard University Press, Cambridge. 223 pp.

Hessler, R.B., J.D. Isaacs & E.L. Mills. 1972. Giant amphipod from the abyssal Pacific Ocean. Science 175: 636–637.

Hill, L.G. 1966. Studies on the biology of the spring cavefish, *Chologaster agassizi* Putnam. Ph.D. Dissertation, University of Louisville, Louisville. 112 pp.

Huppop, K. 1985. The role of metabolism in the evolution of cave animals. Bull. Nat. Speleol. Soc. 47: 136–146.

Jegla, T.C. & T.L. Poulson. 1970. Circannian rhythms. I. Reproduction in a cave crayfish, *Orconectes inermis*. Comp. Biochem. Phys. 33: 347–355.

Koslow, J.A. 1996. Energetic and life history patterns of deep-sea benthic, benthopelagic, and seamount-associated fish. J. Fish Biol. 49 (Suppl. A): 54–74.

Koslow, J.A. 1997. Seamounts and the ecology of deep-sea fisheries. Amer. Sci. 85: 168–176.

Krebs, C.J. 1994.. Ecology: the experimental analysis of distribution and abundance. Harper-Collins, New York. 801 pp.

Marshall, N.B. 1960. Swimbladder structure of deep-sea fishes in relation to their systematics and biology. Disc. Rept. 31: 1–122.

Marshall, N.B. 1964. Bathypelagic macrourid fishes. Copeia 1964: 86–93.

Marshall, N.B. 1984. Progenetic tendencies in deep-sea fishes. pp. 91–101. *In:* G.W. Potts & R.J. Wooton (ed.) Fish Reproduction: Strategies and Tactics, Academic Press, London.

Mead, G.W., E. Bertelsen & D.M. Cohen. 1964. Reproduction among deep-sea fishes. Deep-Sea Res. 11: 569–596.

Mertz, D.B. 1971. Life history phenomena in increasing and decreasing populations. pp. 361–369. *In:* G.P. Patil, E.C. Pielou & W.E. Waters (ed.) Sampling and Modeling Biological Populations and Population Dynamics, Pennsylania State University, University Park.

Murphy, G.I. 1968. Patterns in life history and the environment. Amer. Nat. 102: 391–403.

Parzefall, J. 1985. On the heredity of behavior patterns in cave animals and their epigean relatives. Bull. Nat. Speleol. Soc. 47: 128–135.

Poulson, T.L. 1961. Cave adaptation in amblyopsid fishes. Ph.D. Dissertaion, University Michigan, Ann Arbor. 684 pp.

Poulson, T.L. 1963. Cave adaptation in amblyopsid fishes. Amer. Midl. Nat. 70: 257–290.

Poulson, T.L. 1964. Animals in aquatic environments: animals in caves. pp. 749–771. *In:* D.B. Dill (ed.) Handbook of Physiology: Environment, Amer. Physiol. Soc., Washington, D.C.

Poulson, T.L. 1971. Biology of cave and deep sea organisms: a comparison. Bull. Nat. Speleol. Soc. 33: 51–61.

Poulson, T.L. 1985. Evolutionary reduction by neutral mutations: plausibility arguments and data from amblyopsid fishes and linyphiid spiders. Bull. Nat. Speleol. Soc. 47: 109–117.

Poulson, T.L. 1992. The Mammoth Cave ecosystem. pp. 569–611. *In:* A. Camacho (ed.) The Natural History of Biospeleology, Museo Nacional de Ciencias Naturales, Madrid.

Poulson, T.L. & D.C. Culver. 1969. Diversity in terrestrial cave communities. Ecology 50: 153–158.

Poulson, T.L & K.H. Lavoie. 2000. The trophic basis of subsurface ecosystems. pp. 118–136. *In:* D.C. Culver, W. Humphreys & H. Wilkens (ed.) Ecosystems of the World: Caves, Elsevier, Amsterdam.

Poulson, T.L. & W.B. White. 1969. The cave environment. Science 165: 971–981.

Priede, J.G., P.M. Bagley, J.D. Armstrong, K.L. Smith & N.R. Merrett. 1991. Direct documentation of active dispersal of food-falls by deep-sea demersal fishes. Nature 351: 647–649.

Sanders, H.L. & R.B. Hessler. 1969. Ecology of the deep-sea benthos. Science 163: 1419–1424.

Stearns, S.C. 1992. The evolution of life histories. Oxford University Press, Oxford. 249 pp.

Wilkens, H. 1985. The evolution of polygenic systems, studied on epigean and cave populations of *Astyanax fasciatus* (Characidae, Pisces). Bull. Nat. Speleol. Soc. 47: 101–108.

Wilkens, H. 1988. Evolution and genetics of epigean and cave *Astyanax fasciatus* (Characidae, Pisces). Evol. Biol. 23: 271–367.

Species and subject index

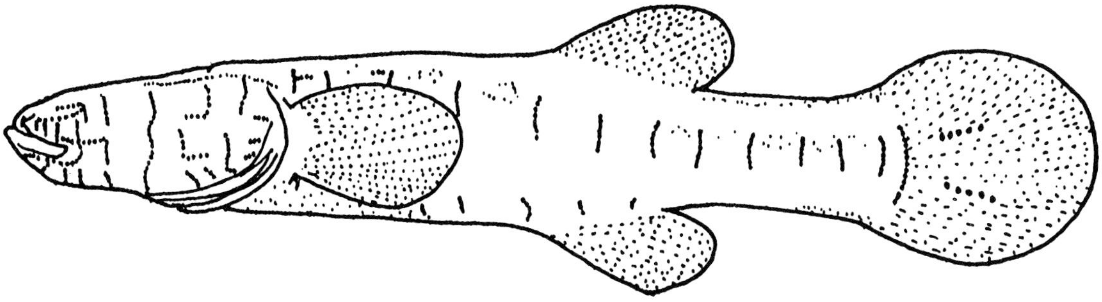